TABLE OF SYMBOLS

 P9-CEH-691

$v_{ss}(t)$	Steady-state portion of $v(t)$
$w(x,t)$, $w(\eta,\tau)$	Transverse displacement of beam
$x(0)$, $\dot{x}(0)$, $x_i(0)$, $\dot{x}_i(0)$	Initial displacement, initial velocity
$x(t)$, $x(\tau)$, $x_i(t)$	Displacement
x_{max}	Magnitude of $x(\tau_m)$
x_o	Static equilibrium position
$x_{ss}(t)$	Steady-state portion of $x(t)$
$\dot{x}(t)$, $\dot{x}(\tau)$, $\dot{x}_i(t)$	Velocity
$\ddot{x}(t)$, $\ddot{x}(\tau)$, $\ddot{x}_i(t)$	Acceleration
$y(t)$	Displacement of base or nondimensional displacement
$\ddot{y}(t)$	Acceleration of base
$z(t)$	Relative displacement between mass and base
A	Cross-sectional area of beam
A_o	Magnitude of displacement due to initial conditions
$[A]$	State matrix
B_i	Nondimensional torsion stiffness
B_w	Bandwidth of a filter
C_n	Beam eigenfunction parameter
$[C]$	Damping matrix
D	Rayleigh dissipation function
$D(\Omega)$	Denominator of $H^2(\Omega)$
E	Young's modulus
E_d, E_{diss}	Dissipation energy
E_T	Total energy in a signal
$E(\omega)$	Signal energy as a function of frequency
$\mathbf{F}$, $\mathbf{F}_i$	Force vector
$\mathbf{F}_s$	Internal force vector
$F_T(t)$	Force transmitted to the base
F_o	Magnitude of $f(t)$
$F(x)$	Spring force
$F(s)$	Laplace transform of forcing $f(t)$ or $f(\tau)$
G	Shear modulus
G_B	Beam energy function
G_0, G_L	Beam energy function at boundaries
$G_{ij}(j\Omega)$	Frequency-response function of inertial element i to force applied to inertial element j
$G(\Omega)$	Frequency-response function
$H(\Omega)$	Amplitude response function—single-degree-of-freedom system
$H_{ij}(\Omega)$	Amplitude response function of inertial element i to force applied to inertial element j
$H_{st}(\Omega)$	Amplitude response function—system with structural damping
$H_{mb}(\Omega)$	Amplitude response function—system with base excitation
$H_{ub}(\Omega)$	Amplitude response function—system with rotating unbalanced mass
$\mathbf{H}$	Angular momentum
I	Moment of inertia of beam cross-section about bending axis
J	Mass moment of inertia about axis of rotation
J_G	Mass moment of inertia about center of mass
J_O	Mass moment of inertia about point "o"

THOMSON
BROOKS/COLE
™

isher: *Bill Stenquist*

nology Project Manager: *Burke Taft*

keting Manager: *Tom Ziolkowski*

keting Assistant: *Jennifer Gee*

ect Manager, Editorial Production: *Mary Vezilich*

t/Media Buyer: *Judy Inouye*

nissions Editor: *Bob Kauser*

Production Service: *RPK Editorial Services*

Text Designer: *Roy R. Neuhaus*

Illustrator: *Wellington Studios*

Cover Designer: *Lisa Henry*

Cover Image: *Glen Allison/Getty Images*

Compositor: *ATLIS Graphics*

Printer: *Phoenix Book Technology Park*

For more information about our products, contact us at:
Thomson Learning Academic Resource Center
1-800-423-0563
For permission to use material from this text, contact us by:
Phone: 1-800-730-2214   **Fax:** 1-800-730-2215
Web: http://www.thomsonrights.com

TLAB is a registered trademark of The MathWorks, Inc.
MATLAB product information, please contact:
MathWorks, Inc.
pple Hill Drive
ick, MA, 01760-2098 USA
 508-647-7000
: 508-647-7101
ail: info@mathworks.com
: www.mathworks.com

rary of Congress Control Number: 2003102724

N: 0-534-39510-4

Brooks/Cole–Thomson Learning
20 Davis Drive
Belmont, CA 94002
USA

Asia
Thomson Learning
5 Shenton Way #01-01
UIC Building
Singapore 068808

Australia/New Zealand
Thomson Learning
102 Dodds Street
Southbank, Victoria 3006
Australia

Canada
Nelson
1120 Birchmount Road
Toronto, Ontario M1K 5G4
Canada

Europe/Middle East/Africa
Thomson Learning
High Holborn House
50/51 Bedford Row
London WC1R 4LR
United Kingdom

THIS BOOK IS PRINTED ON ACID-FREE RECYCLED PAPER

VIBRATION

Balakumar Balachandran
University of Maryland

Edward B. Magrab
University of Maryland

THOMSON
™
BROOKS/COLE

Australia • Canada • Mexico • Singapore • Spain • United Kingdom •

To Malini, Ragini and Nitin

To June Coleman Magrab
and

In memory of
Jon Grabowski
(10/17/67–9/11/01)

Contents

6 Single Degree-of-Freedom Systems Subjected to Transient Excitations   260

7 Multiple Degree-of-Freedom Systems: Governing Equations and Characteristics of Free Responses   308

Preface

Vibration principles have been known and studied for many centuries. Over the years, the use of these principles to understand and design systems has seen considerable growth in the diversity of systems that are designed with vibrations in mind: mechanical, aerospace, electromechanical and microelectromechanical devices and systems, biomechanical and biomedical systems, ships and submarines, and civil structures. As the performance envelope of an engineered system is pushed to higher limits, nonlinear effects also have to be taken into account.

This book has been written to enable the use of vibration principles in a broad spectrum of applications and to meet the wide range of challenges faced by system analysts and designers. To this end, the authors have the following goals: a) to provide an introduction to the subject of vibrations for undergraduate students in engineering and the physical sciences; b) to present vibration principles in a general context and illustrate the use of these principles through carefully chosen examples from different disciplines; c) to use a balanced approach that integrates principles of linear and nonlinear vibrations with modeling, analysis, prediction, and measurement so that physical understanding of the vibratory phenomena and their relevance for engineering design can be emphasized; and d) to deduce design guidelines that are applicable to a wide range of vibratory systems.

In writing this book, the authors have used the following guidelines. The material presented should, to the extent possible, have a physical relevance to justify its introduction and development. The examples should be relevant and wide ranging and they should be drawn from different areas such as biomechanics, electronic circuit boards and components, machines, machining (cutting) processes, microelectromechanical devices, and structures. There should be a natural integration and progression between linear and nonlinear systems, between the time domain and the frequency domain, among the responses of systems to harmonic and transient excitations, and between discrete and continuous system models. There should be a minimum emphasis placed on the discussion of numerical methods and procedures per se and, instead, advantage should be taken of tools such as MATLAB® for generating the numerical solutions and complementing analytical solutions. The algorithms for generating numerical solutions should be presented external to the chapters, as they tend to break the flow of the material being presented. (The MATLAB algorithms used to construct and generate all solutions can be found at http:/engineering.brookscole.com/vibrations) Further advantage should be taken of tools such as MATLAB in concert with analysis, so that linear systems can be extended to include nonlinear elements. Finally, there should be a natural and integrated interplay and presentation among analysis, modeling, measurement, prediction, and design so that a reader does not develop artificial distinctions among them.

Many parts of this book have been used for classroom instruction in a Vibrations course offered at the junior level at the University of Maryland. Typically, students in this course have had a sophomore level course on dynamics and a course on ordinary differential equations, including Laplace transforms. Beyond that, some fundamental material on complex numbers and linear algebra is introduced at the appropriate places in the course.

This book has the following features. Both Newton's laws and Lagrange's equations are used to develop models of systems. Since an important part of this development requires kinematics, this material is reviewed in Chapter 1. We use Laplace transforms to develop analytical solutions for linear vibratory systems, and from the Laplace domain, extend these results to the frequency domain. The responses of these systems are discussed in both the time and frequency domains to emphasize their duality. Notions of transfer functions and frequency-response functions are also used throughout the book to help the reader develop a comprehensive picture of vibratory systems. We have introduced design for vibration (DFV) guidelines that are based on vibration principles developed throughout the book. The guidelines appear at appropriate places in each chapter. These design guidelines serve the additional function of summarizing the preceding material by encapsulating the most important elements as they relate to some aspect of vibration design. Many examples are included from the area of microelectromechanical systems throughout the book to provide a physical context for the application of principles of vibrations at "small" length scales. In addition, there are several examples of vibratory models from biomechanics. Throughout the book, extensive use has been made of MATLAB, and in doing so we have been able to include a fair amount of new numerical results which were not accessible—or not easily accessible—to analysis previously. These results reveal many interesting features which the authors believe help expand our understanding of vibrations.

The book is organized into nine chapters, with the topics covered ranging from pendulum systems and spring-mass-damper prototypes to beams. In mechanics, the subject of vibrations is considered a subset of dynamics, in which one is concerned with the motion of bodies subjected to forces and moments. For much of the material covered in this book, a background in dynamics on the plane is sufficient. In Chapter 1, a summary is provided of concepts such as degrees of freedom and principles such as Newton's linear momentum principle and Euler's angular momentum principle.

In Chapter 2, the elements that are used to construct a vibratory system model are introduced and discussed. The notion of equivalent spring stiffness is presented in different physical contexts. Different damping models that can be used in modeling vibratory systems are also presented in this chapter. Subsequently, in Chapter 3, the derivation of the equation governing a single degree-of-freedom vibratory system is addressed. For this purpose, principles of linear momentum balance and angular momentum balance and Lagrange's equations are used. Notions such as natural frequency and damping factor are introduced here. Linearization of nonlinear systems is also illustrated. In Chapter 4, the solution for the response of a single degree-of-freedom system is obtained by using Laplace transforms. A primary motivation for using the Laplace transform approach for finding the solutions of linear vibratory systems is the fact that this approach is used in the study of control systems. Responses to different initial conditions, including impact, are examined in this chapter. In addition, systems with linear and nonlinear springs are compared, the notion of stability is briefly addressed, and the important phenomenon of machine-tool chatter is introduced.

In Chapter 5, the responses of single degree-of-freedom systems subjected to periodic excitations are considered. The notions of resonance, frequency-response functions, and transfer functions are discussed in detail. The responses of linear and nonlinear vibratory systems subjected to harmonic excitations are also examined. The Fourier transform is introduced, and considerable attention is paid to relating the information in the time domain to the frequency domain and vice versa. For different excitations, sensitivity of frequency-response functions with respect to the system parameters is also examined for design purposes. Accelerometer design is discussed and the notion of equivalent damping is presented. In Chapter 6, the responses of single degree-of-freedom systems to different types of external transient excitations are addressed and analyzed in terms of their frequency spectra. The notion of spectral energy is used to study vibratory responses in this chapter.

Multi-degree-of-freedom systems are treated in Chapters 7 and 8, leading up to systems with infinite number of degrees of freedom in Chapter 9. In Chapter 7, derivation of governing equations of motion of a system with multiple degrees of freedom is addressed by using principles of linear momentum balance and angular momentum balance and Lagrange's equations. Characteristics of free oscillations of undamped and damped systems are also studied in this chapter, and the notion of a vibratory mode is explained. Linearization of nonlinear multi-degree-of-freedom systems is addressed and systems with gyroscopic forces are treated. Stability notions discussed in Chapter 4 for a single degree-of-freedom system are extended to multi-degree-of-freedom systems in Chapter 7. Conservation of energy and momentum are also studied.

In Chapter 8, different approaches that can be used to obtain the response of a multi-degree-of-freedom system are presented. These approaches include the direct approach for harmonic excitation, the normal-mode approach, the Laplace transform approach, and the state-space formulation. Explicit solution forms for responses of multi-degree-of-freedom systems are obtained and used to find the response to initial conditions and different types of forcing. The importance of the normal-mode approach to carry out modal analysis of vibratory systems with special damping properties is addressed in this chapter. The state-space formulation is used to show how vibratory systems with arbitrary forms of damping can be treated. The notion of resonance in a multi-degree-of-freedom system is also addressed. Notions of frequency-response functions and transfer functions, introduced in Chapter 5 for single degree-of-freedom systems, are revisited and the relevance of these notions for system identification and design of vibration absorbers, mechanical filters, and vibration isolation systems is brought forth. The vibration-absorber material includes the traditional treatment of linear vibration absorbers and a brief introduction to the design of nonlinear vibration absorbers. Tools based on optimization techniques are also introduced for tailoring vibration absorbers and vibration isolation systems.

In Chapter 9, the subject of beam vibrations is treated at length as a representative example of vibrations of systems with infinite number of degrees of freedom. Derivation of governing equations of motion for isotropic beams is addressed and both free and forced oscillations of beams are studied for a variety of boundary conditions. In particular, considerable attention is paid to free-oscillation characteristics such as mode shapes, and the effects of axial forces, elastic foundation, and beam geometry on these characteristics. This book also includes five appendices, one on Laplace transform pairs (A), one on Fourier series (B), one on the decibel (C), one on solutions to ordinary differential equations (D), and one on matrices (E).

We express our sincere thanks to our former students for their spirited participation with earlier versions of this book and for providing feedback; to the many anonymous reviewers of this manuscript for their constructive suggestions; to our colleagues Professor Bruce Berger for his careful reading of Chapter 1; Professor Amr Baz for suggesting material and examples for inclusion in several chapters; Professor Donald DeVoe for pointing us to some of the literature on microelectromechanical systems; Dr. Henry Haslach for reading and commenting on parts of Chapter 9; and to Professor Jae-Eung Oh of Hanyang University, South Korea, for spending a generous amount of time in reading the early versions of Chapters 1 to 6 and for providing feedback for this material as well as suggestions for the exercises and their solutions. We would also like to thank Bingen Yang, University of Southern California; R. G. Parker, Ohio State University; and Kon-Well Wang, Pennsylvnia State University for their helpful comments and suggestions. We are also thankful to Mr. William Stenquist of Brooks-Cole for supporting and encouraging this book project. Last, but not least, we are grateful to our families for their support and understanding throughout this undertaking.

B. Balachandran
E. B. Magrab

College Park, MD

Table of Examples

Chapter 4

Chapter 5

Two important contributors to the field of vibrations: Galileo Galilei made the first pendulum frequency measurements and Sir Isaac Newton discovered the laws of motion. (*Source:* Galileo image courtesy of Taxi/Getty Images; Newton image courtesy of The Bridgeman Art Library/Getty Images.)

1

Introduction

1.1 INTRODUCTION

Vibrations occur in many aspects of our life. For example, in the human body, there are low-frequency oscillations of the lungs and the heart, high-frequency oscillations of the ear, oscillations of the larynx as one speaks, and oscillations induced by rhythmical body motions such as walking, jumping, and dancing. Many man-made systems also experience or produce vibrations. For example, any unbalance in machines with rotating parts such as fans, ventilators, centrifugal separators, washing machines, lathes, centrifugal pumps, rotary presses, and turbines, can cause vibrations. For these machines, vibrations are generally undesirable. Buildings and structures can experience vibrations due to operating machinery; passing vehicular, air, and rail traffic; or natural phenomena such as earthquakes and winds. Pedestrian bridges and floors in buildings also experience vibrations due to human movement on them. In structural systems, the fluctuating stresses due to vibrations can result in fatigue failure. Vibrations are also undesirable when performing measurements with precision instruments such as an electron microscope and when fabricating microelectromechanical systems. In vehicle design, noise due to vibrating panels must be reduced. Vibrations, which can be responsible for unpleasant sounds called noise, are also responsible for the music that we hear.

Vibrations are also beneficial for many purposes such as atomic clocks that are based on atomic vibrations, vibratory parts feeders, paint mixers, ultrasonic instrumentation used in eye and other types of surgeries, sirens and alarms for warnings, determination of fundamental properties of thin films from an understanding of atomic vibrations, and stimulation of bone growth.

The word *oscillations* is often used synonymously with *vibrations* to describe to and fro motions; however, in this book, the word *vibrations* is used in the context of mechanical and biomechanical systems, where the system energy components are kinetic energy and potential energy.

It is likely that the early interest in vibrations was due to development of musical instruments such as whistles and drums. As early as 4000 B.C., it is believed that in India and China there was an interest in understanding music, which is described as a pulsating effect due to rapid change in pitch. The origin of the harmonica can be traced back to 3000 B.C., when in China, a bamboo reed instrument called a "sheng" was introduced. From archeological studies of the royal tombs in Egypt, it is known that stringed instruments have also been around from about 3000 B.C. A first scientific study into such instruments is attributed to the Greek philosopher and mathematician Pythagoras (582–507 B.C.). He showed that if two like strings are subjected to equal tension, and if one is half the length of the other, the tones they produce are an octave (a factor of two) apart. It is interesting to note that although music is considered a highly subjective and personal art, it is closely governed by vibration principles such as those determined by Pythagoras and others who followed him.

The vibrating string was also studied by Galileo Galilei (1564–1642), who was the first to show that pitch is related to the frequency of vibration. Galileo also laid the foundations for studies of vibrating systems through his observations made in 1583 regarding the motions of a lamp hanging from a cathedral in Pisa, Italy. He found that the period of motion was independent of the amplitude of the swing of the lamp. This property holds for all vibratory systems that can be described by linear models. The pendulum system studied by Galileo has been used as a paradigm to illustrate the principles of vibrations for many centuries. Galileo and many others who followed him have laid the foundations for vibrations, which is a discipline that is generally grouped under the umbrella of mechanics. A brief summary of some of the major contributors and their contributions is provided in Table 1.1. The biographies of many of the individuals listed in this table can be found in the *Dictionary of Scientific Biography*.[1] It is interesting to note from Table 1.1 that the early interest of the investigators was in pendulum and string vibrations, followed by a phase where the focus was on membrane, plate, and shell vibrations, and a subsequent phase in which vibrations in practical problems and nonlinear oscillations received considerable attention.

Lord Rayleigh's book *Theory of Sound*, which was first published in 1877, is one of the early comprehensive publications on vibrations. In fact, many of the mathematical developments that are commonly taught in a vibrations course can be traced back to the 1800s and before. However, since then, the use of these principles to understand and design systems has seen considerable growth in the diversity of systems that are designed with vibrations in mind: mechanical, electromechanical and microelectromechanical devices and systems, biomechanical and biomedical systems, ships and submarines, and civil structures.

---

[1]C. C. Gillispie, ed., *Dictionary of Scientific Biography*, 18 Vols., Scribner, New York (1970–1990).

TABLE 1.1 Major Contributors to the Field of Vibrations and Their Contributions	Contributor	Area of Contributions
	Galileo Galilei (1564–1612)	Pendulum frequency measurement, vibrating string
	Marin Mersenne (1588–1648)	Vibrating string
	John Wallis (1616–1703)	String vibration: observations of modes and harmonics
	Christian Huygens (1629–1695)	Nonlinear oscillations of pendulum
	Robert Hooke (1635–1703)	Pitch–frequency relationship; Hooke's law of elasticity
	Isaac Newton (1642–1727)	Laws of motion, calculus
	Gottfried Leibnitz (1646–1716)	Calculus
	Joseph Sauveur (1653–1716)	String vibration: coined the name "fundamental harmonic" for lowest frequency and "harmonics" for higher frequency components
	Brook Taylor (1685–1731)	Vibrating string frequency computation; Taylor's theorem
	Daniel Bernoulli (1700–1782)	Principle of linear superposition of harmonics; string and beam vibrations
	Leonhard Euler (1707–1783)	Angular momentum principle, complex number, Euler's equations; beam, plate, and shell vibrations
	Jean d'Alembert (1717–1783)	D'Alembert's principle; equations of motion; wave equation
	Charles Coulomb (1736–1806)	Torsional vibrations; friction
	Joseph Lagrange (1736–1813)	Lagrange's equations; frequencies of open and closed organ pipes
	E. F. F. Chladni (1756–1824)	Plate vibrations: nodal lines
	Jacob Bernoulli (1759–1789)	Beam, plate, and shell vibrations
	J. B. J. Fourier (1768–1830)	Fourier series
	Sophie Germain (1776–1831)	Equations governing plate vibrations
	Simeon Poisson (1781–1840)	Plate, membrane, and rod vibrations; Poisson's effect
	G. R. Kirchhoff (1824–1887)	Plate and membrane vibrations
	R. F. A. Clebsch (1833–1872)	Vibrations of elastic media
	Lord Rayleigh (1842–1919)	Energy methods: Rayleigh's method; Strutt diagram; vibration treatise
	Gaston Floquet (1847–1920)	Stability of periodic oscillations: Floquet theory
	Henri Poincaré (1854–1912)	Nonlinear oscillations; Poincaré map; stability; chaos
	A. M. Liapunov (1857–1918)	Stability of equilibrium
	Aurel Stodola (1859–1943)	Beam, plate, and membrane vibrations; turbine blades
	C. G. P. De Laval (1845–1913)	Vibrations of unbalanced rotating disk: practical solutions
	Stephen Timoshenko (1878–1972)	Beam vibrations; vibration problems in electric motors, steam turbines, and hydropower turbines
	Balthasar van der Pol (1889–1959)	Nonlinear oscillations: van der Pol oscillator
	Jacob Pieter Den Hartog (1901–1989)	Nonlinear systems with Coulomb damping; vibration of rotating and reciprocating machinery; vibration textbook

1.2 PRELIMINARIES FROM DYNAMICS

Dynamics can be thought as having two parts, one being kinematics and the other being kinetics. While kinematics deals with the mathematical description of motion, kinetics deals with the physical laws that govern a motion. Here, first, particle kinematics and rigid-body kinematics are reviewed. Then, the notions of generalized coordinates and degrees of freedom are discussed. Following that, particle dynamics and rigid-body dynamics are addressed and the principles of linear momentum and angular momentum are presented. Finally, work and energy are discussed.

1.2.1 Kinematics of Particles and Rigid Bodies

Particle Kinematics

In Figure 1.1, a particle in free space is shown. In order to study the motions of this particle, a reference frame R and a set of unit vectors[2] i, j, and k fixed in this reference frame are considered. These unit vectors, which are orthogonal with respect to each other, are assumed fixed in time; that is, they do not change direction and length with time t. A set of orthogonal coordinate axes pointing along the X, Y, and Z directions is also shown in Figure 1.1. The origin O of this coordinate system is fixed in the reference frame R for all time t. The position vector $r^{P/O}$ from the origin O to the particle P is written as

$$r^{P/O} = x_p i + y_p j + z_p k \tag{1.1}$$

where the superscript in the position vector is used to indicate that the vector runs from point O to point P. When there is no ambiguity about the position

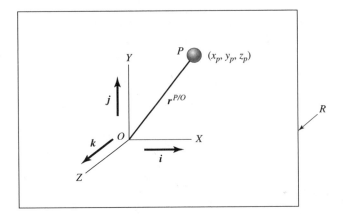

FIGURE 1.1
Particle kinematics. R is reference frame in which the unit vectors i, j, and k are fixed.

[2]As a convention throughout the book, bold and italicized letters represent vectors.

vector in question, as in Figure 1.1, this superscript notation is dropped for convenience; that is, $r^{P/O} \equiv r$. The particle's velocity v and acceleration a, which are both vector quantities, are defined as, respectively,

$$v = \frac{dr}{dt}$$

$$a = \frac{dv}{dt} = \frac{d^2r}{dt^2} \tag{1.2}$$

where the derivatives are with respect to time.[3] Using Eqs. (1.1) and (1.2) and noting that the unit vectors do not change with time, one obtains

$$v = \dot{r} = \dot{x}_p i + \dot{y}_p j + \dot{z}_p k$$

$$a = \ddot{r} = \ddot{x}_p i + \ddot{y}_p j + \ddot{z}_p k \tag{1.3}$$

where the over dot is used to indicate differentiation with respect to time.

Planar Rigid-Body Kinematics

A rigid body is shown in the X-Y plane in Figure 1.2 along with the coordinates of the center of mass of this body. In this figure, the unit vectors i, j, and k are fixed in reference frame R, while the unit vectors e_1 and e_2 are fixed

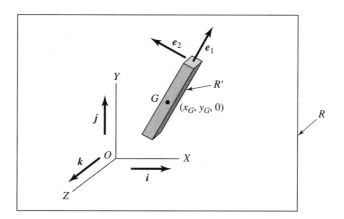

FIGURE 1.2
Planar rigid-body kinematics.

[3]The time-derivative operator d/dt is defined only when a reference frame is specified. To be precise, one would write Eqs. (1.2) as follows:

$$^R v^{P/O} = \frac{^R dr^{P/O}}{dt}$$

$$^R a^{P/O} = \frac{^R dv^{P/O}}{dt} = \frac{^R d^2 r^{P/O}}{dt^2}$$

However, for notational ease, the superscripts have been dropped in this book. Throughout this book, although the presence of a reference frame such as R in Figure 1.1 is not made explicit, it is assumed that the time-derivative operator has been defined in a specific reference frame.

in the reference frame R' and they are defined such that the vector cross product $e_1 \times e_2 = k$. While the unit vectors i, j, and k are fixed in time, the unit vectors e_1 and e_2 share the motion of the body and, therefore, change with respect to time. The velocity and acceleration of the center of mass of the rigid body with respect to the point O fixed in the reference frame R is determined in a manner similar to that previously discussed for the particle, and they are given by

$$v^{G/O} = \dot{r}^{G/O} = \dot{x}_G i + \dot{y}_G j$$
$$a^{G/O} = \ddot{r}^{G/O} = \ddot{x}_G i + \ddot{y}_G j \tag{1.4}$$

To determine the velocity and acceleration of any other point P on the body in terms of the corresponding quantities of the center of mass of the body, one needs to consider the angular velocity ω and the angular acceleration α of the rigid body.[4] Since the motion of the rigid body is confined to the plane in this case, both of these vectors point in the Z direction and they are each parallel to the unit vector k. In terms of the velocity and acceleration of the center of mass of the rigid body, the velocity and acceleration at any other point P on this body is written as

$$v^{P/O} = v^{G/O} + \omega \times r^{P/G}$$
$$a^{P/O} = a^{G/O} + \omega \times v^{P/G} + \alpha \times r^{P/G} \tag{1.5}$$

where $r^{P/G}$ is the position vector from point G to point P, $v^{P/O}$ and $a^{P/O}$ are the velocity and acceleration of point P with respect to O, respectively, and again, the symbol "$\times$" refers to the vector cross product. Equations (1.5) are derived by starting from the definitions given by Eqs. (1.2). Noting from the first of Eqs. (1.5) that the relative velocity of point P with respect to G is given by

$$v^{P/G} = v^{P/O} - v^{G/O}$$
$$= \omega \times r^{P/G} \tag{1.6}$$

the acceleration $a^{P/O}$ is written as[5]

$$a^{P/O} = a^{G/O} + \omega \times (\omega \times r^{P/G}) + \alpha \times r^{P/G} \tag{1.7}$$

Equations (1.5) through (1.7) hold for any two arbitrary points on the rigid body. If the body-fixed unit vectors e_1 and e_2 are used to resolve the position and velocity vectors when evaluating the velocities and accelerations using Eqs. (1.2), (1.5), (1.6), and (1.7), then one has to realize that the body-fixed unit vectors change orientation with respect to time; this is given by

$$\frac{de_1}{dt} = \omega \times e_1$$

$$\frac{de_2}{dt} = \omega \times e_2 \tag{1.8}$$

---

[4]Again, for notational ease, instead of the precise notations of $^R\omega^{R'}$ and $^R\alpha^{R'}$ for the angular velocity of the reference frame R' with respect to R and the angular acceleration of the reference frame R' with respect to R, respectively, the left and right superscripts have been dropped.

[5]In arriving at Eqs. (1.7), it has been assumed that all of the time derivatives are being evaluated in reference frame R.

In the definitions and relations provided in this section for velocities and accelerations, if the point O is fixed for all time in what is called an *inertial reference frame*,[6] then the corresponding quantities are referred to as *absolute velocities* and *absolute accelerations*. These quantities are important for determining the governing equations of a system, as discussed in Section 1.2.3. Three examples are provided next to illustrate how particle and rigid-body kinematics can be used to describe the motion of rigid bodies that will be considered in the following chapters.

EXAMPLE 1.1   Kinematics of a planar pendulum

Consider the planar pendulum shown in Figure 1.3, where the orthogonal unit vectors e_1 and e_2 are fixed to the pendulum and they share the motion of the pendulum. The unit vectors i, j, and k, which point along the X, Y and Z directions, respectively, are fixed in time. The velocity and acceleration of the planar pendulum P with respect to point O are the quantities of interest. The position vector from point O to point P is written as

$$r^{P/O} = r^{Q/O} + r^{P/Q}$$
$$= hj - Le_2 \tag{a}$$

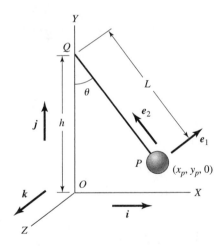

FIGURE 1.3
Planar pendulum. The reference frames are not explicitly shown in this figure, but it is assumed that the unit vectors i, j, and k are fixed in R and that the unit vectors e_1 and e_2 are fixed in R'.

[6]An inertial reference frame is an "absolute fixed" frame, which for our purposes does not share any of the motions of the particles or bodies considered in the vibration problems. (Of course, in a broader context, a true inertial reference frame does not exist.) Throughout this book, the reference frame R is used as an inertial reference frame, but it is not explicitly shown in Figure 1.3 and in any of the following figures in this chapter and later chapters. However, the choice of this frame is assumed in defining the absolute velocities and absolute accelerations.

Making use of Eqs. (1.2) and (1.8) and noting that both h and L are constant with respect to time and the angular velocity $\boldsymbol{\omega} = \dot{\theta}\boldsymbol{k}$, the pendulum velocity is

$$
\begin{aligned}
\boldsymbol{v}^{P/O} &= -L\frac{d\boldsymbol{e_2}}{dt} = -L\boldsymbol{\omega} \times \boldsymbol{e_2} \\
&= -L\dot{\theta}\boldsymbol{k} \times \boldsymbol{e_2} = L\dot{\theta}\boldsymbol{e_1}
\end{aligned} \tag{b}
$$

and the pendulum acceleration is

$$
\begin{aligned}
\boldsymbol{a}^{P/O} &= \frac{d\boldsymbol{v}^{P/O}}{dt} \\
&= \frac{d(L\dot{\theta}\boldsymbol{e_1})}{dt} = L\ddot{\theta}\boldsymbol{e_1} + L\dot{\theta}\boldsymbol{\omega} \times \boldsymbol{e_1} \\
&= L\ddot{\theta}\boldsymbol{e_1} + L\dot{\theta}(\dot{\theta}\boldsymbol{k} \times \boldsymbol{e_1}) \\
&= L\ddot{\theta}\boldsymbol{e_1} + L\dot{\theta}^2\boldsymbol{e_2}
\end{aligned} \tag{c}
$$

In arriving at Eqs. (b) and (c), the following relations have been used.

$$
\begin{aligned}
\boldsymbol{k} \times \boldsymbol{e_2} &= -\boldsymbol{e_1} \\
\boldsymbol{k} \times \boldsymbol{e_1} &= \boldsymbol{e_2}
\end{aligned} \tag{d}
$$

Noting that the unit vectors $\boldsymbol{e_1}$ and $\boldsymbol{e_2}$, which are fixed to the pendulum, are resolved in terms of the unit vectors $\boldsymbol{i}$ and $\boldsymbol{j}$ as follows,

$$
\begin{aligned}
\boldsymbol{e_1} &= \cos\theta\boldsymbol{i} + \sin\theta\boldsymbol{j} \\
\boldsymbol{e_2} &= -\sin\theta\boldsymbol{i} + \cos\theta\boldsymbol{j}
\end{aligned} \tag{e}
$$

one can rewrite the velocity and acceleration of the pendulum given by Eqs. (b) and (c), respectively, as

$$
\begin{aligned}
\boldsymbol{v}^{P/O} &= L\dot{\theta}(\cos\theta\boldsymbol{i} + \sin\theta\boldsymbol{j}) \\
\boldsymbol{a}^{P/O} &= (L\ddot{\theta}\cos\theta - L\dot{\theta}^2\sin\theta)\boldsymbol{i} + (L\ddot{\theta}\sin\theta + L\dot{\theta}^2\cos\theta)\boldsymbol{j}
\end{aligned} \tag{f}
$$

EXAMPLE 1.2   Kinematics of a rolling disc

A disc of radius r rolls without slipping in the X-Y plane, as shown in Figure 1.4. The unit vectors $\boldsymbol{e_1}$ and $\boldsymbol{e_2}$ are fixed to the disc and $\dot{\theta}$ is the angular speed of the disc. To determine the velocity of the center of mass of the disc located at G with respect to point O, the first of Eqs. (1.5) is used; that is,

$$
\boldsymbol{v}^{G/O} = \boldsymbol{v}^{C/O} + \boldsymbol{\omega} \times \boldsymbol{r}^{G/C} \tag{a}
$$

Noting that the instantaneous velocity of the point of contact $\boldsymbol{v}^{C/O}$ is zero, the angular velocity $\boldsymbol{\omega} = -\dot{\theta}\boldsymbol{k}$, and $\boldsymbol{r}^{G/C} = r\boldsymbol{j}$, the velocity of point G with respect to the origin O can be evaluated. In addition, from the definitions given by Eqs. (1.2), the acceleration can be evaluated. The resulting expressions are

$$
\begin{aligned}
\boldsymbol{v}^{G/O} &= -\dot{\theta}\boldsymbol{k} \times r\boldsymbol{j} = r\dot{\theta}\boldsymbol{i} \\
\boldsymbol{a}^{G/O} &= r\ddot{\theta}\boldsymbol{i}
\end{aligned} \tag{b}
$$

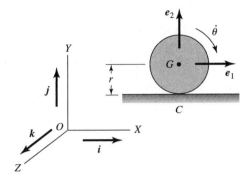

FIGURE 1.4
Rolling disc.

As expected, the center of mass of a disc rolling along a straight line does not experience any vertical acceleration due to this motion.

EXAMPLE 1.3 Kinematics of a particle in a rotating frame[7]

In this example, a particle is free to move in a plane, which rotates about an axis normal to the plane as shown in Figure 1.5. The x-y-z frame rotates with respect to the X-Y-Z frame, which is fixed in time. The rotation takes place about the z-axis with an angular speed ω. The unit vector k is oriented along the axis of rotation, which points along the z direction. The unit vectors e_1 and e_2 are fixed to the rotating frame, and they point along the x and y directions, respectively.

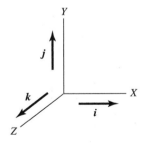 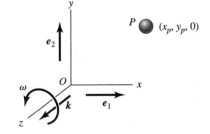

FIGURE 1.5
Particle in a rotating frame.

[7]In Chapters 7 and 8, the particle kinematics discussed here will be used in examining vibrations of a gyro-sensor.

The unit vectors i and j, which point along the X and Y directions, respectively, have fixed directions for all time. Here, we are interested in determining the velocity of the particle P located in the x-y plane.

In the rotating frame, the position vector from the fixed point O to the point P is

$$r = x_p e_1 + y_p e_2 \qquad\qquad\qquad (a)$$

Making use of the first of Eqs. (1.2) for the velocity of a particle, noting Eqs. (1.8), recognizing that $\boldsymbol{\omega} = \omega k$, and making use of Eqs. (d) of Example 1.1, the velocity of the particle is

$$
\begin{aligned}
v = \frac{dr}{dt} &= \frac{d}{dt}(x_p e_1) + \frac{d}{dt}(y_p e_2) \\
&= \dot{x}_p e_1 + x_p \frac{de_1}{dt} + \dot{y}_p e_2 + y_p \frac{de_2}{dt} \\
&= (\dot{x}_p - \omega y_p)e_1 + (\dot{y}_p + \omega x_p)e_2
\end{aligned}
\qquad (b)
$$

1.2.2   Generalized Coordinates and Degrees of Freedom

To describe the physical motion of a system, one needs to choose a set of variables or coordinates, which are referred to as *generalized coordinates.*[8] They are commonly represented by the symbol q_k.

The motion of a free particle, which is shown in Figure 1.6a, is described by using the generalized coordinates $q_1 = x_p$, $q_2 = y_p$, and $q_3 = z_p$. Here, all three of these coordinates are needed to describe the motion of the system. The minimum number of independent coordinates needed to describe the motion of a system is called the *degrees of freedom* of a system. Any free particle in space has three degrees of freedom.

In Figure 1.6b, a planar pendulum is shown. The pivot point of this pendulum is fixed at $(x_t, y_t, 0)$ and the pendulum has a constant length L. For this case, the coordinates are chosen as x_p and y_p. However, since the pendulum length is constant, these coordinates are not independent of each other because

$$(x_p - x_t)^2 + (y_p - y_t)^2 = L^2 \qquad\qquad (1.9)$$

Equation (1.9) is an example of a *constraint equation,* which in this case is a geometric constraint.[9] The motion of the pendulum in the plane can be described by using either x_p or y_p. Since $x_p = L \sin\theta$ and $y_p = L - L \cos\theta$, one

---

[8]In a broader context, the term "generalized coordinates" is also used to refer to any set of parameters that can be used to specify the system configuration. There are subtle distinctions in the definitions of generalized coordinates used in the literature. Here, we refer to the generalized coordinates as the coordinates that form the minimum set or the smallest possible number of variables needed to describe a system (J. L. Synge and B. A. Griffith, *Principles of Mechanics,* Section 10.6, McGraw Hill, New York, 1959).

[9]A geometric constraint is an example of a holonomic constraint, which can be expressed in the form $\phi(q_1, q_2, \ldots, q_n; t) = 0$, where q_i are the generalized coordinates and t is time.

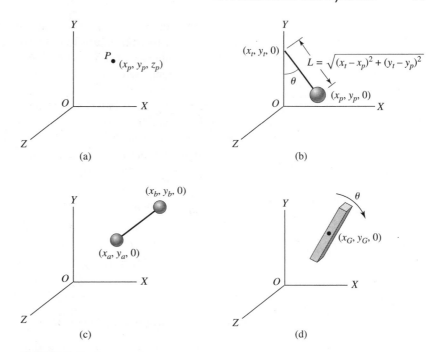

FIGURE 1.6
(a) Free particle in space; (b) planar pendulum; (c) dumbbell in plane; and
(d) free rigid body in plane.

can also use the variable θ to describe the motion of the pendulum, which is an
independent coordinate that qualifies as a generalized coordinate. Since only
one independent variable or coordinate is needed to describe the pendulum's
motion, a planar pendulum of constant length has one degree of freedom.

As a third example, a dumbbell in the plane is considered in Figure 1.6c.
In this system of particles, a massless rod of constant length connects two
particles. Here, the coordinates are chosen as x_a, y_a, x_b, and y_b, where this set
includes the coordinates of each of the two particles in the plane. Since the
length of the dumbbell is constant, only three of these coordinates are independent. A minimum of three of the four coordinates is needed to describe
the motion of the dumbbell in the plane. Hence, a dumbbell in the plane has
three degrees of freedom. The generalized coordinates in this case are chosen as x_G, y_G, and θ, where x_G and y_G are the coordinates of the center of
mass of the dumbbell and θ represents the rotation about an axis normal to
the plane of the dumbbell.

In Figure 1.6d, a rigid body that is free to move in the X-Y plane is
shown. The generalized coordinates $q_1 = x_G$ and $q_2 = y_G$ specify the translation of the center of mass of the rigid body. Apart from these two generalized coordinates, another generalized coordinate θ, which represents the rotation about the Z-axis, is also needed. Hence, a rigid body that is free to
move in the plane has three degrees of freedom. It is not surprising that the
dumbbell in the plane has the same number of degrees of freedom as the rigid

body shown in Figure 1.6d, since the collection of two particles, which are a fixed distance L apart in Figure 1.6c, is a rigid body.

In the examples shown in Figures 1.6b and 1.6c, one needed to take into account the constraint equations in determining the number of degrees of freedom of a system. For a system configuration specified by n coordinates, which are related by m independent constraints, the number of degrees of freedom N is given by

$$N = n - m \qquad\qquad (1.10)$$

The dynamics of vibratory systems with a single degree of freedom is treated in Chapters 1 through 6, the dynamics of vibratory systems with finite but more than one degree of freedom is treated in Chapters 7 and 8, and the dynamics of vibratory systems with infinite number of degrees of freedom is treated in Chapter 9. For much of the material presented in these chapters, the physical systems move in the plane and the systems are subjected only to geometric constraints.

1.2.3   Particle and Rigid-Body Dynamics

In the previous two subsections, it was shown how one can determine the velocities and accelerations of particles and rigid bodies and the number of independent coordinates needed to describe the motion of a system. Here, a discussion of the physical laws governing the motion of a system is presented. The velocities and accelerations determined from kinematics will be used in applying these laws to a system.

For all systems treated in this book, the speeds at which the systems travel will be much less than the speed of light. In the area of mechanics, the dynamics of such systems are treated under the area broadly referred to as *Newtonian Mechanics*. Two important principles that are used in this area are the principle of linear momentum and the principle of angular momentum, which are due to Newton and Euler, respectively.

Principle of Linear Momentum

The Newtonian principle of linear momentum states that in an inertial reference frame, the rate of change of the linear momentum of a system is equal to the total force acting on this system. This principle is stated as

$$\boldsymbol{F} = \frac{d\boldsymbol{p}}{dt} \qquad\qquad (1.11)$$

where $\boldsymbol{F}$ represents the total force vector acting on the system and $\boldsymbol{p}$ represents the total linear momentum of the system. Again, it is important to note that the time derivative in Eq. (1.11) is defined in an inertial reference frame and that the linear momentum is constructed based on the absolute velocity of the system of interest.

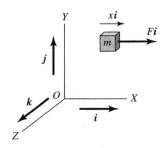

FIGURE 1.7
Free particle of mass m translating along the i direction.

Particle Dynamics

For a particle, the linear momentum is given by

$$p = mv \tag{1.12}$$

where m is the mass of the particle and v is the absolute velocity of the particle. When the mass m is constant, Eq. (1.11) takes the familiar form

$$F = \frac{d(mv)}{dt} = ma \tag{1.13}$$

which is referred to as *Newton's second law of motion.* The velocity in Eq. (1.12) and the acceleration in Eq. (1.13) are determined from kinematics. Therefore, for the particle shown in Figure 1.7, it follows from Eq. (1.13) that

$$F i = m\ddot{x} i$$

or

$$F = m\ddot{x}$$

Dynamics of a System of n Particles

For a system of n particles, the principle of linear momentum is written as

$$F = \sum_{i=1}^{n} F_i = \frac{dp}{dt} = \sum_{i=1}^{n} \frac{dp_i}{dt}$$
$$= \sum_{i=1}^{n} m_i \frac{dv_i}{dt} \tag{1.14}$$

where the subscript i refers to the ith particle in the collection of n particles, F_i is the external force acting on particle i, p_i is the linear momentum of this particle, m_i is the constant mass of the ith particle, and v_i is the absolute velocity of the ith particle. For the jth particle in this collection, the governing equation takes the form

$$F_j + \sum_{\substack{i=1 \\ i \neq j}}^{n} F_{ij} = \frac{dp_j}{dt} = \frac{d(m_j v_j)}{dt} \tag{1.15}$$

where F_{ij} is the internal force acting on particle j due to particle i. Note that in going from the equation of motion for an individual particle given by Eq. (1.15) to that for a system of particles given by Eq. (1.14), it is assumed that all of the internal forces satisfy Newton's third law of motion; that is, the assumption of equal and opposite internal forces.

If the center of mass of the system of particles is located at point G, then Eq. (1.14) can be shown to be equivalent to

$$F = \frac{d(mv_G)}{dt} \tag{1.16}$$

where m is the total mass of the system and v_G is the absolute velocity of the center of mass of the system. Equation (1.16) is also valid for a rigid body.

It is clear from Eq. (1.11) that *in the absence of external forces, the linear momentum of the system is conserved; that is, the linear momentum of the system is constant for all time.* This is an important conservation theorem, which is used, when applicable, to examine the results obtained from analysis of vibratory models.

Principle of Angular Momentum

The principle of angular momentum states that the rate of change of the angular momentum of a system with respect to the center of mass of the system or a fixed point is equal to the total moment about this point. This principle is stated as

$$M = \frac{dH}{dt} \tag{1.17}$$

where the time derivative is evaluated in an inertial reference frame, M is the net moment acting about a fixed point O in an inertial reference frame or about the center of mass G, and H is the total angular momentum of the system about this point.

For a particle, the angular momentum is given by

$$H = r \times p \tag{1.18}$$

where r is the position vector from the fixed point O to the particle and p is the linear momentum of this particle based on the absolute velocity of this particle. For a system of n particles, the principle given by Eq. (1.17) can be applied with respect to either the center of mass G of the system or a fixed point O. The angular momentum takes the form

$$H = \sum_{i=1}^{n} H_i = \sum_{i=1}^{n} r_i \times p_i \tag{1.19}$$

where r_i is the position vector from either point O or point G to the ith particle and p_i is the linear momentum of the particle based on the absolute velocity of the particle. For a rigid body moving in the plane as in Figure 1.6d, the angular momentum about the center of mass is given by

$$H_G = J_G \dot{\theta} k \tag{1.20}$$

where J_G is the mass moment of inertia about the center of mass and the unit vector k points along the Z-direction. The angular momentum of the rigid body about the fixed point O is given by

$$H_O = J_O \dot{\theta} k \tag{1.21}$$

where J_O is the mass moment of inertia about the fixed point.

From Eq. (1.17), it is clear that *in the absence of external moments, the angular momentum of the system is conserved; that is, the angular momentum of the system is constant for all time.* This is another important conservation theorem, which is used, when applicable, to examine the results obtained from analysis of vibratory models.

The governing equations derived for vibratory systems in the subsequent chapters are based on Eqs. (1.11) and (1.17). Specific examples are not provided here, but the material provided in later chapters are illustrative of how these important principles are used for developing mathematical models of a system.

1.2.4 Work and Energy

The definition for *kinetic energy T* of a system is provided and the relation between work done on a system and kinetic energy is presented. The kinetic energy of a system is a scalar quantity. For a system of *n* particles, the kinetic energy is defined as

$$T = \frac{1}{2} \sum_{i=1}^{n} m_i(\dot{r}_i \cdot \dot{r}_i) = \frac{1}{2} \sum_{i=1}^{n} m_i(v_i \cdot v_i) \tag{1.22}$$

where v_i is the absolute velocity of the *i*th particle and the "·" symbol is the scalar dot product of two vectors.

For a rigid body, the kinetic energy is written in the following convenient form, which is due to König,[10]

$$T = T_{(\text{translational})} + T_{(\text{rotational})} \tag{1.23}$$

where the first term on the right-hand side represents the kinetic energy associated with the translation of the system and the second term on the right-hand side represents the kinetic energy associated with rotation about the center of mass of the system. For a rigid body moving in the plane as shown in Figure 1.6d, the kinetic energy takes the form

$$T = \frac{1}{2} m(v_G \cdot v_G) + \frac{1}{2} J_G \dot{\theta}^2 \tag{1.24}$$

where the first term on the right-hand side is the translational part based on the velocity v_G of the center of mass of the system and the second term on the right-hand side is associated with rotation about an axis, which passes through the center of mass and is normal to the plane of motion. For a rigid

[10]D. T. Greenwood, *Principles of Dynamics*, Prentice Hall, Upper Saddle River, NJ, Chapter 4 (1988).

body rotating in the plane about a fixed point O, the kinetic energy is determined from

$$T = \frac{1}{2} J_O \dot{\theta}^2 \tag{1.25}$$

Work-Energy Theorem

According to the *work-energy theorem*, the work done in moving a system from a point A to point B is equal to the change in kinetic energy of the system, which is stated as

$$W_{AB} = T_B - T_A \tag{1.26}$$

where W_{AB} is the work done in moving the system from the initial point A to the final point B, T_B is the kinetic energy of the system at point B, and T_A is the kinetic energy of the system at point A. The work done W_{AB} is a scalar quantity.

Another form of energy called potential energy of a system is addressed in Chapter 2. Specific examples are not provided here, but the use of kinetic energy and work done in developing mathematical models of vibratory systems is illustrated in the subsequent chapters.

1.3 SUMMARY

In this chapter, a review from dynamics has been presented in the spirit of summarizing material that is typically a prerequisite to the study of vibrations. In carrying out this discussion, attention has been paid to kinematics, the notion of degree of freedom, and the principles of linear and angular momentum. This background material provides a summary of the foundation for the material presented in the subsequent chapters.

EXERCISES

1.1 Choose any two contributors from Table 1.1, study their contributions, and write a paragraph about each of them.

1.2 Consider the planar pendulum kinematics discussed in Example 1.1, start with position vector $r^{P/O}$ resolved in terms of the unit vectors i and j, and verify the expressions obtained for the acceleration and velocity given by Eqs. (f) of Example 1.1.

1.3 Consider the kinematics of the rolling disc considered in Example 1.2, and verify that the instantaneous acceleration of the point of contact is not zero.

1.4 Show that the acceleration of the particle in the rotating frame of Example 1.3 is

$$a = (\ddot{x}_p - 2\omega \dot{y}_p - \omega^2 x_p - \alpha y_p)e_1 \\ + (\ddot{y}_p + 2\omega \dot{x}_p - \omega^2 y_p + \alpha x_p)e_2$$

where α is the magnitude of the angular acceleration of the rotating frame about the z axis.

1.6 Determine the kinetic energy of the planar pendulum of Example 1.1.

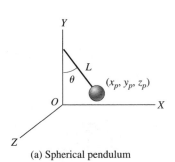

(a) Spherical pendulum

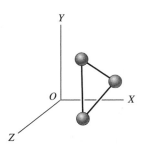

(b) System of three particles

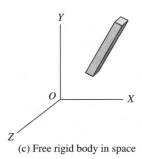

(c) Free rigid body in space

FIGURE E1.5

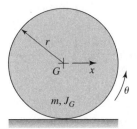

FIGURE E1.7

1.7 Consider the disc rolling along a line in Figure E1.7. The disc has a mass m and a rotary inertia J_G about the center of mass G. Answer the following: a) How many degrees of freedom does this system have? and b) Determine the kinetic energy for this system.

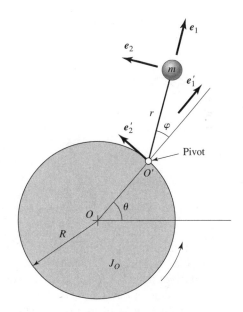

FIGURE E1.8

1.5 Determine the number of degrees of freedom for the systems shown in Figure E1.5. Assume that the length L of the pendulum shown in Figure E1.5a is constant and that the length between each pair of particles in Figure E1.5b is constant. *Hint:* For Figure E1.5c, the rigid body can be thought of as a system of particles where the length between each pair of particles is constant.

1.8 Determine the absolute velocity of the pendulum attached to the rotating disc shown in Figure E1.8; this system is representative of a centrifugal pendulum absorber treated in Chapter 8. Assume that the point O is fixed in an inertial reference frame, the unit vectors e_1 and e_2 are fixed to the pendulum, and that the orthogonal unit vectors e_1' and e_2' are fixed to the rotating disc. The motions of the pendulum are restricted to the plane containing the unit vectors e_1 and e_2, the angle θ describes the rotation of the disc with respect to the horizontal, and the angle φ describes the position of the pendulum relative to the disc.

1.9 In the system shown in Figure E1.8, if the mass of the pendulum is m, the length of the pendulum is r, and the rotary inertia of the disc about the point O is J_O, determine the system kinetic energy.

1.10 Determine the angular momentum of the system shown in Figure E1.8 about the point O and discuss if it is conserved.

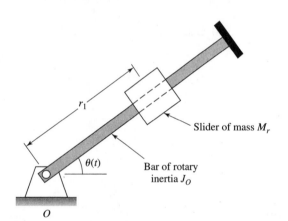

FIGURE E1.11

1.11 In Figure E1.11, a slider of mass M_r is located on a bar whose angular displacement in the plane is described by the coordinate θ. The motion of the slider from the pivot point is measured by the coordinate r_1. The acceleration due to gravity acts in a direction normal to the plane of motion. Assume that point O is fixed in an inertial reference frame and determine the absolute velocity and absolute acceleration of the slider.

1.12 Determine the kinetic energy of the system shown in Figure E1.11.

1.13 Determine the linear momentum for the system shown in Figure E1.11 and discuss if it is conserved. Assume that the mass of the bar is M_{bar} and the distance from the point O to the center of the bar is L_{bar}.

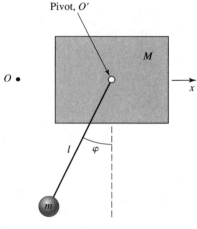

FIGURE E1.14

1.14 A pendulum of mass m attached to a moving pivot of mass M is shown in Figure E1.14. Assume that the pivot cannot translate in the vertical direction. If the horizontal translation of the pivot point from the fixed point O is measured by the coordinate x and the angle φ is used to describe the angular displacement of the pendulum from the vertical, determine the absolute velocity of the pendulum.

1.15 Referring to Figure E1.14 and assuming that the bar to which the pendulum mass m is connected is massless, determine the kinetic energy for the system.

1.16 Draw free-body diagrams for each of the masses shown in Figure E1.14 and obtain the equations of motion along the horizontal direction by using Eq. (1.15).

1.17 Draw the free-body diagram for the whole system shown in Figure E1.14 along the horizontal and obtain the system equation of motion by using Eq. (1.14) and verify that this equation can be obtained from Eq. (1.15).

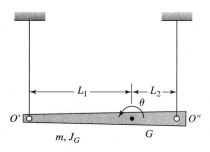

FIGURE E1.18

1.18 A rigid body is suspended from the ceiling by two elastic cables that are attached to the body at the points O' and O'', as shown in Figure E1.18. Point G is the center of mass of the body. Which of these points would you choose to carry out an angular-momentum balance based on Eq. (1.17)?

1.19 Consider the rigid body shown in Figure E1.19. This body has a mass m and rotary inertia J_G about the center of mass G. It is suspended from a point O on the ceiling by using an elastic suspension. The point of attachment O' is at a distance l from the center of mass G of this body. $M(t)$ is an external moment applied to the system along an axis normal to the plane of the body. Use the generalized coordinates x, which describes the up and down motions of point O' from point O, and θ, which describes the angular oscillations about an axis normal to the plane of the rigid body, and construct the system kinetic energy.

1.20 For the system shown in Figure E1.19, use the principle of angular-momentum balance given by Eq. (1.17) and obtain an equation of motion for the system. Assume that gravity loading is present.

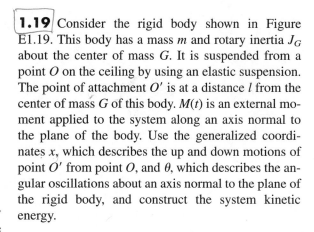

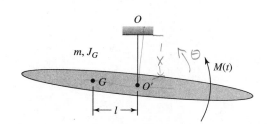

FIGURE E1.19

Components of vibrating systems: an air spring and a steel helical spring. (*Source:* Airspring courtesy of Goodyear, Inc.; steel spring courtesy of Earthworks/Getty Images.)

2

Modeling of Vibratory Systems

2.1 INTRODUCTION

In this chapter, the elements that comprise a vibratory system model are described and the use of these elements to construct models is illustrated with examples. There are, in general, three elements that comprise a vibrating system: i) inertia elements, ii) stiffness elements, and iii) dissipation elements. In addition to these elements, one must also consider externally applied forces and moments and external disturbances from prescribed initial displacements and/or initial velocities.

The *inertia element* stores and releases kinetic energy, the *stiffness element* stores and releases potential energy, and the *dissipation* or *damping element* is

TABLE 2.1	Quantity	Units
Units of Components Comprising a Vibrating Mechanical System and Their Customary Symbols	Translational motion	
	Mass, m	kg
	Stiffness, k	N/m
	Damping, c	N·s/m
	External force, F	N
	Rotational motion	
	Mass moment of inertia, J	kg·m^2
	Stiffness, k_t	N·m/rad
	Damping, c_t	N·m·s/rad
	External moment, M	N·m

used to express energy loss in a system. Each of these elements has different excitation-response characteristics and the excitation is in the form of either a force or a moment and the corresponding response of the element is in the form of a displacement, velocity, or acceleration. The inertia elements are characterized by a relationship between an applied force (or moment) and the corresponding acceleration response. The stiffness elements are characterized by a relationship between an applied force (or moment) and the corresponding displacement (or rotation) response. The dissipation elements are characterized by a relationship between an applied force (or moment) and the corresponding velocity response. The nature of these relationships, which can be linear or nonlinear, are presented in this chapter. The units associated with these elements and the commonly used symbols for the different elements are shown in Table 2.1.

2.2   INERTIA ELEMENTS

Translational motion of a mass is described as motion along the path followed by the center of mass. The associated inertia property depends only on the total mass of the system and is independent of the geometry of the mass distribution of the system. The inertia property of a mass undergoing rotational motions, however, is a function of the mass distribution, specifically the mass moment of inertia, which is usually defined about its center of mass or a fixed point O. When the mass oscillates about a fixed point O or a pivot point O, the rotary inertia J_O is given by

$$J_O = J_G + md^2 \tag{2.1}$$

where m is the mass of the element, J_G is the mass moment of inertia about the center of mass, and d is the distance from the center of gravity to the point O. In Eq. (2.1), the mass moment of inertias J_G and J_O are both defined with respect to axes normal to the plane of the mass. This relationship between the mass moment of inertia about an axis through the center of mass G and a parallel axis through another point O follows from the *parallel-axes theorem.* The mass moments of inertia of some common shapes are given in Table 2.2.

The questions of how the inertia properties are related to forces and moments and how these properties affect the kinetic energy of a system are examined next. In Figure 2.1, a mass m translating with a velocity of magnitude

TABLE 2.2
Mass Moments of Inertia about Axis z Normal to the x-y Plane and Passing Through the Center of Mass

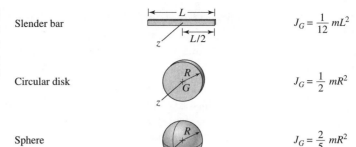

Slender bar		$J_G = \frac{1}{12} mL^2$
Circular disk		$J_G = \frac{1}{2} mR^2$
Sphere		$J_G = \frac{2}{5} mR^2$

$\dot{x}$ in the X-Y plane is shown. The direction of the velocity vector is also shown in the figure, along with the direction of the force acting on this mass.

In stating the principles of linear momentum and angular momentum in Chapter 1, certain relationships between inertia properties and forces and moments were assumed. These relationships are revisited here. Based on the principle of linear momentum, which is given by Eq. (1.11), the equation governing the motion of the mass is

$$F\mathbf{i} = \frac{d}{dt}(m\dot{x}\mathbf{i})$$

which, when m and $\mathbf{i}$ are independent of time, simplifies to

$$F = m\ddot{x} \tag{2.2}$$

On examining Eq. (2.2), it is evident that *for translational motion, the inertia property m is the ratio of the force to the acceleration.* The units for mass shown in Table 2.1 should also be evident from Eq. (2.2). From Eq. (1.22), it follows that the kinetic energy of mass m is given by

$$T = \frac{1}{2} m(\dot{x}\mathbf{i}\cdot\dot{x}\mathbf{i}) = \frac{1}{2} m\dot{x}^2 \tag{2.3}$$

and we have used the identity $\mathbf{i} \cdot \mathbf{i} = 1$. From the definition given by Eq. (2.3), it is clear that *the kinetic energy of translational motion is linearly proportional to the mass.* Furthermore, the kinetic energy is proportional to the second power of the velocity magnitude. To arrive at Eq. (2.3) in a different manner, let us consider the work-energy theorem discussed in Section 1.2.4. We assume that the mass shown in Figure 2.1 is translated from an initial rest state, where the velocity is zero at time t_o, to the final state at time t_f. Then, it follows from Eq. (1.26) that the work W done under the action of a force $F\mathbf{i}$ is

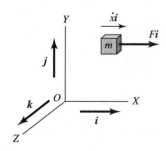

FIGURE 2.1
Mass in translation.

$$W = \int_0^x F\mathbf{i}\cdot d\mathbf{x}\mathbf{i} = \int_0^x m\ddot{x}\mathbf{i}\cdot d\mathbf{x}\mathbf{i} = \int_0^x m\ddot{x}dx$$

$$= \int_{t_o}^{t_f} m\ddot{x}\dot{x}dt = \int_0^{\dot{x}} m\dot{x}d\dot{x} = \frac{1}{2} m\dot{x}^2 \tag{2.4}$$

where we have used the relation $dx = \dot{x}dt$. Hence, the kinetic energy is

$$T(t = t_f) - T(t = t_0)^{=0} = W = \frac{1}{2}m\dot{x}^2 \tag{2.5}$$

which is identical to Eq. (2.3).

For a rigid body undergoing only rotation in the plane with an angular speed $\dot{\theta}$, one can show from the principle of angular momentum discussed in Section 1.2.3 that

$$M = J\ddot{\theta} \tag{2.6}$$

where M is the moment acting about the center of mass G or a fixed point O as shown in Figure 2.2 along the direction normal to the plane of motion and J is the associated mass moment of inertia. From Eq. (2.6), it follows that *for rotational motion, the inertia property J is the ratio of the moment to the angular acceleration.* Again, one can verify that the units of J shown in Table 2.1 are consistent with Eq. (2.6). This inertia property is also referred to as *rotary inertia.* Furthermore, to determine how the inertia property J affects the kinetic energy, we use Eq. (1.25) to show that the kinetic energy of the system is

$$T = \frac{1}{2}J\dot{\theta}^2 \tag{2.7}$$

Hence, *the kinetic energy of rotational motion only is linearly proportional to the inertia property J, the mass moment of inertia.* Furthermore, the kinetic energy is proportional to the second power of the angular velocity magnitude.

In the discussions of the inertia properties of vibratory systems provided thus far, the inertia properties are assumed to be independent of the displacement of the motion. This assumption is not valid for all physical systems. For a slider mechanism discussed in Example 2.2, the inertia property

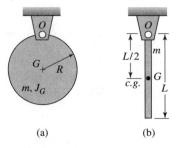

(a)   (b)

FIGURE 2.2
(a) Uniform disk hinged at a point on its perimeter and (b) bar of uniform mass hinged at one end.

is a function of the angular displacement. Other examples can also be found in the literature.[1]

EXAMPLE 2.1   Determination of mass moments of inertia

We shall illustrate how the mass moments of inertia of several different rigid body distributions are determined.

Uniform Disk

Consider the uniform disk shown in Figure 2.2a. If J_G is the mass moment of inertia about the disk's center, then from Table 2.2

$$J_G = \frac{1}{2} mR^2$$

and, therefore, the mass moment of inertia about the point O, which is located a distance R from point G, is

$$J_O = J_G + mR^2 = \frac{1}{2} mR^2 + mR^2 = \frac{3}{2} mR^2 \tag{a}$$

Uniform Bar

A bar of length L is suspended as shown in Figure 2.2b. The bar's mass is uniformly distributed along its length. Then the center of gravity of the bar is located at $L/2$. From Table 2.2, we have that

$$J_G = \frac{1}{12} mL^2$$

and therefore, after making use of the parallel axis theorem, the mass moment of inertia about the point O is

$$J_O = J_G + m\left(\frac{L}{2}\right)^2 = \frac{1}{12} mL^2 + \frac{m}{4} L^2 = \frac{1}{3} mL^2 \tag{b}$$

EXAMPLE 2.2   Slider mechanism: system with varying inertia property

In Figure 2.3, a slider mechanism with a pivot at point O is shown. A slider of mass m_s slides along a uniform bar of mass m_l. Another bar, which is pivoted at point O', has a portion of length b that has a mass m_b and another portion of length e that has a mass m_e. We will determine the rotary inertia J_O of this system and show its dependence on the angular displacement coordinate φ.

[1]J. P. Den Hartog, *Mechanical Vibrations,* Dover, NY, p. 352 (1985).

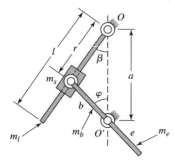

FIGURE 2.3
Slider mechanism.

If a_e is the distance from the midpoint of bar of mass m_e to O and a_b is the distance from the midpoint of bar of mass m_b to O, then from geometry we find that

$$r^2(\varphi) = a^2 + b^2 - 2ab \cos \varphi$$
$$a_b^2(\varphi) = (b/2)^2 + a^2 - ab \cos \varphi$$
$$a_e^2(\varphi) = (e/2)^2 + a^2 - ae \cos(\pi - \varphi) \qquad \text{(a)}$$

and hence, all motions of the system can be described in terms of the angular coordinate φ. The rotary inertia J_O of this system is given by

$$J_O = J_{m_l} + J_{m_s}(\varphi) + J_{m_b}(\varphi) + J_{m_e}(\varphi) \qquad \text{(b)}$$

where

$$J_{m_l} = \frac{1}{3} m_l l^2, \qquad J_{m_s}(\varphi) = m_s r^2(\varphi)$$

$$J_{m_b}(\varphi) = m_b \frac{b^2}{12} + m_b a_b^2 = m_b \left[\frac{b^2}{3} + a^2 - ab \cos \varphi \right]$$

$$J_{m_e}(\varphi) = m_e \frac{e^2}{12} + m_e a_e^2 = m_e \left[\frac{e^2}{3} + a^2 - ae \cos(\pi - \varphi) \right] \qquad \text{(c)}$$

In arriving at Eqs. (b) and (c), the parallel-axes theorem has been used in determining the bar inertias J_{m_b}, J_{m_e}, and J_{m_l}. From Eqs. (b) and (c), it is clear that the rotary inertia J_O of this system is a function of the angular displacement φ.

2.3 STIFFNESS ELEMENTS

2.3.1 Introduction

The stiffness elements store and release the potential energy of a system. To examine how the potential energy is defined, let us consider the illustration shown in Figure 2.4, in which a spring is held fixed at end O, and at the other end, a force of magnitude F is directed along the direction of the unit vector j. Under the action of this force, let the element stretch from an initial or unstretched length L_o to a length $L_o + x$ along the direction of j. In undergoing this deformation, the relationship between F and x can be linear or nonlinear as discussed subsequently.

If F_s represents the internal force acting within the stiffness element, as shown in the free-body diagram in Figure 2.4b, then in the lower spring portion this force is equal and opposite to the external force F; that is,

$$F_s = -Fj$$

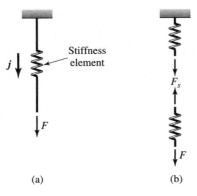

FIGURE 2.4
(a) Stiffness element with a force acting on it and (b) its free-body diagram.

Since the force F_s tries to restore the stiffness element to its undeformed configuration, it is referred to as a *restoring force*. As the stiffness element is deformed, energy is stored in this element, and as the stiffness element is undeformed, energy is released. *The potential energy V is defined[2] as the work done to take the stiffness element from the deformed position to the undeformed position; that is, the work needed to undeform the element to its original shape.* For the element shown in Figure 2.4, this is given by

$$V(x) = \int_x^0 F_s \cdot dx$$

$$= \int_x^0 -Fj \cdot dxj = \int_0^x F dx \tag{2.8}$$

where we have used the identity $j \cdot j = 1$ and $F_s = -Fj$. Like the kinetic energy T, the potential energy V is a scalar-valued function.

The relationship between the deformation experienced by a spring and an externally applied force may be linear as discussed in Section 2.3.2 or nonlinear as discussed in Section 2.3.3. The notion of an equivalent spring element is also introduced in Section 2.3.2.

---

[2]A general definition of potential energy V takes the form

$$V(x) = \int_{\text{deformed position}}^{\text{initial or referene position}} F_s \cdot dx$$

where the force F_s is a conservative force. The work done by a conservative force is independent of the path followed between the initial and final positions.

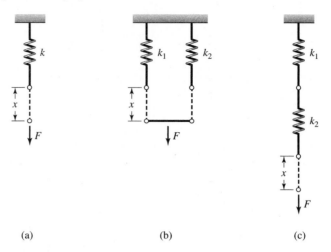

(a)   (b)   (c)

FIGURE 2.5
Various spring configurations: (a) single spring, (b) two springs in parallel, and (c) two springs in series.

2.3.2 Linear Springs

Translation Spring

If a force F is applied to a linear spring as shown in Figure 2.5a, this force produces a deflection x such that

$$F(x) = kx \tag{2.9}$$

where the coefficient k is called the *spring constant* and there is a linear relationship between the force and the displacement. Based on Eqs. (2.8) and (2.9), the potential energy V stored in the spring is given by

$$V(x) = \int_0^x F(x)dx = \int_0^x kxdx = k\int_0^x xdx = \frac{1}{2}kx^2 \tag{2.10}$$

Hence, for a linear spring, the associated potential energy is linearly proportional to the spring stiffness k and proportional to the second power of the displacement magnitude.

Torsion Spring

If a linear torsion spring is considered and if a moment τ is applied to the spring at one end while the other end of the spring is held fixed, then

$$\tau(\theta) = k_t\theta \tag{2.11}$$

where k_t is the spring constant and θ is the deformation of the spring. The potential energy stored in this spring is

$$V(\theta) = \int_0^\theta \tau(\theta)d\theta = \int_0^\theta k_t\theta d\theta = \frac{1}{2}k_t\theta^2 \tag{2.12}$$

Combinations of Linear Springs

Different combinations of linear spring elements are now considered and the equivalent stiffness of these combinations is determined. First, combinations of translation springs shown in Figures 2.5b and 2.5c are considered and following that, combinations of torsion springs shown in Figures 2.6a and 2.6b are considered.

When there are two springs in parallel as shown in Figure 2.5b and the bar on which the force F acts remains parallel to its original position, then the displacements of both springs are equal and, therefore, the total force is

$$
\begin{aligned}
F(x) &= F_1(x) + F_2(x) \\
&= k_1x + k_2x = (k_1 + k_2)x = k_ex
\end{aligned} \tag{2.13}
$$

where $F_j(x)$ is the resulting force in spring k_j, $j = 1, 2$, and k_e is the equivalent spring constant for two springs in parallel given by

$$k_e = k_1 + k_2 \tag{2.14}$$

When there are two springs in series, as shown in Figure 2.5c, the force on each spring is the same and the total displacement is

$$
\begin{aligned}
x &= x_1 + x_2 \\
&= \frac{F}{k_1} + \frac{F}{k_2} = \left(\frac{1}{k_1} + \frac{1}{k_2}\right)F = \frac{F}{k_e}
\end{aligned} \tag{2.15}
$$

where the equivalent spring constant k_e is

$$k_e = \left(\frac{1}{k_1} + \frac{1}{k_2}\right)^{-1} \tag{2.16}$$

In general, for N springs in parallel, we have

$$k_e = \sum_{i=1}^{N} k_i \tag{2.17}$$

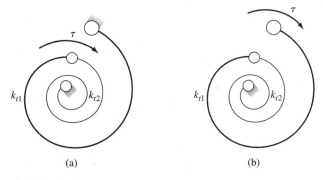

(a) (b)

FIGURE 2.6
Two torsion springs: (a) parallel combination and (b) series combination.

and for N springs in series, we have

$$k_e = \left[\sum_{i=1}^{N} \frac{1}{k_i} \right]^{-1} \tag{2.18}$$

For two torsion springs in series and parallel combinations, we refer to Figure 2.6. From Figure 2.6a, the rotation θ of each spring is the same and, therefore,

$$\tau(\theta) = \tau_1(\theta) + \tau_2(\theta)$$
$$= k_{t1}\theta + k_{t2}\theta = (k_{t1} + k_{t2})\theta = k_{te}\theta \tag{2.19}$$

where τ_j is the resulting moment in spring $k_{tj}, j = 1,2$, and k_{te} is the equivalent torsional stiffness given by

$$k_{te} = k_{t1} + k_{t2} \tag{2.20}$$

For torsion springs in series, as shown in Figure 2.6b, the torque on each spring is the same, but the rotations are unequal. Thus,

$$\theta = \theta_1 + \theta_2 = \frac{\tau}{k_{t1}} + \frac{\tau}{k_{t2}} = \left(\frac{1}{k_{t1}} + \frac{1}{k_{t2}} \right)\tau = \frac{\tau}{k_{te}} \tag{2.21}$$

where the equivalent stiffness k_{te} is

$$k_{te} = \left(\frac{1}{k_{t1}} + \frac{1}{k_{t2}} \right)^{-1} \tag{2.22}$$

Equivalent Spring Constants of Common Structural Elements Used in Vibration Models

To determine the spring constant for numerous elastic structural elements one can make use of known relationships between force and displacement. Several such spring constants that have been determined for different geometry and loading conditions are presented in Table 2.3. For modeling purposes, the inertias of the structural elements such as the beams of cases 4 to 6 in Table 2.3 are usually ignored. In Chapter 9, it is shown under what conditions it is reasonable to make such an assumption.[3]

Since it may not always be possible to obtain a spring constant for a given system through analysis, often one has to experimentally determine this constant. As a representative example, let us return to Figure 2.4a, and consider the experimental determination of the spring constant for this system. The loading F is gradually increased to a chosen value and the resulting deflection x from the unstretched position of the element is recorded for each value of F. These data are plotted in Figure 2.7, where open squares are used to denote the values of the experimentally obtained data. Then, assuming that the stiffness element is linear, curve fitting is done to estimate the unknown parameter k in Eq. (2.9). The resulting value of the spring constant is also shown in Figure 2.7. *Note that the stiffness constant k is a static concept, and hence, a static loading is sufficient to determine this parameter.*

---

[3]See Eqs. (9.119) and (9.178).

TABLE 2.3
Spring Constants for Some
Common Elastic Elements

1	Axially loaded rod or cable		$k = \dfrac{AE}{L}$
2	Axially loaded tapered rod		$k = \dfrac{\pi E d_1 d_2}{4L}$
3	Hollow circular rod in torsion		$k_t = \dfrac{GI}{L}$ $\qquad$ $I = \dfrac{\pi(d_{\text{out}}^4 - d_{\text{in}}^4)}{32}$
4	Cantilever beam		$k = \dfrac{3EI}{a^3};\ 0 < a \le l$
5	Pinned-pinned beam (Hinged, simply supported)		$k = \dfrac{3EI(a+b)}{a^2 b^2}$
6	Clamped-clamped beam (Fixed-fixed beam)		$k = \dfrac{3EI(a+b)^3}{a^3 b^3}$
7	Two circular rods in torsion		$k_{te} = k_{t1} + k_{t2}$ $\qquad$ $k_{t_i} = \dfrac{G_i I_i}{L_i}$
8	Two circular rods in torsion		$k_{te} = \left(\dfrac{1}{k_{t1}} + \dfrac{1}{k_{t2}} \right)^{-1}$ $\qquad$ $k_{t_i} = \dfrac{G_i I_i}{L_i}$
9	Coil spring		$k = \dfrac{G d^4}{8 n D^3}$

A: area of cross section; *E*: Young's modulus; *G*: shear modulus; *I*: area moment of inertia or polar moment of inertia

Force-displacement relationships other than Eq. (2.9) may also be used to determine parameters such as k that characterize a stiffness element. Determination of parameters for a nonlinear spring is discussed in Section 2.3.3. In a broader context, procedures used to determine parameters such as k of a vibratory system element fall under the area called system identification or parameter identification; identification and estimation of parameters of vibratory systems are addressed in the field of *experimental modal analysis*.[4]

[4]D. J. Ewins, *Modal Testing: Theory and Practice*, John Wiley and Sons, NY (1984).

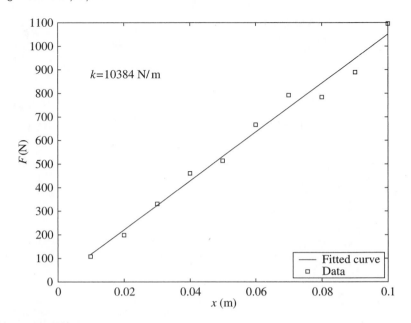

FIGURE 2.7
Experimentally obtained data used to determine the linear spring constant k.

In experimental modal analysis, dynamic loading is used for parameter estimation. A further discussion is provided in Chapter 5, when system input-output relations (transfer functions and frequency response functions) are considered.

Next, some examples are considered to illustrate how the information shown in Table 2.3 can be used to determine equivalent spring constants for different physical configurations.

EXAMPLE 2.3   Equivalent stiffness of a beam-spring combination

Consider the combinations shown in Figure 2.8. In Figures 2.8a and 2.8b, we have a cantilever beam that has a spring attached at its free end. In Figure 2.8a, the force is applied to the free end of the spring. In this case, the forces acting on the cantilever beam and the spring are the same as seen from the associated free-body diagram, and, hence, the springs are in series. Thus, from Eq. (2.18), we arrive at

$$k_e = \left(\frac{1}{k_{beam}} + \frac{1}{k_1} \right)^{-1} \tag{a}$$

where, from Table 2.3,

$$k_{beam} = \frac{3EI}{L^3} \tag{b}$$

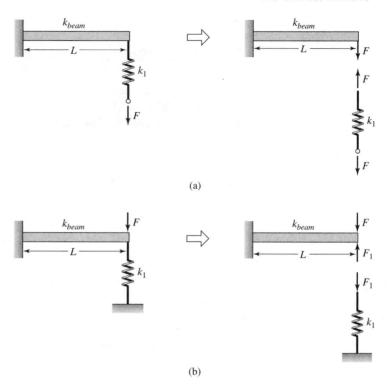

FIGURE 2.8
Spring combinations and free-body diagrams.

In Figure 2.8b, the force is applied simultaneously to the free end of the cantilever beam and a linear spring of stiffness k_1. In this case, the displacements at the attachment point of the cantilever and the spring are equal and the springs are in parallel. Thus, from Eq. (2.17), we find that

$$k_e = k_{beam} + k_1 \qquad\qquad (c)$$

EXAMPLE 2.4 Equivalent stiffness of a beam with a fixed end and a translating support at the other end

In Figure 2.9, a uniform beam of length L and flexural rigidity EI, where E is the Young's modulus of elasticity and I is the area moment of inertia about the bending axis, is shown. This beam is fixed at one end and free to translate along the vertical direction at the other end with the restraint that the beam slope is zero at this end. The equivalent stiffness of this beam is to be determined when the beam is subjected to a transverse loading F at the translating support end.

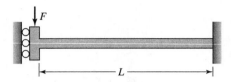

FIGURE 2.9
Beam fixed at one end and free to translate at the other end.

Since the midpoint of a fixed-fixed beam with a transverse load at its middle behaves like a beam with a translating support end, we first determine the equivalent stiffness of a fixed-fixed beam of length $2L$ that is loaded at its middle. To this end, we use Case 6 of Table 2.3 and set $a = b = L$ and obtain

$$k_{fixed} = \frac{3EI(a+b)^3}{a^3 b^3}\bigg|_{a=b=L} = \frac{3EI(L+L)^3}{L^3 L^3} = \frac{24EI}{L^3} \tag{a}$$

Recognizing that the equivalent stiffness of a fixed-fixed beam of length $2L$ loaded at its middle is equal to the total equivalent stiffness of a parallel-spring combination of two end loaded beams of the form shown in Figure 2.9, we obtain from Eq. (2.17) and Eq. (a) that

$$k_e = \frac{1}{2} k_{fixed} = \frac{1}{2} \frac{24EI}{L^3} = \frac{12EI}{L^3} \tag{b}$$

EXAMPLE 2.5 Equivalent stiffness of springs in parallel: removal of a restriction

Let us reexamine the pair of springs in parallel shown in Figure 2.5b. Now, however, we remove the restriction that the bar to which the force is applied has to remain parallel to its original position. Then, we have the configuration shown in Figure 2.10. The equivalent spring constant for this configuration will be determined.

From similar triangles, we see that

$$x = x_2 + \frac{b}{a+b}(x_1 - x_2)$$

$$= \frac{b}{a+b}x_1 + \frac{a}{a+b}x_2 \tag{a}$$

If F_1 is the force acting on k_1 and F_2 is the force acting on k_2, then from the summation of forces and moments on the bar we obtain, respectively,

$$F = F_1 + F_2$$

$$bF_2 = aF_1 \tag{b}$$

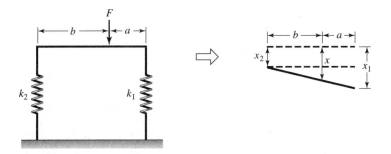

FIGURE 2.10
Parallel springs subjected to unequal forces.

Thus, Eqs. (b) lead to

$$F_1 = \frac{bF}{a + b}$$

$$F_2 = \frac{aF}{a + b} \tag{c}$$

Therefore,

$$x_1 = \frac{F_1}{k_1} = \frac{bF}{k_1(a + b)}$$

$$x_2 = \frac{F_2}{k_2} = \frac{aF}{k_2(a + b)} \tag{d}$$

From Eqs. (a) and (d), we obtain

$$x = \frac{b}{(a + b)}\left[\frac{bF}{k_1(a + b)}\right] + \frac{a}{(a + b)}\left[\frac{aF}{k_2(a + b)}\right]$$

$$= \frac{F}{(a + b)^2}\left[\frac{k_2b^2 + k_1a^2}{k_1k_2}\right] \tag{e}$$

Comparing Eq. (e) to the form

$$x = \frac{F}{k_e} \tag{f}$$

we find that the equivalent spring constant k_e is given by

$$k_e = \frac{k_1k_2(a + b)^2}{k_2b^2 + k_1a^2} \tag{g}$$

We see that even when the force is applied at the center of the bar (i.e., $a = b$) and $k_1 \neq k_2$ this result does not reduce to the result for the system of Figure 2.5b, where the loaded bar remains parallel to its original position. These two cases are identical only when $k_2b = k_1a$.

EXAMPLE 2.6 Equivalent stiffness of a cantilever beam with a transverse end load

A cantilever beam, which is made of an alloy with the Young's modulus of elasticity $E = 72 \times 10^9$ N/m^2, is loaded transversely at its free end. If the length of the beam is 750 mm and the beam has an annular cross-section with inner and outer diameters of 110 mm and 120 mm, respectively, then determine the equivalent stiffness of this beam.

For the given loading, the equivalent stiffness of the cantilever beam is found from Case 4 of Table 2.3 to be

$$k = \frac{3EI}{L^3} \tag{a}$$

where the area moment of inertia I about the bending axis is determined as

$$\begin{aligned} I &= \frac{\pi}{32}(d_{outer}^4 - d_{inner}^4) \\ &= \frac{\pi}{32}[(120 \times 10^{-3})^4 - (110 \times 10^{-3})^4] \\ &= 5.98 \times 10^{-6} \text{ m}^4 \end{aligned} \tag{b}$$

Then, from Eq. (a)

$$\begin{aligned} k &= \frac{3 \times (72 \times 10^9) \times (5.98 \times 10^{-6})}{(750 \times 10^{-3})^3} \text{ N/m} \\ &= 3.06 \times 10^6 \text{ N/m} \end{aligned} \tag{c}$$

When the length is increased from 750 mm to twice its value—that is, to 1500 mm—the stiffness decreases by eightfold from 3.06×10^6 N/m to 0.39×10^6 N/m.

EXAMPLE 2.7 Equivalent stiffness of a microelectromechanical system (MEMS) fixed-fixed flexure[5]

A microelectromechanical sensor system (MEMS) consisting of four flexures is shown in Figure 2.11. Each of these flexures is fixed at one end and connected to a mass at the other end. The length of each flexure is L, the thickness of each flexure is h, and the width of each flexure is b. A transverse loading acts on the mass along the Z-direction, which is normal to the X-Y plane. Each flexure is

[5]G. K. Fedder, "Simulation of Microelectromechanical Systems," Ph.D. dissertation, Department of Electrical Engineering and Computer Sciences, University of California, Berkeley, CA (1994).

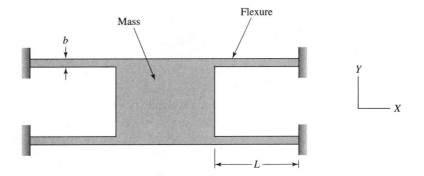

FIGURE 2.11
Fixed-fixed flexure used in a microelectromechanical system.
Source: From G. K. Fedder, "Simulation of Microelectromechanical Systems,"
Ph.D. Dissertation, Department of Electrical Engineering and Computer
Sciences, University of California, Berkeley, CA, (1994). Reprinted with
permission of the author.

fabricated from a polysilicon material, which has a Young's modulus of elasticity $E = 150$ GPa. If the length of each flexure is 100 μm and the width and
thickness are each 2 μm, then determine the equivalent stiffness of the system.

Each of the four flexures can be treated as a beam that is fixed at one end
and free to translate only at the other end, similar to the system shown in Figure 2.9. This means that the equivalent stiffness of each flexure is given by
Eq. (b) of Example 2.4 as

$$k_{flexure} = \frac{12EI}{L^3} \tag{a}$$

where the area moment of inertia is given by

$$I = \frac{bh^3}{12} \tag{b}$$

because the bending axis is along the Y direction. Since each of the four flexures experiences the same displacement at its end in the Z direction, this is a
combination of four stiffness elements in parallel; hence, the equivalent stiffness of the system is given by

$$k_e = 4 \times k_{flexure}$$
$$= \frac{48\,EI}{L^3} \tag{c}$$

Thus,

$$k_e = \frac{48(150 \times 10^9)(2 \times 10^{-6})(2 \times 10^{-6})^3}{12(100 \times 10^{-6})^3} \text{ N/m}$$
$$= 9.6 \text{ N/m} \tag{d}$$

2.3.3 Nonlinear Springs

Nonlinear stiffness elements appear in many applications, including leaf springs in vehicle suspensions and uniaxial microelectromechanical devices in the presence of electrostatic actuation.[6] For a nonlinear spring, the spring force $F(x)$ is a nonlinear function of the displacement variable x. A series expansion of this function can be interpreted as a combination of linear and nonlinear spring components.

For a stiffness element with a linear spring element and a cubic nonlinear spring element, the force-displacement relationship is written as

$$F(x) = \underbrace{kx}_{\substack{\text{Linear spring} \\ \text{element}}} + \underbrace{\alpha k x^3}_{\substack{\text{Nonlinear spring} \\ \text{element}}} \tag{2.23}$$

where α is used to express the stiffness coefficient of the nonlinear term in terms of the linear spring constant k. (This notation will be used later in Sections 4.6.1 and 5.10.) The quantity α can be either positive or negative. A spring element for which α is positive is called a *hardening spring,* and a spring element for which α is negative is called a *softening spring.* From Eq. (2.23), the potential energy V is

$$V(x) = \int_0^x F(x)dx = \frac{1}{2}kx^2 + \frac{1}{4}\alpha k x^4 \tag{2.24}$$

Hence, for a nonlinear spring, the potential energy is proportional to powers of displacement magnitude higher than two.

For a linear stiffness element, the force versus displacement graph is a straight line, and the slope of this line gives the stiffness constant k. For a nonlinear stiffness element described by Eq. (2.23), the graph is no longer a straight line. The slope of this graph at a location $x = x_l$ is given by

$$\left.\frac{dF}{dx}\right|_{x=x_l} = (k + 3\alpha k x^2)|_{x=x_l}$$
$$= (k + 3\alpha k x_l^2) \tag{2.25}$$

Thus, in the vicinity of displacements in a neighborhood of $x = x_l$, the cubic nonlinear stiffness element may be replaced by a linear stiffness element with a stiffness constant given by Eq. (2.25).

The constant of proportionality αk for the nonlinear cubic spring is determined experimentally in the same manner that the constant k was determined for a linear spring, as discussed in Section 2.3.2. The results of a typical experiment are shown in Figure 2.12. The fitted line is determined using standard nonlinear curve fitting techniques.[7]

---

[6]S. G. Adams et al., "Independent Tuning of Linear and Nonlinear Stiffness Coefficients," *J. Microelectromechanical Systems,* Vol. 7, No. 2 (June 1998).

[7]The MATLAB function `lsqcurvefit` from the Optimization Toolbox was used.

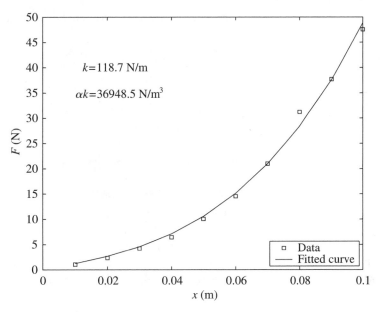

FIGURE 2.12
Experimentally obtained data used to determine the nonlinear spring constant αk.

EXAMPLE 2.8 Nonlinear stiffness due to geometry

In this example, we illustrate that for certain geometric conditions, even a linear spring can give rise to nonlinear stiffness in a vibratory system. One such system is shown in Figure 2.13, where the stretching of the linear spring is described by using Eq. (2.9). We assume that when $\gamma = 0$, the spring is under an initial deflection δ_o, and the spring, therefore, is initially under a tension force $T_o = k\delta_o$. When the spring is moved up or down an amount x in the vertical direction, the force in the spring is

$$F_s(x) = k\delta_o + k(\sqrt{L^2 + x^2} - L) \qquad \text{(a)}$$

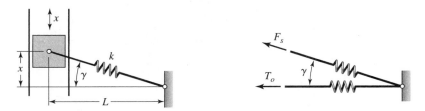

FIGURE 2.13
Nonlinear stiffness due to geometry: spring under an initial tension, one end of which is constrained to move in the vertical direction.

The force in the x-direction is obtained from Eq. (a) as

$$F_x(x) = F_s \sin \gamma = \frac{F_s x}{\sqrt{L^2 + x^2}}$$

$$= \frac{x k \delta_o}{\sqrt{L^2 + x^2}} + \frac{kx(\sqrt{L^2 + x^2} - L)}{\sqrt{L^2 + x^2}} \qquad \text{(b)}$$

which clearly shows that the spring force opposing the motion is a nonlinear function of the displacement x. Hence, a vibratory model of the system shown in Figure 2.13 will have nonlinear stiffness.

Cubic Springs and Linear Springs

If, in Eq. (b), we assume that $|x/L| \ll 1$ and expand the denominator of each term on the right-hand side of Eq. (b) as a binomial expansion and keep only the first two terms, we obtain

$$F_x(x) = k\delta_o \frac{x}{L} + \frac{k}{2}(L - \delta_o)\left(\frac{x}{L}\right)^3 \qquad \text{(c)}$$

When the nonlinear term is negligible, Eq. (c) leads to the following linear relationship

$$F_x(x) = k\delta_o \frac{x}{L} = T_o \frac{x}{L} \qquad \text{(d)}$$

From Eq. (d), it is seen that the spring constant is proportional to the initial tension in the spring.

Another example of a nonlinear spring is one that is piecewise linear as shown in Figure 2.14. Here, each spring is linear; however, as the deflection increases, another linear spring comes into play and the spring constant changes (increases). An illustration of the effects of this type of spring on a vibrating system is given in Section 4.6.1.

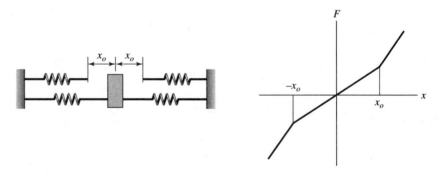

FIGURE 2.14
Nonlinear spring composed of a set of linear springs.

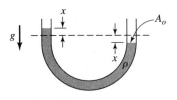

FIGURE 2.15
Manometer.

2.3.4 Other Forms of Potential Energy Elements

In the previous two subsections, we saw stiffness elements in which the source of the restoring force is a structural element. Here, we consider other stiffness elements in which there is a mechanism for storing and releasing potential energy. The source of the restoring force is a fluid element or a gravitational loading.

Fluid Element

As an example of a fluid element, consider the manometer shown in Figure 2.15 in which the fluid is displaced by an amount x in one leg of the manometer.[8] Consequently, the fluid has been displaced a total of $2x$. If the fluid has a mass density ρ and the manometer has an area A_o, then the magnitude of the total force of the displaced fluid acting on the rest of the fluid is

$$F_m(x) = 2\rho g A_o x \tag{2.26}$$

Consequently, the equivalent spring constant of this fluid system is

$$
\begin{aligned}
k_e &= \frac{dF_m}{dx} \\
&= 2\rho g A_o
\end{aligned}
\tag{2.27}
$$

from which it is clear that the fluid-element stiffness depends on the mass density ρ, manometer cross-section area A_o, and the acceleration due to gravity g. The corresponding potential energy is

$$V(x) = \frac{1}{2} k_e x^2 = \rho g A_o x^2 \tag{2.28}$$

Alternatively, the potential energy can also be obtained directly from the work done

$$
\begin{aligned}
V(x) &= \int_0^x F_m(x)dx \\
&= 2\rho g A_o \int_0^x x\,dx = \rho g A_o x^2
\end{aligned}
\tag{2.29}
$$

Compressed Gas

Consider the piston shown in Figure 2.16, in which gas is stored at a pressure P_o and entrained in the volume $V_o = A_o L_o$, where A_o is the area of the piston and L_o is the original length of the cylindrical cavity. Thus, when the piston

[8]For a proposed practical application of this type of system see: S. D. Xue, "Optimum Parameters of Tuned Liquid Column Damper for Suppressing Pitching Vibration of an Undamped Structure," *J. Sound Vibration,* Vol. 235, No. 2, pp. 639–653 (2000).

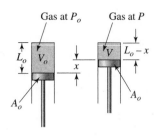

FIGURE 2.16
Gas compression with a piston.

moves by an amount x along the axis of the piston, V_o decreases to a volume V_c, where

$$V_c(x) = V_0 - A_0 x$$

$$= A_0 L_0 \left(1 - \frac{x}{L_0}\right) \tag{2.30}$$

The equation of state for the gas is

$$PV_c^n = P_o V_o^n = c_o = \text{constant} \tag{2.31}$$

When a gas is compressed slowly, the compression is isothermal and $n = 1$. When it is compressed rapidly, the compression is adiabatic and $n = c_p/c_v$, the ratio of specific heats of the gas, which for air is 1.4. To determine the spring constant, we note that the magnitude of the force on the piston is

$$F = A_o P = A_o c_o V_c^{-n} \tag{2.32}$$

Thus, upon substitution of V_c given by Eq. (2.30), we obtain

$$F(x) = A_o c_o V_o^{-n}(1 - x/L_o)^{-n} = A_o P_o (1 - x/L_o)^{-n} \tag{2.33}$$

Thus, the pressure-filled gas element provides the stiffness. Equation (2.33) describes a nonlinear force versus displacement relationship. As discussed earlier, this relationship may be replaced in the vicinity of $x = x_l$ by a straight line with a slope k_e, where k_e is the stiffness of an equivalent linear stiffness element. This equivalent stiffness is given by

$$k_e = \left. \frac{dF}{dx} \right|_{x=x_l}$$

$$= \frac{n A_o P_o}{L_o}(1 - x_l/L_o)^{-n-1} \tag{2.34a}$$

For $|x_l/L_o| << 1$, Eq. (2.34a) is simplified to

$$k_e = \frac{n A_o P_o}{L_o} \tag{2.34b}$$

For arbitrary x/L_o, the potential energy $V(x)$ is determined by using Eq. (2.33) as

$$V(x) = \int_0^x F(x)dx = A_o P_o \int_0^x (1 - x/L_o)^{-n}dx$$

$$= -A_o P_o L_o \ln(1 - x/L_o) \qquad n = 1$$

$$= \frac{A_o P_o L_o}{n - 1}[(1 - x/L_o)^{1-n} - 1] \qquad n \neq 1 \tag{2.35}$$

Pendulum System

Figure 2.17(a) Consider the bar of uniformly distributed mass m shown in Figure 2.17a that is pivoted at its top. Let the reference position be located at a distance $L/2$ below the pivot point, where the center of mass is located

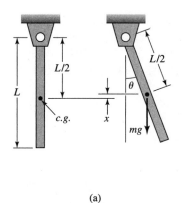

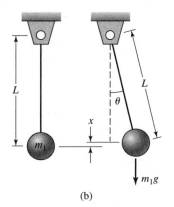

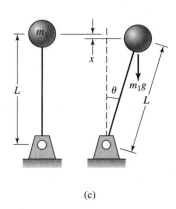

(a)   (b)   (c)

FIGURE 2.17
Pendulum systems: (a) bar with uniformly distributed mass; (b) mass on a weightless rod; and (c) inverted mass on a weightless rod.

when the pendulum is at $\theta = 0$. When the bar rotates either clockwise or counterclockwise, the vertical distance through which the center of gravity of the bar moves up from the reference position is

$$x = \frac{L}{2} - \frac{L}{2}\cos\theta = \frac{L}{2}(1 - \cos\theta) \tag{2.36}$$

Since $F = mg$, the increase in the potential energy is

$$V(x) = \int_0^x F(x)dx = \int_0^x mgdx = mgx \tag{2.37a}$$

or, from Eq. (2.36),

$$V(\theta) = \frac{mgL}{2}(1 - \cos\theta) \tag{2.37b}$$

When the angle of rotation θ about the upright position $\theta = 0$ is "small," we can use the Taylor series approximation[9]

$$\cos\theta = 1 - \frac{\theta^2}{2} + \cdots \tag{2.38}$$

and substitute this expression into Eq. (2.37b) to obtain

$$V(\theta) \approx \frac{1}{2}\left(\frac{mgL}{2}\right)\theta^2 = \frac{1}{2}k_e\theta^2 \tag{2.39}$$

where the equivalent spring constant is

$$k_e = \frac{mgL}{2} \tag{2.40}$$

[9]T. B. Hildebrand, *Advanced Calculus for Applications*, Prentice Hall, Englewood Cliffs NJ (1976).

Figure 2.17(b)   In a similar manner, for "small" rotations about the upright position $\theta = 0$ in Figure 2.17b, we obtain the increase in potential energy for the system. Here it is assumed that a weightless bar supports the mass m_1. Choosing the reference position as the bottom position, we obtain

$$V(\theta) \approx \frac{1}{2} m_1 g L \theta^2 = \frac{1}{2} k_e \theta^2 \tag{2.41}$$

where the equivalent spring constant is

$$k_e = m_1 g L \tag{2.42}$$

In the configuration shown in Figure 2.17b, if the weightless bar is replaced by one that has a uniformly distributed mass m, then the total potential energy of the bar and the mass is

$$V(\theta) \approx \frac{1}{4} m g L \theta^2 + \frac{1}{2} m_1 g L \theta^2 = \frac{1}{2} \left(\frac{m}{2} + m_1 \right) g L \theta^2 = \frac{1}{2} k_e \theta^2 \tag{2.43}$$

where the equivalent spring constant is

$$k_e = \left(\frac{m}{2} + m_1 \right) g L \tag{2.44}$$

Figure 2.17(c)   When the pendulum is inverted as shown in Figure 2.17c, then there is a decrease in potential energy; that is,

$$V(\theta) \approx -\frac{1}{2} m_1 g L \theta^2 \tag{2.45}$$

EXAMPLE 2.9   Equivalent stiffness due to gravity loading

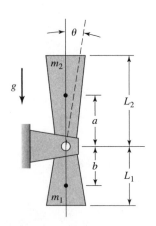

FIGURE 2.18
Pivoted bar with unequally distributed mass along its length.

Consider the pivoted bar shown in Figure 2.18. The lower portion of the bar has a mass m_1 and the upper portion a mass m_2. The distance of the center of gravity of the upper portion of the bar to the pivot is a and the distance of the center of gravity of the lower portion of the bar to the pivot is b. For "small" rotations about the upright position $\theta = 0$, the potential energy is

$$V(\theta) = \frac{1}{2} m_1 g b \theta^2 - \frac{1}{2} m_2 g a \theta^2 = \frac{1}{2} (m_1 b - m_2 a) g \theta^2 \tag{a}$$

There is a gain or loss in potential energy depending on whether $m_1 b > m_2 a$ or *vice versa*.

A special case of Eq. (a) is when the bar has a uniformly distributed mass m so that

$$m_1 = \frac{m L_1}{L}$$

$$m_2 = \frac{m L_2}{L} \tag{b}$$

where $L = L_1 + L_2$. Then, since $b = L_1/2$ and $a = L_2/2$, Eq. (a) becomes

$$V(\theta) = \left(\frac{L_1^2}{4(L_1 + L_2)} - \frac{L_2^2}{4(L_1 + L_2)} \right) mg\theta^2$$

$$= \frac{1}{2} \frac{(L_1 - L_2)}{2} mg\theta^2 = \frac{1}{2} k_e \theta^2 \qquad \text{(c)}$$

where the equivalent stiffness k_e is

$$k_e = \frac{(L_1 - L_2)}{2} mg \qquad \text{(d)}$$

Hence, the stiffness is due to gravity loading. When $L_1 = L_2$, $V = 0$. It is mentioned that if the systems were to lie in the plane of the page, that is, normal to the direction of gravity, then there is no restoring force when the bar is rotated an amount θ and these relationships are not applicable. Furthermore, when $L_2 = 0$, Eq. (c) reduces to Eq. (2.39) with L replaced by L_1.

2.4 DISSIPATION ELEMENTS

Damping elements are assumed to have neither inertia nor the means to store or release potential energy. The mechanical motion imparted to these elements is converted to heat or sound and, hence, they are called *nonconservative* or *dissipative* because this energy is not recoverable by the mechanical system. There are four common types of damping mechanisms used to model vibratory systems. They are (i) viscous damping, (ii) Coulomb or dry friction damping, (iii) material or solid or hysteretic damping, and (iv) fluid damping. In all these cases, the damping force is usually expressed as a function of velocity.

2.4.1 Viscous Damping

When a viscous fluid flows through a slot or around a piston in a cylinder, the damping force generated is proportional to the relative velocity between the two boundaries confining the fluid. A common representation of a viscous damper is a cylinder with a piston head, as shown in Figure 2.19. In this case, the piston head moves with a speed $\dot{x}$ relative to the cylinder housing, which is fixed. The damper force magnitude F always acts in a direction opposite to that of velocity. Depending on the damper construction and the velocity range, the magnitude of the damper force $F(\dot{x})$ is a nonlinear function of velocity or can be approximated as a linear function of velocity. In the linear case, the relationship is expressed as

$$F(\dot{x}) = c\dot{x} \qquad (2.46)$$

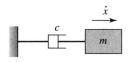

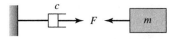

FIGURE 2.19

Representation of a viscous damper.

where the constant of proportionality denoted by c is called the *damping coefficient*. The damping coefficient c has units of N/(m/s). Viscous damping of the form given by Eq. (2.46) is also called *slow-fluid damping*.

In the case of a nonlinear viscous damper described by a function $F(\dot{x})$, the equivalent linear viscous damping around an operating speed $\dot{x} = \dot{x}_l$ is determined as follows:

$$c_e = \left. \frac{dF(\dot{x})}{d\dot{x}} \right|_{\dot{x}=\dot{x}_l} \tag{2.47}$$

Linear viscous damping elements can be combined in the same way that linear springs are, except that the forces are proportional to velocity instead of displacement.

Energy Dissipation

The energy dissipated by a linear viscous damper is given by

$$E_d = \int F dx = \int F \dot{x} dt = \int c\dot{x}^2 dt = c \int \dot{x}^2 dt \tag{2.48}$$

Parallel-Plate Damper

An example of a viscous damper is shown in Figure 2.20, which is used to illustrate how the damping coefficient c depends on fluid viscosity denoted by μ. The system consists of two parallel plates with a viscous fluid layer of height h confined between them. The top plate, which moves with a speed $\dot{x}$ relative to the lower plate, has a surface area A. For the damper construction shown in Figure 2.20, and assuming that the fluid behaves like a Newtonian fluid, the shear force acting on the bottom plate is written as

$$F(\dot{x}) = \underbrace{\left(\frac{\mu \dot{x}}{h} \right)}_{\substack{\text{Shear stress} \\ \text{acting on} \\ \text{top plate}}} A = \frac{\mu A}{h} \dot{x} \tag{2.49}$$

Equation (2.49) has the same form as Eq. (2.46), where the damping coefficient c for the parallel-plate construction of Figure 2.20 is given by

$$c = \frac{\mu A}{h} \tag{2.50}$$

From Eq. (2.50), it is clear that the damping coefficient is linearly proportional to fluid viscosity μ, linearly proportional to the surface area A of the top plate, and inversely proportional to the separation h between the two plates. Thus, as

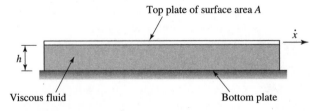

FIGURE 2.20
Parallel-plate viscous damper construction.

the area A is increased and/or the separation h is decreased, the damping coefficient is increased. However, there are limits to the area A and the separation h that one can realize. Hence, alternate damper designs are often used.

EXAMPLE 2.10   Design of a parallel-plate damper

A parallel-plate damper with a top plate of dimensions 100 mm $\times$ 100 mm is to be pulled across an oil layer of thickness 0.2 mm, which is confined between the moving plate and a fixed plate. We are given that this oil is SAE 30 oil, which has a viscosity of 345 mPa·s (345×10^{-3} N·s/m^2).

We shall determine the viscous damping coefficient of this system. To this end, we make use of Eq. (2.50) and substitute the given values into this expression and find that

$$c = \frac{(345 \times 10^{-3} \text{ N/m}^2\text{·s}) \times (100 \times 10^{-3}\text{m} \times 100 \times 10^{-3} \text{ m})}{0.2 \times 10^{-3}\text{m}}$$

$$= 17.25 \text{ N·s/m}$$

EXAMPLE 2.11   Equivalent damping coefficient and equivalent stiffness of a vibratory system

Consider the vibratory system shown in Figure 2.21a in which the motion of mass m is restrained by a set of linear springs and linear viscous dampers. A free-body diagram of this system is shown in Figure 2.21b. In determining

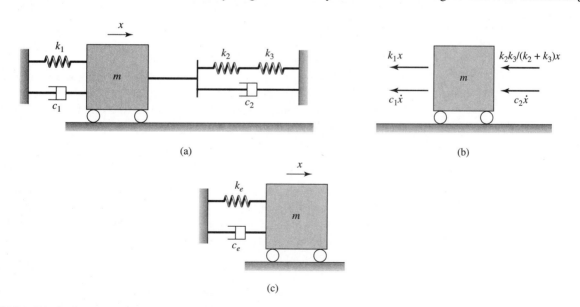

(a)

(b)

(c)

FIGURE 2.21
(a) Linear vibratory system; (b) free-body diagram of mass m; and (c) equivalent system.

the spring force generated by the springs k_2 and k_3, we have made use of the fact these springs are in series and used Eq. (2.18).

In Figure 2.21c, the springs and dampers shown in Figure 2.21a have been collected and expressed as an equivalent spring and equivalent damper combination. Thus, we have

$$k_e = k_1 + \frac{k_2 k_3}{k_2 + k_3}$$

$$c_e = c_1 + c_2$$

EXAMPLE 2.12   Equivalent linear damping coefficient of a nonlinear damper

It has been experimentally determined that the damper force-velocity relationship is given by the function

$$F(\dot{x}) = (4 \text{ N·s/m})\dot{x} + (0.3 \text{ N·s}^3/\text{m}^3)\dot{x}^3 \tag{a}$$

We shall determine the equivalent linear damping coefficient around an operating speed of 3 m/s. To determine this damping coefficient, we use Eq. (2.47) and Eq. (a) and arrive at

$$
\begin{aligned}
c_e = \left.\frac{dF(\dot{x})}{d\dot{x}}\right|_{\dot{x}=3 \text{ m/s}} &= 4 \text{ N·s/m} + (0.9 \text{ N·s}^3/\text{m}^3)\dot{x}^2\big|_{\dot{x}=3 \text{ m/s}} \\
&= 4 \text{ N·s/m} + (0.9 \text{ N·s}^3/\text{m}^3) \times (3^2 \text{ m}^2/\text{s}^2) \\
&= 12.1 \text{ N·s/m}
\end{aligned}
\tag{b}
$$

2.4.2   Other Forms of Dissipation

Coulomb Damping or Dry Friction

This type of damping is due to the force caused by friction between two solid surfaces. The force acting on the system must oppose the motion; therefore, the sign of this force must have the opposite sense (direction) of velocity as shown in Figure 2.22. If the kinetic coefficient of friction is μ and the force compressing the surfaces is N, then

$$F(\dot{x}) = \mu N \, \text{sgn}(\dot{x}) \tag{2.51}$$

where sgn is the signum function, which takes on the value of $+1$ for positive values of the argument ($\dot{x}$ in this case), -1 for negative values of the argument, and 0 when the argument is zero. If the normal force is due to the system weight, then $N = mg$ and we have

$$F(\dot{x}) = \mu mg \, \text{sgn}(\dot{x}) \tag{2.52}$$

The energy dissipated in this case is given by

$$
\begin{aligned}
E_d &= \int F dx = \int F \dot{x} dt \\
&= \mu mg \int \text{sgn}(\dot{x}) \dot{x} dt
\end{aligned}
\tag{2.53}
$$

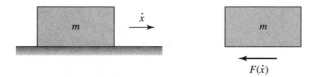

FIGURE 2.22
Dry friction.

Although dry friction can result in loss of efficiency of internal combustion engines, wear on contacting parts, and the loss of position accuracy in servomechanisms, it has been used to enhance the performance of turbomachinery blades, certain built-up structures, and earthquake isolation.[10]

Fluid Damping

This type of damping is associated with a system whose mass is vibrating in a fluid medium. It is often referred to as *velocity-squared damping*.[11] This force always acts in a direction opposite to that of the velocity of the mass. The magnitude of the damping force is given by

$$F(\dot{x}) = c_d \dot{x}^2 \, \text{sgn}(\dot{x}) = c_d \mid \dot{x} \mid \dot{x} \tag{2.54}$$

where

$$c_d = \frac{1}{2} \, C\rho A \tag{2.55}$$

and C is a drag coefficient, A is the projected area of the mass in a direction normal to $\dot{x}$, and ρ is the mass density of the fluid. In Example 6.5, the fluid-damping model is used to study a car seat.

The energy dissipated is

$$\begin{aligned} E_d &= \int F dx = \int F\dot{x} dt \\ &= c_d \int \text{sgn}(\dot{x})\dot{x}^3 dt \end{aligned} \tag{2.56}$$

Fluid damping of the form given by Eq. (2.54) is often referred to as *fast-fluid damping*.

Structural or Solid or Hysteretic Damping

This type of damping describes the losses in materials due to internal friction. The damping force is a function of displacement and velocity and is of the form

$$F = k\pi\beta_h \, \text{sgn}(\dot{x})|x| \tag{2.57}$$

where β_h is an empirically determined constant. The energy dissipated is

$$\begin{aligned} E_d &= \int F dx = \int F\dot{x} dt \\ &= k\pi\beta_h \int \text{sgn}(\dot{x})|x|\dot{x} dt \end{aligned} \tag{2.58}$$

---

[10]A. A. Ferri, "Friction Damping and Isolation Systems," *ASME J. Vibrations Acoustics, Special 50th Anniversary Design Issue,* Vol. 117, pp. 196–206 (June 1995).

[11]This type of damping is also referred to as *turbulent-water damping;* see J. P. Hartog, *ibid.*

From the discussions of Sections 2.4.1 and 2.4.2, it is clear that the damping force is linearly proportional to velocity for linear viscous damping, and the damping force is a nonlinear function of velocity for nonlinear viscous damping, Coulomb damping, and fluid damping. The structural damping model is used only in the presence of harmonic excitation, as discussed further in Section 5.8. Although damping models were presented only for translational motions, they are equally valid for rotational motions.

2.5 MODEL CONSTRUCTION

2.5.1 Introduction

In this section, four examples are provided to illustrate how the previously described inertia, stiffness, and damping elements are used to construct system models. Modeling is an art, and often experience serves as a guide in model construction. In this section, the examples are drawn from different areas, and are presented in a progressive order proceeding from the use of discrete inertia, stiffness, and damping elements in a model to distributed elements, and finally, to a combination of distributed and discrete elements.

As discussed in the subsequent chapters, the mass, stiffness, and damping of a system appear as parameters in the governing equations of the system. When only discrete elements are used to model a physical system, the associated system of governing equations is referred to as a *discrete system* or a *lumped-parameter system*. In these cases, as will become evident in later chapters, since a finite number of independent displacement or rotation coordinates suffice to describe the position of a physical system, discrete systems are also referred to as *finite degree-of-freedom systems*. When a distributed element is used to model a physical system, the associated system of governing equations is referred to as a *distributed-parameter system* or a *continuous system*. In this case, one or more displacement functions are needed to describe the position of a physical system. Since a function is equivalent, in a sense, to specifying the displacement at every point of the physical system or the displacements at an infinite number of points, distributed-parameter systems are also referred to as *infinite degree-of-freedom systems*.

2.5.2 A Microelectromechanical System

In Figure 2.23, a microelectromechanical accelerometer[12] is shown along with the vibratory model of this sensor. In this sensor, the dimensions of the cantilevered structure are of the order of micrometers and the weight of the end mass is of the order of micrograms. A coating on top of the structure serves as one of the faces of a capacitor and another layer below the structure serves as another face of the capacitor. The gap between the capacitor faces changes in

---

[12]K. E. Petersen, A. Shartel, and N. F. Raley, "Micromechanical accelerometer integrated with MOS detection circuitry," *IEEE Transactions of Electronic Devices,* Vol. ED-29, No. 1, pp. 23–27 (1982).

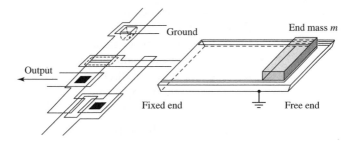

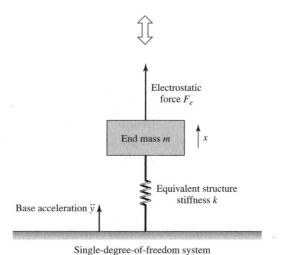

FIGURE 2.23
Microelectromechanical accelerometer and a vibratory model of this sensor.
Source: From *Systems Dynamics and Control,* by Umez-Eronini. © 1999.
Reprinted with permission of Brooks/Cole, a division of Thomson Learning:
www.thomsonrights.com.

response to the accelerations experienced by the sensor, and the change in voltage across this capacitor is sensed to determine the acceleration.

In constructing the vibratory model, the inertia of the cantilevered structure is ignored and this structure is represented by an equivalent spring with stiffness k. The mass of the cantilevered structure is assumed to be negligible and the end mass is modeled as a point mass of mass m. Consequently, the motion of this inertial element is described by a single generalized coordinate x, and the model is an example of a single degree-of-freedom system. The electrostatic force due to the capacitor acts directly on the mass, while the acceleration to be measured $\ddot{y}$ acts at the base of the vibratory model. The electrostatic force that acts directly on the inertial element is an example of a direct excitation, while the acceleration acting at the base is an example of a base excitation. In a refined model of the system, the mass of the cantilevered structure can also be lumped together with that of the end mass to obtain an effective point mass. No damping elements are used in constructing the vibratory model because the physical system has "very low" damping levels.

Single degree-of-freedom systems are treated at length in Chapters 3 to 6. In particular, the response of a single degree-of-freedom system subjected to a base excitation or direct excitation such as that shown in Figure 2.23 is discussed in Section 5.5.

2.5.3 The Human Body

In Figure 2.24, the human body and a vibratory model used to study the response of this physical system when subjected to vertical excitations are shown. While the vibratory model used in the previous section has only one discrete inertia element and one discrete spring element, the model[13] shown in Figure 2.24 has many inertial, spring, and damper elements.

Since many independent displacement variables are needed to describe the motion of this physical system, this vibratory model is an example of a system with multiple degrees of freedom. The response of systems with multiple degrees of freedom is treated in Chapters 7 and 8.

The human body is highly sensitive to vibration levels. While the body may sense displacements with amplitudes in the range of a hundredth of a mm, some of the components of the ear can sense even smaller displacements. In the low-frequency range from 1 Hz to 10 Hz, the perception of motion is said to be proportional to acceleration, and in the mid-frequency range from 10 Hz to 100 Hz, the perception of motion is said to be proportional to velocity. In addition, the level of stimulation also needs to be considered. The response of different parts of a human body is also dependent upon the frequency content of the excitation. For example, the thorax-abdomen system is highly responsive to vibrations in the range of 3 Hz to 6 Hz, the head-neck shoulder system to vibrations in the range of 20 Hz to 30 Hz, and the eyeball to vibrations in the range of 60 Hz to 90 Hz. In terms of modeling, the detailed model shown in Figure 2.24 can be further simplified based on the frequency content of the excitation. If, for example, the excitation has no frequency content below 20 Hz, then it is not necessary to have a detailed spring-mass-damper model for the thorax-abdomen system.

In biomechanics and biomedical engineering, there is an area called *whole-body vibration* where one is concerned with the response of a human body to different types of vibratory excitations and medical aspects of occupational exposure to these excitations. There are detailed international standards[14] that provide acceptable vibration levels in terms of acceleration magnitudes for horizontal and vertical vibrations based on exposure times and

---

[13]M. P. Norton, *Fundamentals of Noise and Vibration Analysis for Engineers,* Cambridge University Press, New York (1989).

[14]ISO 2631/1, "Evaluation of human exposure to whole-body vibration: General requirements," ISO 2631, Part 1, International Standards Organization, Geneva, Switzerland (1985) and ISO 2631/2, "Evaluation of human exposure to whole-body vibration: Continuous and shock-induced vibration in buildings (1 to 80 Hz)," ISO 2631, Part 2, International Standards Organization, Geneva, Switzerland (1985).

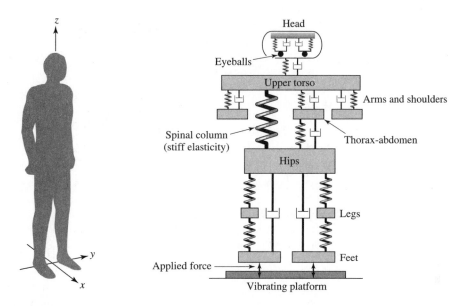

FIGURE 2.24

Human body and a vibratory model. *Source:* From M. P. Norton, *Fundamentals of Noise and Vibration Analysis for Engineers,* Cambridge University Press, New York (1989). Copyright © 1989 Cambridge University Press. Reprinted with the permission of Cambridge University Press.

frequency content. In another area, called *hand-arm vibration,* one is concerned with the response of the hand-arm system to vibratory excitations and medical aspects of occupational exposure to such excitations. These are also covered by international standards.[15]

2.5.4 A Ski

A cross-country ski is shown in Figure 2.25. The corresponding vibratory model is usually a system with an infinite number of degrees of freedom. In the model shown here, the ski is modeled as a collection of discrete inertial elements. In order to take into account that each end or boundary of the ski is free, the inertial elements at the boundaries are only constrained by a spring element on one side in the *X*-direction. Furthermore, each of these inertial elements is allowed two translational degrees of freedom, one along the *X*-direction and another along the *Y*-direction. If these elements are considered as point masses only, then they do not have any rotational degrees of freedom. However, if the inertial elements are treated as rigid bodies, then rotational degrees of freedom about the axis oriented along the *Z*-direction also need to be taken into account.

[15]ISO 5349/1, "Measurement and evaluation of human exposure to hand-transmitted vibration: General requirements," ISO 5349, Part 1, International Standards Organization, Geneva, Switzerland (1999) and ISO 5349/2, "Measurement and evaluation of human exposure to hand-transmitted vibration: Practical guidance for measurement at the workplace," ISO 5349, Part 2, International Standards Organization, Geneva, Switzerland (2001).

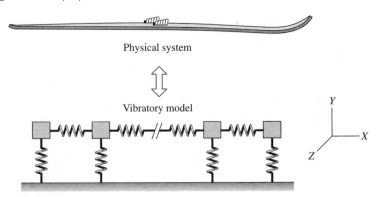

FIGURE 2.25
Cross-country ski, which is a physical system with distributed stiffness and inertia properties, and its vibratory model.

In the limit, as the ski is broken up into a collection of infinitesimal segments, where each segment is modeled as either a point mass or a rigid body, one ends up with a vibratory model with an infinite number of degrees of freedom. The process of discretizing a spatially distributed system into a collection of inertial elements can be realized through various means. This aspect is not addressed in detail in this book; however, if the model of the ski is a beam, then the results of Chapter 9 may be applicable.

2.5.5 Cutting Process

Consider the cutting process model[16] shown in Figure 2.26. The physical system consists of different components of the turret, lathe bed, and the tool used for cutting the work piece. Unlike the previous vibratory models, a more complex model is constructed. While previously only lumped masses or discrete inertia elements were used, here, a distributed inertia element, a beam, is also used in the model.

In the vibratory model, the bed is modeled as an elastic beam with a length L_b, mass per unit length m_b, area moment of inertia I_b about the bending axis, and Young's modulus of elasticity E_b. The turret is modeled as rigid body with a rotational degree of freedom θ and a translational degree of freedom along the vertical direction. The mass moment of inertia of the turret is represented by J_θ and the mass of the turret and the tool together is represented by M_m. Spring elements are introduced between the turret and the bed, and a damper element is introduced to model damping in the turret. A

---

[16]M. U. Jen and E. B. Magrab, "The dynamic interaction of the cutting process, work piece, and lathe's structure in facing," *ASME Journal of Manufacturing Science and Engineering,* Vol. 118, pp. 348–358 (1996).

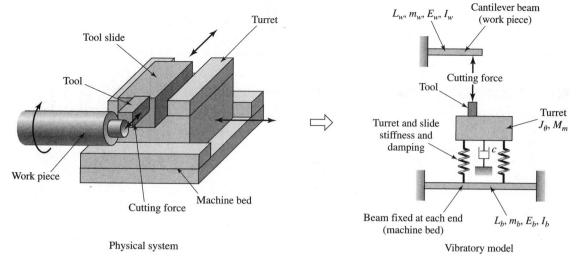

FIGURE 2.26
Work-piece-tool turning system and vibratory model of this system.

viscous-damping element c is used to model the damping experienced by the turret and the damping experienced in the tool-slide system. The work piece is also modeled as a distributed inertia element with a length L_w, mass per unit length m_w, area moment of inertia I_w about the bending axis, and Young's modulus of elasticity E_w. Since the model shown in Figure 2.26 consists of distributed inertial elements, this model has infinite number of degrees of freedom. In the modeling, the beam model used for the turret is considered fixed at both ends, while the beam model used for the work piece is considered fixed at one end and hinged or free at the other end where it is being cut.

In Section 4.5, the stability of a machine tool is determined from a vibratory model to avoid undesirable cutting conditions called chatter. This type of analysis can be used to choose parameters such as width of cut, spindle rpm, etc. In Chapter 9, vibrations of beams used in the model in Figure 2.26 are discussed at length.

2.6 SUMMARY

In this chapter, the inertia elements, stiffness elements, and dissipation elements that are used to construct a vibratory model were discussed. An ideal inertia element, which does not have any stiffness or dissipation characteristics, can store or release kinetic energy. An ideal stiffness element, which does not have any inertia or dissipation characteristics, can store or release potential energy. The stiffness element can be due to a structure, a fluid element, or gravity loading. The notion of equivalent stiffness was also introduced. An ideal dissipation element, which does not have any stiffness or inertia characteristics, is used to represent energy losses of the system. The notion of equivalent damping was also introduced.

EXERCISES

2.1 Examine Eqs. (2.1) and (2.5) and verify that the units (dimensions) of the different terms in the respective equations appearing in these equations are consistent.

2.2 Extend the spring combinations shown in Figures 2.5b and 2.5c to cases with three springs, and verify that the equivalent stiffness of these spring combinations is consistent with Eqs. (2.14) and (2.16), respectively.

(a)

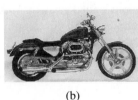

(b)

(c)

FIGURE E2.3
(a) Car; (b) motorcycle; and (c) cable car.

2.3 Construct vibratory models for each of the three systems shown in the Figure E2.3. Discuss the number of degrees of freedom and the associated generalized coordinates in each case.

2.4 Consider the slider mechanism of Example 2.2 and show that the rotary inertia J_O' about the pivot point O' is also a function of the angular displacement φ.

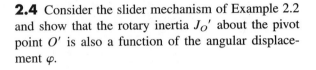

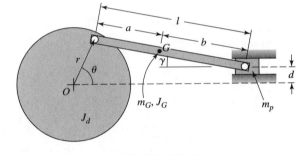

FIGURE E2.5

2.5 Consider the crank-mechanism system shown in Figure E2.5 and determine the rotary inertia of this system about the point O, and express it as a function of the angular displacement θ. The disc has a rotary inertia J_d about the point O, the crank has a mass m_G and rotary inertia J_G about the point G, the center of mass of the crank. The mass of the slider is m_p.

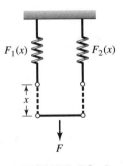

FIGURE E2.6

2.6 Consider the two nonlinear springs in parallel shown in Figure E2.6. The force-displacement relations for each spring is,

$$F_j(x) = k_j x + k_j \alpha x^3 \qquad j = 1, 2$$

a) Obtain the expressions from which the equivalent spring constant can be determined.
b) If $F = 1000$ N, $k_1 = k_2 = 50{,}000$ N/m, $\alpha = 2$ m^{-2}, then determine the equivalent spring constant.

$F_1(x)$

$F_2(x)$

x

F

FIGURE E2.7

2.7 Consider the two nonlinear springs in series shown in Figure E2.7. The force-displacement relations for each spring is,

$$F_j(x) = k_j x + k_j \alpha x^3 \qquad j = 1, 2$$

a) Obtain the expressions from which the equivalent spring constant can be determined.
b) If $F = 1000$ N, $k_1 = 50{,}000$ N/m, $k_2 = 25{,}000$ N/m, $\alpha = 2$ m^{-2}, then determine the equivalent spring constant.

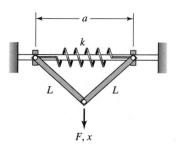

a

k

L L

F, x

FIGURE E2.8

2.8 Consider the mechanical spring system shown in Figure E2.8. Assume that the bars are rigid and determine the equivalent spring constant k_e.

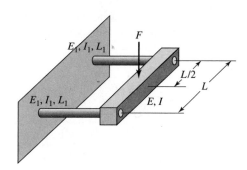

E_1, I_1, L_1

F

$L/2$

L

E_1, I_1, L_1

E, I

FIGURE E2.9

2.9 Consider the three beams connected as shown in Figure E2.9. The beam that is attached to the ends of the two cantilever beams is pinned so that its ends can rotate unimpeded. Determine the system's equivalent spring constant for the transverse loading shown.

2.10 Consider the manometer shown in Figure 2.15 and seal the ends. Assume that the initial gas pressure of the sealed system is P_o and L_o is the initial length of the cavity. Determine the equivalent spring constant of the system when the column of liquid undergoes "small" displacements.

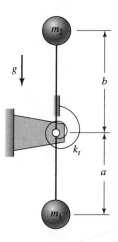

m_2

g

b

k_t

a

m_1

FIGURE E2.11

2.11 Consider "small" amplitude angular oscillations of the pendulum shown in Figure E2.11. Considering the gravitational loading and the torsion spring k_t at the pivot point, determine the expression for the system's equivalent spring constant. The masses are held with rigid, weightless rods for the loading shown.

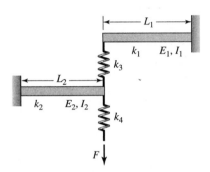

FIGURE E2.12

2.12 For the two cantilever beams whose free ends are connected to springs as shown in Figure E2.12, give the expressions for the spring constants k_1 and k_2 and determine the equivalent spring constant k_e for the system.

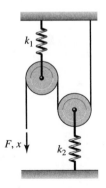

FIGURE E2.13

2.13 For the weightless pulley system shown in Figure E2.13, determine the equivalent spring con-

stant. Recall that when the center of the pulley moves an amount x, the rope moves an amount $2x$.

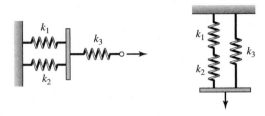

FIGURE E2.14

2.14 Determine the equivalent stiffness for each of the systems shown in Figure E2.14. Each system consists of three linear springs with stiffness k_1, k_2, and k_3.

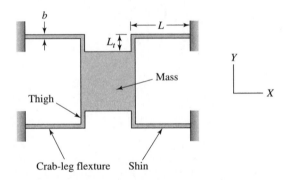

FIGURE E2.15
Source: From G. K. Fedder, "Simulation of Microelectromechanical Systems," Ph.D. Dissertation, Department of Electrical Engineering and Computer Sciences, University of California, Berkeley, CA, (1994). Reprinted with permission of the author.

2.15 Consider the system shown Figure E2.15, which lies in the X-Y plane. This system, which is called a crab-leg flexure, is used in microelectromechanical sensors.[17] A load along the X direction is applied to the mass to which all of the four crab-leg flexures are attached. Each flexure has a shin of length L along the X direction, width b along the Y direction, and thickness h along the Z direction. The thigh of each flexure has a length L_t along the Y di-

[17]G. K. Fedder, *ibid.*

rection, a width b_t along the X direction, and a thickness h along the Z direction.

The equivalent stiffness of each of these flexures in the X direction can be shown to be

$$k_{flexure} = \frac{Ehb^3 \left(4L + \left(\frac{b}{b_t}\right)^3 L_t\right)}{L^3 \left(L + \left(\frac{b}{b_t}\right)^3 L_t\right)}$$

where the Young's modulus of elasticity is E, which has a value of 160 GPa for the polysilicon material from which the flexure is fabricated. The dimensions of each crab leg are as follows: $L = 100$ μm and $b = b_t = h = 2$ μm. For the thigh length L_t spanning the range from 10 μm to 75 μm, plot the graph of the equivalent stiffness of the system versus L_t.

2.16 Based on the expression for $k_{flexure}$ provided in Exercise 2.15, the sensitivity of the equivalent stiffness of each flexure with respect to the flexure width b and the thigh width b_t can be assessed by determining the derivatives $dk_{flexure}/db$ and $dk_{flexure}/db_t$, respectively. Carry out these operations and discuss the expressions obtained.

TABLE E2.17
Tire Load Versus Deflection Data

Tire Load (lbs)	Tire Deflection	
	Dual Tires (in)	Single Wide-Base Tire (in)
0	0	0
2000	0.30	0.40
4000	0.55	0.75
6000	0.75	1.10
8000	0.95	1.40
10,000	1.10	1.65

2.17 Consider the data in Table E2.17, in which the experimentally determined tire loads versus tire deflections have been recorded. These data are for a set

of dual tires and a single wide-base tire.[18] The inflation pressure for all tires is 105 psi (pounds/square inch). Examine the stiffness characteristics of the two different tire systems and discuss them.

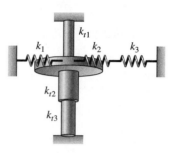

FIGURE E2.18

2.18 For the system of linear and torsion springs shown in Figure E2.18, determine the equivalent spring constant for torsional oscillations. The disc has a radius b and the translation springs are tangential to the disc at the point of attachment.

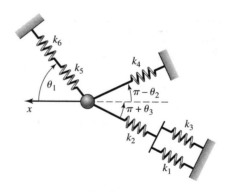

FIGURE E2.19

2.19 For the system of translation springs shown in Figure E2.19, determine the equivalent spring constant for motion in the horizontal (x) direction only.

[18]J. C. Tielking, "Conventional and wide base radial tyres," in Proceedings of the Third International Symposium on Heavy Vehicle Weights and Dimensions, D. Cebon and C. G. B. Mitchell, eds., Cambridge, UK, 28 June–2 July 1992, pp. 182–190.

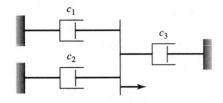

FIGURE E2.20

2.20 Determine the equivalent damping of the system shown in Figure E2.20.

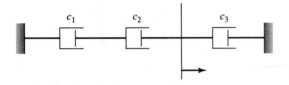

FIGURE E2.21

2.21 Determine the equivalent damping for the system shown in Figure E2.21.

TABLE E2.22
Racecar Damper

Damper Force Magnitude (pounds)	Damper Rod Speed (inches/s)
0	0
13	0.1
26	0.2
40	0.3
60	0.4
80	0.5
100	0.6
110	0.7
120	0.8
130	0.9
140	1.0
150	1.1
160	1.2
170	1.3
180	1.4
190	1.5
200	1.6

2.22 Representative damping force magnitudes versus speed data are given in Table E2.22 for a racecar damper in compression.[19] Examine these data and discuss the type of damping model that can be used to represent them.

2.23 Determine the viscosity of the fluid that one needs to use to realize a parallel-plate damper when the top plate has an area of $0.02 \ m^2$, the gap between the parallel plates $h = 0.2$ mm, and the required damping coefficient is 20 N·s/m.

2.24 Determine the equivalent damping coefficient for the following nonlinear damper

$$F(\dot{x}) = c_1 \dot{x} + c_3 \dot{x}^3$$

where $c_1 = 5$ N·s/m and $c_3 = 0.6$ N·s³/m³, if the damper is to be operated around a speed of 5 m/s.

2.25 Consider the viscous-damping model given by Eq. (2.46) and the dry-friction model given by Eq. (2.52) and sketch the force versus velocity graphs in each case for the following parameter values: $c = 100$ N/(m/s), $m = 100$ kg, and $\mu = 0.1$.

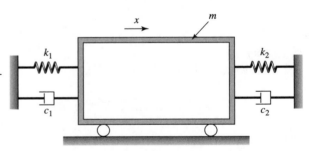

FIGURE E2.26

2.26 Represent the vibratory system given in Figure E2.26 as an equivalent vibratory system with mass m, equivalent stiffness k_e, and equivalent damping coefficient c_e.

---

[19]Racecar dampers are called "shocks" in the United States, and they have different damping characteristics during compression (called "bump") and expansion (called "rebound") phases. For material on racecar damping, see, for example, P. Haney and J. Braun, *Inside Racing Technology,* (ISBN 0-9646414-0-2).

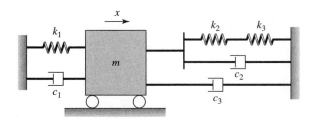

FIGURE E2.27

$$E_k = \int_{x(0)}^{x(2\pi/\omega)} kx\,dx = \int_{0}^{2\pi/\omega} kx\dot{x}\ dt$$

$$E_d = \int_{x(0)}^{x(2\pi/\omega)} c\dot{x}\,dx = \int_{0}^{2\pi/\omega} c\dot{x}^2\,dt$$

respectively, and show that $E_k = 0$ and $E_d = c\pi\omega X_o^2$.

2.27 Represent the vibratory system given in Figure E2.27 as an equivalent vibratory system with mass m, equivalent stiffness k_e, and equivalent damping coefficient c_e.

2.28 For a vibratory system with a mass $m = 10$ kg, $k = 1500$ N/m, and $c = 2500$ N/(m/s), and given that the displacement response has the form $x(t) = 0.2\sin(9t)$ m, plot the graphs of $m\ddot{x}$, the spring force kx, and the damper force $c\dot{x}$ versus time and discuss them.

2.29 The vibrations of a system with stiffness k and damping coefficient c is of the form $x(t) = X_o\sin\omega t$. This type of response, which is called a harmonic response, is possible when a vibratory system is excited by a harmonic force. Evaluate the work done by the spring force and the work done by the damper force, which are given by the integrals

2.30 Normalize the linear viscous damper force given by Eq. (2.46) using the damping coefficient c, the dry friction force given by Eq. (2.52) using μmg, the fluid damping force given by Eq. (2.54) using the damping coefficient c_d, and the hysteretic force given by Eq. (2.57) using $k\pi\beta_h$, and plot the time histories of the normalized damper forces versus time for harmonic displacement of the form $x(t) = 0.4\sin(2\pi t)$ m. Discuss the characteristics of these plots.

2.31 For the system of Exercise 2.29, assume $k = 1000$ N/m, $c = 2500$ N/(m/s), $X_o = 0.1$ m, and $\omega = 9$ rad/s, and plot the graph of the sum of the spring force and damper force versus the displacement; that is, $kx + c\dot{x}$ versus x.

2.32 For the system of Exercise 2.29, assume that $k > 0$, $c > 0$, and $\omega > 0$; show that the graph of $kx + c\dot{x}$ versus x will have the form of an ellipse.

Springs, masses, and dampers are used to model vibrations systems. In the motorcycle, the coil spring in parallel with a viscous damper is attached to a mass composed of the tire and brake assembly. In the wind turbine, the mass of the propellers is supported by the column, which acts as the spring. (*Source:* Images courtesy of EyeWire/Getty Images.)

3

Single Degree-of-Freedom Systems:
Governing Equations

3.1   INTRODUCTION

In this chapter, two approaches are illustrated for deriving the equation governing the motion of a single degree-of-freedom system. The momentum principles of Chapter 1 form the basis of one approach, comprising force-balance and moment-balance methods. The second approach is based on Lagrange's equations, first introduced in this chapter and extensively used throughout the remaining chapters. Based on the parameters that appear in the governing equation, the expressions for the natural frequency and damping factor are defined. The full physical significance of the natural frequency and damping factor is brought forth in Section 4.3, when free oscillations of single degree-of-freedom systems are considered.

Furthermore, system stability can also be assessed from the system parameters, as discussed in Section 4.4.

When either an analytical or a numerical solution of the governing equation is obtained, one can determine the relative effects that the system's components have on the system's response to various externally applied forces and initial conditions. One can also study vibratory systems by determining the transfer function of the system (Section 5.3) or the system's impulse response (Section 6.2). In this chapter, we will only derive the governing equations for various systems. We shall then obtain the general solutions in Chapter 4 and illustrate numerous applications and interpretations of the solutions in Chapters 4, 5, and 6.

3.2   FORCE-BALANCE AND MOMENT-BALANCE METHODS

In this section, we illustrate the use of force-balance and moment-balance methods for deriving governing equations of motion of single degree-of-freedom systems, show how the static-equilibrium position of a vibratory system can be determined, and carry out linearization of a nonlinear system for "small" amplitude oscillations about a system's equilibrium position.

3.2.1   Force-Balance Methods

Consider the principle of linear momentum, which is Newton's second law of motion. The statement of dynamic equilibrium given by Eq. (1.11) is recast in the form

$$F - \dot{p} = 0 \qquad (3.1a)$$

where F is the net external force vector acting on the system, p is the absolute linear momentum of the considered system and the overdot indicates the derivative with respect to time. For a system of constant mass m whose center of mass is moving with absolute acceleration a, the rate of change of linear momentum $\dot{p} = ma$ and Eq. (3.1a) lead to

$$F - ma = 0 \qquad (3.1b)$$

The term $-ma$ is referred to as the *inertia force*. The interpretation of Eq. (3.1b) is that the sum of the external forces and inertial forces acting on the system is zero; that is, the system is in equilibrium under the action of external and inertial forces.[1]

Vertical Vibrations of a Spring-Mass-Damper System

In Figure 3.1, a spring-mass-damper model is shown. A linear spring with stiffness k and a viscous damper with damping coefficient c are connected in parallel to the inertia element m. In addition to the three system elements that were discussed in Chapter 2, an external forcing is also considered. We

---

[1]This statement is also referred to as *D'Alembert's principle.* According to the generalized form of this principle, when a set of so-called *virtual displacements* are imposed on the system of interest, the net work done by the external forces and inertial forces is zero. [See, for example, D. T. Greenwood, *Principles of Dynamics,* Chapter 8, Prentice Hall, Upper Saddle River, NJ (1988).]

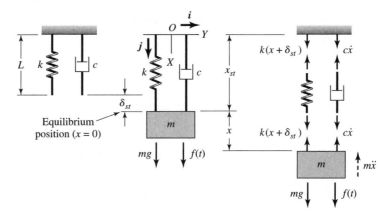

FIGURE 3.1
Vertical vibrations of a spring-mass-damper system.

would like to obtain an equation to describe the motions of the system in the vertical direction. In order to derive such an equation for translation motions, the force-balance method is used. Prior to obtaining the governing equation of motion for the system of Figure 3.1, we choose a set of orthogonal unit vectors i and j fixed in an inertial reference frame and a coordinate system with X and Y axes and an origin O that is fixed. Since the mass m translates only along the j direction, the force balance is considered only along this direction.

Let the unstretched length of the spring shown in Figure 3.1 be L. Then the mass is located at the position $(L + \delta_{st} + x)j$ from the fixed surface, where the term δ_{st} will be determined shortly and explained. After determining δ_{st}, the equation of motion will be developed in terms of the displacement variable x. The position vector of the mass from the fixed point O is given by

$$\boldsymbol{r} = rj = (L + \delta_{st} + x)j \tag{3.2}$$

The directions of different forces along with their magnitudes are shown in Figure 3.1. Note that the inertia force $-m\ddot{x}j$ is also shown along with the free-body diagram of the inertia element. Since the spring force is a restoring force and the damper force is a resistive force, they oppose the motion as shown in Figure 3.1. Based on Eq. (3.1b), we can carry out a force balance along the j direction and obtain the resulting equation

$$\underbrace{f(t)j + mgj}_{\substack{\text{External forces acting} \\ \text{on system}}} - \underbrace{(kx + k\delta_{st})j}_{\substack{\text{Spring force acting} \\ \text{on mass}}} - \underbrace{c\frac{dr}{dt}j}_{\substack{\text{Damping force} \\ \text{acting on mass}}} - \underbrace{m\frac{d^2r}{dt^2}j}_{\text{Inertia force}} = 0 \tag{3.3}$$

Upon making use of Eq. (3.2), noting that L and δ_{st} are constants, and rearranging terms, Eq. (3.3) reduces to the following scalar differential equation

$$m\frac{d^2x}{dt^2} + c\frac{dx}{dt} + k(x + \delta_{st}) = f(t) + mg \tag{3.4}$$

Static-Equilibrium Position

The *static-equilibrium position* of a system is the position that corresponds to the system's rest state; that is, a position with zero velocity and zero acceleration. Dropping the time-dependent forcing term $f(t)$ and setting the velocity and acceleration terms in Eq. (3.4) to zero, we find that the static-equilibrium position is the solution of

$$k(x + \delta_{st}) = mg \tag{3.5}$$

If, in Eq. (3.5), we choose

$$\delta_{st} = \frac{mg}{k} \tag{3.6}$$

we find that $x = 0$ is the static-equilibrium position of the system. Equation (3.6) is interpreted as follows. Due to the weight of the mass m, the spring is stretched an amount δ_{st}, so that the spring force balances the weight mg. For this reason, δ_{st} is called the *static displacement*. Recalling that the spring has an unstretched length L, the static-equilibrium position measured from the origin O is given by

$$\boldsymbol{x}_{st} = x_{st}\boldsymbol{j} = (L + \delta_{st})\boldsymbol{j} \tag{3.7}$$

which is the rest position of the system. For the vibratory system of Figure 3.1, it is clear from Eq. (3.6) that the static-equilibrium position is determined by the spring force and gravity loading. An example of another type of static loading is provided in Example 3.1.

Equation of Motion for Oscillations about the Static-Equilibrium Position

Upon substituting Eq. (3.6) into Eq. (3.4), we obtain

$$m\frac{d^2x}{dt^2} + c\frac{dx}{dt} + kx = f(t) \tag{3.8}$$

Equation (3.8) is the governing equation of motion of a single degree-of-freedom system for oscillations about the static-equilibrium position given by Eq. (3.7). Note that the gravity loading does not appear explicitly in Eq. (3.8). For this reason, in development of models of vibratory systems, the measurement of displacement from the static-equilibrium position turns out to be a convenient choice, since one does not have to explicitly take static loading into account.

The left-hand side of Eq. (3.8) describes the forces from the components that comprise a single degree-of-freedom system. The right-hand side represents the external force acting on the mass. Examples of external forces acting on a mass are fluctuating air pressure loading such as that on the wing of an aircraft, fluctuating electromagnetic forces such as in a loudspeaker coil, electrostatic forces that appear in some microelectromechanical devices, forces caused by an unbalanced mass in rotating machinery (see Section 3.5), and buoyancy forces on floating systems. The system represented by Eq.

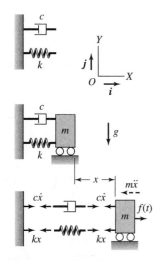

FIGURE 3.2
Horizontal vibrations of a spring-mass-damper system.

(3.8) is a linear, ordinary differential equation with constant coefficients m, c, and k. As mentioned in Sections 2.2, 2.3, and 2.4, these quantities are also referred to as *system parameters*.

Horizontal Vibrations of a Spring-Mass-Damper System

In Figure 3.2, a mass moving in a direction normal to the direction of gravity is shown. It is assumed that the mass moves without friction. The unstretched length of the spring is L, and a fixed point O is located at the unstretched position of the spring, as shown in the figure. Noting that the spring does not undergo any static deflection and carrying out a force balance along the i direction gives Eq. (3.8) directly. Here, the static-equilibrium position $x = 0$ coincides with the position corresponding to the unstretched spring.

Force Transmitted to Fixed Surface

From Figure 3.1, we see that the total reaction force due to the spring and the damper on the fixed surface is the sum of the static and dynamic forces. Thus,

$$F_R = \underbrace{k\delta_{st}}_{\substack{\text{Static} \\ \text{component}}} + \underbrace{kx + c\frac{dx}{dt}}_{\substack{\text{Dynamic} \\ \text{component}}} \tag{3.9}$$

If we consider only the dynamic part of the reaction force—that is, only those forces created by the motion $x(t)$ from its static equilibrium position—then Eq. (3.9) leads to

$$F_{Rd} = c\frac{dx}{dt} + kx \tag{3.10}$$

where $x(t)$ is the solution of Eq. (3.8). Equation (3.10) is used in later chapters to determine the force transmitted to the ground (Section 5.4) or the force transmitted to the mass when the base is in motion (Sections 5.5 and 6.7).

EXAMPLE 3.1   Wind-driven oscillations about a system's static-equilibrium position

In examining wind flow across civil structures such as buildings, water towers, and lampposts, it has been found[2] that the wind typically generates a force on the structure that consists of a steady-state part and a fluctuating part. In such cases, the excitation force $f(t)$ is represented as

$$f(t) = f_{ss} + f_d(t) \tag{a}$$

[2]E. Naudascher and D. Rockwell, *Flow-Induced Vibrations: An Engineering Guide,* A. A. Balkema, Rotterdam, p. 103 (1994).

where f_{ss} is the time-independent steady-state force and $f_d(t)$ is the fluctuating time-dependent portion of the force. A single degree-of-freedom model of the vibrating structure has the form of Eq. (3.8), where $x(t)$ is the displacement of the structure in the direction of the wind loading and the forcing is given by Eq. (a); that is,

$$m \frac{d^2x}{dt^2} + c \frac{dx}{dt} + kx = f_{ss} + f_d(t) \qquad \text{(b)}$$

We shall determine the static-equilibrium position of this system and obtain the equation of motion governing oscillations about this static-equilibrium position. To this end, we assume a solution of the following form for Eq. (b)

$$x(t) = x_o + x_d(t) \qquad \text{(c)}$$

where x_o is determined by the static loading and $x_d(t)$ determines motions about the static position. Thus, after substituting Eq. (c) into Eq. (b) and noting that x_o is independent of time, we find that

$$x_o = f_{ss}/k \qquad \text{(d)}$$

and that $x_d(t)$ is the solution to the equation

$$m \frac{d^2x_d}{dt^2} + c \frac{dx_d}{dt} + kx_d = f_d(t) \qquad \text{(e)}$$

which is the governing equation of motion for oscillations about the static-equilibrium position $x_o = f_{ss}/k$.

EXAMPLE 3.2   Eardrum oscillations: nonlinear oscillator[3] and linearized systems

In this example, we consider a nonlinear oscillator used to study eardrum oscillations. We shall determine the static-equilibrium positions of this system and illustrate how the governing nonlinear equation can be linearized to study oscillations local to an equilibrium position. The governing nonlinear equation has the form

$$m \frac{d^2x}{dt^2} + kx + \underbrace{kx^2}_{\substack{\text{Quadratic} \\ \text{nonlinearity}}} = 0 \qquad \text{(a)}$$

The restoring force of the eardrum has a component with a quadratic nonlinearity.

[3]R. E. Mickens, *Oscillations in Planar Dynamic Systems,* World Scientific, Singapore (1996).

Static-Equilibrium Positions

Noting that there are no external time-dependent forcing terms in this problem and setting the acceleration term to zero, we find that the equilibrium positions $x = x_o$ are solutions of the algebraic equation

$$k(x_o + x_o^2) = 0 \qquad (b)$$

The solutions of Eq. (b) provide us two static-equilibrium positions for the eardrum, namely,

$$
\begin{aligned}
x_{o1} &= \quad 0 \\
x_{o2} &= -1
\end{aligned}
\qquad (c)
$$

Linearization

Next, we substitute

$$x(t) = x_o + \hat{x}(t) \qquad (d)$$

into Eq. (a) and linearize the nonlinear stiffness term for oscillations about one of the equilibrium positions in terms of the variable $\hat{x}(t)$. To this end, we note that

$$x^2(t) = (x_o + \hat{x}(t))^2 \approx x_o^2 + 2x_o\hat{x}(t) + \ldots \qquad (e)$$

In addition,

$$
\begin{aligned}
\frac{dx}{dt} &= \frac{d(x_o + \hat{x})}{dt} = \frac{d\hat{x}}{dt} \\
\frac{d^2x}{dt^2} &= \frac{d^2(x_o + \hat{x})}{dt^2} = \frac{d^2\hat{x}}{dt^2}
\end{aligned}
\qquad (f)
$$

Linearized system for "small" amplitude oscillations around $x_{o1} = 0$
Making use of Eqs. (e) and (f) in Eq. (a), and noting that $x_o = x_{o1} = 0$, we arrive at the linear equation

$$m\frac{d^2\hat{x}}{dt^2} + k\hat{x} = 0 \qquad (g)$$

Linearized system for "small" amplitude oscillations around $x_{o2} = -1$
Making use of Eqs. (e) and (f) in Eq. (a), and noting that $x_o = x_{o2} = -1$, we arrive at the linear equation

$$m\frac{d^2\hat{x}}{dt^2} - k\hat{x} = 0 \qquad (h)$$

Comparing Eqs. (g) and (h), it is clear that the equations have different stiffness terms.

3.2.2 Moment-Balance Methods

For single degree-of-freedom systems that undergo rotational motion, such as the system shown in Figure 3.3, the moment balance method is useful in deriving the governing equation. A shaft with torsional stiffness k_t is attached to a disc with rotary inertia J_G about the axis of rotation, which is directed along the k direction. An external moment $M(t)$ acts on the disc, which is immersed in an oil-filled housing. Let the variable θ describe the rotation of the disc, and let the rotary inertia of the shaft be negligible in comparison to that of the disk.

The principle of angular momentum given by Eq. (1.17) is applied to obtain the equation governing the disc's motion. First the angular momentum H of the disc is determined. Since the disk is a rigid body undergoing rotation in the plane, Eq. (1.20) is used to write the angular momentum about the center of mass of the disc as

$$H = J_G \dot{\theta} k$$

Thus, since the rotary inertia J_G and the unit vector k do not change with time, Eq. (1.17) is rewritten as

$$M - J_G \ddot{\theta} k = 0 \tag{3.11}$$

where M is the total external moment acting on the free disk. Based on the free-body diagram shown in Figure 3.3, which also includes the inertial moment $-J_G \ddot{\theta} k$, the governing equation of motion is

$$\underbrace{M(t)k}_{\substack{\text{External moment}\\\text{acting on disk}}} - \underbrace{k_t\theta k}_{\substack{\text{Restoring moment due}\\\text{to shaft stiffness}}} - \underbrace{c_t \frac{d\theta}{dt} k}_{\substack{\text{Damping moment due}\\\text{to oil in housing}}} - \underbrace{J_G \frac{d^2\theta}{dt^2} k}_{\substack{\text{Inertial moment}}} = 0 \tag{3.12}$$

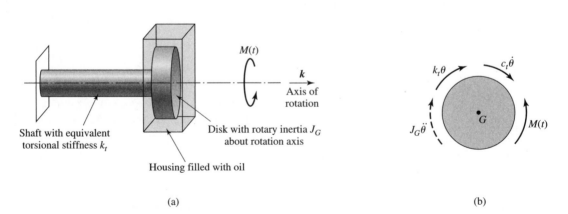

(a)   (b)

FIGURE 3.3
(a) A disc undergoing rotational motions and (b) free-body diagram of this disc in the plane normal to the axis of rotation.

After collecting the scalar coefficients of the different vector terms in Eq. (3.12) and setting them to zero, we arrive at the following scalar equation:

$$J_G \frac{d^2\theta}{dt^2} + c_t \frac{d\theta}{dt} + k_t\theta = M(t) \tag{3.13}$$

The form of Eq. (3.13) is identical to Eq. (3.8), which was obtained for a translating vibratory system; that is, the first term on the left-hand side is determined by the inertia element J_G, the second term on the left-hand side is determined by the damping element c_t, the third term on the left-hand side is determined by the stiffness element k_t, and the right-hand side contains the external forcing, the moment $M(t)$. *All linear single degree-of-freedom systems are governed by a linear second-order ordinary differential equation with an inertia term, a stiffness term, a damping term, and a term related to the external forcing imposed on the system.*

EXAMPLE 3.3   Hand biomechanics[4]

Consider the rotational motion of the hand in the *X-Y* plane shown in Figure 3.4. This motion is described by the angle θ. An object of mass *M* is held in the hand, and the forearm has a mass *m* and a length *l*. If we use simplified models for the forces generated by the muscles, then the biceps provide a force of magnitude $F_b = -k_b\theta$, where k_b is a constant, and the triceps provide a force of magnitude $F_t = K_t v$, where K_t is a constant and *v* is the magnitude of the velocity with

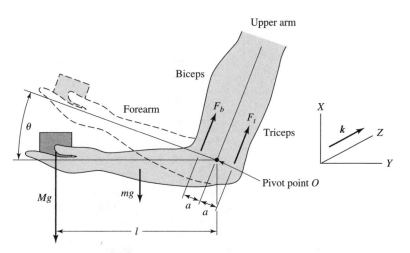

FIGURE 3.4
Hand motion.
Source: From P. Maróti, L. Berkes, and F. Tölgyesi, *Biophysics Problems,* Akademiai Kiado, Budapest, Hungary (1998). Copyright © 1998 P. Maroti, L. Berkes, and F. Tolgyesi. Reprinted with permission.

[4]P. Maróti, L. Berkes, and F. Tölgyesi, *Biophysics Problems,* Akademiai Kiado, Budapest, Hungary (1998).

which the triceps are stretched. It is further assumed that the forearm can be treated as a uniform rigid beam. The governing nonlinear equation of motion is obtained, and following along the lines of Example 3.2, the nonlinear system is linearized for "small" oscillations about the system equilibrium position.

The equation of motion of the system of Figure 3.4 is derived by carrying out a moment balance about either the center of the forearm or the pivot point O (the elbow) if the pivot point is a fixed point. Assuming that the pivot point is a fixed point, Eq. (1.17) is used to carry out a moment balance about point O. Here, Eq. (1.17) takes the form

$$\mathbf{M} - J_o\ddot{\theta}\mathbf{k} = 0 \tag{a}$$

where J_o is the rotary inertia of the forearm and the object held in the hand. The net moment $\mathbf{M}$ acting about the point O due to gravity loading and the forces due to the biceps and triceps is given by

$$\mathbf{M} = -Mgl\cos\theta\mathbf{k} - mg\frac{l}{2}\cos\theta\mathbf{k} + F_b a\mathbf{k} - F_t a\mathbf{k} \tag{b}$$

Therefore, from Eqs. (a) and (b), we find that the governing equation takes the form

$$-Mgl\cos\theta\mathbf{k} - mg\frac{l}{2}\cos\theta\mathbf{k} + F_b a\mathbf{k} - F_t a\mathbf{k} - J_o\ddot{\theta}\mathbf{k} = 0 \tag{c}$$

Recognizing that

$$F_b = -k_b\theta$$
$$F_t = K_t v = K_t a\dot{\theta}$$
$$J_o = \frac{1}{3}m\,l^2 + Ml^2 \tag{d}$$

and collecting the scalar coefficients of all the vector terms in Eq. (c) and setting them to zero, we obtain the scalar equation

$$\left(M + \frac{m}{3}\right)l^2\,\ddot{\theta} + K_t a^2\dot{\theta} + k_b a\theta + \left(M + \frac{m}{2}\right)gl\cos\theta = 0 \tag{e}$$

In this equation, the inertia term is due to the rotary inertia of the forearm and rotary inertia of the end mass, the damping term is due to the triceps, the stiffness term is due to the biceps, and there is an additional term due to gravity that makes the equation nonlinear due to the presence of the $\cos\theta$ term. This latter term influences the static equilibrium position and the stiffness of the system, and this influence depends on the magnitude of θ.

Static-Equilibrium Position

Noting that time dependent external moment terms are absent in Eq. (e) and setting the velocity term and the acceleration term to zero, we find that the equilibrium position $\theta = \theta_o$ is a solution of the transcendental equation

$$k_b a\theta_o + \left(M + \frac{m}{2}\right)gl\cos\theta_o = 0 \tag{f}$$

Linear System Governing "Small" Oscillations about the Static-Equilibrium Position

We now consider oscillations about the static-equilibrium position and expand the angular variable $\theta(t)$ in the form

$$\theta(t) = \theta_o + \hat{\theta}(t) \tag{g}$$

and linearize the nonlinear term $\cos \theta$ in Eq. (e). To do this, we carry out the Taylor-series expansions[5] of this term and retain only linear terms in $\hat{\theta}$. This leads to

$$\cos \theta = \cos(\theta_o + \hat{\theta}) \approx \cos \theta_o - \hat{\theta} \sin \theta_o + \dots \tag{h}$$

Evaluating the time derivatives of $\theta(t)$, we find that

$$\ddot{\theta}(t) = \frac{d^2}{dt^2}(\theta_o + \hat{\theta}) = \ddot{\hat{\theta}}(t)$$

$$\dot{\theta}(t) = \frac{d}{dt}(\theta_o + \hat{\theta}) = \dot{\hat{\theta}}(t) \tag{i}$$

On substituting Eqs. (i) and (h) into Eq. (e) and making use of Eq. (f), we arrive at the following linear equation of motion governing "small" oscillations of the forearm about the system-equilibrium position:

$$\left(M + \frac{m}{3}\right) l^2 \ddot{\hat{\theta}} + K_t a^2 \dot{\hat{\theta}} + \left[k_b a - \left(M + \frac{m}{2}\right) gl \sin \theta_o\right] \hat{\theta} = 0 \tag{j}$$

It is noted out that the "linear" stiffness of the linearized system is influenced by the gravity loading.

3.3 NATURAL FREQUENCY AND DAMPING FACTOR

In this section, we define the natural frequency and damping factor for a vibratory system and explore how these quantities depend on various system properties. These quantities are discussed without explicitly determining the solution of the system. They are only a function of the system's inertia, stiffness, and damping parameters and independent of the external time-dependent forcing imposed on a system. The solutions for the responses of vibratory systems represented by Eqs. (3.8) and (3.13), which are discussed in Chapters 4 to 6, can be characterized in terms of the system's damping factor.

[5]T. B. Hildebrand, *ibid.*

3.3.1   Natural Frequency

Translation Vibrations: Natural Frequency

For translation oscillations of a single degree-of-freedom system, the *natural frequency* ω_n of the system is defined as

$$\omega_n = 2\pi f_n = \sqrt{\frac{k}{m}} \ \ \text{rad/s} \tag{3.14}$$

where k is the stiffness of the system and m is the system mass. The quantity f_n, which is also referred to as the natural frequency, has the units of Hz.

For the configuration shown in Figure 3.1, the vibratory system exhibits vertical oscillations. For such oscillations, we make use of Eq. (3.6) and Eq. (3.14) and obtain

$$\omega_n = 2\pi f_n = \sqrt{\frac{g}{\delta_{st}}} \ \ \text{rad/s} \tag{3.15}$$

where δ_{st} is the static deflection of the system.

Rotational Vibrations: Natural Frequency

Drawing a parallel to the definition of natural frequency of translation motions of a single degree-of-freedom system, the natural frequency for rotational motions is defined as

$$\omega_n = 2\pi f_n = \sqrt{\frac{k_t}{J}} \ \ \text{rad/s} \tag{3.16}$$

where k_t is the torsion stiffness of the system and J is the mass moment of inertia of the system.

> **Design Guideline:**   For single degree-of-freedom systems, an increase in the stiffness or a decrease in the mass or mass moment of inertia increases the natural frequency, whereas a decrease in the stiffness and/or an increase in the mass or mass moment of inertia decreases the natural frequency. Equivalently, when applicable, the greater the static displacement the lower the natural frequency; however, from practical considerations too large of a static displacement may be undesirable.

Period of Undamped Free Oscillations

For an unforced and undamped system, the *period of free oscillation* of the system is given by

$$T = \frac{1}{f_n} = \frac{2\pi}{\omega_n} \tag{3.17}$$

Thus, increasing the natural frequency decreases the period and vice versa.

EXAMPLE 3.4   Natural frequency from static deflection of a machine system

For a particular choice of machinery mounting, the static deflection of a piece of machinery is found to be $\delta_{st1} = 0.1$ mm. For two other choices of machinery mounting, this deflection is found to be $\delta_{st2} = 1$ mm and $\delta_{st3} = 10$ mm. Based on the static deflection, we will determine the natural frequency for vertical vibrations for each of the three machinery mountings. To this end, we make use of Eq. (3.15). Noting that the acceleration due to gravity $g = 9.81$ m/s^2, we arrive at the following for the natural frequencies of the three machinery mountings:

$$f_{n1} = \frac{1}{2\pi} \sqrt{\frac{g}{\delta_{st1}}} = \frac{1}{2\pi} \sqrt{\frac{9.81 \text{ m/s}^2}{0.1 \times 10^{-3} \text{m}}} = 49.85 \text{ Hz}$$

$$f_{n2} = \frac{1}{2\pi} \sqrt{\frac{g}{\delta_{st2}}} = \frac{1}{2\pi} \sqrt{\frac{9.81 \text{ m/s}^2}{1 \times 10^{-3} \text{m}}} = 15.76 \text{ Hz}$$

$$f_{n3} = \frac{1}{2\pi} \sqrt{\frac{g}{\delta_{st3}}} = \frac{1}{2\pi} \sqrt{\frac{9.81 \text{ m/s}^2}{10 \times 10^{-3} \text{m}}} = 4.98 \text{ Hz}$$

EXAMPLE 3.5   Static deflection and natural frequency of the tibia bone in a human leg

Consider a person of 100 kg mass standing upright. We shall determine the static deflection in the tibia bone and an estimate of the natural frequency of axial vibrations. The tibia has a length of 40 cm, and it is modeled as a hollow tube with an inner diameter of 2.4 cm and an outer diameter of 3.4 cm. The Young's modulus of elasticity of the bone material is 2×10^{10} N/m^2.

The static deflection will be determined by using Eq. (3.6), and Eq. (3.15) will be used to determine the natural frequency. We assume that both legs support the weight of the person equally, so that the weight supported by the tibia is

$$mg = \left(\frac{100}{2} \text{ kg}\right) \times (9.81 \text{ m/s}^2) = 490.5 \text{ N} \tag{a}$$

To determine the stiffness k of the tibia, we use Case 1 of Table 2.3 to obtain

$$k = \frac{EA}{L} = \frac{(2 \times 10^{10} \text{ N/m}^2) \times \frac{\pi}{4}[(3.4 \times 10^{-2})^2 - (2.4 \times 10^{-2})^2]\text{m}^2}{40 \times 10^{-2} \text{ m}}$$

$$= 22.78 \times 10^6 \text{ N/m} \tag{b}$$

Hence, from Eqs. (3.6), (a), and (b), the static deflection is given by

$$\delta_{st} = \frac{mg}{k} = \frac{490.5 \text{ N}}{22.78 \times 10^6 \text{ N/m}} = 21.53 \text{ } \mu\text{m} \tag{c}$$

and, from Eqs. (3.15) and (c), the natural frequency is

$$f_n = \frac{1}{2\pi}\sqrt{\frac{g}{\delta_{st}}} = \frac{1}{2\pi}\sqrt{\frac{9.81 \text{ m/s}^2}{21.53 \text{ }\mu\text{m}}} = 107.4 \text{ Hz} \tag{d}$$

EXAMPLE 3.6   System with a constant natural frequency

In many practical situations, different pieces of machinery are used with a single spring-mounting system. Under these conditions, one would like for the system natural frequency to be constant for the different machinery-mounting-system combinations; that is, we are looking for a system whose natural frequency does not change as the system mass is changed. Through this example, we examine how the spring-mounting system can be designed and discuss a realization of this spring in practice.

From Eq. (3.14), it is clear that the natural frequency depends on the mass of the system. In order to realize the desired objective of constant natural frequency regardless of the system weight, we need a nonlinear spring whose equivalent spring constant is given by

$$k = AW \tag{a}$$

where A is a constant, the weight $W = mg$, and g is the gravitational constant. Then, from Eqs. (3.15) and (a), we arrive at

$$f_n = \frac{1}{2\pi}\sqrt{\frac{k}{m}} = \frac{1}{2\pi}\sqrt{\frac{kg}{W}} = \frac{1}{2\pi}\sqrt{Ag} \text{ Hz} \tag{b}$$

from which it is clear that the natural frequency is constant irrespective of the weight of the mass.

Nonlinear Spring Mounting

When the side walls of a rubber cylindrical tube are compressed into the non-linear region,[6] the equivalent spring stiffness of this system approximates the characteristic given by Eq. (a). For illustrative purposes, consider a spring that has the general force-displacement relationship

$$F(x) = a\left(\frac{x}{b}\right)^c \tag{c}$$

where a and b are scale factors and c is a shape factor. Noting that for a machinery of weight W, the static deflection x_o is determined by using Eq. (c) as

$$x_o = b\left(\frac{W}{a}\right)^{1/c} \tag{d}$$

---

[6]E. I. Riven, *Stiffness and Damping in Mechanical Design*, Marcel Dekker, NY, pp. 58–61 (1999).

For "small" amplitude vibrations about x_o, the linear equivalent stiffness of this spring is determined from Eqs. (c) and (d) to be

$$k_{eq} = \left.\frac{dF(x)}{dx}\right|_{x=x_o} = \frac{ac}{b}\left(\frac{x_o}{b}\right)^{c-1}$$

$$= \frac{ac}{b}\left(\frac{W}{a}\right)^{(c-1)/c} \tag{e}$$

Then, from Eqs. (3.14) and (e), we determine the natural frequency of this system as

$$f_n = \frac{1}{2\pi}\sqrt{\frac{k_{eq}}{W/g}} = \frac{1}{2\pi}\sqrt{\frac{gc}{b}\left(\frac{W}{a}\right)^{-1/c}}$$

$$= \frac{1}{2\pi}\sqrt{\frac{gc}{b}}\left(\frac{W}{a}\right)^{-1/(2c)} \text{ Hz} \tag{f}$$

Representative Spring Data

We now consider the representative data of a nonlinear spring shown in Figure 3.5a (on next page). By using standard curve-fitting procedures,[7] we find that $a = 2500$ N, $b = 0.011$ m, and $c = 2.77$. After substituting these values into Eq. (f), we arrive at the natural frequency values shown in Figure 3.5b (on next page). It is seen that over a sizable portion of the load range, the natural frequency of the system varies within the range of ±8.8%. The natural frequency of a system with a linear spring whose static displacement ranges from 12 mm to 5 mm varies approximately 55%.

3.3.2 Damping Factor

Translation Vibrations: Damping Factor

For translating single degree-of-freedom systems, such as those described by Eq. (3.8), the *damping factor* or *damping ratio* ζ is defined as

$$\zeta = \frac{c}{2m\omega_n} = \frac{c}{2\sqrt{km}} = \frac{c\omega_n}{2k} \tag{3.18}$$

where c is the system damping coefficient with units of N·s/m, k is the system stiffness, and m is the system mass. The damping factor is a nondimensional quantity.

Critical Damping, Underdamping, and Overdamping

Defining the quantity c_c, called the *critical damping*, as

$$c_c = 2m\omega_n = 2\sqrt{km} \tag{3.19}$$

---

[7]The MATLAB function `lsqcurvefit` from the Optimization Toolbox was used.

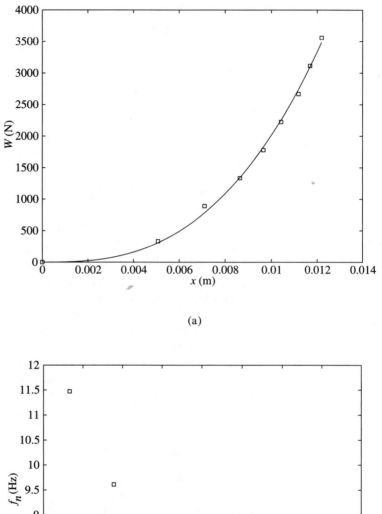

(a)

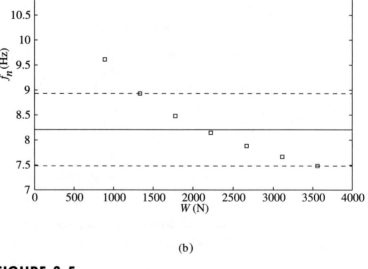

(b)

FIGURE 3.5
(a) Curve fit of nonlinear spring data: squares—experimental data values; solid line—fitted curve; (b) natural frequency for data values in (a) above: horizontal broken lines are within ±8.8% from the solid horizontal line.

the damping ratio is rewritten in the form

$$\zeta = \frac{c}{c_c} \tag{3.20}$$

When $c = c_c$, $\zeta = 1$. The significance of c_c is discussed in Sections 4.2 and 4.3, where free oscillations of vibratory systems are considered. A system for which $0 < \zeta < 1$ is called an *underdamped* system and a system for which $\zeta > 1$ is called an *overdamped* system. A system for which $\zeta = 1$ is called a *critically* damped system.

Rotational Vibrations: Damping Factor

For rotating single degree-of-freedom systems such as those described by Eq. (3.13), the *damping ratio* ζ is defined as

$$\zeta = \frac{c_t}{2J\omega_n} = \frac{c_t}{2\sqrt{k_t J}} \tag{3.21}$$

where the damping coefficient c_t has the units N·m·s/rad.

From Eqs. (3.14) and (3.16), we see that the stiffness and inertia properties affect the natural frequency. From Eqs. (3.18) and (3.21), we see that the damping ratio is affected by any change in the stiffness, inertia, or damping property. However, for both ω_n and ζ, one can change more than one system parameter in such a way that the net effect on ω_n and ζ remains unchanged. This is shown in Example 3.7.

Governing Equation of Motion in Terms of Natural Frequency and Damping Factor

Introducing the definitions given by Eqs. (3.14) and (3.18) into Eq. (3.8), we obtain

$$\frac{d^2x}{dt^2} + 2\zeta\omega_n \frac{dx}{dt} + \omega_n^2 x = \frac{f(t)}{m} \tag{3.22}$$

The significance of the quantities ω_n and ζ will become apparent when the solution to Eq. (3.22) is discussed in detail in the subsequent chapters. If we introduce the dimensionless time $\tau = \omega_n t$, then Eq. (3.22) can be written as

$$\frac{d^2x}{d\tau^2} + 2\zeta \frac{dx}{d\tau} + x = \frac{f(\tau)}{k} \tag{3.23}$$

It is seen from Eq. (3.23) that the natural frequency associated with the nondimensional system is always unity (one), and that the damping factor ζ is the only system parameter that appears explicitly on the left-hand side of the equation. We shall use both forms of Eqs. (3.22) and (3.23) in subsequent chapters.

In the absence of forcing, that is, when $f(\tau) = 0$, the motion of a vibratory system expressed in terms of nondimensional quantities can be described by just one system parameter. This fact is further elucidated in Section 4.3, where free oscillations are considered and it is shown that the qualitative nature of these oscillations can be completely characterized by the damping factor. In the presence of forcing, that is, $f(\tau) \neq 0$, both the damping factor ζ and the natural frequency ω_n are important for characterizing the nature of the response. This is further addressed when the forced responses of single degree-of-freedom systems are considered in Chapters 5 and 6.

Since the damping coefficient is one of the most important descriptors of a vibratory system, it is important to understand its interrelationships with the component's parameters m (or J), c (or c_t), and k (or k_t). We shall illustrate some of these relationships with the following example.

EXAMPLE 3.7   Effects of system parameters on the damping ratio

An engineer finds that a single degree-of-freedom system with mass m, damping c, and spring constant k has too much static deflection δ_{st}. The engineer would like to decrease δ_{st} by a factor of 2, while keeping the damping ratio constant. Through this example, we shall determine what options the engineer has.

Noting that this is a problem involving vertical vibrations, it is seen from Eqs. (3.6), (3.15), and (3.18) that

$$\delta_{st} = \frac{mg}{k}$$

$$2\zeta = \frac{c}{m}\sqrt{\frac{\delta_{st}}{g}} = c\sqrt{\frac{\delta_{st}}{gm^2}} = \frac{1}{m}\sqrt{\frac{c^2\delta_{st}}{g}} \tag{a}$$

From Eqs. (a), we see that there are three ways that one can achieve the goal.

First Choice

For the first choice, let c remain constant. Then, when δ_{st} is reduced by one-half,

$$2\zeta = c\sqrt{\frac{\delta_{st}}{gm^2}} \rightarrow c\sqrt{\frac{\delta_{st}}{2gm^2}} \tag{b}$$

and, therefore, the mass has to be reduced by a factor of $\sqrt{2}$ and the stiffness has to be increased by a factor of $\sqrt{2}$; that is,

$$m \rightarrow \frac{m}{\sqrt{2}} \quad \text{and} \quad k \rightarrow k\sqrt{2} \tag{c}$$

since,

$$\delta_{st} = \frac{mg}{k} \rightarrow \frac{m}{\sqrt{2}}\frac{g}{k\sqrt{2}} = \frac{\delta_{st}}{2} \tag{d}$$

Thus, for $\delta_{st} \rightarrow \delta_{st}/2$ and c held constant, $m \rightarrow m/\sqrt{2}$ and $k \rightarrow k\sqrt{2}$.

Second Choice

For the second choice, let m remain constant. Then, when δ_{st} is reduced by one-half,

$$2\zeta = \frac{1}{m}\sqrt{\frac{c^2\delta_{st}}{g}} \rightarrow \frac{1}{m}\sqrt{\frac{c^2\delta_{st}}{2g}} \tag{e}$$

and, therefore, the damping coefficient has to be increased by a factor of $\sqrt{2}$ and the stiffness has to be increased by a factor of 2; that is,

$$c \rightarrow c\sqrt{2} \quad \text{and} \quad k \rightarrow 2k \tag{f}$$

since,

$$\delta_{st} = \frac{mg}{k} \rightarrow \frac{mg}{2k} = \frac{\delta_{st}}{2} \tag{g}$$

Thus, for $\delta_{st} \rightarrow \delta_{st}/2$ and m held constant, $k \rightarrow 2k$ and $c \rightarrow c\sqrt{2}$.

Third Choice

For the last choice, let k remain constant. Then, when δ_{st} is reduced by one-half,

$$\delta_{st} = \frac{mg}{k} \rightarrow \frac{m}{2}\frac{g}{k} = \frac{\delta_{st}}{2} \tag{h}$$

Thus, the mass has to be reduced by a factor of 2; that is,

$$m \rightarrow \frac{m}{2} \tag{i}$$

Furthermore, since

$$2\zeta = \frac{c}{\sqrt{km}} \tag{j}$$

the damping coefficient has to be reduced by a factor of $\sqrt{2}$; that is,

$$c \rightarrow \frac{c}{\sqrt{2}} \tag{k}$$

Thus, for $\delta_{st} \rightarrow \delta_{st}/2$ and k held constant, $m \rightarrow m/2$ and $c \rightarrow c/\sqrt{2}$.

The results of this example can be generalized to a design guideline (see Exercises 3.7 and 3.8).

EXAMPLE 3.8   Effect of mass on the damping factor

A system is initially designed to be critically damped—that is, with a damping factor of $\zeta = 1$. Due to a design change, the mass of the system is increased 20%—that is, from m to $1.2m$. Will the system still be critically damped if the stiffness and the damping coefficient of the system are kept the same?

The definition of the damping factor is given by Eq. (3.18) and that for the critical damping factor is given by Eq. (3.19). Then, the damping factor of the system after the design change is given by

$$\zeta_{new} = \frac{c}{2\sqrt{k(1.2m)}} = 0.91\frac{c}{2\sqrt{km}} = 0.91\frac{c}{c_c} = 0.91 \times 1 = 0.91$$

Therefore, the system with the increased mass is no longer critically damped; rather, it is now underdamped.

In the next two subsections, the governing equations for different types of damping models and forcing conditions are presented. For all of these cases, translational motions are considered for illustrative purposes, and the equations are obtained by carrying out a force balance along the direction of motion. The form of the governing equations will be similar for systems involving rotational motions.

3.4   GOVERNING EQUATIONS FOR DIFFERENT TYPES OF DAMPING

The governing equations of motion for systems with different types of damping are obtained by replacing the term corresponding to the force due to viscous damping with the force due to either the fluid, structural, or dry friction type damping. Solutions for different periodically forced systems are given in Section 5.8, where equivalent viscous damping coefficients for different damping models are obtained.

Coulomb or Dry Friction Damping

After using Eq. (2.52) to replace the $c\dot{x}$ term in Eq. (3.8), the governing equation of motion takes the form

$$m\,\frac{d^2x}{dt^2} + kx + \underbrace{\mu mg\,\text{sgn}(\dot{x})}_{\substack{\text{Nonlinear dry} \\ \text{friction force}}} = f(t) \tag{3.24}$$

which is a nonlinear equation because the damping characteristic is piece-wise linear. This piece-wise linear property can be used to find the solution of this system.

Fluid Damping

After using Eq. (2.54) to replace the $c\dot{x}$ term in Eq. (3.8), the governing equation takes the form

$$m\,\frac{d^2x}{dt^2} + \underbrace{c_d\,|\dot{x}|\,\dot{x}}_{\substack{\text{Nonlinear fluid} \\ \text{damping force}}} + kx = f(t) \tag{3.25}$$

which is a nonlinear equation due to the nature of the damping.

Structural Damping

After using Eq. (2.57) to replace the $c\dot{x}$ term in Eq. (3.8), we arrive at the governing equation

$$m\,\frac{d^2x}{dt^2} + k\beta\pi\,\text{sgn}(\dot{x})|x| + kx = f(t) \tag{3.26}$$

Equation (3.26) is further addressed in Section 5.8.

3.5 GOVERNING EQUATIONS FOR DIFFERENT TYPES OF APPLIED FORCES

In Section 3.2, we addressed governing equations of single degree-of-freedom systems whose inertial elements were subjected to direct excitations. Here, we address governing equations of single degree-of-freedom systems subjected to base excitations, excited by rotating unbalance, and immersed in a fluid.

3.5.1 System with Base Excitation

The base-excitation model is a prototype that is useful for studying buildings subjected to earthquakes, packaging during transportation, vehicle response, and for designing accelerometers (see Section 5.6). Here, the physical system of interest is represented by a single degree-of-freedom system whose base is subjected to a displacement disturbance $y(t)$, and an equation governing the motion of this system is sought to determine the response of the system $x(t)$.

If the system of interest is an automobile, then the road surface on which it is traveling can be a source of the disturbance $y(t)$ and the vehicle response $x(t)$ is to be determined. To avoid failure of electronic components during transportation, a base-excitation model is used to predict the vibration response of the electronic components. For buildings located above or adjacent to subways or above ground railroad tracks, the passage of trains can act as a source of excitation to the base of the building. In designing accelerometers, the accelerometer responses to different base excitations are studied to determine the appropriate accelerometer system parameters, such as the damping factor.

A prototype of a single degree-of-freedom system subjected to a base excitation is illustrated in Figure 3.6. The system represents an instrumentation package being transported in a vehicle. The vehicle provides the base excitation $y(t)$ to the instrumentation package modeled as a single degree-of-freedom system. The displacement response $x(t)$ is measured from the system's static-equilibrium position. In the system shown in Figure 3.6, it is assumed that no external force is applied directly to the mass; that is, $f(t) = 0$. Based on the free-body diagram shown in Figure 3.6, we use Eq. (3.1b) to obtain the following governing equation of motion

$$m \frac{d^2x}{dt^2} + c \frac{dx}{dt} + kx = c \frac{dy}{dt} + ky \qquad (3.27)$$

which, on using Eqs. (3.14) and (3.18), takes the form

$$\frac{d^2x}{dt^2} + 2\zeta\omega_n \frac{dx}{dt} + \omega_n^2 x = 2\zeta\omega_n \frac{dy}{dt} + \omega_n^2 y \qquad (3.28)$$

The displacements $y(t)$ and $x(t)$ are measured from a fixed point O located in an inertial reference frame and a fixed point located at the system's

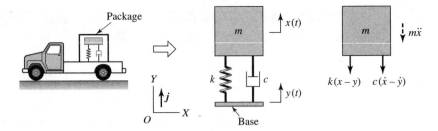

FIGURE 3.6
Base excitation and the free-body diagram of the mass.

static-equilibrium position, respectively. If the relative displacement is desired, then we let

$$z(t) = x(t) - y(t) \tag{3.29}$$

and Eq. (3.27) is written as

$$m \frac{d^2z}{dt^2} + c \frac{dz}{dt} + kz = -m \frac{d^2y}{dt^2} \tag{3.30}$$

while Eq. (3.28) becomes

$$\frac{d^2z}{dt^2} + 2\zeta\omega_n \frac{dz}{dt} + \omega_n^2 z = -\frac{d^2y}{dt^2} \tag{3.31}$$

where $\ddot{y}(t)$ is the acceleration of the base.

3.5.2 System with Unbalanced Rotating Mass

As discussed in Chapter 1, many rotating machines such as fans, clothes dryers, internal combustion engines, and electric motors, have a certain degree of unbalance. In modeling such systems as single degree-of-freedom systems, it is assumed that the unbalance generates a force that acts on the system's mass. This force, in turn, is transmitted through the spring and damper to the fixed base. The unbalance is modeled as a mass m_o that rotates with an angular speed ω, and this mass is located a fixed distance ϵ from the center of rotation as shown in Figure 3.7. Note that in Figure 3.7, M does not include the unbalance m_o.

For deriving the governing equation, only motions along the vertical direction are considered, since the presence of the lateral supports restrict motion in the j direction. The displacement of the system $x(t)$ is measured from the system's static-equilibrium position. The fixed point O is chosen to coincide with the vertical position of the static-equilibrium position. Based on the discussion in Section 3.2, gravity loading is not explicitly taken into account.

From the free-body diagram of the unbalanced mass m_o, we find that the reactions at the point O' are given by

$$N_x = -m_o(\ddot{x} - \epsilon\omega^2 \sin \omega t)$$
$$N_y = m_o\epsilon\omega^2 \cos \omega t \tag{3.32}$$

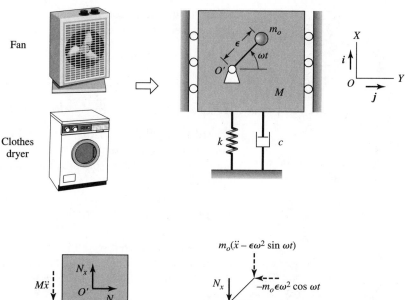

Fan

Clothes dryer

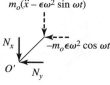

FIGURE 3.7
System with unbalanced rotating mass and free-body diagrams.

and from the free-body diagram of mass M we find that

$$M \frac{d^2x}{dt^2} + c \frac{dx}{dt} + kx = N_x \tag{3.33}$$

Then, substituting for N_x from Eqs. (3.32) into Eq. (3.33), we arrive at the equation of motion

$$(M + m_o) \frac{d^2x}{dt^2} + c \frac{dx}{dt} + kx = m_o \epsilon \omega^2 \sin \omega t \tag{3.34}$$

which is rewritten as

$$\frac{d^2x}{dt^2} + 2\zeta\omega_n \frac{dx}{dt} + \omega_n^2 x = \frac{F(\omega)}{m} \sin \omega t \tag{3.35}$$

where

$$m = M + m_o$$
$$\omega_n = \sqrt{\frac{k}{m}}$$
$$F(\omega) = m_o \epsilon \omega^2 \tag{3.36}$$

In Eqs. (3.35) and (3.36), $F(\omega)$ is the magnitude of the unbalanced force. This magnitude depends on the unbalanced mass m_o and it is proportional to the

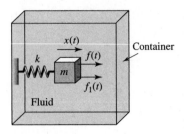

FIGURE 3.8
Vibrations of a system immersed in a fluid.

square of the excitation frequency. From Eq. (3.6) it follows that the static displacement of the spring is

$$\delta_{st} = \frac{(M + m_o)g}{k} = \frac{mg}{k} \tag{3.37}$$

3.5.3 System with Added Mass Due to a Fluid

Consider a rigid body that is connected to a spring as shown in Figure 3.8. The entire system is immersed in a fluid. From Eq. (3.8) and Figure 3.8, and noting that $c = 0$ because there is no damper, the equation of motion of the system is

$$m \frac{d^2x}{dt^2} + kx = f(t) + f_1(t) \tag{3.38}$$

where $x(t)$ is measured from the unstretched position of the spring, $f(t)$ is the externally applied force, and $f_1(t)$ is the force exerted by the fluid on the mass due to the motion of the mass. The force generated by the fluid on the rigid body is[8]

$$f_1(t) = -K_o M \frac{d^2x}{dt^2} - C_f \frac{dx}{dt} \tag{3.39}$$

where M is the mass of the fluid displaced by the body, K_o is an added mass coefficient that is a function of the shape of the rigid body and the shape and size of the container holding the fluid, and C_f is a positive fluid damping coefficient that is a function of the shape of the rigid body, the kinematic viscosity of the fluid, and the frequency of oscillation of the rigid body.

After substituting Eq. (3.39) into Eq. (3.38), we obtain the governing equation of motion

$$(m + K_o M) \frac{d^2x}{dt^2} + C_f \frac{dx}{dt} + kx = f(t) \tag{3.40}$$

where $K_o M$ is the *added mass* due to the fluid. From Eq. (3.40), we see that placing a single degree-of-freedom system in a fluid increases the total mass of the system and adds viscous damping to the system.[9] A practical application of the fluid mass loading is in modeling offshore structures.[10]

In this section, the use of force-balance and moment-balance methods for deriving the governing equation of a single degree-of-freedom system

[8]K. G. McConnell and D. F. Young, "Added mass of a sphere in a bounded viscous fluid," *J. Engrg. Mech. Div.,* Proc. ASCE, Vol. 91, No. 4, pp. 147–164 (1965).

[9]A similar result is obtained when one considers the acoustic radiation loading on the mass, when the surface of the mass is used as an acoustically radiating surface. See L. E. Kinsler and A. R. Frey, *Fundamentals of Acoustics,* 2nd ed., John Wiley and Sons, NY, pp.180–183 (1962).

[10]A. Uścilowska and J. A. Kołodzeij, "Free vibration of immersed column carrying a tip mass," *J. Sound Vibration,* Vol. 216, No. 1, pp. 147–157 (1998).

was illustrated. In the next section, a different method to obtain the governing equation of a single degree-of-freedom system is presented. This method is based on Lagrange's equations, where one makes use of scalar quantities such as kinetic energy and potential energy for obtaining the equation of motion.

3.6 LAGRANGE'S EQUATIONS[11]

Lagrange's equations can be derived from differential principles such as the principle of virtual work and integral principles such as those discussed in Chapter 9. We will not derive Lagrange's equations here, but use these equations to obtain governing equations of holonomic[12] systems. We first present Lagrange's equations for a system with multiple degrees of freedom and then apply them to vibratory systems modeled with a single degree of freedom. In Chapter 7, we illustrate how the governing equations of systems with multiple degrees of freedom are determined by using Lagrange's equations.

Let us consider a system with N degrees of freedom that is described by a set of N generalized coordinates q_i, $i = 1, 2, ..., N$. These coordinates are unconstrained, independent coordinates; that is, they are not related to each other by geometrical or kinematical conditions. Then, in terms of the chosen generalized coordinates, Lagrange's equations have the form

$$\frac{d}{dt}\left(\frac{\partial T}{\partial \dot{q}_j}\right) - \frac{\partial T}{\partial q_j} + \frac{\partial D}{\partial \dot{q}_j} + \frac{\partial V}{\partial q_j} = Q_j \qquad j = 1, 2, \ldots, N \qquad (3.41)$$

where $\dot{q}_j$ are the generalized velocities, T is the kinetic energy of the system, V is the potential energy of the system, D is the Rayleigh dissipation function, and Q_j is the generalized force that appears in the jth equation. The generalized forces Q_j are given by

$$Q_j = \sum_l \boldsymbol{F}_l \cdot \frac{\partial \boldsymbol{r}_l}{\partial q_j} + \sum_l \boldsymbol{M}_l \cdot \frac{\partial \boldsymbol{\omega}_l}{\partial \dot{q}_j} \qquad (3.42)$$

where $\boldsymbol{F}_l$ and $\boldsymbol{M}_l$ are the vector representations of the externally applied forces and moments, respectively, the index l indicates which external force or moment is being considered, $\boldsymbol{r}_l$ is the position vector to the location where the force is applied, and $\boldsymbol{\omega}_l$ is the system angular velocity about the axis along which the considered moment is applied. The "·" symbol in Eq. (3.42) indicates the scalar dot product of two vectors.

---

[11]For a derivation of the Lagrange equations see D. T. Greenwood, *Principles of Dynamics,* Prentice Hall, Upper Saddle River, NJ, 1988, Section 6-6.

[12]As discussed in Chapter 1, holonomic systems are systems subjected to holonomic constraints, which are integrable constraints. Geometric constraints discussed in Chapter 1 are in this category.

Linear Vibratory Systems

For vibratory systems with linear inertial characteristics, linear stiffness characteristics, and linear viscous damping characteristics, the quantities T, V, and D take the following form:[13]

$$T = \frac{1}{2} \sum_{j=1}^{N} \sum_{n=1}^{N} m_{jn} \dot{q}_j \dot{q}_n$$

$$V = \frac{1}{2} \sum_{j=1}^{N} \sum_{n=1}^{N} k_{jn} q_j q_n$$

$$D = \frac{1}{2} \sum_{j=1}^{N} \sum_{n=1}^{N} c_{jn} \dot{q}_j \dot{q}_n \tag{3.43}$$

In Eqs. (3.43), each of the summations is carried out over the number of degrees of freedom N and the inertia coefficients m_{jn}, the stiffness coefficients k_{jn}, and the damping coefficients c_{jn} are positive quantities.

Single Degree-of-Freedom Systems

As stated previously, in this chapter we shall limit the discussion to the case when $N = 1$, which is the case of a single degree-of-freedom system. Then Eqs. (3.41) reduce to

$$\frac{d}{dt}\left(\frac{\partial T}{\partial \dot{q}_1}\right) - \frac{\partial T}{\partial q_1} + \frac{\partial D}{\partial \dot{q}_1} + \frac{\partial V}{\partial q_1} = Q_1 \tag{3.44}$$

where the generalized force is obtained from Eq. (3.42), which reduces to

$$Q_1 = \sum_l \boldsymbol{F}_l \cdot \frac{\partial \boldsymbol{r}_l}{\partial q_1} + \sum_l \boldsymbol{M}_l \cdot \frac{\partial \boldsymbol{\omega}_l}{\partial \dot{q}_1} \tag{3.45}$$

Linear Single Degree-of-Freedom Systems

For linear single degree-of-freedom systems, the expressions for the system kinetic energy, the system potential energy, and the system dissipation function given by Eqs. (3.43) reduce to

$$T = \frac{1}{2} \sum_{j=1}^{1} \sum_{n=1}^{1} m_{jn} \dot{q}_j \dot{q}_n = \frac{1}{2} m_{11} \dot{q}_1^2$$

$$V = \frac{1}{2} \sum_{j=1}^{1} \sum_{n=1}^{1} k_{jn} q_j q_n = \frac{1}{2} k_{11} \dot{q}_1^2$$

$$D = \frac{1}{2} \sum_{j=1}^{1} \sum_{n=1}^{1} c_{jn} \dot{q}_j \dot{q}_n = \frac{1}{2} c_{11} \dot{q}_1^2 \tag{3.46a}$$

---

[13]The quadratic forms shown for T, V, and D in Eqs. (3.43) are strictly valid for systems with linear characteristics. In addition, the kinetic energy T is not always a function of only velocities as shown here. Systems in which the kinetic energy has the quadratic form shown in Eqs. (3.43) are called *natural systems*. For a general system with holonomic constraints, Eqs. (3.41) will be used directly in this book to obtain the governing equations.

Comparing the forms of the kinetic energy T, the potential energy and V, and the dissipation function D with the standard forms given in Chapter 2, we find that

$$T = \frac{1}{2} m_e \dot{q}_1^2$$

$$V = \frac{1}{2} k_e q_1^2$$

$$D = \frac{1}{2} c_e \dot{q}_1^2 \qquad (3.46b)$$

where m_e is the *equivalent mass*, k_e is the *equivalent stiffness*, and c_e is the *equivalent viscous damping;* they are given by

$$m_e = m_{11}$$
$$k_e = k_{11}$$
$$c_e = c_{11} \qquad (3.46c)$$

On substituting Eqs. (3.46) into Eq. (3.44), the result is

$$\frac{d}{dt}\left(\frac{\partial}{\partial \dot{q}_1}\left(\frac{1}{2} m_e \dot{q}_1^2\right)\right) - \frac{\partial}{\partial q_1}\left(\frac{1}{2} m_e \dot{q}_1^2\right) + \frac{\partial}{\partial \dot{q}_1}\left(\frac{1}{2} c_e \dot{q}_1^2\right) + \frac{\partial}{\partial q_1}\left(\frac{1}{2} k_e q_1^2\right) = Q_1$$

$$\frac{d}{dt}(m_e \dot{q}_1) - 0 + c_e \dot{q}_1 + k_e q_1 = Q_1$$

$$m_e \ddot{q}_1 + c_e \dot{q}_1 + k_e q_1 = Q_1$$
$$(3.47)$$

Thus, to obtain the governing equation of motion of a linear vibrating system with viscous damping, one first obtains expressions for the system kinetic energy, system potential energy, and system dissipation function. If these quantities can be grouped so that an equivalent mass, equivalent stiffness, and equivalent damping can be identified, then, after the determination of the generalized force, the governing equation is given by the last of Eqs. (3.47). We see further from the definitions Eqs. (3.14) and (3.18) that

$$\omega_n = \sqrt{\frac{k_e}{m_e}}$$

$$\zeta = \frac{c_e}{2 m_e \omega_n} = \frac{c_e}{2\sqrt{k_e m_e}} \qquad (3.48)$$

It is noted that depending on the choice of the generalized coordinate, the determined equivalent inertia, equivalent stiffness, and equivalent damping properties of a system will be different. In the rest of this section, the use of the Lagrange equations is illustrated with nine examples. As illustrated in these examples, we use the last of Eqs. (3.47) to obtain the governing equations of motion if the system kinetic energy, system potential energy, and dissipation function are in the form of Eqs. (3.46b); otherwise, we use Eq. (3.44) directly to obtain the governing equation of motion.

It is noted that only the system displacements and velocities are needed from the kinematics to use Lagrange's method whereas, to use the force-balance and moment-balance methods, one also needs system accelerations and has to deal with internal forces. In addition, with the increasing use of symbolic manipulation programs, it has become more common to have these programs derive the governing equations directly from the Lagrange's equations.

EXAMPLE 3.9   Equation of motion for a linear single degree-of-freedom system

For the linear system of Figure 3.1, the equation of motion is derived by using Lagrange's equations. After choosing the generalized coordinate to be x, we determine the system kinetic energy, the system potential energy, and the dissipation function for the system. From these quantities and the determined generalized force, the governing equation of motion of the system is established for motions about the static equilibrium position.

First, we identify the following

$$q_1 = x, \qquad \boldsymbol{F}_l = f(t)\boldsymbol{j}, \qquad \boldsymbol{r}_l = x\boldsymbol{j}, \qquad \text{and} \qquad \boldsymbol{M}_l = 0 \qquad \text{(a)}$$

where $\boldsymbol{j}$ is the unit vector along the vertical direction. Making use of Eqs. (3.45) and (a), we determine the generalized force as

$$Q_1 = \sum_l \boldsymbol{F}_l \cdot \frac{\partial \boldsymbol{r}_l}{\partial q_j} + 0 = f(t)\boldsymbol{j} \cdot \frac{\partial x\boldsymbol{j}}{\partial x} = f(t) \qquad \text{(b)}$$

From Eqs. (2.3) and (2.10), we find that the system kinetic energy and potential energy are, respectively,

$$T = \frac{1}{2}m\dot{x}^2$$

$$V = \frac{1}{2}kx^2 \qquad \text{(c)}$$

and, from Eqs. (3.46), the dissipation function is

$$D = \frac{1}{2}c\dot{x}^2 \qquad \text{(d)}$$

Comparing Eqs. (c) and (d) to Eqs. (3.46), we recognize that

$$m_e = m, \qquad k_e = k, \qquad \text{and} \qquad c_e = c \qquad \text{(e)}$$

Hence, from Eqs. (e) and the last of Eqs. (3.47), the governing equation of motion has the form

$$m\frac{d^2x}{dt^2} + c\frac{dx}{dt} + kx = f(t) \qquad \text{(f)}$$

which is identical to Eq. (3.8).

EXAMPLE 3.10   Equation of motion for a system that translates and rotates

In Figure 3.9, a system that translates and rotates is illustrated. After choosing a generalized coordinate, we construct the system kinetic energy, the system potential energy, and the dissipation function, and then noting that they are in the form of Eqs. (3.46), we determine the equivalent inertia, equivalent stiffness, and equivalent damping coefficient. Based on these equivalent system properties and the last of Eqs. (3.47), we obtain the governing equation of motion of this system. We also determine the expressions for the natural frequency and the damping factor.

As shown in Figure 3.9, the disc has a mass m and a mass moment of inertia J_G about its center G. The disc rolls without slipping. The horizontal location of the fixed point O is chosen to coincide with the unstretched length of the spring, and the horizontal translations of the center of mass of the disc are measured from this point O. When the center of the disc translates an amount x along the horizontal direction i, then $x = r\theta$, where θ is the corresponding rotation of the disc about an axis parallel to k. We can choose either x or θ as the generalized coordinate and express the one that is not chosen in terms of the other. Here, we choose θ as the generalized coordinate. Furthermore, we recognize that

$$q_1 = \theta, \qquad F_l = 0, \qquad M_l = M(t)k, \qquad \text{and} \qquad \omega_l = \dot\theta k \qquad \text{(a)}$$

Then making use of Eqs. (3.45) and (a), we determine the generalized force to be

$$Q_1 = \sum_l M_l \cdot \frac{\partial \omega_l}{\partial \dot q_1} = M(t)k \cdot \frac{\partial \dot\theta}{\partial \dot\theta} k = M(t) \qquad \text{(b)}$$

To determine the potential energy, we note that we have a linear spring. Therefore, we make use of Eq. (2.10) and arrive at

$$V = \frac{1}{2} kx^2 = \frac{1}{2} k(r\theta)^2 = \frac{1}{2} kr^2\theta^2 \qquad \text{(c)}$$

From Eq. (c), we recognize the equivalent stiffness of the system to be

$$k_e = kr^2 \qquad \text{(d)}$$

To determine the kinetic energy of the system, we make use of Eq. (1.23); that is, the kinetic energy of the disc is the sum of the kinetic energy due to translation of the center of mass of the disc and the kinetic energy due to rotation about the center of mass. Thus, we find that

$$T = \underbrace{\frac{1}{2} m\dot x^2}_{\substack{\text{Translation} \\ \text{kinetic energy}}} + \underbrace{\frac{1}{2} J_G \dot\theta^2}_{\substack{\text{Rotational} \\ \text{kinetic energy}}}$$

$$= \frac{1}{2} [mr^2 + J_G]\dot\theta^2 = \frac{1}{2}\left[\frac{3}{2} mr^2\right]\dot\theta^2 \qquad \text{(e)}$$

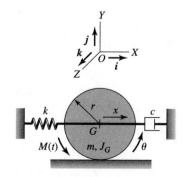

FIGURE 3.9
Disc rolling and translating.

where from Table 2.2, we have used $J_G = mr^2/2$. From Eq. (e), we recognize that the equivalent mass of the system is

$$m_e = \frac{3}{2} mr^2 \tag{f}$$

In this case, the dissipation function takes the form

$$D = \frac{1}{2} c\dot{x}^2 = \frac{1}{2} c(r\dot{\theta})^2 = \frac{1}{2} (cr^2)\dot{\theta}^2 \tag{g}$$

from which we identify the equivalent damping coefficient to be

$$c_e = cr^2 \tag{h}$$

Hence, from the determined generalized force and the equivalent inertia, equivalent stiffness, equivalent damping properties, and the last of Eqs. (3.47), we obtain the governing equation of motion

$$\frac{3}{2} mr^2 \ddot{\theta} + cr^2 \dot{\theta} + kr^2 \theta = M(t) \tag{i}$$

Natural Frequency and Damping Factor

To determine the natural frequency and the damping factor, we make use of Eqs. (3.48) and find that

$$\omega_n = \sqrt{\frac{k_e}{m_e}} = \sqrt{\frac{kr^2}{3mr^2/2}} = \sqrt{\frac{2k}{3m}}$$

$$\zeta = \frac{c_e}{2m_e\omega_n} = \frac{cr^2}{2(3mr^2/2)\sqrt{2k/3m}} = \frac{c}{\sqrt{6km}} \tag{j}$$

EXAMPLE 3.11   Governing equation for an inverted pendulum

For the inverted pendulum shown in Figure 3.10, we obtain the governing equation of motion for "small" oscillations about the upright position. The natural frequency of the inverted pendulum is also determined and the natural frequency of a related pendulum system is examined. In the system of Figure 3.10, the bar, which carries the sphere of mass m_1, has a mass m_2 that is uniformly distributed along its length. A linear spring of stiffness k and a linear viscous damper with a damping coefficient c are attached to the sphere.

Before constructing the system kinetic energy, we determine the mass moments of inertia of the sphere of mass m_1 and the bar of mass m_2 about the point O. The total rotary inertia of the system is given by

$$J_O = J_{O1} + J_{O2} \tag{a}$$

where J_{O1} is the mass moment of inertia of m_1 about point O and J_{O2} is the mass moment of inertia of the bar about point O. Making use of Table 2.2 and the parallel-axes theorem, we find that

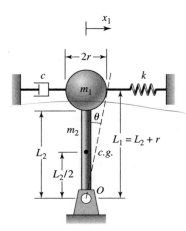

FIGURE 3.10
Inverted planar pendulum restrained by a spring and a viscous damper.

$$J_{O1} = \frac{2}{5} m_1 r^2 + m_1 L_1^2$$

$$J_{O2} = \frac{1}{12} m_2 L_2^2 + m_2 \left(\frac{L_2}{2}\right)^2 = \frac{1}{3} m_2 L_2^2 \qquad (b)$$

After choosing $q_1 = \theta$ as the generalized coordinate, and making use of Eqs. (a) and (b), we find that the system kinetic energy takes the form

$$T = \frac{1}{2} J_O \dot{\theta}^2 = \frac{1}{2} [J_{O1} + J_{O2}] \dot{\theta}^2$$

$$= \frac{1}{2} \left[\frac{2}{5} m_1 r^2 + m_1 L_1^2 + \frac{1}{3} m_2 L_2^2 \right] \dot{\theta}^2 \qquad (c)$$

For "small" rotations about the upright position, we can express the translation of mass m_1 as

$$x_1 \approx L_1 \theta \qquad (d)$$

Then, making use of Eqs. (2.10), (2.39), (2.45), and (d), the system potential energy is constructed as

$$V = \frac{1}{2} k x_1^2 - \frac{1}{2} m_1 g L_1 \theta^2 - \frac{1}{2} m_2 g \frac{L_2}{2} \theta^2$$

$$= \frac{1}{2} \left[k L_1^2 - m_1 g L_1 - m_2 g \frac{L_2}{2} \right] \theta^2 \qquad (e)$$

The dissipation function takes the form

$$D = \frac{1}{2} c \dot{x}_1^2 = \frac{1}{2} c L_1^2 \dot{\theta}^2 \qquad (f)$$

Comparing Eqs. (c), (e), and (f) to Eqs. (3.46), we find that the equivalent inertia, the equivalent stiffness, and the equivalent damping properties of the system are given by, respectively,

$$m_e = \frac{2}{5} m_1 r^2 + m_1 L_1^2 + \frac{1}{3} m_2 L_2^2$$

$$k_e = k L_1^2 - m_1 g L_1 - m_2 g \frac{L_2}{2}$$

$$c_e = c L_1^2 \qquad (g)$$

Noting that the only external force acting on the system is gravity loading, and that this has already been taken into account, the governing equation of motion is obtained from the last of Eqs. (3.47) as

$$m_e \ddot{\theta} + c_e \dot{\theta} + k_e \theta = 0 \qquad (h)$$

Then, from the first of Eqs. (3.48) and (g), we find that

$$\omega_n = \sqrt{\frac{k_e}{m_e}} = \sqrt{\frac{k L_1^2 - m_1 g L_1 - m_2 g \frac{L_2}{2}}{J_{O1} + J_{O2}}} \qquad (i)$$

It is pointed out that k_e can be negative, which affects the stability of the system as discussed in Section 4.4. The equivalent stiffness k_e is positive when

$$kL_1^2 > m_1gL_1 + m_2g\frac{L_2}{2} \tag{j}$$

that is, when the net moment created by the gravity loading is less than the restoring moment of the spring.

Natural Frequency of Pendulum System

In this case, we locate the pivot point O in Figure 3.10 on the top, so that the sphere is now at the bottom. The spring combination is still attached to the sphere. Then this pendulum system resembles the combination of the systems shown in Figures 2.17a and 2.17b. The equivalent stiffness of this system takes the form

$$k_e = kL_1^2 + m_1gL_1 + m_2g\frac{L_2}{2} \tag{k}$$

Noting that the equivalent inertia of the system is the same as in the inverted-pendulum case, we find the natural frequency of this system is

$$\omega_n = \sqrt{\frac{k_e}{m_e}} = \sqrt{\frac{kL_1^2 + m_1gL_1 + m_2g\dfrac{L_2}{2}}{J_{O1} + J_{O2}}} \tag{1}$$

If $m_2 << m_1$, $r << L_1$, and $k = 0$, then from Eqs. (b) and (1), we arrive at

$$\omega_n = \sqrt{\frac{m_1gL_1\left(1 + \dfrac{m_2L_2}{m_1L_1}\right)}{m_1L_1^2\left(1 + \dfrac{2r^2}{5L_1^2}\right)}} \rightarrow \sqrt{\frac{g}{L_1}} \tag{m}$$

which is the natural frequency of a pendulum composed of a rigid, weightless rod carrying a mass a distance L_1 from its pivot. We see that the natural frequency is independent of the mass and inversely proportional to the length L_1.

EXAMPLE 3.12   Governing equation for motion of a disk segment

For the half-disk shown in Figure 3.11, we will choose the coordinate θ as the generalized coordinate and establish the governing equation for the disc. Through this example, we illustrate how the system kinetic energy and the system potential energy can be approximated for "small" amplitude angular oscillations, so that the final form of the governing equation is linear. During the course of obtaining the governing equation, we determine the equivalent mass and equivalent stiffness of this system. The natural frequency of the disc

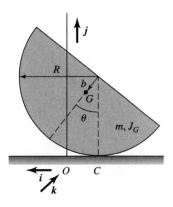

FIGURE 3.11
Half-disk rocking on a surface.

is determined and it is shown that disc can be treated as a pendulum with a certain effective length. After determining the equivalent system properties, we determine the governing equation of motion based on the last of Eqs. (3.47).

As shown in Figure 3.11, the half-disk has a mass m and a mass moment of inertia J_G about the center of mass G. The system is assumed to oscillate without slipping. The point O is a fixed point, and the point of contact C is a distance $R\theta$ from the fixed point for an angular motion θ. The orthogonal unit vectors i, j, and k are fixed in an inertial reference frame. The position vector from the fixed point O to the center of mass G is given by

$$r = (-R\theta + b \sin \theta)i + (R - b \cos \theta)j \tag{a}$$

and the absolute velocity of the center of mass is determined from Eq. (a) to be

$$\dot{r} = -(R - b \cos \theta)\dot{\theta}i + b \sin \theta \dot{\theta}j \tag{b}$$

Then, using Eq. (1.24) and selecting the generalized coordinate $q_1 = \theta$, the system kinetic energy takes the form

$$
\begin{aligned}
T &= \frac{1}{2} J_G \dot{\theta}^2 + \frac{1}{2} m(\dot{r} \cdot \dot{r}) \\
&= \frac{1}{2} J_G \dot{\theta}^2 + \frac{1}{2} m[(R - b \cos \theta)^2 + b^2 \sin^2 \theta]\dot{\theta}^2 \\
&= \frac{1}{2} J_G \dot{\theta}^2 + \frac{1}{2} m[R^2 + b^2 - 2bR \cos \theta]\dot{\theta}^2
\end{aligned}
\tag{c}
$$

Choosing the fixed ground as the datum, the system potential energy takes the form

$$V = \frac{1}{2} mg(R - b \cos \theta) \tag{d}$$

Note that the system kinetic energy and the system potential energy, which are given by Eqs. (c) and (d), respectively, are not in the form of Eqs. (3.46a).

Small Oscillations about the Upright Position

If we use the expressions for the system kinetic energy and system potential energy in Eq. (3.44), then the resulting equation of motion will be a nonlinear equation. Since our final objective is to have a linear equation that can be used to describe "small" amplitude angular oscillations about the upright position of the disk (i.e., $\theta_o = 0$), we express the angular displacement as

$$\theta(t) = \theta_o + \hat{\theta}(t) \tag{e}$$

and expand the trigonometric terms $\sin \theta$ and $\cos \theta$ as the following Taylor-series expansions:

$$\cos \theta = \cos(\theta_o + \hat{\theta}) \approx \cos \theta_o - \hat{\theta} \sin \theta_o - \frac{1}{2} \hat{\theta}^2 \cos \theta_o + \ldots$$

$$\sin \theta = \sin(\theta_o + \hat{\theta}) \approx \sin \theta_o + \hat{\theta} \cos \theta_o - \frac{1}{2} \hat{\theta}^2 \sin \theta_o + \ldots \tag{f}$$

Since $\theta_o = 0$, Eqs. (f) become

$$\cos\theta \approx 1 - \frac{1}{2}\hat{\theta}^2$$

$$\sin\theta \approx \hat{\theta} \tag{g}$$

In the expansions given by Eqs. (g), we have kept up to the quadratic terms in $\hat{\theta}$, since quadratic terms in the expressions for kinetic energy and potential energy lead to linear terms in the governing equations. On substituting Eqs. (e) and (f) into the expressions (c) and (d), noting that $\dot{\theta} = \dot{\hat{\theta}}$, and retaining up to quadratic terms in $\hat{\theta}$ and $\dot{\hat{\theta}}$ in these expressions, we arrive at

$$T \approx \frac{1}{2}[J_G + m(R - b)^2]\dot{\hat{\theta}}^2$$

$$V \approx \frac{1}{2}mg(R - b) + \frac{1}{2}mgb\hat{\theta}^2 \tag{h}$$

On comparing Eqs. (h) with Eqs. (3.46), we see that while the kinetic energy is in the standard form, the potential energy is not in standard form because of the constant term. However, since the datum for the potential energy is not unique, we can shift the datum for the potential energy from the fixed ground of Figure 3.11 to a distance $(R-b)$ above the ground. When this is done, this constant term does not appear and, from Eqs. (h) and (3.46), we identify that the equivalent inertia and stiffness properties of the system are

$$m_e = J_G + m(R - b)^2$$

$$k_e = mgb \tag{i}$$

Since the gravity loading has already been taken into account, the generalized force is zero. Furthermore, since there is no damping, the equivalent damping coefficient c_e is zero. Hence, from Eqs. (3.47) and (i), we arrive at the governing equation

$$[J_G + m(R - b)^2]\ddot{\hat{\theta}} + mgb\hat{\theta} = 0 \tag{j}$$

Natural Frequency

From Eq. (j), we find that the natural frequency is

$$\omega_n = \sqrt{\frac{mgb}{J_G + m(R - b)^2}}$$

$$= \sqrt{\frac{g}{[J_G + m(R - b)^2]/mb}} \tag{k}$$

On comparing the form of Eq. (k) to the form of the equation for the natural frequency of a planar pendulum of length L_1 given by Eq. (m) of Example 3.11, we note that Eq. (k) is similar in form to the natural frequency of a pendulum with an effective length

$$L_e = \frac{J_G + m(R - b)^2}{mb} \tag{l}$$

EXAMPLE 3.13 Governing equation for a translating system with a pretensioned spring

We revisit Example 2.8, and use Lagrange's equations to derive the governing equation of motion for vertical translations x of the mass about the static-equilibrium position of the system. Through this process, we will examine how the horizontal spring with linear stiffness k_1 affects the vibrations. The natural frequency of this system is also determined. The equation of motion will be derived for "small" amplitude vertical oscillations; that is, $|x/L| \ll 1$.

In the initial position, the horizontal spring is pretensioned with a tension T_1 as shown in Figure 3.12, which is produced by an initial extension of the spring by an amount δ_o; that is,

$$T_1 = k_1 \delta_o \tag{a}$$

The kinetic energy of the system is

$$T = \frac{1}{2} m \dot{x}^2 \tag{b}$$

Next, we note that the potential energy is given by

$$V = V_1 + V_2 \tag{c}$$

where V_i, $i = 1, 2$, is the potential energy associated with the spring of stiffness k_i. Note that gravitational loading is not taken into account because we are considering oscillations about the static-equilibrium position. On substituting for V_1 and V_2 in Eq. (b), we arrive at

$$V(x) = \frac{1}{2} k_1 (\delta_o + \Delta L)^2 + \frac{1}{2} k_2 x^2 \tag{d}$$

where ΔL is the change in the length of the spring with stiffness k_1 due to the motion x of the mass. For $|x/L| \ll 1$, as discussed in Example 2.8, this change is

$$\Delta L = \sqrt{L^2 + x^2} - L = L\sqrt{1 + (x/L)^2} - L$$
$$\approx L\left(1 + \frac{1}{2}\left(\frac{x}{L}\right)^2\right) - L = \frac{L}{2}\left(\frac{x}{L}\right)^2 \tag{e}$$

From Eqs. (d) and (e), the system potential energy is

$$V(x) = \frac{1}{2} k_1 \left(\delta_o + \frac{L}{2}\left(\frac{x}{L}\right)^2\right)^2 + \frac{1}{2} k_2 x^2 \tag{f}$$

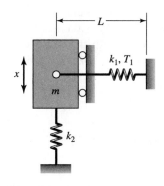

FIGURE 3.12
Single degree-of-freedom system with the horizontal spring under an initial tension T_1.

The expression for potential energy contains terms up to the fourth power of the displacement x, whereas the standard form given in Eqs. (3.46) contains only a quadratic term. However, the kinetic energy is of the form given in Eqs. (3.46). Hence, we will need to use Eq. (3.44) directly to obtain the governing equation. To this end, we recognize that $q_1 = x$ and find that

$$\frac{\partial V}{\partial x} = k_1 \left(\delta_o + \frac{L}{2} \left(\frac{x}{L} \right)^2 \right) \left(\frac{x}{L} \right) + k_2 x$$

$$= \left(k_2 + \frac{k_1 \delta_o}{L} \right) x + \frac{k_1}{2} \frac{x^3}{L^2}$$

$$\cong \left(k_2 + \frac{T_1}{L} \right) x \tag{g}$$

where we have made use of Eq. (a) and we have dropped the cubic term in x since we have assumed that $|x/L| \ll 1$.

Noting that the dissipation function $D = 0$ and that the generalized force $Q_1 = 0$, we substitute Eqs. (b) and (g) into Eq. (3.44) to obtain the following governing equation of motion

$$m\ddot{x} + \left(k_2 + \frac{T_1}{L} \right) x = 0 \tag{h}$$

From Eq. (h) we recognize the natural frequency to be

$$\omega_n = \sqrt{\frac{k_2 + T_1/L}{m}} \tag{i}$$

It is seen that the effect of a spring under tension, which is initially normal to the direction of motion, is to increase the natural frequency of the system.

EXAMPLE 3.14   Equation of motion for a disk with an extended mass

We shall determine the governing equation of motion and the natural frequency for the system shown in Figure 3.13, for "small" angular motions of the pendulum. The system shown in Figure 3.13 is similar to the system shown in Figure 3.9, except that there is an additional pendulum of length L and rigid mass m that is attached to the disk. The disk rolls without slipping. The position of fixed point O is chosen to coincide with the unstretched length of the spring, the coordinate θ is chosen as the generalized coordinate, and the translation $x = -R\theta$.

The kinetic energy of the system is given by

$$T = T_{\text{disk}} + T_{\text{pendulum}} \tag{a}$$

where the kinetic energy of the disk is given by Eq. (e) of Example 3.10. The kinetic energy of the pendulum mass m is given by

$$T_{\text{pendulum}} = \frac{1}{2} m \, (V_m \cdot V_m) \tag{b}$$

where, based on the particle kinematics discussed in Section 1.2 and the first of Eqs. (f) of Example 1.1, we have

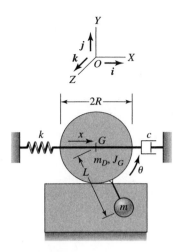

FIGURE 3.13
Disk that is rolling and translating and has a rigidly attached extended mass.

$$V_m = \frac{d\mathbf{r}_m}{dt} = \frac{d}{dt}[(x + L\sin\theta)\mathbf{i} + (L - L\cos\theta)\mathbf{j}]$$

$$V_m = (-R\dot\theta + L\dot\theta\cos\theta)\mathbf{i} + L\dot\theta\sin\theta\mathbf{j} \tag{c}$$

On substituting for the velocity vector from Eq. (c) into Eq. (b) and executing the scalar dot product, we obtain

$$T_{\text{pendulum}} = \frac{1}{2}m[(-R\dot\theta + L\dot\theta\cos\theta)^2 + L^2\dot\theta^2\sin^2\theta]$$

$$= \frac{1}{2}m[R^2\dot\theta^2 + L^2\dot\theta^2 - 2LR\dot\theta^2\cos\theta]$$

$$= \frac{1}{2}m[(R^2 + L^2 - 2LR\cos\theta)]\dot\theta^2 \tag{d}$$

Since the objective is to obtain the governing equation for "small" angular oscillations of the pendulum about the position $\theta = 0$, we retain up to quadratic terms in Eq. (d). To this end, we expand the $\cos\theta$ term as

$$\cos\theta \approx 1 - \frac{1}{2}\theta^2 + \cdots \tag{e}$$

substitute Eq. (e) into Eq. (d), and retain up to quadratic terms to obtain

$$T_{\text{pendulum}} = \frac{1}{2}m[L^2 + R^2 - 2LR]\dot\theta^2 = \frac{1}{2}m(L - R)^2\dot\theta^2 \tag{f}$$

Making use of Eq. (e) of Example (3.10) and Eq. (f), we construct the system kinetic energy from Eq. (a) as

$$T = \frac{1}{2}m(L - R)^2\dot\theta^2 + \frac{1}{2}m_D\dot{x}^2 + \frac{1}{2}J_G\dot\theta^2$$

$$= \frac{1}{2}[m(L - R)^2 + m_D R^2 + J_G]\dot\theta^2 \tag{g}$$

The potential energy of the system is constructed as

$$V = \frac{1}{2}kx^2 + mg(L - L\cos\theta) = \frac{1}{2}kR^2\theta^2 + mgL(1 - \cos\theta) \tag{h}$$

where the datum for the potential energy of the pendulum is located at the bottom position and we have used Eq. (2.36) with $L/2$ replaced by L. To describe small oscillations of the pendulum, we use the expansion for the $\cos\theta$ term given by Eq. (e) and retain up to quadratic terms in Eq. (h) to obtain

$$V = \frac{1}{2}kR^2\theta^2 + \frac{1}{2}mgL\theta^2$$

$$= \frac{1}{2}(kR^2 + mgL)\theta^2 \tag{i}$$

In this case, the dissipation function is given by

$$D = \frac{1}{2}c\dot{x}^2 = \frac{1}{2}cR^2\dot\theta^2 \tag{j}$$

Comparing Eqs. (g), (i), and (j) to Eqs. (3.46), we find that the equivalent system properties are given by

$$m_e = m(L - R)^2 + m_D R^2 + J_G$$
$$k_e = kR^2 + mgL$$
$$c_e = cR^2 \qquad\qquad (k)$$

Thus, making use of the last of Eqs. (3.47) and Eqs. (k) and noting that the generalized force $Q_1 = 0$, we arrive at the governing equation

$$m_e \ddot{\theta} + c_e \dot{\theta} + k_e \theta = 0 \qquad\qquad (l)$$

From Eqs. (k) and the first of Eqs. (3.48), we determine that the system natural frequency is

$$\omega_n = \sqrt{\frac{k_e}{m_e}} = \sqrt{\frac{kR^2 + mgL}{m(L - R)^2 + m_D R^2 + J_G}} \qquad\qquad (m)$$

EXAMPLE 3.15   Lagrange formulation for a microelectromechanical system (MEMS) device

We shall determine the governing equation of motion and the natural frequency for the microelectromechanical system[14] shown in Figure 3.14. The mass m_2 is the scanning micro mirror whose typical dimensions are 300 μm × 400 μm. This mass is modeled as a rigid bar. The torsion springs are rods that are 50 μm in length and 4 μm^2 in area and are collectively modeled by an equivalent torsion spring of stiffness k_t in the figure. Mass m_1 is the mass of the electrostatic comb drive, which is comprised of 100 interlaced "fingers." The comb fingers are 2 μm wide and 40 μm long. The comb drive is connected to the displacement drive through an elastic member that has a spring constant k. The mass m_1 is connected to the bar m_2 by a rigid, weightless rod.

We will use the angular coordinate ϕ as the generalized coordinate, and derive the equation of motion for "small" angular oscillations. The translation $x_o(t)$ is prescribed, and the translations x_1 and x_2 are approximated as

$$x_1 = L_2 \phi$$
$$x_2 = L_1 \phi \qquad\qquad (a)$$

The system potential energy is constructed as

$$V = V_1 + V_2 + V_3 \qquad\qquad (b)$$

[14]M.-H. Kiang, O. Solgaard, K. Y. Lau, and R. S. Muller, "Electrostatic Comb-Drive-Actuated Micromirrors for Laser-Beam Scanning and Positioning," *J. Microelectromechanical Systems,* Vol. 7, No. 1, pp. 27–37 (March 1998).

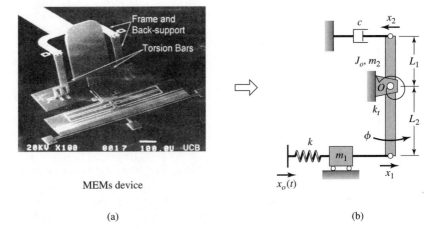

MEMs device

(a)   (b)

FIGURE 3.14
(a) MEMS device and (b) single degree-of-freedom model. *Source:* From
M. H. Kiang, O. Solgaard, K. Y. Lau, and R. S. Muller, "Electrostatic
Comb-Drive-Actuated Micromirrors for Laser-Beam Scanning and Positioning,"
Journal of Microelectromechanical Systems, Vol. 7, No. 1, pp. 27–37 (March
1998). Copyright © 1998 IEEE. Reprinted with permission.

where V_1 is the potential energy of the torsion spring, V_2 is the potential energy of the translation spring, and V_3 is the gravitational potential energy of
the bar. For "small" angular oscillations of the bar, Eq. (c) of Example 2.9 is
used to describe the bar's potential energy. Thus, we arrive at

$$V = \frac{1}{2} k_t \phi^2 + \frac{1}{2} k(x_o(t) - x_1)^2 + \frac{1}{4} m_2 g(L_2 - L_1)\phi^2$$

$$= \frac{1}{2} k_t \phi^2 + \frac{1}{2} k(x_o(t) - L_2\phi)^2 + \frac{1}{4} m_2 g(L_2 - L_1)\phi^2 \tag{c}$$

where we have made use of Eqs. (a). When $L_2 = L_1$, the effects of the increase and decrease in the potential energy of each portion of the bar of mass
m_2 cancel.

Next, the system's kinetic energy is determined as

$$T = \frac{1}{2} J_o \dot{\phi}^2 + \frac{1}{2} m_1 \dot{x}_1^2 = \frac{1}{2} J_o \dot{\phi}^2 + \frac{1}{2} m_1 L_2^2 \dot{\phi}^2$$

$$= \frac{1}{2} (J_o + m_1 L_2^2) \dot{\phi}^2 \tag{d}$$

The system dissipation function is given by

$$D = \frac{1}{2} c \dot{x}_2^2 = \frac{1}{2} c L_1^2 \dot{\phi}^2 \tag{e}$$

where we have again made use of Eqs. (a). Comparing the forms of Eqs. (c),
(d), and (e) to Eqs. (3.46), we find that the potential energy is not in the

standard form. Thus, we will make use of Eq. (3.44) to determine the governing equation of motion. To this end, we find from Eq. (c) that

$$\frac{\partial V}{\partial \phi} = \frac{\partial}{\partial \phi} \left[\frac{1}{2} k_t \phi^2 + \frac{1}{2} k \left(x_o(t) - L_2 \phi \right)^2 + \frac{1}{4} m_2 g (L_2 - L_1) \phi^2 \right]$$

$$= k_t \phi - k L_2 \left(x_o(t) - L_2 \phi \right) + \frac{1}{2} m_2 g (L_2 - L_1) \phi$$

$$= \left[k_t + k L_2^2 + \frac{1}{2} m_2 g (L_2 - L_1) \right] \phi - k L_2 x_o(t) \tag{f}$$

To obtain the governing equation of motion we recognize that $q_1 = \phi$, substitute for the system kinetic energy and the dissipation function from Eqs. (d) and (e), respectively, into Eq. (3.44), make use of Eq. (f), and note that the generalized force $Q_1 = 0$. Thus,

$$(J_o + m_1 L_2^2) \ddot{\phi} + c L_1^2 \dot{\phi} + (k_t + k L_2^2 + m_2 g(L_2 - L_1)/2) \phi = k L_2 x_o(t) \tag{g}$$

From the inertia and stiffness terms of Eq. (g), we find that the system natural frequency is given by

$$\omega_n = \sqrt{\frac{k_t + k L_2^2 + m_2 g (L_2 - L_1)/2}{J_o + m_1 L_2^2}} \tag{h}$$

An experimentally determined value for the natural frequency of a typical system is 2,400 Hz. If the rod connecting the mass m_1 to the bar were not assumed rigid and weightless, one would need to consider additional coordinates to describe the system of Figure 3.14. Systems with more than one degree of freedom are treated in Chapters 7 to 9.

EXAMPLE 3.16 Equation of motion of a slider mechanism[15]

We revisit the slider mechanism system of Example 2.2 and obtain the equation of motion of this system by using Lagrange's equation. Gravity loading is assumed to act normal to the plane of the system shown in Figure 3.15. The mass m_s slides along a uniform bar of mass m_l that is pivoted at the point O. A linear spring of stiffness k restrains the motions of the mass m_s. Another uniform bar of mass $(m_b + m_e)$ is pivoted at O', which is attached to a linear spring of stiffness k_d at one end and attached to the mass m_s at the other end. An excitation $d(t)$ is imposed at one end of the spring with stiffness k_d.

We choose the angular coordinate φ as the generalized coordinate, and we will determine the equation of motion in terms of this coordinate. The geometry imposes the following constraints on the motion of the system:

---

[15]J. Pedurach and B. H. Tongue, "Chaotic Response of a Slider Crank Mechanism," *J. Vibration Acoustics,* Vol. 113, pp. 69–73 (January 1991).

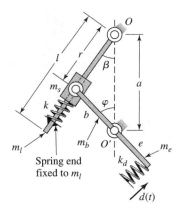

FIGURE 3.15
Slider mechanism.

$$r^2(\varphi) = a^2 + b^2 - 2ab \cos \varphi$$
$$r(\varphi) \sin \beta = b \sin \varphi$$
$$a = r(\varphi) \cos \beta + b \cos \varphi \qquad \text{(a)}$$

At a first glance, although the slider mechanism appears to be a system that would need more than one coordinate for its description, due to the constraints given by Eqs. (a), the system is single degree-of-freedom system that can be described by the independent coordinate φ.

System Kinetic Energy

The total kinetic energy of the system is

$$T = \frac{1}{2} [J_{mb} + J_{me}] \dot{\varphi}^2 + \frac{1}{2} J_{ml} \dot{\beta}^2 + \frac{1}{2} m_s \dot{r}^2(\varphi) \qquad \text{(b)}$$

where J_{mb} and J_{me} are the mass moments of inertia of bar of mass m_b and bar of mass m_e about the fixed point O', respectively, and J_{ml} is the mass moment of inertia of bar m_l about the fixed point O. Then, making use of Eq. (b) of Example 2.1 and Figure 3.15, we find that

$$J_{mb} = \frac{1}{3} m_b b^2, \qquad J_{me} = \frac{1}{3} m_e e^2, \qquad \text{and} \qquad J_{ml} = \frac{1}{3} m_l l^2 \qquad \text{(c)}$$

Since the angle β and the length $r(\varphi)$ are each related to φ by Eqs. (a), we proceed to obtain expressions for $\dot{\beta}$ and $\dot{r}(\varphi)$ in terms of $\dot{\varphi}$. We differentiate the first of Eqs. (a) with respect to time to obtain

$$2r(\varphi)\dot{r}(\varphi) = 2ab\dot{\varphi} \sin \varphi$$

which leads to

$$\dot{r}(\varphi) = \frac{ab}{r(\varphi)} \dot{\varphi} \sin \varphi \qquad \text{(d)}$$

Upon differentiating the third of Eqs. (a) with respect to time, we arrive at

$$\dot{r}(\varphi) \cos \beta - r(\varphi)\dot{\beta} \sin \beta - b\dot{\varphi} \sin \varphi = 0$$

which results in

$$\dot{\beta} = \frac{\dot{r}(\varphi) \cos \beta - b\dot{\varphi} \sin \varphi}{r(\varphi) \sin \beta} \qquad \text{(e)}$$

We now use Eqs. (a) and Eq. (d) in Eq. (e) to obtain

$$\begin{aligned} \dot{\beta} &= \frac{1}{b \sin \varphi} \left\{ \frac{ab}{r(\varphi)} \dot{\varphi} \sin \varphi \left[\frac{a - b \cos \varphi}{r(\varphi)} \right] - b\dot{\varphi} \sin \varphi \right\} \\ &= \frac{\dot{\varphi}}{r^2(\varphi)} [a^2 - ab \cos \varphi - r^2(\varphi)] \\ &= \frac{\dot{\varphi}}{r^2(\varphi)} [ab \cos \varphi - b^2] \qquad \text{(f)} \end{aligned}$$

After substituting Eqs. (d) and (f) into Eq. (b), we arrive at the following expression for the total kinetic energy in terms of $\dot{\varphi}$

$$T = \frac{1}{2} m(\varphi)\dot{\varphi}^2 \tag{g}$$

where

$$m(\varphi) = J_{mb} + J_{me} + J_{ml} \left(\frac{ab \cos \varphi - b^2}{r^2(\varphi)} \right)^2 + m_s \left(\frac{ab}{r(\varphi)} \sin \varphi \right)^2 \tag{h}$$

System Potential Energy

The system potential energy is given by

$$V = \frac{1}{2} k \left[r(\varphi) - r(\varphi_o) \right]^2 + \frac{1}{2} k_d \left[d(t) - e\varphi \right]^2 \tag{i}$$

where φ_o is the static-equilibrium position of the system.

Equation of Motion

Since the expressions for kinetic energy and potential energy are not in the standard form of Eqs. (3.46), we will make use of the Lagrange equation given by Eq. (3.44) to obtain the equation of motion; that is,

$$\frac{d}{dt} \left(\frac{\partial T}{\partial \dot{\varphi}} \right) - \frac{\partial T}{\partial \varphi} + \frac{\partial D}{\partial \dot{\varphi}} + \frac{\partial V}{\partial \varphi} = 0 \tag{j}$$

where we have used the fact that the generalized force is zero. Noting that there is no dissipation in the system—that is, $D = 0$—we substitute for the kinetic energy and potential energy from Eqs. (g) and (i), respectively, into Eq. (j), and carry out the differentiation operations to obtain the following nonlinear equation

$$m(\varphi)\ddot{\varphi} + \frac{1}{2} m'(\varphi)\dot{\varphi}^2 + k \left[r(\varphi) - r(\varphi_o) \right] r'(\varphi) + k_d e^2 \varphi = k_d e d(t) \tag{k}$$

where the prime denotes the derivative with respect to φ.

EXAMPLE 3.17   Oscillations of a crankshaft[16]

Consider the model of a crankshaft shown in Figure 3.16 where gravity is acting in the k direction. The crank of mass m_G and mass moment of inertia J_G about its center of mass is connected to a slider of mass m_p at one end and to a disk of mass moment of inertia J_d about the fixed point O. Choosing the angle θ as the generalized coordinate, we will first derive the governing equa-

[16]G. Genta, *Vibration of Structures and Machines: Practical Aspects,* 2nd ed., Springer-Verlag, NY, pp. 338–341 (1995); and E. Brusa, C. Delprete, and G. Genta, "Torsional Vibration of Crankshafts: Effects of Non-Constant Moments of Inertia," *J. Sound Vibration,* Vol. 205, No. 2, pp. 135–150 (1997).

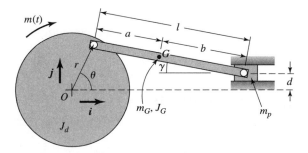

FIGURE 3.16
Crankshaft model.

tion of motion of the system, and then from this equation, determine the equation governing oscillations about a steady rotation rate.

Kinematics

From Figure 3.16, we see that the position vector of the slider mass m_p with respect to point O is

$$\boldsymbol{r}_p = (r \cos \theta + l \cos \gamma)\boldsymbol{i} + d\boldsymbol{j} \tag{a}$$

and that the position vector of the center of mass G of the crank with respect to point O is

$$\boldsymbol{r}_g = (r \cos \theta + a \cos \gamma)\boldsymbol{i} + (r \sin \theta - a \sin \gamma)\boldsymbol{j} \tag{b}$$

Furthermore, from geometry, the angle γ and the angle θ are related by the relation

$$r \sin \theta = d + l \sin \gamma \tag{c}$$

To determine the slider velocity, we differentiate the position vector $\boldsymbol{r}_p$ with respect to time and obtain

$$\boldsymbol{v}_p = (-r\dot{\theta} \sin \theta - l\dot{\gamma} \sin \gamma)\boldsymbol{i} \tag{d}$$

By differentiating Eq. (c) with respect to time, we obtain the following relationship between $\dot{\gamma}$ and $\dot{\theta}$:

$$\dot{\gamma} = \frac{r}{l} \frac{\cos \theta}{\cos \gamma} \dot{\theta} \tag{e}$$

After substituting Eq. (e) into Eq. (d), we obtain the slider velocity to be

$$\boldsymbol{v}_p = -r\dot{\theta}(\sin \theta + \tan \gamma \cos \theta)\boldsymbol{i} \tag{f}$$

The velocity of the center of mass G of the crank is obtained in a similar manner. We differentiate Eq. (b) with respect to time to obtain

$$\boldsymbol{v}_G = (-r\dot{\theta} \sin \theta - a\dot{\gamma} \sin \gamma)\boldsymbol{i} + (r\dot{\theta} \cos \theta - a\dot{\gamma} \cos \gamma)\boldsymbol{j} \tag{g}$$

After substituting Eq. (e) into Eq. (g) and noting that $a + b = l$, we obtain the velocity of the crank's center of mass to be

$$v_G = -\left(\sin\theta + \frac{a}{l}\tan\gamma\cos\theta\right)r\dot{\theta}\mathbf{i} + \left(\frac{b}{l}\cos\theta\right)r\dot{\theta}\mathbf{j} \tag{h}$$

System Kinetic Energy

The total kinetic energy of the system is given by

$$T = \frac{1}{2}J_d\dot{\theta}^2 + \frac{1}{2}m_G(v_G \cdot v_G) + \frac{1}{2}J_G\dot{\gamma}^2 + \frac{1}{2}m_p(v_p \cdot v_p) \tag{i}$$

We now substitute Eqs. (e), (f), and (h) into Eq. (i) to obtain

$$T = \frac{1}{2}J(\theta)\dot{\theta}^2 \tag{j}$$

where

$$J(\theta) = J_d + r^2 m_G\left\{\left(\sin\theta + \frac{a}{l}\tan\gamma\cos\theta\right)^2 + \left(\frac{b}{l}\cos\theta\right)^2\right\} + J_G\left(\frac{r}{l}\frac{\cos\theta}{\cos\gamma}\right)^2$$
$$+ r^2 m_p(\sin\theta + \tan\gamma\cos\theta)^2 \tag{k}$$

and from Eq. (c)

$$\gamma = \sin^{-1}\left\{\frac{r}{l}\sin\theta - \frac{d}{l}\right\} \tag{l}$$

Equation of Motion

Noting that the generalized coordinate $q_1 = \varphi$, the system potential energy is zero, the system dissipation function is zero, and that the generalized moment $Q_1 = -M(t)$, Eq. (3.44) takes the form

$$\frac{d}{dt}\left(\frac{\partial T}{\partial\dot{\theta}}\right) - \frac{\partial T}{\partial\theta} = -M(t) \tag{m}$$

Upon substituting Eq. (j) into Eq. (m) and performing the differentiation operations, we obtain

$$J(\theta)\ddot{\theta} + \frac{1}{2}J'(\theta)\dot{\theta}^2 = -M(t) \tag{n}$$

where the prime denotes the derivative with respect to θ.

The angle θ can be expressed as the superposition of a rigid-body motion at a constant angular velocity ω and an oscillatory rotation ϕ; that is,

$$\theta(t) = \omega t + \phi(t) \tag{o}$$

Then, from Eqs. (n) and (o), we arrive at

$$J(\theta)\ddot{\phi} + \frac{1}{2}J'(\theta)(\omega + \dot{\phi})^2 = -M(t) \tag{p}$$

3.7   SUMMARY

In this chapter, the use of two different methods to derive the governing equation of motion of a single degree-of-freedom system was illustrated. One of these methods is based on applying force and/or moment balance and the other method is based on Lagrange's equations. The underlying approach for each of these methods will be used again for deriving governing equations of systems with multiple degrees of freedom in Chapter 7. Definitions of natural frequency and damping factor were also introduced. It was shown how the static-equilibrium position of a vibratory system can be determined and how nonlinear systems can be linearized to describe small oscillations about a system's equilibrium position.

EXERCISES

3.1 A cylindrical buoy with a radius of 1.5m and a mass of 1000 kg floats in salt water ($\rho = 1026$ kg/m^3). Determine the natural frequency of this system.

3.2 A 10 kg instrument is to be mounted at the end of a cantilever arm of annular cross-section. The arm has a Young's modulus of elasticity $E = 72 \times 10^9$ N/m^2 and a mass density $\rho = 2800$ kg/m^3. If this arm is 500 mm long, determine the cross-section dimensions of the arm so that the first natural frequency of the system is above 50 Hz.

3.3 The static displacement of a system with a motor weight 850 lbs is found to be 1 mil. Determine the natural frequency of vertical vibrations of this system.

3.4 A rotor is attached to one end of a shaft that is fixed at the other end. Let the rotary inertia of the rotor be J_G and assume that the rotary inertia of the shaft is negligible compared to that of the rotor. The shaft has a diameter d, a length L, and it is made from material with a shear modulus G. Determine an expression for the natural frequency of torsional oscillations.

3.5 Rewrite the second-order system given by Eq. (3.8) as a system of two first-order differential equations, by introducing the new variables $x_1 = x$ and $x_2 = \dot{x}$. The resulting system of equations is said to be

in state-space form, a useful form for numerically determining the solutions of vibrating systems.

3.6 Consider the hand motion discussed in Example 3.3 and let the hand move in the horizontal plane; that is, gravity force acts normal to this plane. Assume that the length of the forearm l is 25 cm, the mass of the forearm m is 1.5 kg, the object being carried in the hand has a mass $M = 5$ kg, the constant k_b associated with the restoring force of the biceps is 2×10^3 N/rad, the constant K_t associated with the triceps is 2×10^3 N/rad, and the spacing $a = 4$ cm. Determine the equation of motion of this system, and from this governing equation, find the natural frequency and damping factor of the system.

3.7 Formulate a design guideline for Example 3.7 that would enable a vibratory system designer to decrease the static deflection by a factor n while holding the damping ratio and the damping coefficient constant.

3.8 Formulate a design guideline for Example 3.7 that would enable a vibratory system designer to decrease the static deflection by a factor n while keeping the damping ratio and the mass m constant.

3.9 Obtain the equation of motion for the system with rotating unbalance shown in Figure 3.7 by using Lagrange's equations.

3.10 Obtain the equation of motion for the system shown in Figure 3.10 by using moment balance and compare it to the results obtained by using Lagrange's equation.

3.12 A mass m is attached to the free end of a thin cantilever beam of length L as shown in Figure E3.12. The fixed end of this beam is attached to a shaft of radius r that is rotating about its axis at a speed of Ω rad/s. Derive the governing equation of motion for the transverse vibrations of the beam in terms of the variable x and obtain an expression for the system's natural frequency.

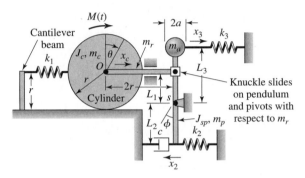

FIGURE E3.11

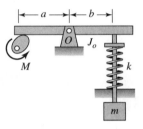

FIGURE E3.13

3.11 Derive the governing equation for the single degree-of-freedom system shown in Figure E3.11 in terms of θ when θ is small, and obtain an expression for its natural frequency. The top mass of the pendulum is a sphere and the mass m_r of the horizontal rod and the mass m_p of the rod that is supporting m_a are each uniformly distributed. The cylinder rolls without slipping. The rotational inertia J_c of the cylinder is about the point "o" and J_{sp} is the total rotational inertia of the rod about the point "s." Assume that these rotational inertias are known.

3.13 Derive the governing equation of motion for the rocker arm-valve assembly shown in Figure E3.13. The quantity J_o is the mass moment of inertia about point O of the rocker arm, k is the stiffness of the linear spring that is fixed at one end, and M is the external moment imposed by the cam on the system.

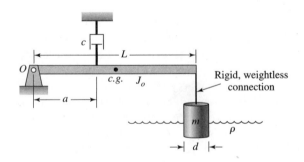

FIGURE E3.14

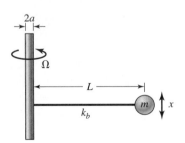

FIGURE E3.12

3.14 For the fluid-float system shown in Figure E3.14, J_o is the mass moment of inertia about point "O." Answer the following: (a) For "small" angular oscillations, derive the governing equation of motion for the fluid float system; and (b) what is the value of the damping coefficient c for which the system is critically damped?

FIGURE E3.17

3.17 Determine the natural frequency of the angle bracket shown in Figure E3.17. Each leg of the bracket has a uniformly distributed mass m and a length L.

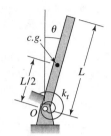

FIGURE E3.15

3.15 Derive the governing equation of motion for the system shown in Figure E3.15. The mass moment of inertia of the bar about the point "O" is J_O and the torsion stiffness of the spring attached to the pivot point is k_t. Assume that there is gravity loading.

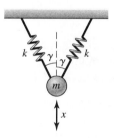

FIGURE E3.18

3.18 Determine the natural frequency for the vertical oscillations of the system shown in Figure E3.18. Let L be the static equilibrium length of the spring and let $|x/L| \ll 1$. The angle γ is arbitrary.

3.19 A vibratory system with a hardening nonlinear spring is governed by the following equation

$$m\ddot{x} + c\dot{x} + k(x + x^3) = 0$$

Determine the static-equilibrium position of this system and linearize the system for "small" oscillations about the system static-equilibrium position.

3.20 A vibratory system with a softening nonlinear spring is governed by the following equation

$$m\ddot{x} + c\dot{x} + k(x - x^3) = 0$$

Determine the static-equilibrium positions of this system and linearize the system for "small" oscillations about each of the system static-equilibrium positions.

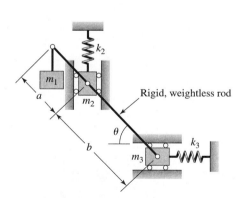

FIGURE E3.16

3.16 Determine the nonlinear governing equation of motion for the kinematically constrained system shown in Figure E3.16.

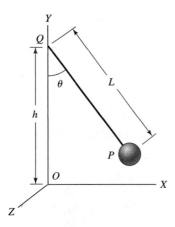

FIGURE E3.21

3.21 A planar pendulum of mass m and constant length l is shown in Figure E3.21. This pendulum is described by the following nonlinear equation

$$ml^2\ddot{\theta} + mgl \sin\theta = 0$$

where θ is the angle measured from the vertical. Determine the static-equilibrium positions of this system and linearize the system for "small" oscillations about each of the system static-equilibrium positions.

3.22 Consider a nonlinear spring that is governed by the force-displacement relationship

$$F(x) = a\left(\frac{x}{b}\right)^c$$

where $a = 3000$ N, $b = 0.015$ m, and $c = 2.80$. If this spring is to be used as a mounting for different machinery systems, obtain a graph similar to that shown in Figure 3.5b and discuss how the natural frequency of this system changes with the weight of the machinery.

3.23 The static deflection in the tibia bone of a 120 kg person standing upright is found to be 25 µm. Determine the associated natural frequency of axial vibrations.

3.24 For the base-excitation prototype shown in Figure 3.6, assume that the base displacement $y(t)$ is

known, choose $x(t)$ as the generalized coordinate, and derive the equation of motion by using Lagrange's equation.

3.25 Determine the natural frequency and damping factor for the system shown in Figure E2.26.

3.26 Determine the natural frequency and damping factor for the system shown in Figure E2.27.

3.27 For the translating and rotating disc system of Figure 3.9, choose the coordinate x measured from the unstretched length of the spring to describe the motion of the system. What are the equivalent inertia, equivalent stiffness, and equivalent damping properties for this system?

3.28 For the inverted pendulum system of Figure 3.10, choose the coordinate x_1 measured from the unstretched length of the spring to describe the motion of the system. What are the equivalent inertia, equivalent stiffness, and equivalent damping properties for this system?

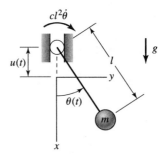

FIGURE E3.29

3.29 Consider a pendulum with an oscillating support as shown in Figure E3.29. The support is oscillating harmonically at a frequency ω; that is,

$$u(t) = U \cos \omega t$$

At the point about which the pendulum rotates, there is a viscous damping moment $cl^2\dot\theta$.

a) Determine expressions for the kinetic energy and the potential energy of the system.

b) Show that the governing equation of motion can be written as

$$\frac{d^2\theta}{d\tau^2} + 2\zeta\frac{d\theta}{d\tau} + [1 - U_o\Omega^2\cos\Omega\tau]\sin\theta = 0$$

where

$$\tau = \omega_o t, \qquad \omega_o^2 = \frac{g}{l}, \qquad \Omega = \frac{\omega}{\omega_o},$$

$$2\zeta = \frac{c}{m\omega_o}, \qquad \text{and} \qquad U_o = \frac{U}{l}$$

c) Linearize the governing equation in (b) for 'small' angular oscillations about $\theta_o = 0$ using a two-term Taylor expansion for $\sin\theta$ and show that the nonlinear stiffness is of the softening type.

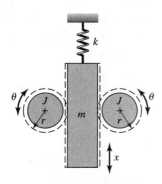

FIGURE E3.30

3.30 Use Lagrange's equation to derive the equation describing the vibratory system shown in Figure E3.30, which consists of two indentical gears, each of radius r and rotary inertia J. They drive an elastically constrained rack of mass m. The elasticity of the constraint is k. From the equation of motion, determine an expression for the natural frequency.

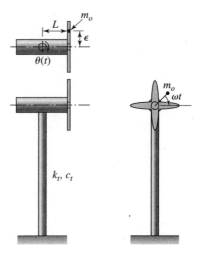

FIGURE E3.31

3.31 Obtain the governing equation of motion in terms of the generalized coordinate θ for torsional oscillations of the wind turbine shown in Figure E3.31. Assume that the turbine blades spin at ω rad/s and that the total mass unbalance is represented by mass m_o located at a distance ε from the axis of rotation. The support for the turbine is a solid circular rod of diameter d, length L, and it is made from a material with a shear modulus G. The turbine body and blades have a rotary inertia J_z. Assume that the damping coefficient for torsional oscillations is c_t.

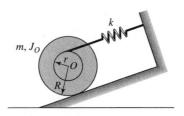

FIGURE E3.32

3.32 The uniform concentric cylinder of radius R rolls without slipping on the inclined surface as shown in Figure E3.32. The cylinder has another cylinder of radius $r < R$ concentrically attached to it. The smaller cylinder has a cable wrapped around it. The other end of the cable is fixed. The cable is parallel to the inclined surface. If the stiffness of the cable is k and the mass and rotary inertia of the two attached cylinders are m and J_O, respectively, then determine an expression for the natural frequency of the system in Hz.

3.33 An instrument's needle indicator has a rotary inertia of 1.4×10^6 kg·m^2. It is attached to a torsion spring whose stiffness is 1.1×10^{-5} N·m/rad and a viscous damper of coefficient c. What is the value of c needed so that the needle is critically damped?

3.34 A solid wooden cylinder of radius r, height h, and specific gravity s_w is placed in a container of tap water such that the axis of the cylinder is perpendicular to the surface of the water. It is assumed the wooden cylinder stays upright under small oscillations.
(a) If the cylinder is displaced a small amount, then determine an expression for its natural frequency.
(b) If the tap water is replaced by salt water with specific gravity 1.2, then determine whether the natural frequency of the wooden cylinder increases or decreases and by what percentage.

tions about the static equilibrium position. The springs are stretched by an amount x_o at the static equilibrium position. The rotary inertia of the pulley about its center is J_O, the radius of the pulley is r, and the stiffness of each linear spring is k.

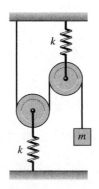

FIGURE E3.36

3.36 Consider the pulley system shown in Figure E3.36. The mass of each pulley is small compared with the mass m and, therefore, can be ignored. Furthermore, the cord holding the mass is inextensible and has negligible mass. Obtain an expression for the natural frequency of the system.

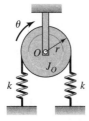

FIGURE E3.35

FIGURE E3.37

3.35 For the pulley shown in Figure E3.35, determine an expression for natural frequency for oscilla-

3.37 The pendulum shown in Figure E3.37 oscillates about the pivot at O. If the mass of the rigid bar of length L_3 can be neglected, then determine the equation of motion and an expression for the natural frequency of the system, for "small" angular oscillations.

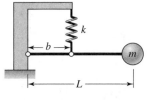

FIGURE E3.39

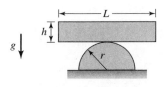

FIGURE E3.38

3.38 A rectangular block of mass m rests on a stationary half-cylinder as shown in Figure E3.38. Find the natural frequency of the block when it undergoes small oscillations about the point of contact with the cylinder.

3.39 Consider the pendulum shown in Figure E3.39. If the bar is rigid and weightless, then determine the natural frequency of the system and compare it to the natural frequency of the pendulum shown in Figure 2.17b. What conclusions can you draw?

Free oscillations of systems are important considerations that must be taken into account in order to obtain effective operations of a system. For a helicopter or a ship crane, the load oscillations must be taken into account to carry out safe load-transfer operations. Stability of vibratory systems such as the machine tool must also be considered in the design of systems subjected to dynamic loads. (*Source:* Image courtesy of PhotoDisc/ Getty Images.)

4

Single Degree-of-Freedom System: Solution for Response and Free-Response Characteristics

4.1   INTRODUCTION

In Chapter 3, we illustrated how the governing equation of a single degree-of-freedom system can be derived. In this chapter, the solution of this governing equation is determined, and based on this solution, the responses of single degree-of-freedom systems subjected to external forcing and different types of initial conditions are discussed. In particular, the Laplace transform method is used to determine the solution of a linear system, and numerical integration is used to determine the solution of a nonlinear system. As pointed out in Chapter 3, it is shown that the free responses can be characterized in terms of the damping factor. The notion of stability of a solution is introduced and is briefly discussed. The problem of machine-tool chatter during turning operations is also considered and numerical determination of stability for this problem is illustrated. The forced responses of single degree-of-freedom systems are addressed in Chapters 5 and 6.

For all linear single degree-of-freedom systems, the governing equation can be put in the form of Eq. (3.22), which is repeated below.

$$\frac{d^2x}{dt^2} + 2\zeta\omega_n \frac{dx}{dt} + \omega_n^2 x = \frac{f(t)}{m} \tag{4.1}$$

A solution is sought for the system described by Eq. (4.1) for a given set of initial conditions. This type of problem is called an *initial-value problem*. Since the system inertia, stiffness, and damping parameters are constant with respect to time, the coefficients in Eq. (4.1) are constant with respect to time. For such linear differential systems with constant coefficients, the solution can be determined by using time-domain methods and the Laplace transform method.[1] The latter has been used here, since a general solution for the response of a forced vibratory system can be determined for arbitrary forms of forcing. However, a price that one pays for generality is that in the Laplace transform method the oscillatory characteristics of the vibratory system are not readily apparent until the final solution is determined. On the other hand, when time-domain methods are used, the explicit forms of the solutions assumed in the initial development allows one to readily see the oscillatory characteristics of a vibratory system. In order to provide a flavor of this complementary approach, time-domain methods are also introduced and discussed at appropriate junctures in this and other chapters and Appendix D.

Laplace Transforms

The Laplace transform of a time-dependent function $g(t)$ is defined as

$$L[g(t)] = G(s) = \int_0^\infty e^{-st} g(t)dt \tag{4.2}$$

where s is the Laplace transform variable. This variable is complex and has the form $s = \sigma + j\omega$, where $j = \sqrt{-1}$ is an imaginary number. The inverse Laplace transform is obtained by carrying out the operation

$$L^{-1}[G(s)] = g(t) = \frac{1}{2\pi j} \int_{\sigma - j\infty}^{\sigma + j\infty} G(s)e^{st}ds \tag{4.3}$$

The integral operation defined in Eq. (4.2), which is a transformation from the time domain to the s domain, is a one-to-one transformation; that is, there is no function $h(t)$, different from $g(t)$, that has $G(s)$ as its Laplace transform. Due to this property, a unique inverse transformation from the s domain to the time domain given by

---

[1]The Laplace transform method is presented here as a tool to determine solutions of linear vibratory systems with constant inertia, stiffness, and damping properties. For a complete discussion of the properties and theorems associated with this transform, the reader is referred to the following books: David L. Widder, *The Laplace Transform*, Princeton University Press, Princeton, N. J. (1941) and Ruel V. Churchill, *Operational Mathematics*, McGraw-Hill, NY (1958).

Eq. (4.3) is also defined. The one-to-one transformation property of the Laplace transform allows one to establish a catalog of Laplace transforms of commonly used time functions, such as those given in Table A of Appendix A. This table is essentially a collection of elementary definite integrals. We shall use this table to obtain the Laplace transform and its inverse.

In the definition given by Eq. (4.2), when the variable s is on the imaginary axis, that is, $s = j\omega$, the resulting transformation takes one from the time domain to the frequency domain. This transformation is used to determine the frequency responses of linear vibratory systems with single degree and multiple degrees of freedom in Chapters 5 and 8, respectively. We shall apply these results to the case of forced vibrations in Chapters 5 and 6. In Chapter 9, the usefulness of Laplace transforms for determining the response of beams is illustrated.

4.2 GENERAL SOLUTION

4.2.1 Introduction

The Laplace transform of Eq. (4.1) is obtained using the Laplace transform pair 2 in Table A of Appendix A. Thus,

$$\underbrace{s^2 X(s) - \dot{x}(0) - sx(0)}_{\text{Laplace transform of } \ddot{x}(t)} + \underbrace{2\zeta\omega_n[sX(s) - x(0)]}_{\text{Laplace transform of } \dot{x}(t)} + \underbrace{\omega_n^2 X(s)}_{\substack{\text{Laplace} \\ \text{transform} \\ \text{of } x(t)}} = \frac{1}{m}\underbrace{F(s)}_{\substack{\text{Laplace} \\ \text{transform} \\ \text{of } f(t)}}$$

which, upon rearrangement, becomes

$$X(s) = \frac{sx(0)}{D(s)} + \frac{2\zeta\omega_n x(0) + \dot{x}(0)}{D(s)} + \frac{F(s)}{mD(s)} \tag{4.4}$$

In Eq. (4.4), $x(0)$ is the initial displacement of the mass, $\dot{x}(0)$ is the initial velocity of the mass, $F(s)$ indicates the Laplace transform of $f(t)$, the forcing directly applied to the inertial element, and

$$D(s) = s^2 + 2\zeta\omega_n s + \omega_n^2 \tag{4.5}$$

Based on the damping factor ζ, the responses of three different types of damped systems and the undamped system are discussed.

4.2.2 Underdamped System: $0 < \zeta < 1$

The vibratory system given by Eq. (4.1) is *underdamped* when $0 < \zeta < 1$. In this case, we see from Eq. (3.20) that the damping coefficient c is less than the critical damping coefficient c_c. To determine the response in this case, we use transform pairs 4, 14, and 16 in Table A of Appendix A to obtain the following inverse transform of Eq. (4.4).

$$x(t) = \underbrace{-x(0)\frac{\omega_n}{\omega_d}e^{-\zeta\omega_n t}\sin(\omega_d t - \varphi) + [2\zeta\omega_n x(0) + \dot{x}(0)]\frac{e^{-\zeta\omega_n t}}{\omega_d}\sin(\omega_d t)}_{\text{Response to initial conditions}}$$

$$\underbrace{+ \frac{1}{m\omega_d}\int_0^t e^{-\zeta\omega_n \eta}\sin(\omega_d \eta)f(t - \eta)d\eta}_{\text{Response to forcing}} \tag{4.6}$$

where

$$\varphi = \cos^{-1}\zeta = \tan^{-1}\frac{\sqrt{1-\zeta^2}}{\zeta} = \sin^{-1}\sqrt{1-\zeta^2}$$

$$\omega_d = \omega_n\sqrt{1-\zeta^2} \tag{4.7}$$

and ω_d is the *damped natural frequency*, which is only defined for $0 < \zeta < 1$. If we let $x(0) = X_o$, $\dot{x}(0) = V_o$, and use the relation

$$-\frac{1}{\sqrt{1-\zeta^2}}\sin(\omega_d t - \varphi) + \frac{\zeta}{\sqrt{1-\zeta^2}}\sin(\omega_d t) = \cos(\omega_d t)$$

then Eq. (4.6) is written as

$$x(t) = \underbrace{X_o e^{-\zeta\omega_n t}\cos(\omega_d t) + \frac{V_o + \zeta\omega_n X_o}{\omega_d}e^{-\zeta\omega_n t}\sin(\omega_d t)}_{\text{Response to initial conditions}}$$

$$+ \underbrace{\frac{1}{m\omega_d}\int_0^t e^{-\zeta\omega_n\eta}\sin(\omega_d\eta)f(t-\eta)d\eta}_{\text{Response to forcing}} \tag{4.8a}$$

or in the form

$$x(t) = A_o e^{-\zeta\omega_n t}\sin(\omega_d t + \varphi_d) + \frac{1}{m\omega_d}\int_0^t e^{-\zeta\omega_n\eta}\sin(\omega_d\eta)f(t-\eta)d\eta \tag{4.8b}$$

or equivalently[2]

$$x(t) = A_o e^{-\zeta\omega_n t}\sin(\omega_d t + \varphi_d) + \frac{e^{-\zeta\omega_n t}}{m\omega_d}\int_0^t e^{\zeta\omega_n\eta}\sin(\omega_d[t-\eta])f(\eta)d\eta \tag{4.8c}$$

where the amplitude A_o and the phase φ_d are given by

$$A_o = \sqrt{X_o^2 + \left(\frac{V_o + \zeta\omega_n X_o}{\omega_d}\right)^2}$$

$$\varphi_d = \tan^{-1}\frac{\omega_d X_o}{V_o + \zeta\omega_n X_o} \tag{4.9}$$

---

[2]In writing Eq. (4.8c), the property of convolution integrals was used: that is,

$$\int_0^t h(\eta)f(t-\eta)d\eta = \int_0^t h(t-\eta)f(\eta)d\eta$$

and we have used the identity

$$a \sin(\omega t + \theta) \pm b \cos(\omega t + \theta) = \sqrt{a^2 + b^2} \sin(\omega t + \theta \pm \psi)$$

$$\psi = \tan^{-1} \frac{b}{a} \tag{4.10}$$

The right-hand side of Eqs. (4.8), which describes the displacement response of the system (4.1), consists of two parts. The first part is determined by the initial displacement and initial velocity, and this part has an oscillatory character. The second part is determined by the forcing $f(t)$ imposed on the system. Responses to different types of excitations are studied in Chapters 5 and 6.

Free Oscillations

In the absence of forcing, that is, $f(t) = 0$, we see from the general solution that the system oscillates with the frequency ω_d and the amplitude decays exponentially as a function of ζ, when $0 < \zeta < 1$. The *period of damped free oscillation* is

$$T_d = \frac{2\pi}{\omega_d} = \frac{2\pi}{\omega_n \sqrt{1 - \zeta^2}} \tag{4.11}$$

Characteristics of free oscillations of underdamped systems are further explored in Section 4.3.

4.2.3 Critically Damped System: $\zeta = 1$

We now consider the solution of Eq. (4.1) for the case $\zeta = 1$. For this value of ζ, the system is *critically damped*. The solution for $\zeta = 1$ can be obtained by taking the limit $\zeta \to 1$ in Eq. (4.8a). Thus, noting that

$$\lim_{\zeta \to 1} \frac{\sin \omega_d t}{\omega_d} = t \lim_{\zeta \to 1} \frac{\sin \omega_d t}{\omega_d t} = t$$

$$\lim_{\zeta \to 1} \cos \omega_d t = 1$$

we arrive at

$$x(t) = \underbrace{X_o e^{-\omega_n t} + [V_o + \omega_n X_o] t e^{-\omega_n t}}_{\text{Response to initial conditions}} + \underbrace{\frac{1}{m} \int_0^t e^{-\omega_n \eta} \eta f(t - \eta) d\eta}_{\text{Response to forcing}} \tag{4.12}$$

The first two terms on the right-hand side of Eq. (4.12) describe the part of the displacement response that is determined by the initial conditions. The third term is present only in the presence of forcing; that is, $f(t) \neq 0$.

Free Response

In the absence of forcing—that is, $f(t) = 0$—the solution given by Eq. (4.12) has exponentially decaying terms, and hence the response decays with time. Furthermore, the motion of the critically damped system is not oscillatory as in the underdamped case; that is, the displacement magnitude does not oscillate between positive and negative values about the system equilibrium position. At best, there may be one zero crossing before the response settles down at the equilibrium position $x(t) = 0$.

4.2.4 Overdamped System: $\zeta > 1$

When the damping factor $\zeta > 1$ the system is *overdamped;* that is, the damping coefficient c is larger than the critical damping coefficient c_c. To determine the response in this case, the polynomial $D(s)$ in Eq. (4.4) is first put in the form

$$D(s) = (s - s_1)(s - s_2)$$

where

$$s_{1,2} = -\zeta\omega_n \pm \omega_n \sqrt{\zeta^2 - 1}$$

Making use of transform pairs 27 and 28 from Table A of Appendix A, the displacement response is written as

$$x(t) = \underbrace{X_o e^{-\zeta\omega_n t}\cosh(\omega'_d t) + \frac{V_o + \zeta\omega_n X_o}{\omega'_d} e^{-\zeta\omega_n t}\sinh(\omega'_d t)}_{\text{Response to initial conditions}}$$

$$+ \underbrace{\frac{1}{m\omega'_d}\int_0^t e^{-\zeta\omega_n \eta}\sinh(\omega'_d \eta)f(t - \eta)d\eta}_{\text{Response to forcing}} \qquad (4.13)$$

where

$$\omega'_d = \omega_n\sqrt{\zeta^2 - 1}$$

In the displacement response of the overdamped system described by Eq. (4.13), the first and second terms on the right-hand side are determined by the initial conditions and the third term is determined by the forcing $f(t)$.

Free Response

Again, as in the other two cases, in the absence of forcing, the response decays with time to the equilibrium position $x(t) = 0$. However, the displacement magnitude does not oscillate about the equilibrium position, as it is approached.

4.2.5 Undamped System: $\zeta = 0$

We now consider the solution of Eq. (4.1) for the case $\zeta = 0$; that is, the system is undamped. In this case, the solution for the response can be obtained by taking the limit $\zeta \to 0$ in Eq. (4.8a). Thus, noting from Eqs. (4.7) that

$$\lim_{\zeta \to 0} \varphi = \frac{\pi}{2}$$

$$\lim_{\zeta \to 0} \omega_d = \omega_n$$

we find that the solution is of the form

$$x(t) = \underbrace{X_o \cos(\omega_n t) + \frac{V_o}{\omega_n} \sin(\omega_n t)}_{\text{Response to initial conditions}} + \underbrace{\frac{1}{m\omega_n} \int_0^t \sin(\omega_n \eta) f(t - \eta) d\eta}_{\text{Response to forcing}} \tag{4.14a}$$

An alternative form is obtained from Eqs. (4.8b) and (4.9) as

$$x(t) = \underbrace{A_o \sin(\omega_n t + \varphi_d)}_{\text{Response to initial conditions}} + \underbrace{\frac{1}{m\omega_n} \int_0^t \sin(\omega_n \eta) f(t - \eta) d\eta}_{\text{Response to forcing}} \tag{4.14b}$$

where the amplitude A_o and the phase φ_d are given by

$$A_o = \sqrt{X_o^2 + \left(\frac{V_o}{\omega_n}\right)^2}$$

$$\varphi_d = \tan^{-1} \frac{\omega_n X_o}{V_o} \tag{4.15}$$

In the displacement response of the undamped system described by Eq. (4.14a), the first and second terms on the right-hand side are determined by the initial conditions and the third term is determined by the forcing $f(t)$.

Free Oscillations

In the absence of forcing, the response does not decay to the equilibrium position but rather the displacement magnitude oscillates sinusoidally about the equilibrium position with an amplitude A_o given by the first of Eqs. (4.15).

4.3 FREE RESPONSES OF UNDAMPED AND DAMPED SYSTEMS

4.3.1 Introduction

In this section, the responses of undamped and damped single degree-of-freedom systems in the absence of forcing—that is, $f(t) = 0$—are explored in detail. These responses are also referred to as *free responses*, and when the system is undamped or underdamped, the responses are referred to as *free oscillations*. In the absence of forcing, the single degree-of-freedom given by Eq. (4.1) reduces to

$$\frac{d^2x}{dt^2} + 2\zeta\omega_n \frac{dx}{dt} + \omega_n^2 x = 0$$

Based on the discussions of Sections 4.2.2 to 4.2.5, we are now in a position to study the differences in the response of the mass for the three different damping levels and to determine the effects of the damping ratio on the rate of decay. To simplify matters, we shall assume that the initial displacement $X_o = 0$ and that the initial velocity $V_o \neq 0$. Then, after introducing the nondimensional time variable $\tau = \omega_n t$, we simplify Eqs. (4.14a), (4.8a), (4.12), and (4.13) to, respectively,

$\zeta = 0$

$$\frac{x(\tau)}{V_o/\omega_n} = \sin(\tau)$$

$\zeta < 1$

$$\frac{x(\tau)}{V_o/\omega_n} = \frac{1}{\sqrt{1 - \zeta^2}} e^{-\zeta\tau} \sin(\tau\sqrt{1 - \zeta^2})$$

$\zeta = 1$

$$\frac{x(\tau)}{V_o/\omega_n} = \tau e^{-\tau}$$

$\zeta > 1$

$$\frac{x(\tau)}{V_o/\omega_n} = \frac{1}{\sqrt{\zeta^2 - 1}} e^{-\zeta\tau} \sinh(\tau\sqrt{\zeta^2 - 1})$$

The time histories for the three damped cases are plotted in Figure 4.1, where it is seen that when $\zeta = 1$, the displacement decays to its equilibrium posi-

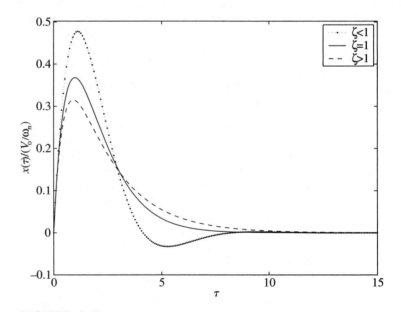

FIGURE 4.1
Response of a single degree-of-freedom system to an initial velocity for three different values of ζ.

tion in the shortest time. This characteristic is made use of, for example, in the design of dampers for doors. In addition, it is seen that for $\zeta < 1$ the response is oscillatory, whereas for $\zeta \geq 1$ the response is not oscillatory. However, as ζ increases, the magnitude of the peak amplitude decreases.

Design Guideline: The free response of a critically damped system reaches its equilibrium or rest position in the shortest possible time.

In the absence of forcing, when $\zeta > 0$, the displacement response always decays to the equilibrium position $x(t) = 0$. However, this is not true when $\zeta < 0$; the response of the system will grow with respect to time. This is an example of an unstable response, which is discussed in Section 4.4.

Next, we present three examples before proceeding to explore the free responses of underdamped and critically damped systems in detail.

EXAMPLE 4.1   Free response of a microelectromechanical system

A microelectromechanical system has a mass of 0.40 μg, a stiffness of 0.08 N/m, and a negligible damping coefficient. The gravity loading is normal to the direction of motion of this mass. We shall determine and discuss the displacement response of this system when there is no forcing acting on this system and when the initial displacement is 2 μm and the initial velocity is zero.

Since $V_o = f(t) = \zeta = 0$, we see from Eq. (4.14a) that the displacement response has the form

$$x(t) = X_o \cos(\omega_n t) \tag{a}$$

where

$$\omega_n = \sqrt{\frac{k}{m}} \tag{b}$$

From Eq. (b), the natural frequency is

$$\omega_n = \sqrt{\frac{0.08 \text{ N/m}}{0.40 \times 10^{-9} \text{ kg}}} = 14142.14 \text{ rad/s}$$

or

$$f_n = \frac{\omega_n}{2\pi} = \frac{14142.14}{2\pi} = 2250.8 \text{ Hz}$$

Substituting this value and the given value of initial displacement 2 μm into Eq. (a) results in

$$x(t) = 2 \cos(14142.14t) \text{ } \mu\text{m} \tag{c}$$

Equation (c) is the displacement response. Based on the form of Eq. (a) or Eq. (c), it is clear that the displacement is a cosine harmonic function that varies periodically with time and has the period

$$T = \frac{2\pi}{\omega_n} = \frac{1}{f_n} = \frac{1}{2250.8} = 444.29 \ \mu s$$

From the form of Eq. (c), it is clear that the response does not decay, and hence, the response does not settle down to the static-equilibrium position. The system, instead, oscillates harmonically about this equilibrium position with a magnitude of 2 μm.

EXAMPLE 4.2   Free response of a car tire

A wide-base truck tire is characterized with a stiffness of 1.23×10^6 N/m, an undamped natural frequency of 30 Hz, and a damping coefficient of 4400 N·s/m. In the absence of forcing, we shall determine the response of the system assuming non-zero initial conditions, evaluate the damped natural frequency of the system, and discuss the nature of the response.

Let the mass of the tire be represented by m. Based on the equation of motion derived in Chapter 3 for the system shown in Figure 3.1, the governing equation of motion of the tire system from the static equilibrium position is given by Eq. (4.1) with zero forcing; that is,

$$\frac{d^2x}{dt^2} + 2\zeta\omega_n \frac{dx}{dt} + \omega_n^2 x = 0 \tag{a}$$

For this case,

$$\omega_n = 2\pi \times 30 = 188.50 \text{ rad/s}$$

$$\zeta = \frac{c}{2m\omega_n} = \frac{c\omega_n}{2k} = \frac{4400 \text{ N·s/m} \times 188.50 \text{ rad/s}}{2 \times 1.23 \times 10^6 \text{ N/m}} = 0.337 \tag{b}$$

Since the damping factor is less than 1, the system is underdamped. Hence, the solution for Eq. (a) is given by Eq. (4.8b) with the forcing set to zero; that is, the displacement response of the tire system about the static-equilibrium position is

$$x(t) = A_o e^{-\zeta\omega_n t} \sin(\omega_d t + \varphi_d) \tag{c}$$

where the constants A_o and φ_d are determined by the initial displacement and initial velocity as indicated by Eq. (4.9). The damping factor ζ and the natural frequency ω_n are given by Eq. (b), and the damped natural frequency ω_d is determined from Eq. (4.7) as

$$\omega_d = \omega_n \sqrt{1 - \zeta^2} = 188.50 \sqrt{1 - 0.337^2} = 177.5 \text{ rad/s}$$

The response given by Eq. (c) has the form of a damped sinusoid with a period

$$T_d = \frac{2\pi}{\omega_d} = \frac{2\pi}{177.5} = 0.0354 \text{ s}$$

Thus, the tire oscillates back and forth about the static-equilibrium position with a period of 35.4 ms. As time unfolds, the amplitude of the displacement response decreases exponentially with time, and in the limit,

$$\lim_{t\to\infty} x(t) = \lim_{t\to\infty} [A_o e^{-\zeta\omega_n t} \sin(\omega_d t + \varphi_d)] = 0$$

because of the exponential term. Thus, after a fast decay, the tire system settles down to the static-equilibrium position.

EXAMPLE 4.3   Free response of a door

A door shown in Figure 4.2 undergoes rotational motions about the vertical axis pointing in the k direction. From Eq. (3.13), the governing equation of motion of this system is

$$J_{door}\ddot{\theta} + c_t\dot{\theta} + k_t\theta = 0 \qquad (a)$$

where the mass moment of inertia $J_{door} = 20 \text{ kg·m}^2$, the viscous damping provided by the door damper is 48 N·m·s/rad, and the rotational stiffness of the door hinge is 28.8 N·m/rad. We shall determine the response of this system when the door is opened with an initial velocity of 4 rad/s from the initial position $\theta = 0$. We shall then plot this response as a function of time and discuss its motion.

Equation (a) is written in the form of Eq. (4.1) by dividing through by the inertia J_{door} to obtain

$$\ddot{\theta} + 2\zeta\omega_n\dot{\theta} + \omega_n^2\theta = 0 \qquad (b)$$

where

$$\zeta = \frac{c_t}{2J_{door}\omega_n}$$

$$\omega_n = \sqrt{\frac{k_t}{J_{door}}} \qquad (c)$$

For the given values of the parameters, the damping factor and the natural frequency are, from Eqs. (c),

$$\omega_n = \sqrt{\frac{28.8 \text{ N·m/rad}}{20 \text{ kg·m}^2}} = 1.2 \text{ rad/s}$$

$$\zeta = \frac{48 \text{ N m·s/rad}}{2 \times 20 \text{ kg·m}^2 \times 1.2 \text{ rad/s}} = 1.0 \qquad (d)$$

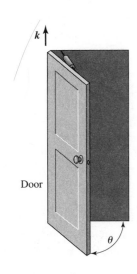

FIGURE 4.2
Door motions.

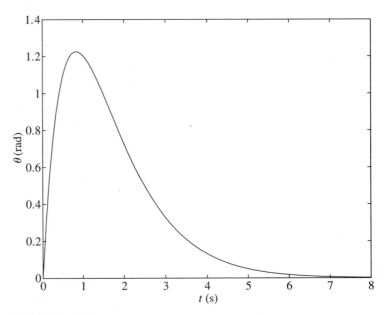

FIGURE 4.3
Displacement time history of the door in Figure 4.2.

Hence, the system is critically damped. The displacement response is given by Eq. (4.12) with the forcing being zero; that is,

$$\theta(t) = \theta(0)e^{-\omega_n t} + [\dot{\theta}(0) + \omega_n \theta(0)]te^{-\omega_n t} \qquad \text{(e)}$$

Upon substituting the given initial conditions, $\theta(0) = 0$ and $\dot{\theta}(0) = 4$ rad/s and the value of the natural frequency from Eq. (d) in Eq. (e), we arrive at the displacement response of the door

$$\theta(t) = 4te^{-1.2t} \text{ rad} \qquad \text{(f)}$$

This response is plotted as a function of time in Figure 4.3. From this figure, it is evident the free-response of this critically damped system quickly reaches the static-equilibrium position $\theta = 0$ after the time exceeds about one period of the undamped oscillation of the system; that is, when $T = 2\pi/\omega_n$ = 5.24 s. The peak displacement amplitude occurs at $\dot{\theta}(t_{peak}) = 0$, or $t_{peak} =$ 1/1.2 = 0.833 s. As expected of a critically damped system, the motion is not periodic and does not oscillate about the equilibrium position.

In the rest of the section, the responses of underdamped single degree-of-freedom systems to certain prescribed initial displacements, initial velocities, or both simultaneously are addressed in detail; that is, systems for which $0 < \zeta < 1$.

From the general form of the solution for $0 < \zeta < 1$, we know that nonzero initial conditions will result in oscillations that decay exponentially with time. In our examination of the responses to initial conditions, it is assumed that the externally applied forces are zero; that is, $f(t) = 0$. Then the general solution given by Eq. (4.8b) becomes

$$x(t) = A_o e^{-\zeta \omega_n t} \sin(\omega_d t + \varphi_d) \tag{4.16}$$

where the amplitude A_o and the phase φ_d are given by Eqs. (4.9) as

$$A_o = \sqrt{X_o^2 + \left(\frac{V_o + \zeta \omega_n X_o}{\omega_d}\right)^2}$$

$$\varphi_d = \tan^{-1} \frac{\omega_d X_o}{V_o + \zeta \omega_n X_o} \tag{4.17}$$

Next, this solution is examined for several situations that occur in the design of single degree-of-freedom systems with a prescribed initial velocity and a prescribed initial displacement.

4.3.2 Initial Velocity

We now examine the free response of a single degree-of-freedom system with a prescribed initial velocity. When a system is subjected to an initial velocity only, we set $X_o = 0$ in Eqs. (4.17). This leads to the following amplitude and phase

$$A_o = \frac{V_o}{\omega_d}$$

$$\varphi_d = 0$$

and, therefore, Eq. (4.16) becomes

$$x(t) = \frac{V_o e^{-\zeta \omega_n t}}{\omega_d} \sin(\omega_d t) \tag{4.18}$$

The velocity and acceleration of the mass are, respectively,

$$\dot{x}(t) = v(t) = -\frac{V_o e^{-\zeta \omega_n t}}{\sqrt{1 - \zeta^2}} \sin(\omega_d t - \varphi)$$

$$\ddot{x}(t) = a(t) = \frac{V_o \omega_n e^{-\zeta \omega_n t}}{\sqrt{1 - \zeta^2}} \sin(\omega_d t - 2\varphi) \tag{4.19}$$

where

$$\varphi = \tan^{-1} \frac{\sqrt{1 - \zeta^2}}{\zeta} \tag{4.20}$$

and Eq. (4.10) has been used.

The displacement, velocity, and acceleration responses given by Eqs. (4.18) and (4.19) are plotted in Figure 4.4. The location of the displacement extrema and the velocity extrema seen in this figure are determined as discussed next.

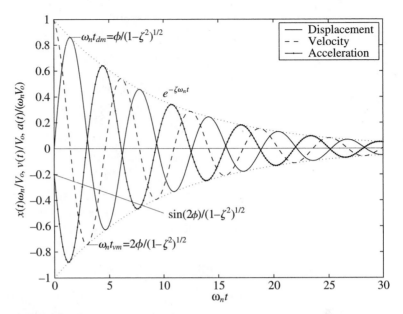

FIGURE 4.4
Time histories of displacement, velocity, and acceleration of a system with prescribed initial velocity V_o.

Extrema of Displacement Response

The displacement has an extremum (maximum or minimum) at those times t_{dm} for which

$$\dot{x}(t_{dm}) = -\frac{V_o e^{-\zeta \omega_n t_{dm}}}{\sqrt{1 - \zeta^2}} \sin(\omega_d t_{dm} - \varphi) = 0 \qquad (4.21)$$

The solution to Eq. (4.21) is

$$\omega_d t_{dm} = \varphi + p\pi \qquad p = 0, 1, 2, \ldots \qquad (4.22)$$

Therefore, from Eq. (4.18), it follows that

$$x(t_{dm}) = x_{\text{max/min}} = \frac{V_o e^{-\zeta \omega_n t_{dm}}}{\omega_d} \sin(\omega_d t_{dm})$$

$$= \frac{V_o e^{-\zeta \omega_d t_{dm}/\sqrt{1-\zeta^2}}}{\omega_d} \sin(\varphi + p\pi)$$

$$= (-1)^p \frac{V_o}{\omega_n} e^{-(\varphi + p\pi)/\tan \varphi} \qquad p = 0, 1, 2, \ldots \qquad (4.23)$$

where we have used Eq. (4.7), that is,

$$\sin \varphi = \sqrt{1 - \zeta^2} \qquad \text{and} \qquad \tan \varphi = \frac{\sqrt{1 - \zeta^2}}{\zeta}$$

The largest displacement occurs when $p = 0$, or at

$$t_{d,\max} = \frac{\varphi}{\omega_d} = \frac{\varphi}{\omega_n\sqrt{1 - \zeta^2}} \tag{4.24}$$

Extrema of Velocity Response

In a similar manner, we find the times t_{vm} at which the velocity is a maximum/minimum, which are determined from the condition that the acceleration is zero; that is, $a(t_{vm}) = 0$. Making use of the second of Eqs. (4.19), these times are found to be

$$\omega_d t_{vm} = 2\varphi + p\pi \qquad p = 0, 1, 2, \ldots \tag{4.25}$$

and, from the first of Eqs. (4.19), the corresponding velocities are determined as

$$\dot{x}(t_{vm}) = (-1)^{p+1}V_o e^{-(2\varphi+p\pi)/\tan\varphi} \qquad p = 0, 1, 2, \ldots \tag{4.26}$$

The use of Eq. (4.26) is illustrated in Example 4.4.

Force Transmitted to Fixed Surface

We shall now determine the dynamic component of the force transmitted to the base of a single degree-of-freedom system such as that shown in Figure 3.1. This force is given by Eq. (3.10); that is,

$$F_R = c\dot{x} + kx \tag{4.27}$$

Upon substituting Eqs. (4.18) and (4.19) into Eq. (4.27), we obtain

$$F_R(t) = \frac{kV_o e^{-\zeta\omega_n t}}{\omega_d}[-2\zeta\sin(\omega_d t - \varphi) + \sin(\omega_d t)] \tag{4.28}$$

At $t = 0$, the reaction force acting on the base is determined from Eq. (4.28) to be

$$F_R(0) = \frac{2\zeta kV_o}{\omega_d}\sin(\varphi) = \frac{2\zeta kV_o}{\omega_n} \tag{4.29}$$

Thus, when the mass of a single degree-of-freedom system is subjected to an initial velocity, the force is instantaneously transmitted to the base. This unrealistic characteristic is a property of modeling the system with a spring and viscous damper combination in parallel. The viscous damper essentially "locks" with the sudden application of the velocity and is thereby momentarily rigid. This temporary rigidity shorts out the spring and instantaneously transmits the force to the base. Representing a support by a combination of a linear spring and linear viscous damper in parallel is called the *Kelvin-Voigt model,* which is one type of elementary *viscoelastic model.* A second type of elementary viscoelastic model, called the *Maxwell model,* consists of a linear spring and a linear viscous damper in series, and this model is discussed in Example 4.7.

State-Space Plot and Energy Dissipation

The values of the displacements and velocities corresponding to these maxima and minima can also be visualized in a state-space plot, which is a graph

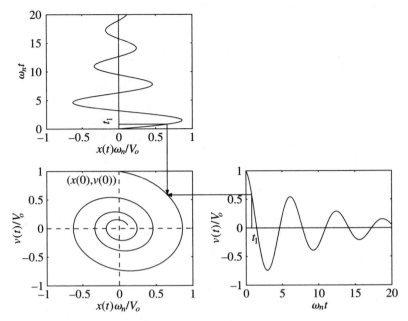

FIGURE 4.5
State-space plot of a single degree-of-freedom system with a prescribed initial velocity V_o.

of the displacement versus the velocity at each instant of time. This graph for the system considered here is shown in Figure 4.5. As time unfolds, the trajectory initiated from a set of initial conditions is attracted to the equilibrium position located at the origin (0, 0).

We now show how the energy dissipated by the system in the time interval $0 \le t \le t_{d,max}$ can be determined. The system of interest is a spring-mass-damper system, as shown in Figure 3.2, which is translating back and forth along the x-axis. The energy dissipated by the system is equal to the difference between the sum of the kinetic energy and the potential energy in the final state and the sum of the kinetic energy and the potential energy in the initial state. Noting that the potential energy in the initial state is zero, and the kinetic energy in the final state is zero, the energy that is dissipated is the difference between the initial kinetic energy and the potential energy that is stored in the spring at $t_{d,max}$. The initial energy is the kinetic energy of the mass, which has been imparted a velocity V_o. The energy stored in the spring is a function of the displacement $x(t_{d,max})$. Thus, the energy E_{diss} that is dissipated is

$$E_{diss} = \underbrace{\frac{1}{2}mV_o^2}_{\substack{\text{Initial kinetic} \\ \text{energy}}} - \underbrace{\frac{1}{2}k[x(t_{d,max})]^2}_{\text{Final potential energy}} = \frac{1}{2}mV_o^2 - \frac{1}{2}k\frac{V_o^2}{\omega_n^2}e^{-2\varphi/\tan\varphi}$$

$$= \frac{1}{2}mV_o^2[1 - e^{-2\varphi/\tan\varphi}] \tag{4.30}$$

where we have used Eqs. (4.23). Thus, the fraction of the total energy that has been dissipated is

$$\frac{E_{diss}}{E_{init}} = [1 - e^{-2\varphi/\tan\varphi}] \tag{4.31}$$

where the initial energy E_{init} is given by

$$E_{init} = \frac{mV_o^2}{2} \tag{4.32}$$

From Eq. (4.31), it is seen that the fraction of the total energy dissipated is only a function of the system's damping factor ζ, since the angle φ is determined only by the damping factor.

The significance of these results is illustrated with several examples, where free oscillations of underdamped systems due to impacts are considered.

EXAMPLE 4.4   Impact of a vehicle bumper

Consider a vehicle of mass m that is travelling at a constant velocity V_o as shown in Figure 4.6a. The bumper is modeled as a spring k and viscous damper c in parallel. If the vehicle's bumper hits a stationary barrier, then after the impact, the displacement and velocity of the mass are those given by Eqs. (4.18) and (4.19), respectively. These results are used to determine the coefficient of restitution of the system and the amount of energy that has been dissipated until the time the bumper is no longer in contact with the barrier.

The bumper is in contact with the barrier only while the sum of the forces

$$kx(t) + c\dot{x}(t) > 0$$

that is, while the spring-damper combination is being compressed. At the instant when they are no longer in compression the acceleration is zero; that is, the time at which the sum of these forces on the mass is zero. The first time instance at which the acceleration is zero is given by Eq. (4.25) for $p = 0$, and the corresponding velocity is given by Eq. (4.26).

Based on Newton's law of impact, the coefficient of restitution ϵ is defined as

$$\epsilon = \frac{-(v_{vehicle} - v_{barrier})_{after\ impact}}{(v_{vehicle} - v_{barrier})_{before\ impact}} = \frac{-(v_{vehicle})_{after\ impact}}{(v_{vehicle})_{before\ impact}} \tag{a}$$

where $v_{vehicle}$ is the vehicle velocity, $v_{barrier}$ is the velocity of the barrier, and the assumption that $v_{barrier}$ is zero has been used. This assumption is valid when the barrier has a "large" inertia. Then, making use of Eq. (a) and Eq. (4.26) with $p = 0$, we find that

$$\epsilon = \frac{-\dot{x}(t_{vm})}{\dot{x}(0)} = \frac{V_o e^{-2\varphi/\tan\varphi}}{V_o} = e^{-2\varphi/\tan\varphi} \tag{b}$$

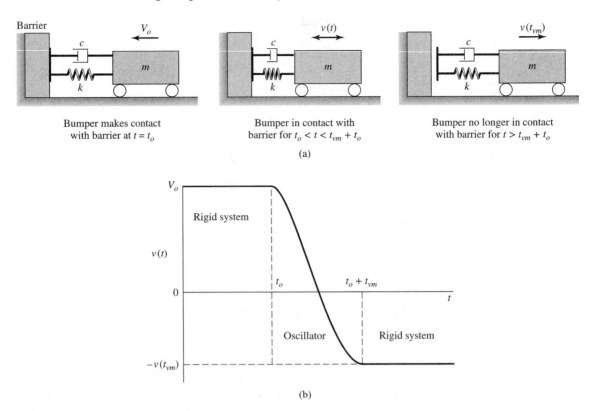

FIGURE 4.6

(a) Model of a car bumper colliding with a stationary barrier and (b) time history of velocity of mass.

We now make use of Eq. (b) to examine how the coefficient of restitution ϵ depends on the damping factor ζ. Considering first the undamped case, we note from Eq. (4.20) that $\varphi/\tan\varphi \to 0$ as $\zeta \to 0$, and, therefore, $\epsilon \to 1$. In other words, there are no losses and the system leaves with the same velocity with which it arrived. This is consistent with the fact that this is an elastic collision. When $\zeta \to 1$, the system becomes critically damped and $\varphi/\tan\varphi \to 1$ and, therefore, from Eq. (b), we find that $\epsilon \to e^{-2}$; that is, the mass leaves the barrier with a velocity of $0.135V_o$.

The amount of energy that the system dissipates during the interval $0 \le t \le t_{vm}$ is the difference between the initial kinetic energy and the kinetic energy at separation. Note that the vehicle does not have any potential energy when it is not in contact with the barrier. Thus,

$$E_{diss} = \frac{1}{2}\,mV_o^2 - \frac{1}{2}\,m[\dot{x}(t_{vm})]^2 = \frac{1}{2}\,mV_o^2 - \frac{1}{2}\,mV_o^2 e^{-4\varphi/\tan\varphi}$$

or

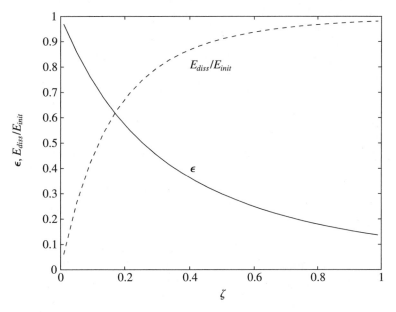

FIGURE 4.7
Coefficient of restitution and fraction of energy dissipation for impacting single degree-of-freedom system.

$$\frac{E_{diss}}{E_{init}} = [1 - e^{-4\varphi/\tan \varphi}]$$

where E_{init} is given by Eq. (4.32). These results are summarized in Figure 4.7. It is noted that these results have been obtained for a collision with a single impact.

EXAMPLE 4.5 Impact of a container housing a single degree-of-freedom system

We shall now consider the effects of dropping onto the floor a system that resides inside a container that has a coefficient of restitution ϵ with respect to the floor. The system is shown in Figure 4.8. If the container falls from a height h, then the magnitude of the velocity at the time of impact with the floor is

$$V_o = \sqrt{2gh}$$

At the instant $t = 0^+$ after impact, the container bounces upwards with a velocity whose magnitude is ϵV_o. Then at $t = 0^+$, the container and the single

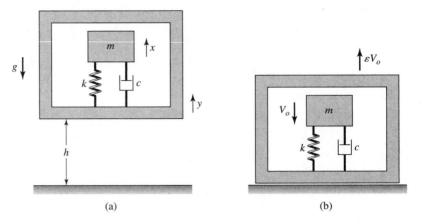

FIGURE 4.8
Single degree-of-freedom system inside a container: (a) dropped from a height h and (b) on rebound immediately after impact with the floor.

degree-of freedom system can be modeled as a single degree-of freedom system with a moving base as discussed in Section 3.5. Thus, if we define the relative displacement

$$z(t) = x(t) - y(t) \tag{a}$$

then, from Eq. (3.30) we have

$$m\frac{d^2z}{dt^2} + c\frac{dz}{dt} + kz = -m\frac{d^2y}{dt^2} \tag{b}$$

However, $\ddot{y} = -g$, since the container is decelerating during the rebound upwards. Then Eq. (b) becomes

$$m\frac{d^2z}{dt^2} + c\frac{dz}{dt} + kz = mg \tag{c}$$

The initial conditions are

$$z(0) = x(0) - y(0) = 0$$
$$\dot{z}(0) = \dot{x}(0) - \dot{y}(0) = -V_o - (\epsilon V_o)$$
$$= -(1 + \epsilon)V_o = -(1 + \epsilon)\sqrt{2gh} \tag{d}$$

The solution to Eq. (c) for $0 < \zeta < 1$ and subject to the initial conditions given by Eq. (d) is determined from Eqs. (4.8), after substituting $f(t) = mg$, to be

$$\frac{z(t)}{\delta_{st}} = \frac{-R}{\sqrt{1 - \zeta^2}} e^{-\zeta\omega_n t} \sin(\sqrt{1 - \zeta^2}\omega_n t)$$

$$+ 1 - \frac{1}{\sqrt{1 - \zeta^2}} e^{-\zeta\omega_n t} \sin(\sqrt{1 - \zeta^2}\omega_n t + \varphi) \tag{e}$$

where φ is given by Eq. (4.20), $\delta_{st} = mg/k$, and the coefficient of restitution-dependent parameter R is

$$R = (1 + \epsilon)\sqrt{\frac{2h}{\delta_{st}}} \tag{f}$$

The corresponding velocity is

$$\frac{\dot{z}(t)}{\delta_{st}\omega_n} = \frac{R}{\sqrt{1 - \zeta^2}}\, e^{-\zeta\omega_n t}\sin(\sqrt{1 - \zeta^2}\,\omega_n t - \varphi)$$

$$+ \frac{1}{\sqrt{1 - \zeta^2}}\, e^{-\zeta\omega_n t}\sin(\sqrt{1 - \zeta^2}\,\omega_n t) \tag{g}$$

The extremum of the relative displacement is determined from $z_{max} = z(t_{max})$ where t_{max} is the time at which $\dot{z}(t_{max}) = 0$. In this particular case, an explicit analytical expression for t_{max} cannot be found, so the maximum displacement is determined numerically from Eq. (e). The magnitude of the maximum displacement is a function of the initial velocity, which is a function of the drop height h, the coefficient of restitution of the container, and the damping ratio and the static displacement of the spring of the single degree-of-freedom system inside the container. The numerically obtained results are shown in Figure 4.9. We see that there are many ways by which one can decrease the maximum relative displacement of the mass, which lead to the following design guidelines.

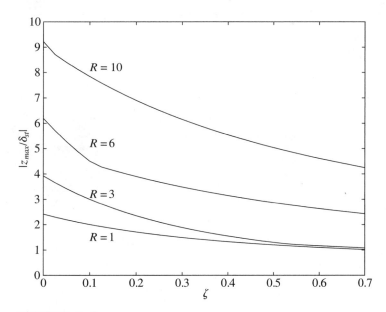

FIGURE 4.9
Normalized maximum relative displacement of a system inside a container that is dropped from a height h as a function of coefficient of restitution of the container and the damping ratio of the single degree-of-freedom system.

Design Guidelines: To minimize the maximum relative displacement of the mass, keep h and, therefore the velocity V_o, as small as possible. Make the container of a material that absorbs the impact, so that the coefficient of restitution is as small as possible. Make the natural frequency of the single degree-of-freedom system as low as possible. Since, in packaging, the mass is usually not a parameter that can be specified, one has to make the equivalent spring as soft as practical. Increase the equivalent damping of the packing material.

Although not explored here, an equally important design goal is to minimize the absolute acceleration of the mass m, as large accelerations can be detrimental to the single degree-of-freedom system.

EXAMPLE 4.6   Collision of two viscoelastic bodies

We use the single degree-of-freedom model to analyze the impact[3] (collision) of two viscoelastic bodies. In Figure 4.10, two bodies m_A and m_B whose relative velocity is V_{AB} just before impact is shown. During contact, the contact force between the masses m_A and m_B is represented by using a Kelvin-Voigt model: that is, a linear spring with stiffness k in parallel with a viscous damper with damping coefficient c.

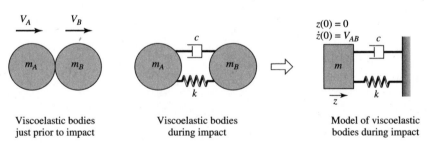

Viscoelastic bodies   Viscoelastic bodies   Model of viscoelastic
just prior to impact   during impact   bodies during impact

FIGURE 4.10
Impact of two viscoelastic bodies.

[3]C. Rajalingham and S. Rakheja, "Analysis of Impact Force Variation During Collision of Two Bodies Using a Single-Degree-of-Freedom System Model," *J. Sound Vibration,* Vol. 229, No. 4, pp. 823–835 (2000). For the case where both masses are traveling in the same direction in contact with each other and impact a rigid wall, see W. J. Stronge, *Impact Dynamics,* Cambridge University Press, Cambridge, U.K., Chapter 5 (2000).

If we let x_A and x_B represent the absolute displacements of m_A and m_B, respectively, then the relative displacement between the two masses, the relative velocity between them, and the relative acceleration between them are given by, respectively,

$$z = x_A - x_B$$
$$V_{AB} = \dot{z} = \dot{x}_A - \dot{x}_B$$
$$\ddot{z} = \ddot{x}_A - \ddot{x}_B \tag{a}$$

The magnitude of the contact force F_{AB} that acts on each mass for the duration of contact is

$$F_{AB} = c\frac{dz}{dt} + kz \tag{b}$$

From the free-body diagram of each mass during impact, we arrive at

$$\ddot{x}_A = -\frac{F_{AB}}{m_A}$$
$$\ddot{x}_B = \frac{F_{AB}}{m_B} \tag{c}$$

Since there are no external forces acting on the system at the time of impact, the system's linear momentum is conserved. Thus, from Eqs. (1.11) and (1.12), we have that

$$\frac{d}{dt}(m_A\dot{x}_A + m_B\dot{x}_B) = 0 \tag{d}$$

and, therefore,

$$m_A\ddot{x}_A + m_B\ddot{x}_B = 0 \tag{e}$$

From Eq. (e) and the last of Eqs. (a), we find that

$$\ddot{x}_A = \frac{m}{m_A}\ddot{z}$$
$$\ddot{x}_B = -\frac{m}{m_B}\ddot{z} \tag{f}$$

where the quantity m is called the *effective mass* and it is given by

$$m = \frac{m_A m_B}{m_A + m_B} \tag{g}$$

Then, from Eqs. (a), (b), (c), (f), and (g), we find that when the masses are in contact, the governing equation for the effective mass of the colliding bodies is given by

$$m\ddot{z} = -F_{AB} = -(c\dot{z} + kz)$$

or

$$\ddot{z} + 2\zeta\omega_n\dot{z} + \omega_n^2 z = 0 \tag{h}$$

where

$$\omega_n^2 = \frac{k}{m} \quad \text{and} \quad \zeta = \frac{c}{2m\omega_n} \tag{i}$$

Equation (h) is valid up until the time that $F_{AB} = 0$, which, from the above equation, is also the time at which $\ddot{z} = 0$. Noting that the initial conditions are $z(0) = 0$ and $\dot{z}(0) = V_{AB}$, the solution for the relative displacement z between the two masses is given by Eq. (4.18) where $V_o = V_{AB}$. The coefficient of restitution can be determined from Eq. (a) of Example 4.4.

EXAMPLE 4.7 Vibratory system employing a Maxwell model

We shall now modify the single degree-of-freedom system shown in Figure 3.1 to obtain a more realistic description of the reaction force transmitted to the fixed support, when the inertial element is given an initial velocity. As noted earlier, if a Kelvin-Voigt model is used, there is an instantaneous reaction force at the base when an initial velocity is imparted to the mass. This unrealistic response is eliminated by using the modified system shown in Figure 4.11a, where we have introduced a linear spring k_1 in series with a lin-

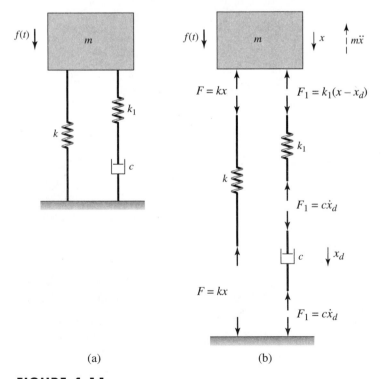

(a)   (b)

FIGURE 4.11
(a) Single degree-of-freedom system with a spring added in series with the damper and (b) forces on the system's elements.

ear viscous damper c. The combination of the linear spring k_1 in series with the linear viscous damper c is called a *Maxwell model*. To describe the motion of the system, we need, in addition to the displacement variable x of the mass m, another displacement variable x_d to describe the displacement at the spring-damper junction in the Maxwell model. Both x and x_d are measured from their respective static-equilibrium positions.

Governing Equations of Motion and Solution for Response

The governing equations are obtained for the general case with forcing and, from this case, the free response of the mass subjected to an initial velocity is determined. Making use of Figure 4.11b and carrying out a force balance along the vertical direction for the mass m and a force balance for the Maxwell element, we arrive at

$$m \frac{d^2x}{dt^2} + kx + k_1 (x - x_d) = f(t)$$

$$k_1 (x - x_d) = c \frac{dx_d}{dt} \tag{a}$$

Since we have an additional first-order equation, apart from the second-order equation typical of a single degree-of-freedom system, the vibratory system of Figure 4.11 is also referred to as a *one and a half degree-of-freedom system*.

Introducing the natural frequency

$$\omega_n = \sqrt{\frac{k}{m}} \tag{b}$$

and the nondimensional quantities

$$\tau = \omega_n t$$

$$\gamma = \frac{k_1}{k} \tag{c}$$

Eqs. (a) are rewritten as

$$\ddot{x} + (1 + \gamma)x - \gamma x_d = f(\tau)/k$$

$$\gamma x - \gamma x_d = 2\zeta \dot{x}_d \tag{d}$$

and the overdot indicates the derivative with respect to τ and

$$2\zeta = \frac{c\omega_n}{k} \tag{e}$$

In the limiting case, when $\gamma \rightarrow \infty$ (i.e., $k_1 \rightarrow \infty$), the second of Eqs. (d) leads to a Kelvin-Voigt model with a linear spring of stiffness k in parallel with a linear damper with damping coefficient c. Therefore, Eqs. (d) can be used to study a vibratory system with a Maxwell model as well as a Kevin-Voigt model.

If we represent the Laplace transform of $x(\tau)$ by $X(s)$, the Laplace transform of $x_d(\tau)$ by $X_d(s)$, and the Laplace transform of $f(\tau)$ by $F(s)$, then, from pair 2 in Table A of Appendix A, the Laplace transforms of Eqs. (d) are

$$(s^2 + 1 + \gamma)X(s) - \gamma X_d(s) = G(s)$$

$$\gamma X(s) - (\gamma + 2\zeta s)X_d(s) = 0 \tag{f}$$

where we have assumed that $x_d(0) = 0$, and used the notation

$$G(s) = \frac{F(s)}{k} + sx(0) + \dot{x}(0) \tag{g}$$

Upon solving for $X(s)$ and $X_d(s)$ from Eqs. (f), we obtain, respectively,

$$X(s) = \frac{G(s)(\gamma + 2\zeta s)}{2\zeta s^3 + \gamma s^2 + 2\zeta(1 + \gamma)s + \gamma}$$

$$X_d(s) = \frac{\gamma G(s)}{2\zeta s^3 + \gamma s^2 + 2\zeta(1 + \gamma)s + \gamma} \tag{h}$$

Force Transmitted to the Fixed Support

From Figure 4.11b, the reaction force on the base is seen to be

$$F_B = F_1 + F = c\frac{dx_d}{dt} + kx \tag{i}$$

which, in terms of the nondimensional quantities given by Eqs. (c), is written as

$$\frac{F_B}{k} = 2\zeta\dot{x}_d + x \tag{j}$$

where the overdot is the derivative with respect to τ. Upon taking the Laplace transform of Eq. (j), again assuming that $x_d(0) = 0$, and using Eqs. (h), we find that

$$\frac{F_B}{k} = \frac{G(s)[\gamma + 2\zeta(1 + \gamma)s]}{2\zeta s^3 + \gamma s^2 + 2\zeta(1 + \gamma)s + \gamma} \tag{k}$$

This expression will be revisited in Example 5.11.

We shall limit the rest of our discussion to the case where the applied force and the initial displacement are zero; that is, $f(\tau) = 0$ and $x(0) = 0$, and the initial velocity is

$$\frac{dx(0)}{dt} = \omega_n\frac{dx(0)}{d\tau} = V_o \tag{l}$$

Therefore, Eq. (g) simplifies to

$$G(s) = \frac{V_o}{\omega_n} \tag{m}$$

Upon substituting Eq. (m) into Eq. (k), we arrive at

$$\frac{F_B}{(kV_o/\omega_n)} = \frac{\gamma + 2\zeta(1 + \gamma)s}{2\zeta s^3 + \gamma s^2 + 2\zeta(1 + \gamma)s + \gamma} \tag{n}$$

Before evaluating Eq. (n), we recall that the limiting case when $\gamma \to \infty$ (i.e., $k_1 \to \infty$) recovers the Kelvin-Voigt model, where a linear spring k is in parallel with a linear viscous damper c. For this limiting case, we divide the nu-

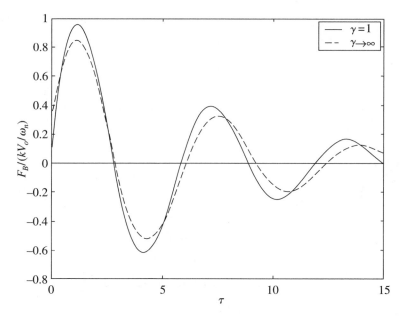

FIGURE 4.12
Reaction force of the system shown in Figure 4.11 for $\zeta = 0.15$.

merator and denominator of Eq. (n) by γ and take the limit as $\gamma \to \infty$. This operation results in

$$\frac{F_B}{(kV_o/\omega_n)} = \frac{1 + 2\zeta s}{s^2 + 2\zeta s + 1} \tag{o}$$

Upon using Laplace transform pairs 14 and 16 in Table A of Appendix A, the inverse Laplace transform of Eq. (o) results in Eq. (4.28). The numerically[4] computed inverse Laplace transforms of Eq. (n) for $\zeta = 0.15$ and $\gamma = 1$ and Eq. (o) for $\zeta = 0.15$ are shown in Figure 4.12. At $\tau = 0$, we see that the reaction force F_B has a discontinuity for the Kevin-Voigt model, while this reaction force is zero for the Maxwell model.

EXAMPLE 4.8 Vibratory system with Maxwell model revisited

As a continuation of Example 4.7, we now consider the case where the support consists only of a Maxwell element; that is, the spring k is absent. In this

[4]The MATLAB function `ilaplace` from the Symbolic Toolbox was used.

case, we again examine the force transmitted to the fixed base. Setting $k = 0$ in Eq. (a) of Example 4.7, we arrive at

$$m \frac{d^2x}{dt^2} + k_1(x - x_d) = f(t)$$

$$k_1(x - x_d) = c \frac{dx_d}{dt} \tag{a}$$

Introducing a new set of quantities

$$\tau' = \omega_{1n}t \quad \text{and} \quad \omega_{1n}^2 = \frac{k_1}{m} \tag{b}$$

Eq. (a) is written as

$$\ddot{x} + x - x_d = f(t)/k_1$$

$$x - x_d = 2\zeta_1\dot{x}_d \tag{c}$$

where the overdot now indicates the derivative with respect to τ' and

$$2\zeta_1 = \frac{c\omega_{1n}}{k_1} \tag{d}$$

If we representing the Laplace transform of $x(\tau)$ by $X(s)$, the Laplace transform of $x_d(\tau)$ by $X_d(s)$, and the Laplace transform of $f(\tau)$ by $F(s)$, then, from pair 2 in Table A of Appendix A, the Laplace transforms of Eqs. (c) are

$$(s^2 + 1)X(s) - X_d(s) = G_1(s)$$

$$X(s) - (1 + 2\zeta_1 s)X_d(s) = 0 \tag{e}$$

where we have assumed that $x_d(0) = 0$ and

$$G_1(s) = \frac{F(s)}{k_1} + sx(0) + \dot{x}(0) \tag{f}$$

Upon solving for $X(s)$ and $X_d(s)$ in Eqs. (e), we obtain

$$X(s) = \frac{G_1(s)(1 + 2\zeta_1 s)}{s(2\zeta_1 s^2 + s + 2\zeta_1)} = \frac{G_1(s)(2\zeta_m + s)}{s(s^2 + 2\zeta_m s + 1)}$$

$$X_d(s) = \frac{G_1(s)}{s(2\zeta_1 s^2 + s + 2\zeta_1)} \tag{g}$$

where

$$2\zeta_m = \frac{1}{2\zeta_1} = \frac{k_1}{c\omega_{1n}} \tag{h}$$

Note that $\zeta_m < 1$ only when $\zeta_1 > 0.25$.

When the spring with stiffness k is absent, the reaction force on the base is

$$F_B(t) = F_1 = c \frac{dx_d}{dt} \tag{i}$$

which is rewritten in terms of the nondimensional quantity given by Eq. (d) as

$$\frac{F_B(\tau')}{k_1} = 2\zeta_1 \dot{x}_d \tag{j}$$

where the overdot is the derivative with respect to τ'. Upon taking the Laplace transform of Eq. (j), assuming that $x_d(0) = 0$, and using the second of Eqs. (g), we find that

$$\frac{F_B(s)}{k_1} = \frac{2\zeta_1 G_1(s)}{s(2\zeta_1 s^2 + s + 2\zeta_1)} \tag{k}$$

We again limit the discussion to the case where the applied force and the initial displacement are zero; that is, $f(\tau) = 0$ and $x(0) = 0$, and the initial velocity is

$$\frac{dx(0)}{dt} = \omega_{1n} \frac{dx(0)}{d\tau'} = V_o \tag{l}$$

Therefore, Eq. (f) simplifies to

$$G_1(s) = \frac{V_o}{\omega_{1n}} \tag{m}$$

Upon substituting Eq. (m) into Eq. (k), we arrive at

$$\frac{F_B(s)}{(k_1 V_o / \omega_{1n})} = \frac{2\zeta_1}{s(2\zeta_1 s^2 + s + 2\zeta_1)}$$
$$= \frac{1}{s(s^2 + 2\zeta_m s + 1)} \tag{n}$$

For $\zeta_m < 1$, the inverse Laplace transform of Eq. (n) is given by Laplace transform pair 15 in Table A of Appendix A with $\omega_n = 1$ and $\zeta = \zeta_m$. The numerically computed[5] inverse Laplace transform of Eq. (n) for $\zeta_m = 0.15$ and $\zeta_m = 1.2$ are shown in Figure 4.13a. This model also exhibits a reaction force $F_B = 0$ at $\tau' = 0$. However, as time increases the model gives the unrealistic result that the force on the base approaches a constant, nonzero value. This limiting value is determined from transform pair 31 in Table A of Appendix A as

$$\lim_{\tau' \to \infty} \frac{F_B(\tau')}{(k_1 V_o / \omega_{1n})} \to \lim_{s \to 0} \frac{sF_B(s)}{(k_1 V_o / \omega_{1n})} = \lim_{s \to 0} \frac{s}{s(s^2 + 2\zeta_m s + 1)} = 1 \tag{o}$$

In Figure 4.13b, we have plotted the displacement response obtained from the numerical inverse Laplace transform of the first of Eqs. (g) with $G_1(s)$ given by Eq. (m) and for $\zeta_m = 0.15$. It is noticed that the nondimensional displacement ratio does not approach zero as time increases. This is because there is no spring in parallel with the viscous damper to restore the

---

[5]The function `ilaplace` from MATLAB's Symbolic Toolbox was used.

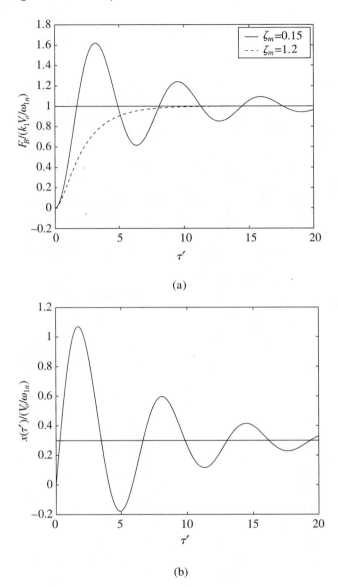

FIGURE 4.13
Maxwell element: (a) reaction force of the system for $\zeta_m = 0.15$ and (b) displacement response of the mass for $\zeta_m = 0.15$.

system to its original equilibrium position and, therefore, the Maxwell element undergoes a permanent deformation. Consequently, based on these observations, the Maxwell model is not used by itself, but in parallel with another spring, as shown in Figure 4.11. The limiting value is determined from transform pair 31 in Table A of Appendix A as

$$\lim_{\tau' \to \infty} \frac{x(\tau')}{(V_o/\omega_{1n})} \to \lim_{s \to 0} \frac{sX(s)}{(V_o/\omega_{1n})} = \lim_{s \to 0} \frac{s(2\zeta_m + s)}{s(s^2 + 2\zeta_m s + 1)} = 2\zeta_m = 0.3 \quad \text{(p)}$$

4.3.3 Initial Displacement

We now examine the free response of an underdamped single degree-of-freedom system with a prescribed initial displacement. When a system is subjected to an initial displacement only, we set $V_o = 0$ and Eq. (4.17) for the amplitude and phase simplify to

$$A_o = \frac{X_o}{\sqrt{1 - \zeta^2}}$$

$$\varphi_d = \tan^{-1} \frac{\sqrt{1 - \zeta^2}}{\zeta} = \varphi$$

Therefore, Eq. (4.16), which describes the displacement response, becomes

$$x(t) = \frac{X_o}{\sqrt{1 - \zeta^2}} e^{-\zeta\omega_n t} \sin(\omega_d t + \varphi) \tag{4.33}$$

and, after using Eq. (4.10), the velocity and acceleration are, respectively,

$$\dot{x}(t) = v(t) = -\frac{X_o \omega_n}{\sqrt{1 - \zeta^2}} e^{-\zeta\omega_n t} \sin(\omega_d t)$$

$$\ddot{x}(t) = a(t) = \frac{X_o \omega_n^2}{\sqrt{1 - \zeta^2}} e^{-\zeta\omega_n t} \sin(\omega_d t - \varphi) \tag{4.34}$$

Equations (4.33) and (4.34) are plotted in Figure 4.14 and the corresponding state space plot is shown in Figure 4.15 along with their respective time histories. As time unfolds, the trajectory is attracted to the equilibrium position located at the origin (0, 0).

Logarithmic Decrement[6]

Consider the displacement response of a single degree-of-freedom system subjected to an initial displacement as shown in Figure 4.16. The logarithmic decrement δ is defined as the natural logarithm of the ratio of any two successive amplitudes of the response that occur a period T_d apart, where T_d is given by Eq. (4.11). For these two amplitudes, it is possible to determine the damping ratio ζ. To this end, we determine a relationship between the logarithmic decrement and the damping factor. We start from

$$\delta = \ln\left(\frac{x(t)}{x(t + T_d)}\right) \tag{4.35}$$

If we let

$$x_p = x(t + pT_d) \qquad p = 0, 1, 2, \ldots \tag{4.36}$$

---

[6]Although the definition of the logarithmic decrement is provided in Section 4.3.3, it applies equally to all free responses considered in Section 4.3.

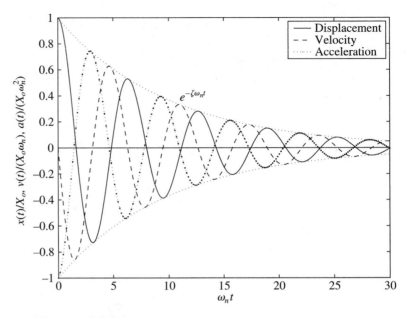

FIGURE 4.14
Time histories of displacement, velocity, and acceleration of a system with a prescribed initial displacement.

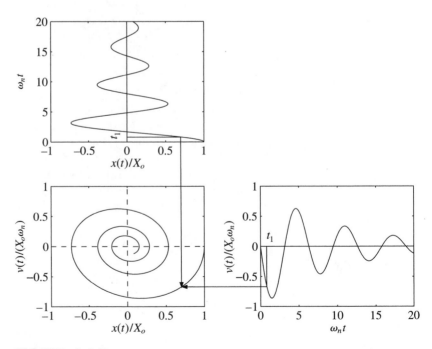

FIGURE 4.15
State-space plot of single degree-of-freedom system with prescribed initial displacement.

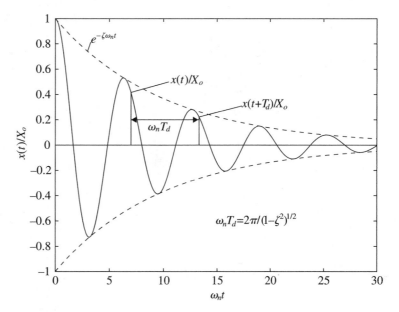

FIGURE 4.16
Quantities used in the definition of the logarithmic decrement.

then, by definition,

$$\frac{x_0}{x_1} = \frac{x_1}{x_2} = \frac{x_2}{x_3} = \cdots \frac{x_{p-1}}{x_p} = e^{\delta} \tag{4.37}$$

Furthermore, we also notice from Eq. (4.37) that

$$\frac{x_0}{x_p} = \frac{x_0}{x_1}\frac{x_1}{x_2}\frac{x_2}{x_3} \cdots \frac{x_{p-1}}{x_p} = e^{p\delta}$$

and, therefore, the logarithmic decrement in terms of two amplitudes measured p cycles apart is expressed as

$$\delta = \frac{1}{p}\ln\left(\frac{x_o}{x_p}\right) = \frac{1}{p}\ln\left(\frac{x(t)}{x(t + pT_d)}\right), \qquad p = 1, 2, \ldots \tag{4.38}$$

Making use of Eq. (4.33) and Eq. (4.11) and substituting for the free response p cycles apart into Eq. (4.38), we obtain

$$\delta = \frac{1}{p}\ln\left(\frac{X_o e^{-\zeta\omega_n t}\sin(\omega_d t + \varphi)/\sqrt{1 - \zeta^2}}{X_o e^{-\zeta\omega_n(t+pT_d)}\sin(\omega_d(t + pT_d) + \varphi)/\sqrt{1 - \zeta^2}}\right)$$

$$= \frac{1}{p}\ln\left(\frac{e^{\zeta\omega_n pT_d}\sin(\omega_d t + \varphi)}{\sin(\omega_d t + \varphi + 2p\pi)}\right) = \frac{1}{p}\ln\left(e^{\zeta\omega_n pT_d}\right) = \frac{1}{p}\zeta\omega_n pT_d = \zeta\omega_n T_d$$

$$= \frac{2\pi\zeta}{\sqrt{1 - \zeta^2}} \tag{4.39}$$

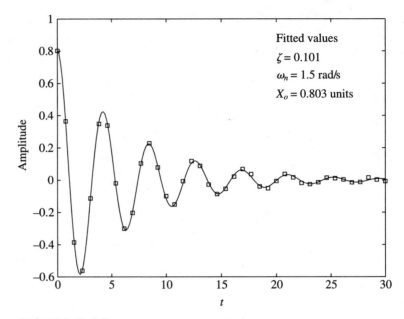

FIGURE 4.17
Curve fit to a set of sampled data from the response of a system with pre-scribed initial displacement.

Thus, from a measurement of the amplitudes x_0 and x_p, one can obtain the damping ratio ζ from

$$\zeta = \frac{1}{\sqrt{1 + (2\pi/\delta)^2}} \tag{4.40}$$

As an alternative to this type of estimation for δ, the free response of a system can also be curve-fitted to determine the damping factor. Based on digitally sampled data, one can use a standard nonlinear curve-fitting procedure for estimating the amplitude, damping ratio, and natural frequency of a system based on Eq. (4.33). A representative set of sampled data and the numerically obtained curve-fit values[7] are shown in Figure 4.17. The open squares represent the data through which the fitted curve is depicted as a solid line. As discussed in Section 5.3.2, the estimation of parameters such as damping factor and natural frequency can be also carried out based on the system transfer function.

EXAMPLE 4.9 Estimate of damping ratio using the logarithmic decrement

It is found from a plot of the response of a single degree-of-freedom system to an initial displacement that at time t_o the amplitude is 40% of its initial

---

[7]These results were obtained using `lsqcurvefit` from the MATLAB Optimization Toolbox.

value. Two periods later the amplitude is 10% of its initial value. We shall determine an estimate of the damping ratio. Thus, from Eq. (4.38)

$$\delta = \frac{1}{2} \ln \left(\frac{0.4}{0.1} \right) = 0.693$$

Then from Eq. (4.40) we find that

$$\zeta = \frac{1}{\sqrt{1 + (2\pi/0.693)^2}} = 0.11$$

4.3.4 Initial Displacement and Initial Velocity

We shall now consider the case when a system is subjected to an initial displacement and an initial velocity simultaneously. The solution is given by Eq. (4.16), which is repeated below for convenience.

$$x(t) = A_o e^{-\zeta \omega_n t} \sin(\omega_d t + \varphi_d) \tag{4.41}$$

where, from Eqs. (4.17), we find that amplitude and phase are given by

$$A_o = \sqrt{X_o^2 + \left(\frac{V_o + \zeta \omega_n X_o}{\omega_d} \right)^2} = X_o \sqrt{1 + \frac{(V_r + \zeta)^2}{1 - \zeta^2}}$$

$$\varphi_d = \tan^{-1} \frac{\omega_d X_o}{V_o + \zeta \omega_n X_o} = \tan^{-1} \frac{\sqrt{1 - \zeta^2}}{\zeta + V_r} \tag{4.42}$$

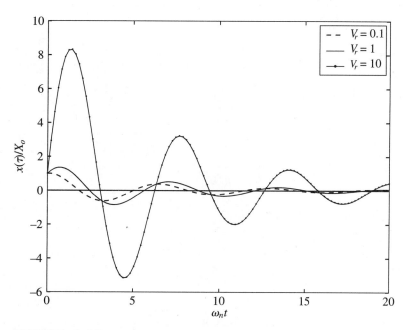

FIGURE 4.18

Displacement response of a system with prescribed initial displacement and prescribed initial velocity.

and $V_r = V_o/(\omega_n X_o)$ is a velocity ratio. The velocity response is determined from Eq. (4.41) to be

$$\dot{x}(t) = -A_o\omega_n e^{-\zeta\omega_n t} \sin(\omega_d t + \varphi_d t + \varphi_d - \varphi) \tag{4.43}$$

The numerically evaluated result for $x(t)/X_o$ is shown in Figure 4.18. For "small" values of V_r, the displacement response is similar to that obtained for a system with a prescribed initial displacement and for "large" values of V_r, the displacement response is similar to that obtained for a system with a prescribed initial velocity.

EXAMPLE 4.10   Inverse problem: information from a state-space plot

Consider the state-space plot shown in Figure 4.19. From this graph, we shall determine the following: (a) the value of the damping ratio and (b) the time $\tau_{max} = \omega_n t_{max}$ at which the maximum displacement occurs.

From the graph, the initial conditions are $x(0) = X_o$ and $v(0) = 1.6 X_o \omega_n$. To determine ζ, the logarithmic decrement is used. For convenience, we select the values of the displacement from Figure 4.19 that are along the line $v(t) = 0$. Then,

$$x(t) \approx 0.95 X_o \qquad \text{and} \qquad x(t + T_d) \approx 0.5 X_o \tag{a}$$

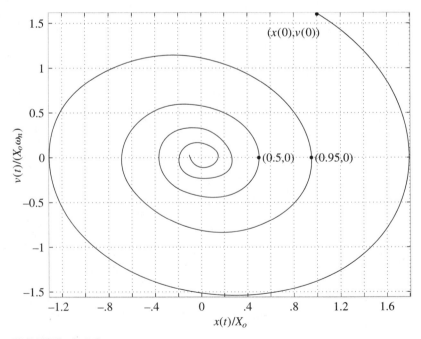

FIGURE 4.19
State-space graph for a system with prescribed initial velocity and prescribed initial displacement.

and from Eq. (4.38) and Eq. (a), we determine the logarithmic decrement

$$\delta = \ln\left(\frac{0.95\,X_o}{0.5X_o}\right) = \ln(1.90) = 0.642 \tag{b}$$

Then, from Eq. (4.40) and Eq. (b), we find the damping factor to be

$$\zeta = \frac{1}{\sqrt{1 + (2\pi/\delta)^2}} = \frac{1}{\sqrt{1 + (2\pi/0.642)^2}} = 0.10 \tag{c}$$

The value of τ_{max} is obtained from Eqs. (4.41), (4.42), (c) and Figure 4.19 in the following manner. We note that

$$V_r = \frac{v(0)}{x(0)\omega_n} = \frac{1.6X_o\omega_n}{X_o\omega_n} = 1.6$$

$$A_o = X_o\sqrt{1 + \frac{(1.6 + 0.1)^2}{1 - (0.1)^2}} = 1.976X_o$$

$$\varphi_d = \tan^{-1}\frac{\sqrt{1 - (0.1)^2}}{1.6 + 0.1} = \tan^{-1}0.5853 = 0.53 \text{ rad} \tag{d}$$

From Figure 4.19 it is seen that the maximum displacement is $1.8X_o$, which occurs when $v(t) = 0$. Then Eq. (4.41) becomes

$$1.8X_o = 1.976X_oe^{-0.1\tau_{max}}\sin(\tau_{max}\sqrt{1 - (0.1)^2} + 0.53)$$

$$0.91 = e^{-0.1\tau_{max}}\sin(0.995\tau_{max} + 0.53) \tag{e}$$

which we solve numerically[8] to obtain $\tau_{max} = 0.945$.

The time τ_{max} can also be obtained from Eq. (4.43), which is the earliest time that $v(\tau_{max}) = 0$; that is, the time at which the argument of the sine function is zero. Thus, we have

$$\omega_d t_{max} + \varphi_d - \varphi = \tau_{max}\sqrt{1 - \zeta^2} + \varphi_d - \varphi = 0 \tag{f}$$

or equivalently

$$\tau_{max} = \frac{\varphi - \varphi_d}{\sqrt{1 - \zeta^2}} \tag{g}$$

Since

$$\varphi = \tan^{-1}\frac{\sqrt{1 - \zeta^2}}{\zeta} = \tan^{-1}\frac{0.995}{0.1} = 1.47 \text{ rad} \tag{h}$$

we find that

$$\tau_{max} = \frac{1.47 - 0.53}{0.995} = 0.945 \tag{i}$$

---

[8]The MATLAB function `fzero` was used.

4.4 STABILITY OF A SINGLE DEGREE-OF-FREEDOM SYSTEM

A linear single degree-of-freedom system is considered stable if, for all selections of finite initial conditions and finite forcing functions,

$$|x(t)| \leq A \qquad t > 0$$

where A has a finite value. This is a boundedness condition, which requires the system response $x(t)$ be bounded for bounded system inputs. If this is not the case, then the system is considered *unstable*.[9] For the systems that are dealt with in this book, the unstable responses grow either linearly with time or exponentially with time.

Instability of Unforced System

We consider an unforced vibratory system subjected to finite initial conditions and study when this system can be unstable. To this end, we start from the solution for the response given in the Laplace domain by Eq. (4.4) and set $F(s) = 0$ to obtain

$$X(s) = \frac{sx(0)}{D(s)} + \frac{2\zeta\omega_n x(0) + \dot{x}(0)}{D(s)} \tag{4.44}$$

Now let us examine the denominator $D(s)$, which is given by Eq. (4.5). We factor $D(s)$ to obtain

$$\begin{aligned} D(s) &= s^2 + 2\zeta\omega_n s + \omega_n^2 = s^2 + (c_e/m_e)s + k_e/m_e \\ &= (s - s_1)(s - s_2) \end{aligned} \tag{4.45}$$

where we have used $\zeta_e = c_e/(2m_e\omega_n)$, $\omega_n = \sqrt{k_e/m_e}$, and

$$s_{1,2} = \frac{1}{2m_e}[-c_e \pm \sqrt{c_e^2 - 4m_e k_e}] \tag{4.46}$$

and we have switched to the more general equivalent forms for the inertia, stiffness and damping. From Eq. (4.44), we see that there are two terms on the right-hand side that involve, respectively, the polynomial ratios

$$\frac{1}{D(s)} = \frac{1}{(s - s_1)(s - s_2)} = \frac{1}{(s_1 - s_2)}\left[\frac{1}{(s - s_1)} - \frac{1}{(s - s_2)}\right] \tag{4.47}$$

and

$$\frac{s}{D(s)} = \frac{s}{(s - s_1)(s - s_2)} = \frac{1}{(s_2 - s_1)}\left[\frac{s_2}{(s - s_2)} - \frac{s_1}{(s - s_1)}\right] \tag{4.48}$$

Since

$$x(t) = L^{-1}[X(s)]$$

---

[9]Other notions of stability can be found in A. H. Nayfeh and B. Balachandran, *Applied Nonlinear Dynamics: Analytical, Computational and Experimental Methods,* John Wiley & Sons, NY (1995).

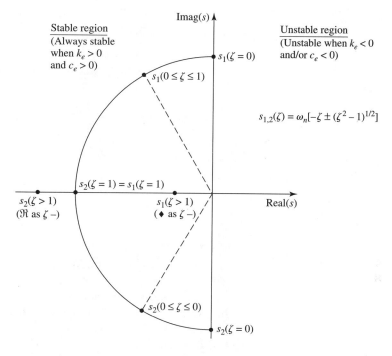

FIGURE 4.20
Root locus diagram.

where L^{-1} indicates the inverse Laplace transform, it is seen that in order to obtain $x(t)$ one needs the inverse Laplace transforms of Eqs. (4.47) and (4.48), which are

$$L^{-1}\left[\frac{1}{s - s_{1,2}}\right] = e^{s_{1,2}t}$$

where we have used Laplace transform pair 7 in Table A of Appendix A.

The condition under which

$$e^{s_{1,2}t}$$

remains finite for $t > 0$ is

$$\text{Re}[s_{1,2}] \leq 0 \tag{4.49}$$

that is, the real parts of the roots have to be less than or equal to zero. From Eq. (4.46), it is seen that this condition is satisfied when $c_e \geq 0$ and $k_e \geq 0$. When either $c_e < 0$ or $k_e < 0$, the system is unstable. These results are usually summarized in a root locus diagram like that shown in Figure 4.20 in which the roots s_1 and s_2 are plotted in the complex plane for different values of the damping parameter ζ.

EXAMPLE 4.11   Instability of inverted pendulum

The inverted pendulum that was examined in Example 3.11 is a system that can be unstable, depending on the values of the parameters. For this system, we have $m_e > 0$, $c_e > 0$, and

$$k_e = kL_1^2 - m_1gL_1 - m_2g\frac{L_2}{2}$$

Thus, the system is stable as long as $k_e > 0$; that is, when

$$kL_1^2 \geq m_1gL_1 + m_2g\frac{L_2}{2}$$

Asymptotic Stability

So far, we have only discussed the notion of bounded stability and how the system parameters such as equivalent stiffness k_e and equivalent damping coefficient c_e can affect the stability of the system. A notion of stability that is useful for studying the oscillations of vibratory systems is asymptotic stability.[10] Instead of defining this motion for a general system, let us consider Eq. (4.1) without any forcing; that is, $f(t) = 0$. Then

$$\frac{d^2x}{dt^2} + 2\zeta\omega_n\frac{dx}{dt} + \omega_n^2x = 0 \tag{4.50}$$

The equilibrium position $x = 0$ of this system is said to be asymptotically stable if

$$\lim_{t \to \infty} x(t) \to 0 \tag{4.51}$$

that is, the equilibrium position is approached as time increases. Since the governing equation is an equation with constant coefficients, a solution to this equation can be written in the form

$$x(t) = Ae^{\lambda t} \tag{4.52}$$

where A is a constant and λ is an unknown quantity. Upon substituting Eq. (4.52) into Eq. (4.50) and requiring that $A \neq 0$, we obtain

$$\lambda^2 + 2\zeta\omega_n\lambda + \omega_n^2 = 0 \tag{4.53}$$

Equation (4.53) is referred to as the *characteristic equation* and the roots of this equation λ_1 and λ_2 are referred to as *characteristic roots* or *eigenvalues*. The eigenvalues are special values for which $x(t)$ given by Eq. (4.52) has a non-zero value. The roots of Eq. (4.53) are given by

$$\lambda_{1,2} = \omega_n\left[-\zeta \pm \sqrt{\zeta^2 - 1}\right] \tag{4.54}$$

---

[10]A. H. Nayfeh and B. Balachandran, *ibid.*

Therefore, the solution given by Eq. (4.52) is written as

$$x(t) = A_1 e^{\lambda_1 t} + A_2 e^{\lambda_2 t} \tag{4.55}$$

Hence, if the real part of the exponents λ_1 and λ_2 are negative, Eq. (4.51) is satisfied, and the equilibrium position is asymptotically stable. It should not be surprising that the polynomial in Eq. (4.53) is identical to Eq. (4.45) and that the requirement that the real parts of the exponents be negative for stability is identical to Eq. (4.49), because they both pertain to the free oscillations of the system described by Eq. (4.50).

It is clear that when the damping factor is positive—that is, $\zeta > 0$—the real parts of the exponents λ_i are negative and hence, this ensures stability, in particular, asymptotic stability. On the other hand, for negative damping factors that are possible in the presence of fluid forces in certain physical systems, the exponents have positive real parts indicating instability.

4.5 MACHINE TOOL CHATTER

In Figure 4.21, a model of a turning operation on a lathe is shown. When the cutting parameters such as spindle speed and width of cut are carefully chosen, the turning operation can produce the desired surface finish on the work piece. However, this turning operation can become unstable for certain values of spindle speed and width of cut. When these undesirable conditions are present, the tool and work piece system "chatters," producing an undesirable surface finish and a shortening of tool life. In this example, we shall explore the loss of stability that leads to the onset of chatter.

For a rigid work piece and a flexible tool, the cutting force acting on the tool due to the uncut material and the associated damping can be modeled as shown in Figure 4.21. The mass m represents the mass of the tool and tool holder, k is the stiffness of the tool holder's support structure, and c is the equivalent viscous damping of the structure. The dynamic cutting force F_c is

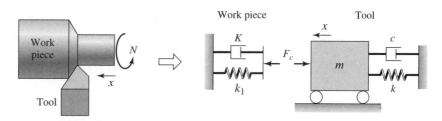

FIGURE 4.21
Model of a tool and work piece during turning.

the sum of the forces due to the change in chip thickness and the change in the penetration rate of the tool. Thus,[11] we have

$$F_c = \underbrace{k_1}_{\substack{\text{Cutting} \\ \text{stiffness}}} \underbrace{[x(t) - \mu x(t - 2\pi/N)]}_{\text{Change in chip thickness}} + \underbrace{K\frac{2\pi}{N}\frac{dx}{dt}}_{\text{Damping}}$$

where μ is the overlap factor ($0 \le \mu \le 1$), k_1 is an experimentally determined dynamic coefficient called the *cutting stiffness, K* is the experimentally determined penetration rate coefficient, and N is the rotational speed of either the tool or the work piece in revolutions per second. Then carrying out a force balance based on Figure 4.21, the tool vibrations can be described by the following equation

$$\frac{d^2x}{d\tau^2} + \left(\frac{1}{Q} + \frac{K}{k\Omega}\right)\frac{dx}{d\tau} + \left(1 + \frac{k_1}{k}\right)x - \underbrace{\mu\frac{k_1}{k}x(\tau - 1/\Omega)}_{\substack{\text{Time-delay effect due to uncut} \\ \text{chip during previous pass}}} = 0 \quad (4.56a)$$

where the nondimensional time $\tau = \omega_n t$, and

$$\Omega = \frac{N}{2\pi\omega_n}, \quad \omega_n = \sqrt{\frac{k}{m}}, \quad Q = \frac{1}{2\zeta}, \quad \text{and} \quad 2\zeta = \frac{c}{m\omega_n} \quad (4.56b)$$

In Eqs. (4.56b), the quantity Q is called the *quality factor,* which is discussed further in Section 5.3.3.

Since the right-hand side of Eq. (4.56a) is zero, the governing equation is similar to that used to study the free response of a single degree-of-freedom system. However, unlike the other systems treated so far, the system described by Eq. (4.56a) has a time delay $1/\Omega$ present. This makes the system, which is a linear, delay-differential equation, more difficult to deal with when compared to the ordinary differential equations we have considered so far. For this system, when the position $x(\tau) = 0$ is stable, the cutting operation is stable. Physically, the position $x(\tau) = 0$ corresponds to cutting at the specified nominal chip thickness. When this position is unstable, the system can start to chatter. The onset of chatter is marked by tool oscillations. To determine the onset of these oscillations, the conditions that can lead to oscillatory solutions of Eq. (4.56a) are examined next.

A solution to Eq. (4.56a) is of the form

$$x = Ae^{\lambda\tau}$$

which, when substituted into Eq. (4.56) gives the characteristic equation

$$\lambda^2 + \left(\frac{1}{Q} + \frac{K}{k\Omega}\right)\lambda + 1 + \frac{k_1}{k}(1 - \mu e^{-\lambda/\Omega}) = 0 \quad (4.57)$$

where, in general, the exponent is complex and of the form $\lambda = \delta + j\omega$. For the system to be stable, the $\text{Re}[\lambda] < 0$, that is, $\delta < 0$. The boundary between

---

[11]S. A. Tobias, *Machine-Tool Vibration,* Blackie & Sons, Ltd., Glasgow, pp. 146–176 (1965).

the stable and unstable regions corresponds to $\delta = 0$. When the exponent is purely imaginary, the response of the tool is oscillatory. Therefore, to find the stability boundary we let $\lambda = j\omega$ and substitute this value into the quasi-polynomial Eq. (4.57). After setting the real and imaginary parts to zero, we obtain two equations from which the chatter frequency ω and parameter of interest are determined as a function of the nondimensional spindle speed Ω. Here, we seek to determine the quality factor Q as a function of Ω, which is indicative of the damping level in the system.

On substituting $\lambda = j\omega$ into Eq. (4.57), and separating the real and imaginary parts, we arrive at

$$\frac{1}{Q} + \frac{K}{k\Omega} + \frac{\mu k_1}{k} \frac{\sin(\omega/\Omega)}{\omega} = 0$$

$$\omega^2 = 1 + \frac{k_1}{k}(1 - \mu \cos(\omega/\Omega)) \tag{4.58}$$

In Eqs. (4.58), the quantities K/k, μ, and k_1/k are known, and the values of the nondimensional spindle speed Ω are varied over a specified range. At each value of Ω, the value of ω is determined numerically[12] from the second of Eqs. (4.58). The values for Ω and ω are then used in the first of Eqs. (4.58)

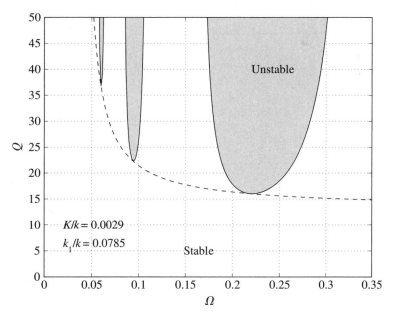

FIGURE 4.22
Stability chart for one set of parameters in turning: $\mu = 1$.

[12]The MATLAB function `fzero` was used.

to determine the positive values of Q that satisfy the equation; that is, those values of Ω and ω for which

$$\frac{1}{Q} = -\frac{K}{k\Omega} - \frac{\mu k_1}{k} \frac{\sin(\omega/\Omega)}{\omega}$$

In the plot of Ω versus Q, we can show the regions for which the system is either stable or unstable. Representative results are shown in Figure 4.22. The shaded regions, which are in the form of lobes, are regions of instability, and are referred to as stability lobes. The asymptote to these lobes is shown in the figure by a dashed line. If one conservatively chooses the cutting parameters so that one is below this asymptote to the lobes, then based on the linear theory presented here, the tool will not chatter. Of course, one can also choose spindle speeds that correspond to regions between the stability lobes as well.

4.6 SINGLE DEGREE-OF-FREEDOM SYSTEMS WITH NONLINEAR ELEMENTS

4.6.1 Nonlinear Stiffness

We illustrate the effects that two different types of nonlinear springs can have on the free response of a system when subjected to either an initial displacement or initial velocity.

System with Hardening Cubic Spring

First, we consider a system that has a spring whose stiffness includes a component that varies as the cube of the displacement. After using Eq. (2.23) for the nonlinear spring force, the governing equation is

$$\frac{d^2x}{d\tau^2} + 2\zeta \frac{dx}{d\tau} + x + \alpha x^3 = 0 \tag{4.59}$$

where the nondimensional time variable $\tau = \omega_n t$. We solve Eq. (4.59) numerically,[13] since it does not have an analytical solution. We assume that $\alpha = 1$ cm^{-2}, $\zeta = 0.15$, and that the initial conditions are $X_o = 2$ cm and $V_o = 0$. The results are shown in Figures 4.23, along with the solution for the system with a linear spring; that is, when $\alpha = 0$.

We see from these results that the response of the system with the nonlinear spring is distinctly different from that with the linear spring. First, the response of the system with the nonlinear spring does not decay exponentially with time and second, the displacement response does not have a constant period of damped oscillation. These differences provide one a means of distinguishing one type of nonlinear system from a linear system based on an examination of the response to an initial displacement. In practice, the nonuniformity of the period is easier to detect, since the dependence of fre-

---

[13]The MATLAB function ode45 was used.

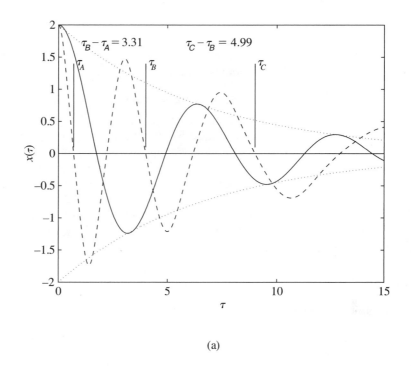

(a)

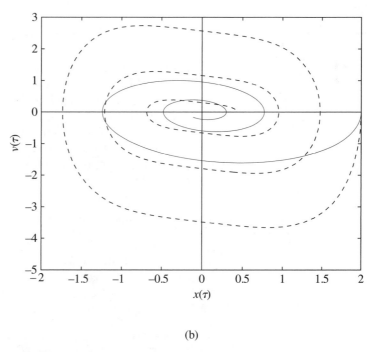

(b)

FIGURE 4.23
Comparison of the responses of linear (solid lines) and nonlinear (dashed lines) systems with prescribed initial displacement: (a) displacement and (b) phase portrait.

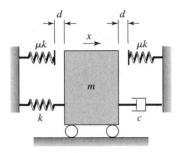

FIGURE 4.24
Single degree-of-freedom system with additional springs that are not contacted until the mass displaces a distance d in either direction.
Source: Reprinted from *Journal of Sound Vibration,* 207, H. Y. Hu. "Primary Resonance of a Harmonically Forced Oscillator with a Pair of Symmetric Set-Up Elastic Stops," pp. 393–401, Copyright © 1997, with permission from Elsevier Science.

quency (or period) on the amplitude of free oscillation is a characteristic of a nonlinear system.

System with Piecewise Linear Springs

We now consider a second nonlinear system shown in Figure 4.24. In this case, the springs are linear; however, the mass is straddled by two additional linear elastic spring-stops that are not contacted until the mass has been displaced by an amount d in either direction. The stiffness of the springs is proportional to the attached spring by a constant of proportionality μ ($\mu \geq 0$). When $\mu = 0$, we have the standard linear single degree-of-freedom system, and when $\mu > 1$, the elastic spring-stops are stiffer than the spring that is permanently attached to the mass. The governing equation describing the motion of the system is[14]

$$\frac{d^2y}{d\tau^2} + 2\zeta \frac{dy}{d\tau} + y + \mu h(y) = 0$$

where

$$
\begin{aligned}
h(y) &= 0 && |y| \leq 1 \\
&= y - \text{sgn}(y) && |y| > 1
\end{aligned}
$$

and, as discussed in Section 2.4.2, the signum function $\text{sgn}(y)$ is $+1$ when $y > 0$ and is -1 when $y < 0$. Furthermore, we have employed the following definitions:

$$\tau = \omega_n t, \quad \omega_n = \sqrt{\frac{k}{m}}, \quad y = \sqrt{\frac{x}{d}}, \quad 2\zeta = \frac{c}{m\omega_n}, \quad \text{and} \quad g(\tau) = \frac{f(t)}{kd}$$

Although it is possible to find a solution for this piecewise linear system, here we obtain a numerical[15] solution for convenience. We shall determine the response of this system when it is subjected to an initial (dimensionless) velocity $dy(0)/d\tau = V_o/(\omega_n d) = 10$, the damping factor $\zeta = 0.15$, and the values of μ are 0, 1, and 10. The results are shown in Figure 4.25. We see that the introduction of the spring-stops decreases the amplitude of the mass. In addition, it has the effect of decreasing the period of oscillation, which is equivalent to increasing its natural frequency.

4.6.2 Nonlinear Damping

We compare the free responses of systems with linear viscous damping, Coulomb damping, and fluid damping, which are obtained from Eqs. (3.23),

---

[14]H. Y. Hu, "Primary Resonance of a Harmonically Forced Oscillator with a Pair of Symmetric Set-up Elastic Stops," *J. Sound Vibration,* Vol. 207, No. 3, pp. 393–401, 1997.

[15]The MATLAB function ode45 was used.

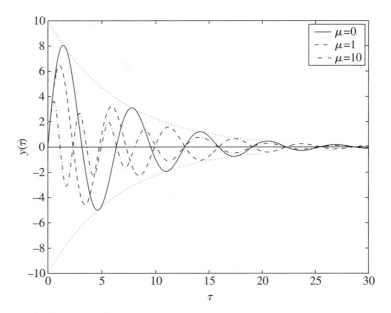

FIGURE 4.25
Response of the system shown in Figure 4.24 with prescribed initial velocity
$V_o/(\omega_n d) = 10$.

(3.24), and (3.25), respectively. They are rewritten here in terms of nondimensional time $\tau = \omega_n t$ as

$$\frac{d^2x}{d\tau^2} + \underbrace{2\zeta\frac{dx}{d\tau}}_{\substack{\text{Linear viscous} \\ \text{damping}}} + x = 0$$

$$\frac{d^2x}{d\tau^2} + \underbrace{d_C\,\text{sgn}(dx/d\tau)}_{\text{Dry friction damping}} + x = 0$$

$$\frac{d^2x}{d\tau^2} + \underbrace{d_F\left|\frac{dx}{d\tau}\right|\frac{dx}{d\tau}}_{\text{Fluid damping}} + x = 0 \qquad (4.60)$$

and the coefficients d_C and d_F are given by

$$d_C = \frac{\mu mg}{k} \qquad \text{and} \qquad d_F = \frac{c_d}{m} \qquad (4.61)$$

The first of Eqs. (4.60) governs a linear vibratory system with viscous damping, the second of Eqs. (4.60) governs a vibratory system with nonlinear damping due to dry friction, and the third of Eqs. (4.60) governs a vibratory system with nonlinear damping due to a fluid.

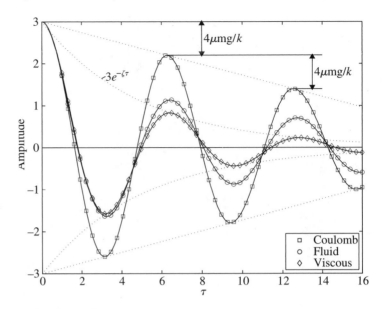

FIGURE 4.26

Comparisons of displacement responses for three different damping models.

We solve Eqs. (4.60) numerically[16] for the case where the initial velocity is zero, the initial displacement is 3 units, and $\zeta = d_C = d_F = d_S = 0.2$. The results are shown in Figure 4.26. The period of damped oscillations about the system equilibrium position is different in all three cases. For viscous damping, we have shown in Section 4.2.2 that the amplitude decays exponentially. For Coulomb damping it can be shown[17] that the amplitude decays linearly with a slope $\pm 2d_C/\pi$, where the plus sign is for the envelope of the negative peaks and the negative sign is for the envelope of positive peaks. The decay of the amplitude for fluid damping does not have any equivalent expression.

Nonlinear System Response Dependence on Initial Conditions

To illustrate that the long-time response of a nonlinear system depends on the initial conditions, the system with dry friction governed by the second of Eqs. (4.60) is revisited. During the free oscillations, the system will come to a stop or reach a rest state when

$$\frac{dx}{d\tau} = 0 \quad \text{and} \quad |x| \leq d_C \tag{4.62}$$

[16]The MATLAB function `ode45` was used.

[17]See, for example, D. J. Inman, *Engineering Vibration*, 2nd ed., Prentice Hall, Upper Saddle River, NJ, pp. 65–68 (2001).

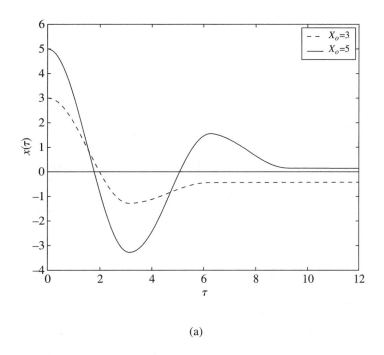

(a)

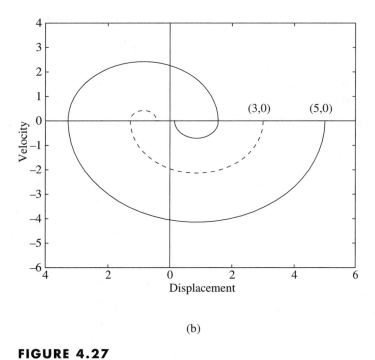

(b)

FIGURE 4.27
(a) Displacement histories and (b) phase portraits for the free response of a system with dry friction subjected to two different initial displacements: $d_C = 0.86$.

In other words, this system has multiple equilibrium positions, and the locus of these positions in the state space is the straight line joining the points $(-d_c, 0)$ and $(d_c, 0)$.

The free responses of the system initiated from two sets of initial conditions $x(0) = 3$ units and $\dot{x}(0) = 0$, and $x(0) = 5$ units and $\dot{x}(0) = 0$, in the state space are shown in Figure 4.27. Again, these responses can be determined analytically by noting that the system is linear in the region $\dot{x}(\tau) > 0$ and linear in the region $\dot{x}(\tau) < 0$. However, the responses shown in Figure 4.27 are determined numerically.[18] As seen from Figure 4.27, for the two different initial conditions, the system comes to a rest in finite time at two different positions, and the respective rest positions are reached at two different times.

4.7   SUMMARY

In this chapter, the solution for the response for a linear vibratory system has been obtained using Laplace transforms and studied. It was shown that the solution is determined by the initial conditions and the forcing imposed on the system. In particular, in the absence of forcing, the responses for different types of initial conditions have been explored and discussed. The notion of logarithmic decrement was introduced and the relationship between the damping factor and the logarithmic decrement was established. In addition, the notion of stability was briefly addressed and the problem of machine-tool chatter was examined. The influence of nonlinear stiffness and nonlinear damping on the free response of a vibratory system was also studied.

EXERCISES

4.1 A 25 kg television set is placed on a light table supported by four cylindrical legs made from a steel alloy material with a Young's modulus of elasticity $E = 400$ MPa. Each of these legs has a length of 0.5 m and a diameter of 10 mm. Consider free motions of this system in the vertical direction and determine the displacement response when the initial displacement is zero and the initial velocity is 0.2 m/s.

4.2 Consider the translational vibrations of a "small" rigid lathe along the cutting direction. A model of this system has a stiffness element $k =$ 20×10^6 N/m, an inertia element $m = 22.5$ kg, and a damping coefficient $c = 21205$ N·s/m. Determine the free response of this system for a 5 mm initial displacement and zero initial velocity. Plot the response and discuss it. How does the nature of the response of the system change when the damping coefficient is increased to 43000 N·s/m?

4.3 A microelectromechanical system has a mass of 0.30 µg and a stiffness of 0.15 N/m. Assume that the damping coefficient for the system is negligible and that the gravity loading is normal to the direction of

---

[18]The MATLAB function ode45 was used.

motion. Determine the displacement response for this system if the system is provided an initial displacement of 2 μm and an initial velocity of 5 mm/s.

4.4 A truck tire is characterized with a stiffness of 1.25×10^6 N/m, a mass of 35 kg, and a damping coefficient of 4200 N·s/m. Determine the natural frequency and damping factor for this system, and obtain the free response of the system initiated from an initial displacement of X_o m and zero initial velocity.

4.6 Determine the characteristics of a damped vibratory system from the time-history data provided in Figure E4.6 for the displacement of a vibratory system.

4.7 In a certain experiment, assume that you can only measure the velocity of a vibratory system. Derive an expression that can be used to determine the logarithmic decrement and the damping factor from velocity data similar to the one derived in Section 4.3 for displacement data.

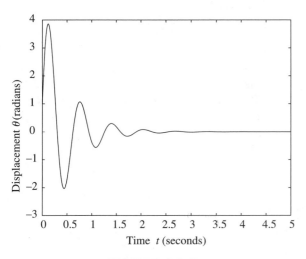

FIGURE E4.5

FIGURE E4.8

4.5 Determine the characteristics of a damped vibratory system from the time-history data provided in Figure E4.5 for the displacement.

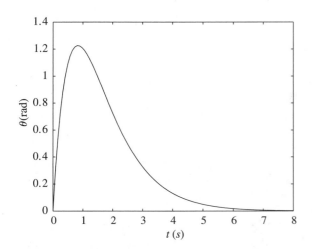

FIGURE E4.6

4.8 Consider the slender tower shown in Figure E4.8, which vibrates in the transverse direction as shown in the figure.[19] It is made from reinforced concrete. An estimate for the first natural frequency of this system is 0.15 Hz. The logarithmic decrement values measured for the tower with uncracked reinforced concrete material and cracked reinforced concrete material are 0.04 and 0.10, respectively. If a wind gust induces an initial displacement of 0.5 m and an initial velocity of 0.2 m/s, determine the peak displacement amplitudes in the cases with uncracked concrete material and cracked concrete material.

[19]H. Bachmann et al., *Vibration Problems in Structures: Practical Guidelines,* Birkhäuser Verlag, Basel, Germany (1995).

4.9 An empirical formula used for determining the natural frequency associated with translational motions of a steel chimney has the form[20]

$$\omega_n = 7100 \frac{d}{h^2} \text{ rad/s}$$

where d is the diameter of the chimney and h is the height of the chimney. The damping factor associated with a chimney of diameter 7 m and height of 100 m is 0.002. This chimney is located in a region where wind gusts are capable of producing initial displacements of 0.1 m and initial velocities of 0.4 m/s. Determine the maximum displacement experienced by the chimney and check if it satisfies the construction codes, which require the maximum displacement to be less than 4% of the diameter. If the present chimney does not satisfy this code, what design changes would you propose so that the construction code is satisfied?

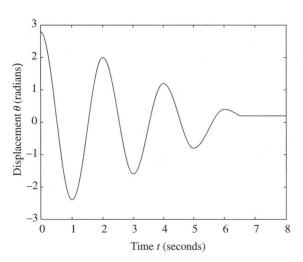

FIGURE E4.10

4.10 Consider the displacement time history shown in Figure E4.10 for free oscillations of a vibratory system. Examine this history and discuss if this system can be characterized as having viscous damping.

[20]Bachmann et al., *ibid.*

4.11 Consider the vibratory model of the micro-electromechanical system treated in Example 3.15. If the stiffness of the translation spring k, the stiffness of the torsion spring k_t, and the system damping coefficient c are all positive values, would it be possible for the system to have unstable behavior?

4.12 When there is a fluid moving with a speed V through a pipe, the transverse vibrations y of a mass m attached to the pipe can be described by the following equation

$$m\ddot{y} + [k_e - cV^2]y = 0$$

where k_e is the equivalent stiffness of the pipe and c is a constant that depends on the pipe cross-section area, the length of the pipe, and the density of the fluid. If m is always positive, k_e is always positive, c is positive, and V is positive, determine when the system can exhibit unstable behavior.

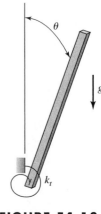

FIGURE E4.13

4.13 A rigid and uniform bar undergoing rotational motions in the vertical plane is restrained by a torsional spring of stiffness k_t at the pivot point O, as shown in Figure E4.13. The bar has a mass m and a length l. Determine what k_t should be, so that small oscillations about the upright position of the bar—that is, $\theta = 0$—are stable.

4.14 In studies of aircraft wing flutter, the following simplified model is used to study the system vibrations

$$m\ddot{x} + (c + a)\dot{x} + kx = 0$$

where the system parameters are m, c, and k. The damping constant a is due to aerodynamic forces, which changes with the angle of attack. Beyond a certain angle of attack, this constant can assume negative values. Determine the conditions on m, c, a, and k for which this system can be unstable.

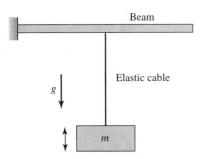

FIGURE E4.15

4.15 A load of mass m is suspended by an elastic cable from the mid-span of a beam, which is held fixed at one end and is free at the other end, as shown in Figure E4.15. The beam has a length L_b of 4 m and a flexural stiffness $EI = 60$ Nm^2. The elastic cable has a length L_c of 8 m, a diameter d of 200 mm, and it is made from material with a Young's modulus of elasticity $E_c = 3 \times 10^9$ N/m^2. The load has a mass of 200 kg. Assume that the damping in this vibratory system is negligible. If the mass is provided an initial velocity of 0.1 m/s, determine the ensuing motion and describe it. If the boundary conditions of the beam were changed so that the beam is now simply supported at both ends, how does the maximum amplitude of motion change?

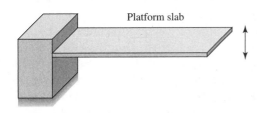

FIGURE E4.16

4.16 A diving platform is shown in Figure E4.16. The damping factor associated with this platform, which is made from reinforced concrete, is 0.012, and the first natural frequency of the platform-slab vibrations is designed to be 12 Hz.[21] When a diver jumps off this platform, an initial displacement of 1 mm and an initial velocity of 0.1 m/s are induced to the platform. Determine the resulting vibrations.

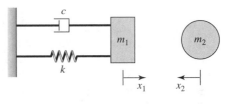

FIGURE E4.17

4.17 Consider two masses shown in Figure E4.17, which are involved in an impact. The mass is $m_1 = 2$ kg, the spring stiffness is $k = 500$ N/m, and the damping coefficient is $c = 15$ N·s/m. This system is initially at rest when it is impacted by a 1 kg mass m_2, which is traveling with a speed of 10 m/s just before impact. The coefficient of restitution e associated with these two impacting bodies is 0.8. Determine the displacement responses of these bodies after impact and plot the time histories of their respective responses. Is there a possibility for another impact?

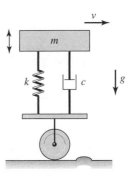

FIGURE E4.18

[21]Typical design rules require this frequency to be greater than or equal to 10 Hz.

4.18 A quarter-car model of a heavy vehicle is shown in Figure E4.18. This vehicle is traveling with a constant speed v on a flat road. It hits a bump, which produces an initial displacement of 0.2 m and an initial velocity of 0.1 m/s at the base of the system. If the mass m of the vehicle is 5000 kg, the stiffness k is 2800 kN/m, and the damping coefficient c is 18 kN·s/m, determine the displacement response of this system and discuss when the system returns to its equilibrium position.

4.19 A shaft undergoing torsional vibrations is held fixed at one end, and it has attached at the other end a disk with a rotary inertia of 4 kg·m² about the rotational axis. The rotary inertia of the shaft is negligible compared to that of the disk and the torsional stiffness of the shaft is 8 N·m/rad. It is proposed to place the disk in an oil housing so that it provides an equivalent torsional viscous damping c_t. What should the value of c_t be so that the system is critically damped?

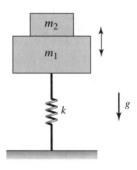

FIGURE E4.20

4.20 In the system shown in Figure E4.20, mass m_2, is resting on mass m_1, which is supported by a spring of stiffness k. The mass m_1 is 1 kg, the mass m_2 is 0.5 kg, and the stiffness k is 1 kN/m. If the spring is pushed down 15 mm below the system's static equilibrium position and released, determine if the mass m_2 will ever lose contact during the subsequent motion.

4.21 Compare the free oscillation characteristics of the following systems, when they are set into motion from the initial displacement of 0.1 m and zero initial velocity.

a) $m\ddot{x} + c\dot{x} + kx = 0$
b) $m\ddot{x} + c\dot{x} + kx + \alpha x^3 = 0$
c) $m\ddot{x} + c\dot{x} + kx - \alpha x^3 = 0$

Assume that the parameter values are as follows: $m = 10$ kg, $c = 10$ N·s/m, $k = 10$ N/m, and $\alpha = 40$ N/m³. Consider another set of free responses, when each of these three systems is set into motion from an initial displacement of 0.6 m. Compare these responses with the previously obtained responses. What can you conclude?

4.22 Replace the Kelvin-Voigt model used in Example 4.6 with a Maxwell model and determine the governing equation of motion when the two viscoelastic bodies are in contact with each other.

4.23 Consider the system with Maxwell model treated in Example 4.7 and treat the case where the spring with stiffness k is absent. Determine the force transmitted to the fixed base and plot the time histories of the normalized force as shown in Figure 4.13 for $\zeta_m = 0.2$ and $\zeta_m = 1.3$.

4.24 For the system of Example 4.5, determine design guidelines that one can use to minimize the relative acceleration $\ddot{z}$; that is, the acceleration of the single degree-of-freedom system inside the container relative to the container. Present these guidelines in terms of the coefficient of restitution ϵ, the undamped natural frequency ω_n, and the damping of the packaging material. This type of design guideline is useful for packaging of electronic components.

4.25 Consider the system with piecewise linear springs given in Figure 4.24 and choose the damping factor ζ to be 0.1. Set the system in motion from the initial conditions $y(0) = 0$ units and the nondimensional initial velocity $dy(0)/d\tau = V_o/(\omega_n d) = 20$. Obtain the displacement time histories of the system, and plot them as shown in Figure 4.25 for the following values of the stiffness coefficient μ:

a) 0
b) 1
c) 20

In each case, tabulate the nondimensional period for each cycle of oscillation and discuss the results.

4.26 Assume that the hinged door of Figure 4.2 is 0.8 m wide and has a mass of 15 kg. It is found that it takes a force of 15 N to keep the door opened at an angle of 90°. When the force is released, the door takes 2.1 s to reach its closed position. What is the damping ratio of the system?

4.27 Consider the mercury-filled (ρ = 13.6 × 10^3 kg/m^3) u-tube manometer shown in Figure 2.15. The total length of the mercury in the manometer is 0.7 m. When the mercury is displaced from its equilibrium position, damped oscillations are observed. The oscillations are such that the peak amplitude nine periods away from the peak amplitude of the first cycle has decreased by a factor of eight. What is the value of the viscous damping coefficient and the damped frequency of oscillation?

Forced periodic vibrations can occur from rotating systems, such as the gears in the gear reduction unit. The transmission of these vibrations to their surroundings can be reduced by various vibration isolation and reduction techniques. These techniques are used to design an air glove to reduce jackhammer vibrations to the hand and soft elastic mounts to reduce rotating machinery vibrations to the foundation. (*Source:* Gear reducer image courtesy of Polman/Bandung Polytechnic for Manufacturing; jackhammer image courtesy of PhotoDisc/Getty Images.)

5

Single Degree-of-Freedom Systems Subjected to Periodic Excitations

5.1 INTRODUCTION

In Chapter 3, the governing equation of a single degree-of-freedom system was derived. From this equation, the solution for the system subjected to forcing and initial

171

conditions was determined in Chapter 4. In the absence of forcing, the responses of a vibratory system subjected to different types of initial conditions were studied. In this chapter, we address those situations in which a physical system is subjected to an external force $f(t)$ and the initial displacement and the initial velocity are zero. In particular, we consider the response to harmonic and other periodic excitations. The notion of frequency-response function is introduced and related to the notion of transfer function, which is used for system design.

In Figure 5.1, a vibratory system subjected to a forcing $f(t)$ is illustrated along with a conceptual illustration of the system's input-output relationship. In this chapter, we determine this relationship in the time domain and in the transformed domain for periodic excitations. In Chapter 6, we determine this relationship in the time domain and in transformed domains for arbitrary excitations.

Response of vibratory systems in the presence of a rotating unbalanced mass is also studied in this chapter. Physical systems subjected to base motions are analyzed, and in this context, the device called an accelerometer, which is used for acceleration measurements, is introduced. Vibration isolation is examined at length. Energy dissipation for different damping models is discussed and the notion of equivalent viscous damping is introduced. The influence of stiffness nonlinearities on the forced response is also treated.

Referring to Figure 5.1, we consider the case where $f(t)$ varies periodically, and in Chapter 6, we consider the case where $f(t)$ is an arbitrary time-varying force. The governing equation of motion is of the form [recall Eq. (3.22)]

$$\frac{d^2x}{dt^2} + 2\zeta\omega_n \frac{dx}{dt} + \omega_n^2 x = \frac{f(t)}{m} \tag{5.1a}$$

The initial conditions are taken to be zero; that is,

$$x(0) = 0 \quad \text{and} \quad \dot{x}(0) = 0 \tag{5.1b}$$

Noting that the initial conditions are zero for this system, the general solution given by Eqs. (4.8) for $0 \le \zeta < 1$ is reduced to

$$x(t) = \frac{1}{m\omega_d} \int_0^t e^{-\zeta\omega_n\eta} \sin(\omega_d\eta) f(t - \eta) d\eta$$

$$= \frac{1}{m\omega_d} \int_0^t e^{-\zeta\omega_n(t - \eta)} \sin(\omega_d[t - \eta]) f(\eta) d\eta \tag{5.2}$$

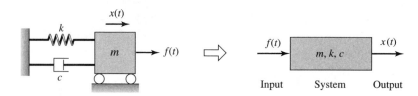

FIGURE 5.1
Vibratory system subjected to forcing and conceptual illustration of system.

where η is the variable of integration[1] and

$$\omega_d = \omega_n\sqrt{1 - \zeta^2}$$

Next, the response of linear vibratory systems subjected to a harmonic excitation is considered. First, the response is studied when the excitation is initiated at time $t = 0$, and then the response is studied when the excitation is present for all time.

5.2 RESPONSE TO HARMONIC EXCITATION

5.2.1 Excitation Applied from $t = 0$

In this section, responses to sine harmonic and cosine harmonic excitation are considered. It will be shown that although the initial conditions are zero, the fact that the excitation is suddenly applied at $t = 0$ results in a response that consists of a transient portion and a steady-state portion. These transients are typical of situations where a motor is started or where an excitation is intermittently turned on and off. In the absence of damping, the response of a vibratory system cannot be characterized as having a transient portion and a steady-state portion.

Case 1: Sine Harmonic Excitations

We first consider the periodic forcing function

$$f(t) = F_o \sin(\omega t)u(t) \tag{5.3}$$

where $u(t)$ is the unit step function[2]

$$\begin{aligned} u(t) &= 0 & t &< 0 \\ u(t) &= 1 & t &\geq 0 \end{aligned} \tag{5.4}$$

When Eq. (5.3) is substituted into Eq. (5.2), the result is

$$x(t) = \frac{F_o e^{-\zeta\omega_n t}}{m\omega_d} \int_0^t e^{\zeta\omega_n \eta} \sin(\omega_d[t - \eta])\sin(\omega\eta)d\eta \tag{5.5}$$

---

[1]In writing Eqs. (5.2), the following property of convolution integrals was used:

$$\int_0^t h(\eta)f(t - \eta)d\eta = \int_0^t h(t - \eta)f(\eta)d\eta$$

[2]The unit step function is used as a compact form that simplifies the way that we can express a function like Eq. (5.3), which is non-zero only in a specific interval. Without using the unit step function, Eq. (5.3) would have been written as

$$\begin{aligned} F(t) &= \sin(\omega t) & t &\geq 0 \\ &= 0 & t &< 0 \end{aligned}$$

If, further, we introduce the dimensionless time $\tau = \omega_n t$, then Eq. (5.5) becomes

$$x(\tau) = \frac{F_o e^{-\zeta \tau}}{k\sqrt{1 - \zeta^2}} \int_0^\tau e^{\zeta \xi} \sin(\sqrt{1 - \zeta^2}[\tau - \xi]) \sin(\Omega \xi) d\xi \tag{5.6}$$

where the nondimensional excitation frequency $\Omega = \omega/\omega_n$, the nondimensional time variable of integration $\xi = \omega_n \eta$, and we have used the definition of ω_d given by the second of Eqs. (4.7). Note that when the excitation frequency is at the natural frequency—that is, $\omega = \omega_n$—$\Omega = 1$.

Solution for forced response   After performing the integration,[3] Eq. (5.6) leads to

$$x(\tau) = [x_{strans}(\tau) + x_{sss}(\tau)]u(\tau) \tag{5.7}$$

where the *steady-state portion* of the response is given by

$$x_{sss}(\tau) = \frac{F_o}{k} H(\Omega) \sin(\Omega \tau - \theta(\Omega))$$

$$D(\Omega) = (1 - \Omega^2)^2 + (2\zeta\Omega)^2$$

$$H(\Omega) = \frac{1}{\sqrt{D(\Omega)}} = \frac{1}{\sqrt{(1 - \Omega^2)^2 + (2\zeta\Omega)^2}}$$

$$\theta(\Omega) = \tan^{-1} \frac{2\zeta\Omega}{1 - \Omega^2} \tag{5.8a}$$

and the *transient portion* of the response is given by

$$x_{strans}(\tau) = \frac{F_o}{k} \frac{H(\Omega)\Omega e^{-\zeta \tau}}{\sqrt{1 - \zeta^2}} \sin(\tau\sqrt{1 - \zeta^2} + \theta_t(\Omega))$$

$$\theta_t(\Omega) = \tan^{-1} \frac{2\zeta\sqrt{1 - \zeta^2}}{2\zeta^2 - (1 - \Omega^2)} \tag{5.8b}$$

After a long period of time, which here means after many cycles of forcing, the response reduces to

$$\lim_{\tau \to \infty} x(\tau) = x_{ss}(\tau) = x_{sss}(\tau) = \frac{F_o}{k} H(\Omega) \sin(\Omega \tau - \theta(\Omega)) \tag{5.9}$$

In Eqs. (5.8a), the quantity $H(\Omega)$ is called the *amplitude response* and the quantity $\theta(\Omega)$ is called the *phase response,* which provides the phase relative to the forcing $f(t)$. We see that the steady-state portion varies periodically at the nondimensional frequency Ω, the frequency of the applied force $f(t)$, and with an amplitude $F_o H(\Omega)/k$. In addition, the displacement response is delayed by an amount $\theta(\Omega)$ with respect to the input. The amplitude response $H(\Omega)$ and phase response $\theta(\Omega)$ are plotted in Figure 5.2 for several values of ζ. We shall discuss the significance of these quantities in Section 5.3.

---

[3]The MATLAB function int from the Symbolic Toolbox was used.

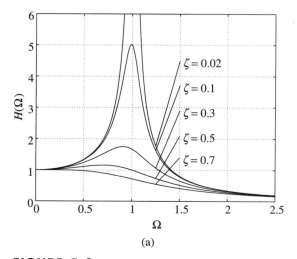

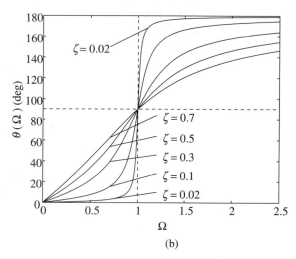

FIGURE 5.2
Harmonic excitation applied directly to the mass of the system: (a) amplitude response and (b) phase response.

The transient response $x_{strans}(\tau)$ varies periodically with a frequency ω_d/ω_n and its amplitude decreases exponentially with time as a function of the damping ratio ζ. In addition, the response is shifted by an amount $\theta_t(\Omega)$ with respect to the input.

Duration of transient response   For practical purposes, we define the time duration τ_d beyond which the system can be considered as having reached steady state; that is,

$$x_{ss}(\tau) \approx x(\tau) \quad \text{for} \quad \tau > \tau_d$$

To obtain an estimate of this duration, let the envelope of the transient decay to a value d at a nondimensional time τ_d, which is given by the relation

$$\frac{\Omega e^{-\zeta \tau_d}}{\sqrt{1 - \zeta^2}} = d \tag{5.10a}$$

If $|d| \ll 1$, that is, when the amplitude of the transient portion of the displacement is much less than the amplitude of the steady-state portion of the displacement, the transient is said to have "died out" and only the steady-state portion of the response remains. Solving for the nondimensional time τ_d, we obtain

$$\tau_d \geq -\frac{1}{\zeta} \ln \left[\frac{d\sqrt{1 - \zeta^2}}{\Omega} \right] \tag{5.10b}$$

We can also express the nondimensional time τ_d, which corresponds to the dimensional time t_d, in terms of the number of periods N_d at the excitation frequency ω that it takes for the normalized displacement

↓ζ/Ω→	0.01	1	2
0.05	0.51	12.5	29.3
0.3	0.09	2.1	4.9
0.7	0.04	0.97	2.3

TABLE 5.1 Some Values of N_d for $d = 0.02$

$x(\tau)/(F_oH(\Omega)/k)$ to decay to d. Since the period of the excitation $t_\omega = 2\pi/\omega$, then the period $t_d = N_d t_\omega = 2\pi N_d/\omega$, and

$$\tau_d = \omega_n t_d = \frac{\omega_n 2\pi N_d}{\omega} = \frac{2\pi N_d}{\Omega} \tag{5.10c}$$

Therefore, Eqs. (5.10) lead to

$$N_d \geq -\frac{\Omega}{2\pi\zeta} \ln\left[\frac{d\sqrt{1-\zeta^2}}{\Omega}\right] \qquad \zeta < 1, \quad |d| << 1 \tag{5.11}$$

Some typical values of N_d obtained from Eq. (5.11) are provided in Table 5.1. It is interesting to note from this table that for a given damping factor ζ, the number of cycles that the transient lasts increases with the excitation frequency. As expected, for a given excitation frequency, the duration of the transient portion of the response decreases for an increase in the damping factor.

Representative system responses for sine harmonic excitation The three sets of graphs shown in Figure 5.3 give the normalized displacement response $x(\tau)/(F_o/k)$ defined by Eq. (5.7) for three values of Ω, and at each value of Ω, the response is obtained for three different values of ζ. The first set shown in Figure 5.3a corresponds to an excitation frequency that is less than the natural frequency: $\Omega < 1$. The second set, which is shown in Figure 5.3b, corresponds to an excitation frequency that is equal to the natural frequency: $\Omega = 1$. The third set, which is shown in Figure 5.3c, corresponds to an excitation frequency that is greater than the natural frequency: $\Omega > 1$. For each of these nine combinations of values, the values of $H(\Omega)$ and N_d are given. As τ increases, the transient portion dies out, and the amplitude of the displacement response approaches the magnitude $H(\Omega)$ (the steady-state value). The response magnitude is within 2% ($d = 0.02$) or less of this steady-state value after N_d periods or, equivalently, when $\tau \geq \tau_d$. It is noted that when $\Omega < 1$ or $\Omega > 1$, the system response decays to within $\pm d$ of its steady-state value. When $\Omega = 1$, the displacement response increases until it reaches within $\pm d$ of its steady-state value. Furthermore, during the portion of the response where the transient portion is pronounced, the response is not periodic. However, when τ_d has elapsed, each of the responses approaches periodicity with the period determined by the excitation frequency; that is, $t_\omega = 2\pi/\omega$.

Case 2: Cosine Harmonic Excitations

For completeness, consider the periodic forcing function

$$f(t) = F_o \cos(\omega t)u(t) \tag{5.12}$$

After substituting Eq. (5.12) into Eq. (5.2), the result is

$$x(\tau) = \frac{F_o e^{-\zeta\tau}}{k\sqrt{1-\zeta^2}} \int_0^\tau e^{\zeta\xi} \sin(\sqrt{1-\zeta^2}[\tau-\xi])\cos(\Omega\xi)d\xi \qquad (5.13)$$

where the nondimensional frequency Ω and the variable of integration ξ are as defined for Case 1.

Solution for forced response Performing the integration[4] in Eq. (5.13), we arrive at the displacement response

$$x(\tau) = [x_{css}(\tau) + x_{ctrans}(\tau)]u(\tau) \qquad (5.14)$$

where the steady-state portion of the response is given by

$$x_{css}(\tau) = \frac{F_o}{k} H(\Omega)\cos(\Omega\tau - \theta(\Omega)) \qquad (5.15a)$$

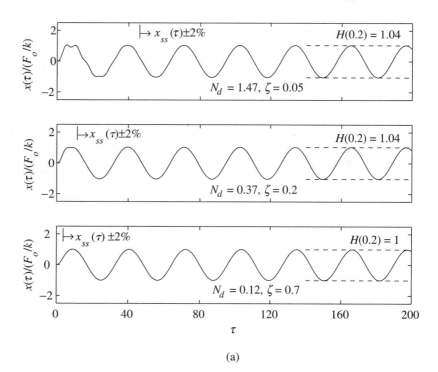

(a)

FIGURE 5.3
Normalized response of a system to a suddenly applied sine wave forcing function when the transient envelope parameter $d = 0.02$ and for different values of ζ: (a) $\Omega = 0.2$.

[4]The MATLAB function `int` from the Symbolic Toolbox was used.

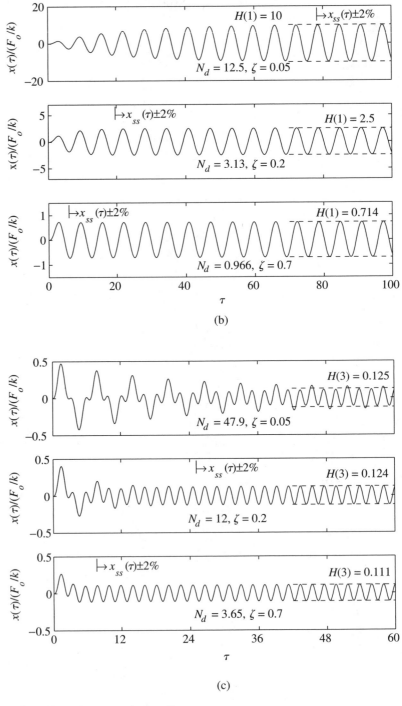

FIGURE 5.3 (*continued*)
(b) $\Omega = 1.0$ and (c) $\Omega = 3.0$.

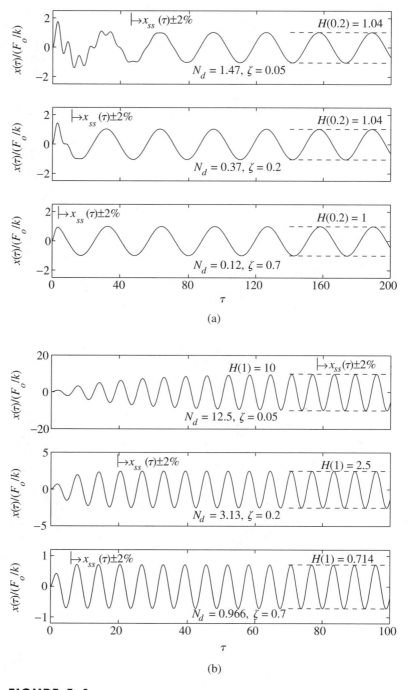

FIGURE 5.4
Response of a system to a suddenly applied cosine wave forcing function when the transient envelope parameter $d = 0.02$ and for different values of ζ: (a) $\Omega = 0.2$ and (b) $\Omega = 1.0$.

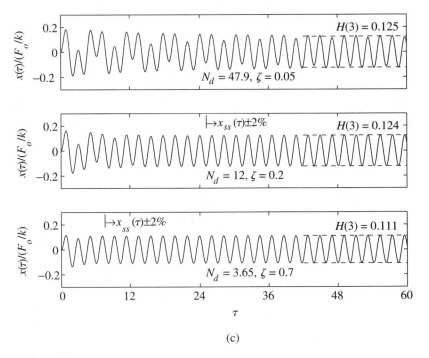

FIGURE 5.4 (continued)
(c) $\Omega = 3.0$.

and the transient portion of the response is given by

$$x_{ctrans}(\tau) = \frac{F_o}{k} \frac{H(\Omega)e^{-\zeta\tau}}{\sqrt{1 - \zeta^2}} \cos(\tau\sqrt{1 - \zeta^2} - \theta_{ct}(\Omega))$$

$$\theta_{ct}(\Omega) = \tan^{-1} \frac{-\zeta(1 + \Omega^2)}{(\Omega^2 - 1)\sqrt{1 - \zeta^2}} \qquad (5.15b)$$

In Eqs. (5.14) and (5.15), the amplitude response $H(\Omega)$ and the phase $\theta(\Omega)$ are given by Eqs. (5.8a), and it is noted that the proper quadrant must be considered for determining $\theta_{ct}(\Omega)$. Again, as in the case with the sine harmonic excitation, after a long time (many cycles of forcing) the response settles down to the steady-state form; that is,

$$\lim_{\tau \to \infty} x(\tau) = x_{ss}(\tau) = x_{css}(\tau) = \frac{F_o}{k} H(\Omega)\cos(\Omega\tau - \theta(\Omega))$$

The displacement response given by Eq. (5.14) is plotted in Figure 5.4 for three values of Ω, and at each value of Ω, the response is determined for three different values of ζ. For each of these nine combinations of values, the values of $H(\Omega)$ and N_d are given. As in the case of the sine harmonic excitation, the transient is initially aperiodic in each of the nine time histories, and then the response becomes periodic after the system settles down. Although the shapes of the transient responses are different than those obtained for the sine wave forcing function, the times it takes for the transients to die out are the same as in the corresponding cases.

EXAMPLE 5.1   Estimation of system damping ratio to tailor transient response

A single degree-of-freedom system with a natural frequency of 66.4 rad/s is intermittently cycled on and off. When it is on, it vibrates at 5.8 Hz. What should the damping ratio be in order for the system to decay to within 5% of its steady-state amplitude in 150 ms each time that the forcing is applied?

Assuming that the system settles down to the rest state in between the forcing cycles, from Eq. (5.10b), which is applicable when the forcing is turned on, we have that

$$\tau_d = \omega_n t_d = -\frac{1}{\zeta} \ln \left[\frac{d\sqrt{1 - \zeta^2}}{\Omega} \right]$$

$$(66.4)(0.15) = -\frac{1}{\zeta} \ln \left[\frac{0.05\sqrt{1 - \zeta^2}}{(2\pi \times 5.8)/66.4} \right]$$

$$9.96 = -\frac{1}{\zeta} \ln \left[0.0911\sqrt{1 - \zeta^2} \right]$$

Solving numerically[5], we obtain $\zeta = 0.244$.

5.2.2   Excitation Present for All Time

In the previous section, it was shown that for a harmonic periodic excitation initiated at time $t = 0$, the response of the vibratory system consists of a transient part and a steady-state part. After a nondimensional time τ_d, only the steady-state part of the response remains. This observation is taken advantage of to characterize linear systems in terms of frequency-response functions and transfer functions. Once a frequency-response function is determined for a linear vibratory system from a harmonic forcing, this frequency-response function can be used to determine the response of a linear vibratory system for any combination of harmonic inputs. In order to proceed in this direction, first, the previously determined results for the steady-state portion of the response found in Section 5.2.1 are revisited.

When the periodic forcing is given by

$$f(t) = F_o \sin(\omega t) \tag{5.16a}$$

or equivalently, in terms of the nondimensional time variable τ, as

$$f(\tau) = F_o \sin(\Omega \tau) \tag{5.16b}$$

---

[5]The MATLAB function `fzero` was used.

that is, the harmonic excitation is present for all time, the associated steady-state portion of the response is given by Eqs. (5.8) and (5.9). Thus,

$$x_{ss}(\tau) = \underbrace{\frac{F_o}{k} H(\Omega)}_{\text{Steady-state amplitude}} \sin(\underbrace{\Omega\tau - \theta(\Omega)}_{\text{Steady-state phase}}) \tag{5.17}$$

where

$$H(\Omega) = \frac{1}{\sqrt{D(\Omega)}} = \frac{1}{\sqrt{(1 - \Omega^2)^2 + (2\zeta\Omega)^2}}$$

$$\theta(\Omega) = \tan^{-1}\frac{2\zeta\Omega}{1 - \Omega^2} \tag{5.18}$$

The steady-state velocity and steady-state acceleration are, respectively, given by

$$v_{ss}(\tau) = \frac{dx(\tau)}{d\tau} = \frac{F_o\Omega}{k} H(\Omega)\cos(\Omega\tau - \theta(\Omega))$$

$$= \frac{F_o\Omega}{k} H(\Omega)\sin(\Omega\tau - \theta(\Omega) + \pi/2)$$

$$a_{ss}(\tau) = \frac{d^2x(\tau)}{d\tau^2} = \frac{F_o\Omega^2}{k} H(\Omega)\sin(\Omega\tau - \theta(\Omega)) = -\Omega^2 x(\tau) \tag{5.19}$$

We see that for harmonic oscillations, the magnitude of the acceleration is equal to the square of the excitation frequency times the displacement magnitude and the acceleration response is 180° out of phase with the displacement response. The magnitude of the velocity is equal to the excitation frequency times the magnitude of the displacement and the velocity response is 90° out of phase with the displacement response.

EXAMPLE 5.2 Forced response of a damped system

Consider the electric motor shown in Figure 5.5a. The output of the motor is connected to two shafts whose opposite ends are fixed. The motor provides a harmonic drive torque directed along the direction of the unit vector k. This

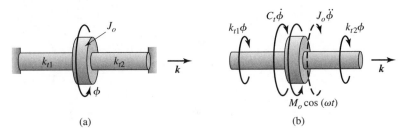

(a)   (b)

FIGURE 5.5
(a) Electric motor driven restrained by two shafts and (b) free-body diagram.

torque has a magnitude $M_o = 100$ N·m and the driving frequency ω is 475 rad/s. The rotary inertia of the electric motor J_o is 0.020 kg·m², the torsional stiffness of the shafts are $k_{t1} = 2500$ N·m/rad and $k_{t2} = 3000$ N·m/rad, and the overall damping experienced by the rotor can be quantified in terms of a torsional damper with the damping coefficient $c_t = 1.25$ N·m·s/rad. We shall determine the form of the steady-state response and the amplitude and the phase of the steady-state response.

The governing equation of the motor is derived based on the principle of angular momentum balance. Consider the free-body diagram shown in Figure 5.5b, which includes the inertial moment $-J_o\ddot{\phi}\mathbf{k}$. The principle of angular momentum applied to the center of the motor leads to the following governing equation

$$J_o\ddot{\phi} + c_t\dot{\phi} + (k_{t1} + k_{t2})\phi = M_o \cos(\omega t) \tag{a}$$

Dividing Eq. (a) by the rotary inertia J_o, we obtain the following equation whose form is similar to Eq. (5.1)

$$\ddot{\phi} + 2\zeta\omega_n\dot{\phi} + \omega_n^2\phi = \frac{M_o}{J_o} \cos(\omega t) \tag{b}$$

where the system natural frequency and damping factor are given by, respectively,

$$\omega_n = \sqrt{\frac{k_{t1} + k_{t2}}{J_o}}$$

$$\zeta = \frac{c_t}{2J_o\omega_n} \tag{c}$$

Based on Eq. (5.17), the solution of Eq. (b) for steady-state motion is given by

$$\phi_{ss}(\tau) = \frac{M_o}{k_{t1} + k_{t2}} H(\Omega)\cos(\Omega\tau - \theta(\Omega)) \tag{d}$$

where the nondimensional excitation frequency $\Omega = \omega/\omega_n$ and the nondimensional time $\tau = \omega_n t$. For the given parameter values, the calculations lead to the following:

$$k_{t1} + k_{t2} = 5500 \quad \text{N·m/rad}$$

$$\omega_n = \sqrt{\frac{5500 \text{ N·m/rad}}{0.020 \text{ kg·m}^2}} = 524.40 \text{ rad/s}$$

$$\zeta = \frac{1.25 \text{ N·m·s/rad}}{2 \times 0.020 \text{ kg·m}^2 \times 524.40 \text{ rad/s}} = 0.06$$

$$\Omega = \frac{\omega}{\omega_n} = \frac{475 \text{ rad/s}}{524.40 \text{ rad/s}} = 0.91$$

$$H(\Omega) = \frac{1}{\sqrt{(1 - 0.91^2)^2 + (2 \times 0.06 \times 0.91)^2}} = 4.91$$

$$\theta(\Omega) = \tan^{-1}\left(\frac{2 \times 0.06 \times 0.91}{1 - 0.91^2}\right) = 0.57 \text{ rad} \tag{e}$$

Hence, from the values provided in Eqs. (e), the steady-state response of the electric motor given by Eq. (d) is written as

$$\phi_{ss}(t) = \frac{100 \text{ N·m}}{5500 \text{ N·m/rad}} \times 4.91 \times \cos(475t - 0.57)$$

$$= 0.09 \cos(475t - 0.57) \text{ rad} \qquad (f)$$

Thus, the amplitude of the harmonic steady-state motion is 0.09 rad at a frequency of 475 rad/s, and the phase relative to the excitation is 0.57 rad.

5.2.3 Response of Undamped System and Resonance

When a linear vibratory system is undamped, $\zeta = 0$, the governing equation given by Eq. (5.1), reduces to

$$\frac{d^2x}{dt^2} + \omega_n^2 x = \frac{f(t)}{m} \qquad (5.20)$$

Response When $\omega \neq \omega_n$ ($\Omega \neq 1$)

For the excitation given by Eq. (5.3), and for zero initial conditions, the response is determined from Eqs. (5.7) and (5.8) with $\zeta = 0$. Thus, we obtain

$$x(\tau) = \frac{F_o}{k(1 - \Omega^2)} \{\sin(\Omega\tau) - \Omega \sin(\tau)\} \qquad (5.21a)$$

or, equivalently,

$$x(t) = \frac{F_o}{k(1 - \Omega^2)} \{\sin(\omega t) - \Omega \sin(\omega_n t)\} \qquad (5.21b)$$

which is valid when the excitation frequency is different from the natural frequency; that is, when $\Omega \neq 1$ (or $\omega \neq \omega_n$). It is clear from the form of Eqs. (5.21) that for an undamped system excited by a sinusoidal forcing at a frequency that is not equal to the natural frequency, the response consists of a frequency component at the excitation frequency ω and a frequency component at the natural frequency ω_n. Unless the ratio of ω/ω_n is a rational number, the displacement response is not periodic.

Response When $\omega = \omega_n$ ($\Omega = 1$)

The response of an undamped system when $\Omega = 1$, is obtained from Eq. (5.6) with $\zeta = 0$ and $\Omega = 1$. Thus, we arrive at

$$x(\tau) = \frac{F_o}{k} \int_0^\tau \sin(\tau - \xi)\sin(\xi)d\xi \qquad (5.22)$$

Evaluating the integral in Eq. (5.22) results in

$$x(\tau) = \frac{F_o}{2k} \{\sin(\tau) - \tau \cos(\tau)\} \qquad (5.23)$$

The displacement response is not periodic, because the amplitude of the second term in Eq. (5.23) increases as time increases.

Resonance and Stability of Response

For the case where $\Omega \neq 1$, the response of the undamped system given by Eq. (5.21) always has a finite magnitude, since $\sin(\Omega\tau)$ and $\sin(\tau)$ have finite values for all time. Thus, it follows that

$$|x(\tau)| \leq A \qquad \tau > 0 \tag{5.24}$$

where A is a positive finite number. Hence, for $\Omega \neq 1$, an undamped system excited by a finite harmonic excitation is stable in the sense of boundedness introduced in Section 4.4. However, this is not true when $\Omega = 1$. When $\Omega = 1$, the term $\tau\cos(\tau)$ in Eq. (5.23) grows linearly in amplitude with time and, hence, it becomes unbounded after a long time. This special ratio $\Omega = 1$ ($\omega = \omega_n$) is called a *resonance relation;* that is, the linear system given by Eq. (5.20) is said to be in *resonance* when the excitation frequency is equal to the natural frequency.

From a practical standpoint, the question of boundedness is important. Since there is always some amount of damping in a system, it is clear from Eqs. (5.7) to (5.9) that the response remains bounded when excited at the natural frequency; that is, for $\Omega = 1$

$$\lim_{\tau \to \infty} x(\tau) = \frac{F_o}{2\zeta k} \sin(\tau - \pi/2) \tag{5.25}$$

The response given by Eq. (5.25) satisfies the boundedness condition given by Eq. (5.24), even though as the damping decreases in magnitude the response increases in magnitude.

EXAMPLE 5.3 Forced response of an undamped system

Consider translational motions of a vibratory system with a mass of 100 kg and a stiffness of 100 N/m. When a harmonic forcing of the form $F_o\sin(\omega t)$ acts on the mass of the system, where $F_o = 1.0$ N, we shall determine the responses of the system and plot them for the following cases: i) $\omega = 0.2$ rad/s, ii) $\omega = 1.0$ rad/s, and iii) $\omega = 2.0$ rad/s. The results are also discussed.

Let the variable x be used to describe the translation of the mass. Then, from Eq. (5.20),

$$\ddot{x} + \omega_n^2 x = \frac{F_o}{m} \sin(\omega t) \tag{a}$$

The natural frequency of the system is

$$\omega_n = \sqrt{\frac{k}{m}} = \sqrt{\frac{100 \text{ N/m}}{100 \text{ kg}}} = 1 \text{ rads} \tag{b}$$

Hence, for cases i) and iii), the excitation frequency is different from the natural frequency, and for case ii), the excitation frequency is equal to the natural frequency of the system.

In cases i) and iii), the solution for the displacement response is given by Eq. (5.21); that is,

$$x(t) = \frac{F_o}{k(1 - \Omega^2)} \{\sin(\omega t) - \Omega \sin(\omega_n t)\} \tag{c}$$

On the other hand, in case ii), the response is given by Eq. (5.23); that is,

$$x(t) = \frac{F_o}{2k} \{\sin(\omega_n t) - \omega_n t \cos(\omega_n t)\} \tag{d}$$

For the given values, the responses are as follows:

Case i

$$x(t) = \frac{1 \text{ N}}{100 \text{ N/m} \times (1 - (0.2/1.0)^2)} \{\sin(0.2t) - (0.2/1.0)\sin(1 \times t)\}$$
$$= 0.01\{\sin(0.2t) - 0.2 \sin t\} \quad \text{m} \tag{e}$$

Case ii

$$x(t) = \frac{1 \text{ N}}{2 \times 100 \text{ N/m}} \{\sin t - t \cos t\}$$
$$= 0.005\{\sin t - t \cos t\} \quad \text{m} \tag{f}$$

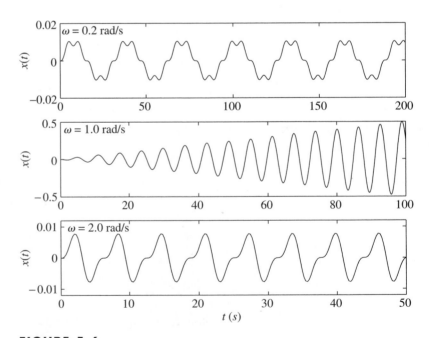

FIGURE 5.6
Displacement response of an undamped system subjected to harmonic forcing at three frequencies: i) $\omega = 0.2$ rad/s; ii) $\omega = 1.0$ rad/s; and iii) $\omega = 2.0$ rad/s.

Case iii

$$x(t) = \frac{1 \text{ N}}{100 \text{ N/m} \times (1 - (0.2/1.0)^2)} \{\sin(2t) - (0.2/1.0)\sin(t)\}$$
$$= -0.003\{\sin(2t) - 2 \sin t\} \quad \text{m} \tag{g}$$

The graphs of Eqs. (e), (f), and (g) are provided in the Figure 5.6. From the graphs, it is clear that the response of the undamped system remains bounded when the excitation frequency is away from the natural frequency, as in cases i) and iii). However, in case ii), the response becomes unbounded after a long period of time, since the amplitude increases with time.

5.2.4 Magnitude and Phase Information

In the case of an undamped linear vibratory system, there is a phase shift of 180° in the response as one goes from an excitation frequency that is less than the natural frequency to an excitation frequency that is greater than the natural frequency. This can be discerned from Eq. (5.21) by noting that the change in sign is brought about by the term $1 - \Omega^2$.

For a linear damped vibratory system excited at the resonance frequency $\Omega = 1$, the response lags the excitation by 90° as shown by Eq. (5.25) and in Figure 5.2b. This observation is used in experiments to determine if the excitation frequency is equal to the undamped natural frequency of the system; that is, $\omega = \omega_n$. The magnitude of the response is also large at $\omega = \omega_n$ ($\Omega = 1$) when ζ is small.

Response Characteristics in Different Excitation Frequency Ranges

Additional characteristics of the response of an underdamped, linear vibratory system can be determined by examining the steady-state response $x_{ss}(\tau)$ in the frequency ranges $\Omega \ll 1$ and $\Omega \gg 1$ and at the frequency location $\Omega = 1$; that is, in a region considerably below the natural frequency, in a region well above the natural frequency, and at the natural frequency, respectively. The examination is performed by studying the values of $H(\Omega)$ and $\theta(\Omega)$ in these three ranges.

$\Omega \ll 1$ In this region, Eqs. (5.18) lead to

$$H(\Omega) \to 1$$
$$\theta(\Omega) \to 0 \tag{5.26}$$

and, therefore, from Eq. (5.17), we obtain

$$x_{ss}(\tau) \cong \frac{F_o}{k} \sin(\Omega\tau) = \frac{1}{k} f(\tau) \tag{5.27}$$

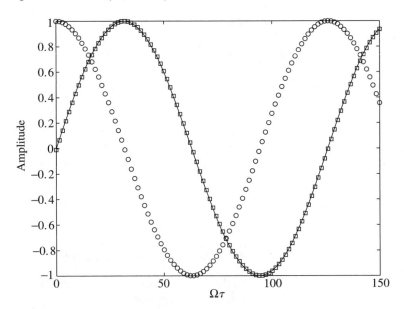

FIGURE 5.7
Phase relationships among displacement, velocity, and force in the stiffness-dominated region: $\Omega \ll 1$. [— $f(\tau)/F_o$; □ $x(\tau)/(F_o/k)$; ○ $v(\tau)/(\Omega F_o/k)$.]

Since the only system parameter that determines the displacement response is the stiffness k, we call this region the *stiffness-dominated region*. In this region, the displacement is in phase with the force.

The velocity response is determined from Eq. (5.27) to be

$$v_{ss}(\tau) = \frac{\Omega F_o}{k} \cos(\Omega\tau) = \frac{\Omega F_o}{k} \sin(\Omega\tau + \pi/2) \tag{5.28}$$

Therefore, as expected, the velocity response leads the displacement response by $\pi/2$; that is, the maximum value of the velocity occurs before the maximum value of the displacement. These results are illustrated in Figure 5.7.

$\Omega = 1$ At this value of Ω, Eqs. (5.18) simplify to

$$H(1) = \frac{1}{2\zeta}$$

$$\theta(1) = \frac{\pi}{2} \tag{5.29}$$

Therefore, the displacement response given by Eq. (5.17) takes the form

$$x_{ss}(\tau) = \frac{F_o}{2k\zeta} \sin(\tau - \pi/2) = \frac{F_o}{c\omega_n} \sin(\tau - \pi/2) \tag{5.30}$$

Thus, for a given excitation amplitude F_o and natural frequency ω_n, the amplitude of the displacement is determined by the damping coefficient c. This region, therefore, is called the *damping-dominated region*. We also see that the displacement lags the force by $\pi/2$ and that this phase lag in independent of ζ. For $\zeta > 0$, it is mentioned that $H(\Omega)$ is not a maximum when $\Omega = 1$. We shall determine this maximum value and the value of Ω at which it occurs subsequently. The amplitude response is characterized by a peak close to the natural frequency for values of ζ in the range $0 < \zeta < 0.3$, as can be seen from Figure 5.2a.

The velocity response is obtained from Eq. (5.30) to be

$$v_{ss}(\tau) = \frac{F_o}{2k\zeta} \cos(\tau - \pi/2) = \frac{F_o}{c\omega_n} \sin(\tau) = \frac{f(\tau)}{c\omega_n} \tag{5.31}$$

Therefore, in this region, the velocity is in phase with the force. These results are illustrated in Figure 5.8.

$\Omega \gg 1$ In this region, the amplitude and phase responses given by Eqs. (5.18) simplify to

$$H(\Omega) \rightarrow \frac{1}{\Omega^2} = \frac{\omega_n^2}{\omega^2}$$

$$\theta(\Omega) \rightarrow \pi \tag{5.32}$$

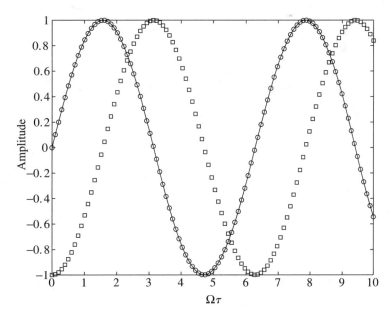

FIGURE 5.8
Phase relationships among displacement, velocity, and force in the damping-dominated region: $\Omega = 1$. [$-$ $f(\tau)/F_o$; $\square$ $x(\tau)/(F_o/(2\zeta k))$; $\bigcirc$ $v(\tau)/(F_o/(2\zeta k))$.]

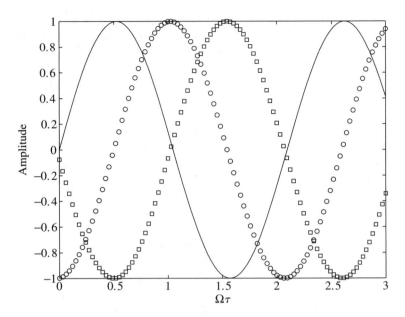

FIGURE 5.9
Phase relationships among displacement, velocity, and force in the inertia-dominated region: $\Omega \gg 1$. [$- f(\tau)/F_o$; $\square\ x(\tau)/(F_o/(k\Omega^2))$; $\bigcirc\ v(\tau)/(F_o/(k\Omega))$.]

and, therefore, the displacement response given by Eq. (5.17) simplifies to

$$x_{ss}(\tau) \cong \frac{F_o \omega_n^2}{k\omega^2} \sin(\Omega\tau - \pi) = -\frac{F_o}{m\omega^2} \sin(\Omega\tau) = -\frac{f(\tau)}{m\omega^2} \qquad (5.33)$$

Thus, the displacement response is 180° out of phase with the applied force. For a given $f(\tau)$, the amplitude of the displacement is inversely proportional to m, the mass. This region, therefore, is called the *inertia-dominated region*.

The velocity is determined from Eq. (5.33) to be

$$v_{ss}(\tau) = \frac{F_o \Omega \omega_n^2}{k\omega^2} \cos(\Omega\tau - \pi) = \frac{\Omega F_o}{m\omega^2} \sin(\Omega\tau - \pi/2) \qquad (5.34)$$

Hence, in this region, the velocity lags the force by $\pi/2$. These results are shown in Figure 5.9.

It can be shown that when $\zeta > 1/\sqrt{2}$, $H(\Omega) < 1$ for $\Omega > 0$, and when $\zeta < 1/\sqrt{2}$, $H(\Omega) < 1$ for

$$\Omega \geq \sqrt{2}\sqrt{1 - 2\zeta^2}$$

or

$$\omega \geq \omega_n \sqrt{2}\sqrt{1 - 2\zeta^2}$$

These response characteristics can also be seen in Figure 5.2.

The dependence of the system response on the different system parameters is summarized as a function of the excitation frequency as shown in Figure 5.10. Further discussion about the response magnitude and the response phase as a function of the excitation frequency is provided in Section 5.3.

Observations

Stiffness-dominated region: When the amplitude of the harmonic exciting force is constant and the excitation frequency is much less than the natural frequency of the system, the magnitude of the displacement is determined by the system's stiffness. The displacement response is in phase with the excitation force.

Damping-dominated region: When the amplitude of the harmonic exciting force is constant and the excitation frequency equals the natural frequency of the system, the magnitude of the displacement response is magnified for $0 < \zeta << 1$, and the amount of magnification is determined by the system's damping coefficient. The displacement response lags the excitation force by 90°.

Inertia-dominated region: When the amplitude of the harmonic exciting force is constant and the excitation frequency is much greater than the natural frequency of the system, the magnitude of the displacement response is determined by the system's inertia. When Ω is greater than $1/\sqrt{2}$, the magnitude of the amplitude response is always less than 1. The displacement response is almost 180° out of phase with the excitation force.

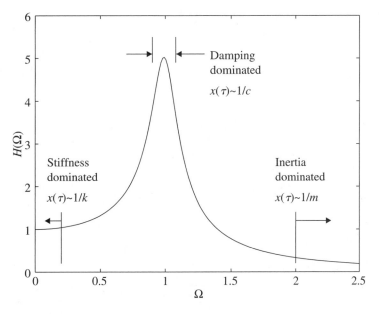

FIGURE 5.10
Three regions of single degree-of-freedom system's amplitude response.

5.3   FREQUENCY-RESPONSE FUNCTION

5.3.1   Introduction

In Section 5.2, the responses of undamped and underdamped systems to harmonic excitations were discussed and the notions of resonance of a linear vibratory system, steady-state motion, and frequency-response function were introduced. In this section, the concept of a frequency-response function is further treated.

As illustrated in Figure 5.11, for a linear vibratory system, a frequency-response function provides a relationship between a system's input and a system's output. The frequency-response function $G(\Omega)$ is a complex-valued function of frequency, and this function provides information about the magnitude and phase of the steady-state response of a linear vibratory system as a function of the excitation frequency.

For the cases treated in Section 5.2, one can define the frequency-response function $G(\Omega)$ in terms of the force input and displacement output of the linear vibratory system. In its general representation, this function has the form

$$G(\Omega) = \frac{1}{k} H(\Omega)e^{-j\theta(\Omega)} \qquad (5.35)$$

where $\Omega = \omega/\omega_n$ is the ratio of the excitation frequency ω to the natural frequency of the system ω_n, k is the stiffness of the vibratory system, the nondimensional function $H(\Omega)$ provides the magnitude, $j = \sqrt{-1}$, and $\theta(\Omega)$ is the phase lag associated with the response. In Eq. (5.35), the frequency-response function has dimensions of displacement/force; that is, m/N. This should not be surprising, since we have considered a force input and a displacement output. If the frequency-response function is desired in nondimensional form, it is defined as

$$\begin{aligned}
\hat{G}(\Omega) &= kG(\Omega) \\
&= H(\Omega)e^{-j\theta(\Omega)}
\end{aligned} \qquad (5.36)$$

where the complex-valued function $\hat{G}(\Omega)$ is now nondimensional.

For a linear vibratory system, where the forcing is applied to the mass directly as in Figure 5.1, the nondimensional function $H(\Omega)$ and the phase $\theta(\Omega)$ are given by Eqs. (5.8). These functions have different forms when the excitation is either applied to the base or the source of the excitation is a rotating unbalance, as discussed in Sections 5.4 and 5.5, respectively.

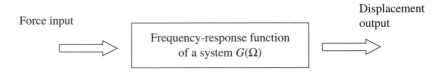

FIGURE 5.11
Conceptual illustration of the frequency-response function.

Based on Eqs. (5.8) and (5.35), the magnitude of the frequency-response function is given by

$$|G(\Omega)| = \frac{H(\Omega)}{k} = \frac{1}{k\sqrt{(1 - \Omega^2)^2 + (2\zeta\Omega)^2}} \qquad (5.37)$$

As evident from Eq. (5.37), this function contains information about the system parameters such as stiffness k, the system characteristics such as the natural frequency ω_n, and the damping factor ζ. In Section 5.3.2, we discuss how the frequency-response function magnitude can be curve-fit to extract information about the characteristics of a linear vibratory system. In Section 5.3.3, the sensitivity of the frequency-response function magnitude to different system parameters is examined, and finally, in Section 5.3.4, the relationship between a frequency-response function and a transfer function is explained.

5.3.2 Curve Fitting and Parameter Estimation

Let us suppose that one conducts a physical experiment where a sinusoidal forcing is applied to a system. After the motions of the system reach steady state, that is, after many cycles of forcing, the forcing and displacement are measured by using appropriate transducers. From these measured quantities, one can determine the ratio of the displacement magnitude to the force magnitude. If this experiment is repeated at different frequencies over a frequency range, and if at each excitation frequency this ratio is determined, the resulting information is plotted as a function of frequency as shown in Figure 5.12a where the open circles represent the experimental data. We assume that the physical system of interest is a linear vibratory system and that a model such as Eq. (5.1) is descriptive of the system. Then, based on Eq. (5.37), one can curve-fit the experimental data and estimate the values of the stiffness k, the natural frequency ω_n, and the damping factor ζ. For the data shown in Figure 5.12a, it is found[6] that $\zeta = 0.15$, $\omega_n = 1.5$ rad/s, and $k = 71.5$ N/m. These values were inserted in Eq. (5.37) to obtain the solid line passing through the data. In Figure 5.12b, the nondimensional function $H(\Omega) = k|G(\omega)|$ is shown along with the experimental data.

In Figure 5.13, the nondimensional function $H(\Omega)$ is plotted on a graph with a logarithmic scale for the frequency axis and a logarithmic scale for the magnitude axis. This type of graph helps highlight the location of a resonance and the overall response characteristics as a function of frequency over several decades of frequency. In such log-log plots, the magnitude is expressed in terms of decibels (dB), which is discussed in Appendix C.

As mentioned in Section 4.3.3, the frequency-response function provides an alternate means to the logarithmic-decrement method, which was based on free-oscillation data for estimating the damping factor in a vibratory system model. In addition, as discussed in Section 5.3.1, and later in

---

[6]The MATLAB function `lsqcurvefit` from the Optimization Toolbox was used.

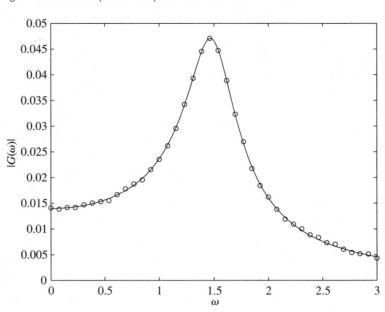

(a)

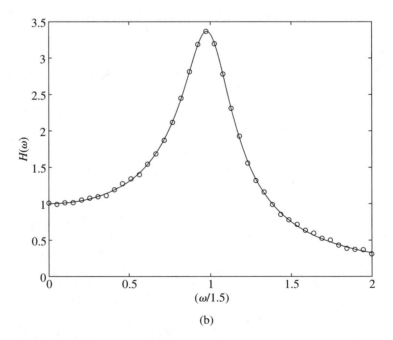

(b)

FIGURE 5.12
Model identification using Eq. (5.37): (a) experimental data along with identified model $|G(\omega)|$ and (b) experimental data along with identified model $H(\Omega) = k|G(\omega)|$.

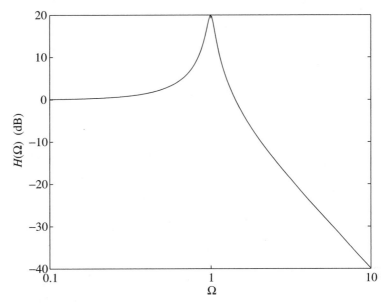

FIGURE 5.13
Logarithmic plot of $H(\Omega)$.

Chapter 6, the frequency-response function of a vibratory system can also be determined for force inputs other than sinusoidal excitations.

5.3.3 Sensitivity to System Parameters and Filter Characteristics

In this section, we illustrate how the sensitivity of a mechanical system's response to an applied forcing is determined and how a mechanical system can be viewed as a filter. The notions of sensitivity and filters are important for the design of many mechanical systems, in particular, sensors used for vibration measurements.

Sensitivity

The amplitude response of a single degree-of-freedom system can be used to determine the response sensitivity. The system's sensitivity is defined as the ratio of the change in the displacement to a change in the applied force. Thus, when $\Omega \ll 1$, we find from Eq. (5.27) that the sensitivity as determined by the response magnitude and force magnitude is

$$S = \left| \frac{\Delta x}{\Delta F} \right| = \frac{1}{k} \tag{5.38}$$

where Δx and ΔF represent the changes in the response and force magnitudes, respectively. Let us suppose that the parameters c and m are held constant and that we consider two systems, one with a spring constant k_1 and another with a spring constant k_2. Then, the sensitivity for the first system is

$$S_1 = \frac{1}{k_1} \tag{5.39}$$

Let

$$a_o = \frac{k_1}{k_2} > 0 \tag{5.40}$$

Furthermore, we have the following expressions for the corresponding natural frequencies, static displacements, and damping factors of the two systems:

$$\omega_{n1} = \sqrt{\frac{k_1}{m}} \qquad\qquad \delta_{st1} = \frac{mg}{k_1}$$

$$\omega_{n2} = \sqrt{\frac{k_2}{m}} = \frac{\omega_{n1}}{\sqrt{a_o}} \qquad \delta_{st2} = \frac{mg}{k_2} = a_o \delta_{st1}$$

$$\zeta_1 = \frac{c}{2\sqrt{k_1 m}} \qquad\qquad \zeta_2 = \frac{c}{2\sqrt{k_2 m}} = \zeta_1 \sqrt{a_o} \tag{5.41}$$

Then, from Eqs. (5.17) and (5.18), the sensitivity S_2 for the modified system is given by

$$S_2 = \left| \frac{\Delta x}{\Delta F} \right| = \frac{a_o}{k_1} \frac{1}{\sqrt{(1 - a_o \Omega^2)^2 + (2 a_o \zeta_1 \Omega)^2}} \tag{5.42a}$$

where the frequency ratio $\Omega = \omega/\omega_{n1}$. When $\Omega \ll 1$, Eq. (5.42a) is approximated as

$$S_2 = \frac{a_0}{k_1} = a_0 S_1 \tag{5.42b}$$

Equation (5.42a) is plotted in Figure 5.14 for $a_o > 1$. We see from this figure and from Eqs. (5.41) that as the sensitivity increases, the natural frequency of the system decreases (and the static displacement increases) and the damping ratio increases. This tradeoff between sensitivity and the location of the natural frequency is important in the design of vibration sensors, where a frequent desire is to have a device with a high sensitivity and a high natural frequency.[7] This discussion presents a broader view of the ideas presented in Example 3.7, where only the influence of the different system parameters on ζ was discussed.

Design Guideline: If c and m are held constant and the stiffness of a single degree-of-freedom system is changed from k_1 to k_2, then the response sensitivity of the system changes from S_1 to $k_1 S_1/k_2$ and the damping ratio changes from ζ_1 to $\zeta_1 \sqrt{k_1/k_2}$.

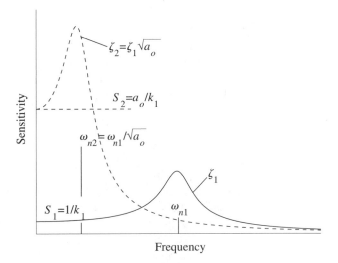

FIGURE 5.14
Effect of change in sensitivity with respect to the natural frequency and damping ratio when c and m remain constant: $a_o = k_1/k_2 > 0$.

EXAMPLE 5.4 Changes in system natural frequency and damping ratio for enhanced sensitivity

A system is to be redesigned so that the modified system's sensitivity is increased by a factor of three. We shall determine the percentage changes in the system natural frequency and the damping ratio.

Since $S_2/S_1 = a_o = 3$, we have that $k_2/k_1 = 1/3$; that is, the stiffness has decreased by a factor of three. Thus, from Eqs. (5.41) we have that $\omega_{n2} = \omega_{n1}/\sqrt{3}$ and $\zeta_2 = \zeta_1\sqrt{3}$. Then the percentage change in the natural frequency $\Delta\omega_n$ is

$$\Delta\omega_n = 100\left(\frac{\omega_{n2} - \omega_{n1}}{\omega_{n1}}\right) = 100\left(\frac{1}{\sqrt{3}} - 1\right) = -42.3\%$$

and the percentage change in damping ratio $\Delta\zeta$ is

$$\Delta\zeta = 100\left(\frac{\zeta_2 - \zeta_1}{\zeta_1}\right) = 100(\sqrt{3} - 1) = 73.2\%$$

[7]See, for example: S. Valoff and W. J. Kaiser, "Presettable Micromachined MEMS Accelerometers," 12th IEEE International Conference on Micro Electrical Mechanical Systems, Orlando, FL, pp. 72–76, (January 1999); and A. Partridge, et al., "A High-Performance Planar Piezoresistive Accelerometer," *J. Microelectromechanical Systems,* Vol. 9, No. 1, pp. 58–65 (March 2000).

Filter

The amplitude-response characteristics of a mechanical system can be compared to that of an electronic band pass filter. A *band pass filter* is a system that lets frequency components in a signal that are within its pass band pass relatively unattenuated or amplified, while frequency components in a signal that are outside the pass band are attenuated. The *pass band* is determined by the cutoff frequencies, which are those frequencies at which

$$H(\Omega) = \frac{H_{max}}{\sqrt{2}} \tag{5.43}$$

In other words, as shown in Figure 5.15, the cutoff frequencies are at amplitudes that are 3 dB below H_{max}. For those systems that have only one cutoff frequency, the system acts as either a *low pass* or a *high pass filter*. A *low pass filter* attenuates frequency components in a signal above the cutoff frequency, and a *high pass filter* attenuates frequency components in a signal below the cutoff frequency.

In the design of mechanical systems, there are two different approaches that are followed, depending on the objective. To reduce vibration levels in a system, one selects the system parameters so that the excitation frequency is not in the system's pass band, because in this region the input is magnified. (Recall the damping-dominated region of Figure 5.10.) On the other hand, there is an application area where mechanical filters are designed for inser-

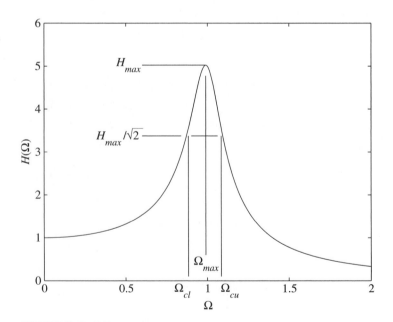

FIGURE 5.15
Definitions of the cutoff frequencies, center frequency, and bandwidth of the amplitude response of a single degree-of-freedom system.

tion in systems. In this case, one designs the resonance frequency of the system to be in the frequency range of interest. This enhances sensitivity as discussed earlier in this section, and is discussed further in Example 8.12.

We now examine the amplitude response $H(\Omega)$ in light of these definitions. The maximum value of $H(\Omega)$ occurs at a frequency ratio Ω_{max}, which is a solution of

$$\frac{dH(\Omega_{max})}{d\Omega} = 0$$

Thus, we find from Eq. (5.18) that for $\zeta \leq 1/\sqrt{2}$, the nondimensional frequency is

$$\Omega_{max} = \sqrt{1 - 2\zeta^2} \tag{5.44a}$$

which means that the corresponding dimensional frequency is

$$\omega_{max} = \omega_n \sqrt{1 - 2\zeta^2} \tag{5.44b}$$

and for $\zeta > 1/\sqrt{2}$, the amplitude response does not have an extremum, as is seen in Figure 5.2a. In this case, the maximum value of $H(\Omega)$ occurs at $\Omega = 0$. Therefore,

$$\begin{aligned} H_{max} = H(\Omega_{max}) &= \frac{1}{2\zeta\sqrt{1 - \zeta^2}} \qquad \zeta \leq 1/\sqrt{2} \\ &= 1 \qquad \zeta > 1/\sqrt{2} \end{aligned} \tag{5.45}$$

Cutoff Frequencies of a Filter   From Eq. (5.18), the location of the *cutoff frequencies* shown in Figure 5.15 is determined by solving

$$\frac{H_{max}}{\sqrt{2}} = \frac{1}{2\zeta\sqrt{2(1 - \zeta^2)}} = \frac{1}{\sqrt{(1 - \Omega^2)^2 + (2\zeta\Omega)^2}} \tag{5.46}$$

Thus, the upper and lower cutoff frequency ratios are given, respectively, by

$$\begin{aligned} \Omega_{cu} &= \sqrt{1 - 2\zeta^2 + 2\zeta\sqrt{1 - \zeta^2}} \\ \Omega_{cl} &= \sqrt{1 - 2\zeta^2 - 2\zeta\sqrt{1 - \zeta^2}} \end{aligned} \tag{5.47}$$

where

$$\Omega_{cu} = \frac{\omega_{cu}}{\omega_n} \qquad \text{and} \qquad \Omega_{cl} = \frac{\omega_{cl}}{\omega_n}$$

and ω_{cu} and ω_{cl} are the respective cutoff frequencies in rad/s. The lower cutoff frequency exists only for those values of ζ for which

$$1 - 2\zeta^2 - 2\zeta\sqrt{1 - \zeta^2} > 0$$

A numerical evaluation[8] of this equation gives that $\zeta < 0.3827$.

---

[8]The MATLAB function `fzero` was used.

Filter Bandwidth The *bandwidth* of the system is

$$B_w = \frac{BW_\omega}{\omega_n} = \frac{\omega_{cu} - \omega_{cl}}{\omega_n} = \Omega_{cu} - \Omega_{cl} \qquad \text{for} \qquad \zeta < 0.3827$$

$$B_w = \Omega_{cu} \qquad \text{for} \qquad \zeta \geq 0.3827 \qquad\qquad (5.48a)$$

where BW_ω is the bandwidth in rad/s. To determine the bandwidth in Hz, we have $BW_f = BW_\omega/2\pi$, and, therefore,

$$B_w = \frac{BW_f}{\omega_n/2\pi} = \frac{f_{cu} - f_{cl}}{f_n} = \Omega_{cu} - \Omega_{cl} \qquad \text{for} \qquad \zeta < 0.3827$$

$$B_w = \Omega_{cu} \qquad \text{for} \qquad \zeta \geq 0.3827 \qquad\qquad (5.48b)$$

since $\Omega_{cu} = \omega_{cu}/\omega_n = (2\pi\omega_{cu})/(2\pi\omega_n) = f_{cu}/f_n$ and $\Omega_{cl} = \omega_{cl}/\omega_n = (2\pi\omega_{cl})/(2\pi\omega_n) = f_{cl}/f_n$.

Quality Factor Another quantity that is often used to define the band pass portion of $H(\Omega)$ when ζ is small is the *quality factor Q*, which is given by

$$Q = \frac{\Omega_c}{B_w} \qquad\qquad (5.49)$$

where Ω_c, which is the center frequency ratio, is defined as the geometric mean

$$\Omega_c = \sqrt{\Omega_{cu}\Omega_{cl}} \qquad\qquad (5.50)$$

The center frequency is not defined for either a low pass or a high pass filter. When $\zeta < 0.1$, it can be shown that

$$Q \cong \frac{1}{2\zeta} \qquad\qquad (5.51)$$

The error made in using this approximation relative to Eq. (5.49) is $< 3\%$, and this error decreases as ζ decreases. The different quantities given above are shown in Figure 5.16 as a function of ζ. We see that for $0 < \zeta < 0.1$, the quality factor of the system is a measure of the maximum magnification of the single degree-of-freedom system.

Design Guideline: The bandwidth of a linear single degree-of-freedom system is a function of the damping ratio and its natural frequency. For a given natural frequency, the smaller the damping ratio, the smaller the bandwidth. For a damping ratio less than 0.1, the maximum amplitude of system's transfer function is approximately inversely proportional to the damping ratio and the corresponding frequency location is slightly less than the system's natural frequency.

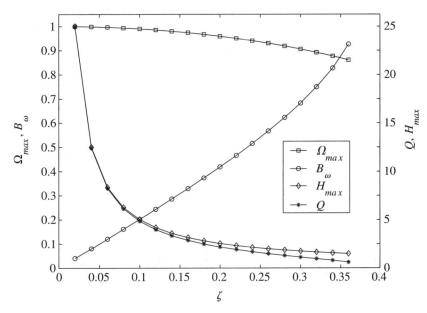

FIGURE 5.16
Quantities used in the definitions of band pass filters as a function of ζ.

EXAMPLE 5.5   Damping ratio and bandwidth to obtain a desired Q

It is desired to have a quality factor of 10 for a single degree-of-freedom system that has a natural frequency of 25.8 Hz. We shall determine the approximate damping ratio and bandwidth of the system. Since $Q = 10$, from Eq. (5.51) we have that $\zeta = 1/(2 \times 10) = 0.05$. Then, from Eqs. (5.47), the upper and lower cutoff frequency ratios are

$$\Omega_{cu} = \sqrt{1 - 2(0.05)^2 + 2(0.05)\sqrt{1 - (0.05)^2}} = 1.05$$
$$\Omega_{cl} = \sqrt{1 - 2(0.05)^2 - 2(0.05)\sqrt{1 - (0.05)^2}} = 0.95$$

and from Eq. (5.48b), the bandwidth is

$$BW_f = f_n B_w = f_n(\Omega_{cu} - \Omega_{cl})$$
$$= 25.8(1.05 - 0.95) = 2.58 \text{ Hz}$$

5.3.4   Relationship of the Frequency-Response Function to the Transfer Function

In Chapter 4, a general solution for the response of a vibratory system governed by Eq. (5.1) was provided. When the initial conditions are zero, it was shown that in the Laplace transform domain

$$X(s) = \frac{F(s)/m}{D(s)} \tag{5.52}$$

which can be determined from Eq. (4.4). The quantity

$$G(s) = \frac{X(s)}{F(s)} \tag{5.53}$$

is called the *transfer function* of the vibratory system. Here, it is defined in terms of a force input and a displacement output. The particular form obtained from Eqs. (5.52), (5.53), and (4.4) is

$$G(s) = \frac{1}{mD(s)} = \frac{1}{m(s^2 + 2\zeta\omega_n s + \omega_n^2)}$$

$$= \frac{1}{k[1 + (s/\omega_n)^2 + 2\zeta(s/\omega_n)]} \tag{5.54}$$

When the complex variable s is set to $j\omega$, we obtain from Eq. (5.54) that

$$G(j\omega) = \frac{1}{k[1 - (\omega/\omega_n)^2 + 2j\zeta(\omega/\omega_n)]} \tag{5.55}$$

where we have used the identity $j^2 = -1$. The function $G(j\omega)$, sometimes referred to as $G(\omega)$, is called the *frequency-response function* of the vibratory system governed by Eq. (5.1). This function can be derived from the transfer function as discussed here or, alternatively, by using Fourier transforms of system input and output signals as discussed later in this section.

On comparing Eq. (5.55) to Eq. (5.35), we recognize that the magnitude of the complex-valued function $kG(j\omega)$—that is, $|kG(j\omega)|$—provides the amplitude response and the phase response associated with $G(j\omega)$. Thus,

$$H(\omega/\omega_n) = |kG(j\omega)| = \frac{1}{\sqrt{(1 - (\omega/\omega_n)^2)^2 + (2\zeta\,\omega/\omega_n)^2}} \tag{5.56a}$$

and

$$\theta(\omega/\omega_n) = \tan^{-1}\frac{2\zeta\,\omega/\omega_n}{1 - (\omega/\omega_n)^2} \tag{5.56b}$$

When $\Omega = \omega/\omega_n$, Eqs. (5.56) for the amplitude and phase response, respectively, take the form

$$H(\Omega) = \frac{1}{\sqrt{(1 - \Omega^2)^2 + (2\zeta\Omega)^2}} \tag{5.57a}$$

and

$$\theta(\Omega) = \tan^{-1}\frac{2\zeta\Omega}{1 - \Omega^2} \tag{5.57b}$$

Equations (5.57) confirm Eqs. (5.18), which were obtained for the amplitude response and the phase response of a linear vibratory system.

Fourier Transforms and Frequency-Response Functions

Instead of determining the frequency response $G(j\omega)$ from the transfer function $G(s)$, the frequency-response function can also be obtained directly based on the system input and output signals. In order to carry this out, one must use the Fourier transform.

The Fourier transform enables conversion of information in the time domain to its frequency domain counterpart. The definitions of the *forward* and *inverse Fourier transforms* are, respectively,

$$F(j\omega) = \int_{-\infty}^{\infty} f(t)e^{-j\omega t}dt$$

$$f(t) = \frac{1}{2\pi} \int_{-\infty}^{\infty} F(j\omega)e^{j\omega t}dt \tag{5.58}$$

where $f(t)$ is the signal in the time domain and $F(j\omega)$ is the corresponding quantity in the frequency domain.

Assuming that the Fourier transforms exist for the input signal $f(t)$ to a vibratory system and the measured output signal $x(t)$, the frequency-response function is defined as

$$G(j\omega) = \frac{X(j\omega)}{F(j\omega)} = \frac{\int_{-\infty}^{\infty} x(t)e^{-j\omega t}dt}{\int_{-\infty}^{\infty} f(t)e^{-j\omega t}dt} \tag{5.59a}$$

Since the displacement and the force signals are present only for $t \geq 0$, Eq. (5.59a) reduces to

$$G(j\omega) = \frac{X(j\omega)}{F(j\omega)} = \frac{\int_{0}^{\infty} x(t)e^{-j\omega t}dt}{\int_{0}^{\infty} f(t)e^{-j\omega t}dt} \tag{5.59b}$$

Usually, in practice, the following approximation is made

$$G(j\omega) = \frac{X(j\omega)}{F(j\omega)} = \frac{\int_{0}^{T} x(t)e^{-j\omega t}dt}{\int_{0}^{T} f(t)e^{-j\omega t}dt} \tag{5.59c}$$

where the integration is carried out over the record length T.

As with the Laplace transform, there are tables from which one can obtain the Fourier transform and its inverse. Here, we shall present only one such Fourier transform pair:

$$f(t) = \frac{e^{-\zeta\omega_n t}}{\omega_d} \sin(\omega_d t)u(t) \Leftrightarrow F(j\omega) = \frac{1}{(\zeta\omega_n + j\omega)^2 + \omega_d^2} \tag{5.60}$$

Returning to Eq. (5.58), we see that when $f(t) \rightarrow f(t)u(t)$, Eq. (5.58) becomes

$$F(j\omega) = \int_{0}^{\infty} f(t)e^{-j\omega t}dt \tag{5.61}$$

which can be obtained from the definition of the Laplace transform by setting $s = j\omega$.

The numerical implementation of the Fourier transform given by Eq. (5.58) is called the *discrete Fourier transform* (DFT), and a very computationally efficient algorithm that performs the DFT is the *fast Fourier transform* (FFT). The DFT is performed over a time interval T_{dft}. Thus, we have two ways to transform the time varying response of a vibratory system to the frequency domain. In the first method, we take the Laplace transform and then set $s = j\omega$. The magnitude of the resulting complex quantity is the amplitude response and the corresponding phase angle is the phase response. This method is limited to linear systems. For those systems in which we obtain the solution numerically or collect digitized signals experimentally, the signals are operated on with the FFT algorithm, which transforms the results to the frequency domain. We shall employ both techniques in this book.

Representative Example

To illustrate the usefulness of the Fourier transform, we return to the system shown in Figure 4.24, for which the corresponding time-domain results are shown in Figure 4.25. It was noticed in Figure 4.25 that the introduction of the spring-stops had the effect of increasing the natural frequency. This was determined by noticing the decrease in the period of the displacement response. Considering the time histories, and using the FFT algorithm,[9] we obtain the corresponding amplitude spectral densities shown in Figure 5.17.

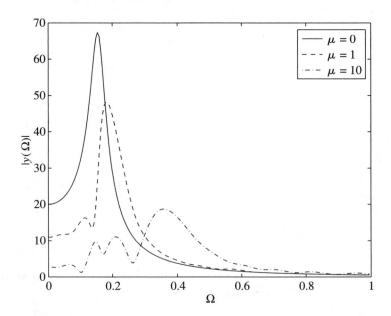

FIGURE 5.17
Amplitude density spectrum of the response of the nonlinear system shown in Figure 4.24 to an initial velocity: $V_o/(\omega_n d) = 10$.

---

[9]The MATLAB function ﬀt was used.

This transformation of information to the frequency domain clearly shows the changes in the effective natural frequency in the three cases considered. For $\mu = 0$, the spectrum indicates a linear response, while for $\mu = 1$ and $\mu = 10$ additional peaks[10] appear in the respective spectra, and there is also a shift in the peak associated with the system's natural frequency.

5.4 SYSTEMS WITH ROTATING UNBALANCED MASS

Many rotating machines such as fans, clothes dryers, internal combustion engines, and electric motors have a certain degree of unbalance. In modeling such systems as single degree-of-freedom systems, it is assumed that the unbalance generates a force that acts on the system's mass. This force, in turn, is transmitted through the spring and damper to the fixed base, as illustrated in Figure 3.7. The unbalance is modeled as a mass m_o rotating with an angular speed ω that is located a fixed distance ϵ from the center of rotation. The equation describing the motion of this system is given by Eq. (3.35), which is repeated below.

$$\frac{d^2x}{dt^2} + 2\zeta\omega_n \frac{dx}{dt} + \omega_n^2 x = \frac{m_o\epsilon\omega^2}{m} \sin \omega t = \frac{F_a}{m} \sin \omega t \tag{5.62a}$$

where $F_a = m_o\epsilon\omega^2$ is the magnitude of the applied (unbalanced) force, and

$$m = M + m_o \quad \text{and} \quad \omega_n = \sqrt{\frac{k}{m}} \tag{5.62b}$$

Rewriting Eq. (5.62a) in terms of the nondimensional time $\tau = \omega_n t$, we have

$$\frac{d^2x}{d\tau^2} + 2\zeta \frac{dx}{d\tau} + x = f(\tau) = M_\epsilon \Omega^2 \sin(\Omega\tau) \tag{5.63a}$$

where

$$M_\epsilon = \frac{m_o\epsilon}{m} \tag{5.63b}$$

Displacement Response

Making use of Eq. (5.17), we see that the solution to Eq. (5.63a) is

$$x(\tau) = \underbrace{M_\epsilon H_{ub}(\Omega)}_{\substack{\text{Displacement} \\ \text{magnitude}}} \sin(\Omega\tau - \underbrace{\theta(\Omega)}_{\text{Phase}}) \tag{5.64}$$

---

[10]Whenever spectra of transient motions are determined, there is a broadening effect in the spectrum, as seen here.

where the amplitude response $H_{ub}(\Omega)$ and the phase response $\theta(\Omega)$ are given by

$$H_{ub}(\Omega) = \Omega^2 H(\Omega) = \frac{\Omega^2}{\sqrt{(1 - \Omega^2)^2 + (2\zeta\Omega)^2}}$$

$$\theta(\Omega) = \tan^{-1}\frac{2\zeta\Omega}{1 - \Omega^2} \tag{5.65}$$

Comparing the forms of the amplitude response and the phase response given by Eqs. (5.65) for the system with the rotating unbalanced mass to those given by Eqs. (5.18) for the system with direct excitation acting on the mass, it is seen that the phase responses are the same and the amplitude responses are different. These similarities and differences can be further seen by comparing the graphs of the amplitude response and phase response of the unbalanced system shown in Figure 5.18 with those for direct excitation of the mass shown in Figure 5.2.

Velocity and Acceleration Responses

The velocity and acceleration responses are determined from the displacement response given by Eq. (5.64) to be, respectively,

$$v(\tau) = \frac{dx(\tau)}{d\tau} = M_e\Omega H_{ub}(\Omega)\cos(\Omega\tau - \theta(\Omega))$$

$$a(\tau) = \frac{d^2x(\tau)}{d\tau^2} = -M_e\Omega^2 H_{ub}(\Omega)\sin(\Omega\tau - \theta(\Omega)) = -\Omega^2 x(\tau) \tag{5.66}$$

We see that for harmonic oscillations, the magnitude of the acceleration is equal to the square of the frequency ratio times the magnitude of the displacement and that the acceleration response lags the displacement response by 180°.

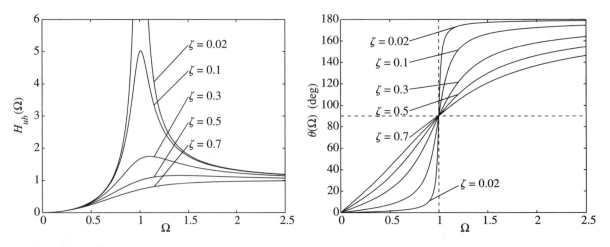

FIGURE 5.18
Harmonic excitation due to rotating unbalance: (a) amplitude response and (b) phase response.

Response Characteristics in Different Excitation Frequency Ranges

The solution is examined in detail in the frequency ranges $\Omega \ll 1$ and $\Omega \gg 1$ and the frequency location $\Omega = 1$. This examination is performed by studying $H_{ub}(\Omega)$ and $\theta(\Omega)$ given by Eqs. (5.65) in these three ranges.

$\Omega \ll 1$ In this region, we find that Eqs. (5.65) simplify to

$$
\begin{aligned}
H_{ub}(\Omega) &\to \Omega^2 \\
\theta(\Omega) &\to 0
\end{aligned}
\tag{5.67}
$$

and, therefore, the displacement response given by Eq. (5.64) simplifies to

$$
x(\tau) \cong M_e \Omega^2 \sin(\Omega\tau) = f(\tau)
\tag{5.68}
$$

Thus, the displacement response is in phase with the unbalanced force $f(\tau)$.

The velocity response is determined from Eq. (5.68) to be

$$
v(\tau) = M_e \Omega^3 \cos(\Omega\tau) = M_e \Omega^3 \sin(\Omega\tau + \pi/2)
\tag{5.69}
$$

The velocity response leads the displacement response by $\pi/2$; that is, the maximum value of the velocity occurs before the maximum value of the displacement. These results are illustrated in Figure 5.19.

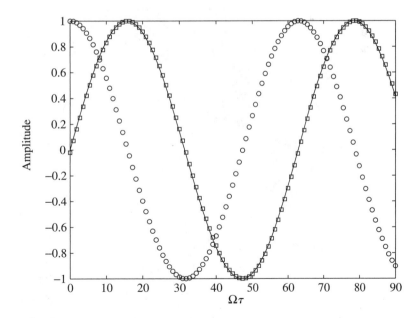

FIGURE 5.19

Phase relationships among displacement, velocity, and force of system with unbalance for $\Omega \ll 1$. [$-$ $f(\tau)/(M_e\Omega^2)$; $\square$ $x(\tau)/(M_e\Omega^2)$; $\bigcirc$ $v(\tau)/(M_e\Omega^3)$.]

$\Omega = 1$ For this value of Ω, we find from Eqs. (5.65) that

$$H_{ub}(1) = \frac{1}{2\zeta}$$

$$\theta(1) = \frac{\pi}{2} \tag{5.70}$$

and, therefore, the displacement response given by Eq. (5.64) reduces to

$$x(\tau) = \frac{M_\epsilon}{2\zeta} \sin(\tau - \pi/2) \tag{5.71}$$

The velocity response follows from Eq. (5.71), and it is given by

$$v(\tau) = \frac{M_\epsilon}{2\zeta} \cos(\tau - \pi/2) = \frac{M_\epsilon}{2\zeta} \sin(\tau) = \frac{1}{2\zeta} f(\tau) \tag{5.72}$$

where we used the fact that $\Omega = 1$. Therefore, in this region, the velocity response is in phase with the unbalanced force. These results are illustrated in Figure 5.20.

$\Omega \gg 1$ In this region, we find from Eqs. (5.65) that

$$\begin{aligned} H_{ub}(\Omega) &\to 1 \\ \theta(\Omega) &\to \pi \end{aligned} \tag{5.73}$$

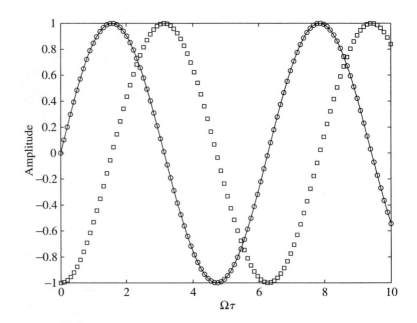

FIGURE 5.20
Phase relationships among displacement, velocity, and force of system with unbalance for $\Omega = 1$. [$-\ f(\tau)/M_\epsilon$; $\square\ x(\tau)/(M_\epsilon/(2\zeta))$; $\bigcirc\ v(\tau)/(M_\epsilon/(2\zeta))$.]

The fact that the amplitude response converges to unity can also be seen from Figure 5.18a. Therefore, the displacement response is approximated from Eq. (5.64) as

$$x(\tau) \cong M_\epsilon \sin(\Omega\tau - \pi) = -\frac{f(\tau)}{\Omega^2} \tag{5.74}$$

Thus, the displacement is 180° out of phase with the applied force. The corresponding velocity response is determined from Eq. (5.74) to be

$$v(\tau) = M_\epsilon \Omega \cos(\Omega\tau - \pi) = M_\epsilon \Omega \sin(\Omega\tau - \pi/2) \tag{5.75}$$

Therefore, in this region the velocity lags the force by $\pi/2$. These results are shown in Figure 5.21. Based on the discussion provided in this section and the graphs shown in Figure 5.18, the following design guideline is proposed.

Design Guideline: In order to reduce the displacement of the mass of a single degree-of-freedom system when the mass is subjected to a harmonic unbalanced force, the natural frequency of the system should be at least twice the excitation frequency or the natural frequency should be at least 50% lower than the excitation frequency. These ranges hold irrespective of the system's damping for $0 < \zeta < 1$.

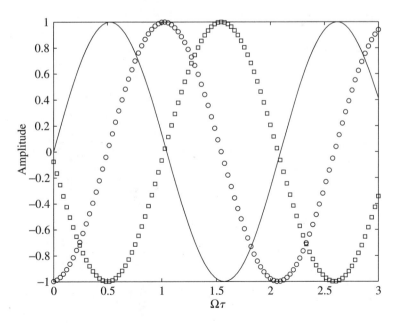

FIGURE 5.21
Phase relationships among displacement, velocity, and force of system with unbalance for $\Omega \gg 1$. [$-$ $f(\tau)/(M_\epsilon\Omega^2)$; $\square$ $x(\tau)/M_\epsilon$; $\bigcirc$ $v(\tau)/(M_\epsilon\Omega)$.]

EXAMPLE 5.6   Locating unbalance on a rotor

A turbine rotor is set up in a balancing machine and operated at the natural frequency of the system composed of the rotor and the supporting spring-damper combination. From an angle indicator attached to the rotor, it was noticed that the maximum displacement of the rotor occurs at an angle of 120° in the direction of rotation. We shall examine where mass must be removed from the rotor in order to reduce the unbalance.

From the graphs shown in Figure 5.18 and the discussion of this section, it is clear that the phase angle between the displacement response and the excitation is 90° when the system is operated at the natural frequency. From Eq. (5.71), it is seen that the response lags the excitation. Hence, the mass must be removed at an angle corresponding to 210° = 90° + 120° on the angle indicator. A shortcoming of this procedure is that for "low" damping levels, the phase response is very sensitive in the vicinity of the natural frequency, as seen in Figure 5.18, and this introduces uncertainty into the procedure.

EXAMPLE 5.7   Fraction of applied force that is transmitted to the base

We shall determine the frequency at which a specified fraction of the magnitude of the applied force F_a due to an unbalanced mass is transmitted to the base. The fraction is denoted by γ, where $\gamma < 1$. The magnitude of the transmitted force F_T is determined from Eq. (3.10) and Figure 3.1 to be

$$F_T = c\frac{dx}{dt} + kx = k\left[2\zeta\frac{dx}{d\tau} + x\right] \tag{a}$$

Substituting Eq. (5.64) into Eq. (a), we obtain[11]

$$\begin{aligned} F_T &= kM_e H_{ub}(\Omega)[2\zeta\Omega\cos(\Omega\tau - \theta(\Omega)) + \sin(\Omega\tau - \theta(\Omega))] \\ &= F_1\sin(\Omega\tau - \theta(\Omega) + \varphi) \end{aligned} \tag{b}$$

where we have used Eq. (4.10) and introduced the amplitude and phase

$$\begin{aligned} F_1 &= kM_e H_{ub}(\Omega)\sqrt{1 + (2\zeta\Omega)^2} \\ \varphi &= \tan^{-1}2\zeta\Omega \end{aligned} \tag{c}$$

respectively.

To determine the frequency at which the magnitude of the force transmitted to the base is a fraction γ of the applied force F_a, we require that $F_1 = \gamma F_a$, or

$$kM_e H_{ub}(\Omega)\sqrt{1 + (2\zeta\Omega)^2} = \gamma m_o\epsilon\omega^2 \tag{d}$$

[11]It is noticed that the transmitted force is at the same frequency as the excitation produced by the rotating unbalance. This captures the transmission force characteristics observed in many industrial applications. However, there are other machines for which the nonharmonic nature of the excitation due to the rotating unbalance becomes important. See Bachmann et al., *ibid.*

Therefore, making use of Eqs. (5.62b), (5.63b), and (5.65), we obtain

$$\frac{\sqrt{1 + (2\zeta\Omega)^2}}{\sqrt{(1 - \Omega^2)^2 + (2\zeta\Omega)^2}} = \gamma \tag{e}$$

Solving for the real positive value of the nondimensional excitation frequency Ω from Eq. (e), we arrive at

$$\Omega = \sqrt{1 - 2\zeta^2(1 - \gamma^{-2}) + \sqrt{(1 - 2\zeta^2(1 - \gamma^{-2}))^2 - 1 + \gamma^{-2}}}$$

$$0 < \gamma \leq 1 \quad \text{(f)}$$

We shall see in the next section that γ, as represented by Eq. (e), is simply the magnitude of the frequency-response function obtained when a harmonic excitation is applied to the base of the system. When $\gamma = 1$, $\Omega = \sqrt{2}$, which is the frequency ratio above which the magnitude of the unbalanced force that is transmitted to the base is less than F_a.

Let us assume that the system has a natural frequency of 35 Hz and determine the angular speed of the rotating mass so that only 15% of the unbalanced force is transmitted to the base when the damping ratio is 0.1. Then, from Eq. (f)

$$\Omega = \frac{\omega}{\omega_n} = \sqrt{1 - 2(0.1)^2(1 - (0.15)^{-2}) + \sqrt{(1 - 2(0.1)^2(1 - (0.15)^{-2}))^2 - 1 + (0.15)^{-2}}}$$

$$= 2.953$$

Since there are 2π rad/rev and 60 s/min, the required angular speed is

$$N = \frac{60}{2\pi}\omega = 2.953\omega_n\frac{60}{2\pi} = 2.953(60)f_n = 2.953(60)(35) = 6200 \text{ rpm}$$

5.5 SYSTEMS WITH BASE EXCITATION

The base excitation model is useful for studying buildings subjected to earthquakes, packaging during transportation, and for designing accelerometers. The equation describing the motion of a single degree-of-freedom system with a vibrating base, which is shown in Figure 3.6, is given by Eq. (3.28). Rewriting Eq. (3.28) in terms of the nondimensional time $\tau = \omega_n t$, we have

$$\frac{d^2x}{d\tau^2} + 2\zeta\frac{dx}{d\tau} + x = 2\zeta\frac{dy}{d\tau} + y \tag{5.76}$$

where it is recalled that $x(\tau)$ is the displacement response of the mass and $y(\tau)$ is the displacement of the base. If the base motion is harmonic—that is,

$$y(\tau) = y_o \sin(\Omega\tau) \tag{5.77}$$

then Eq. (5.76) becomes

$$\frac{d^2x}{d\tau^2} + 2\zeta\frac{dx}{d\tau} + x = 2\zeta y_o\Omega\cos(\Omega\tau) + y_o\sin(\Omega\tau) \tag{5.78}$$

Displacement Response

The right-hand side of Eq. (5.78) contains forcing functions of the form given by Eq. (5.12) and (5.3). Thus, the solution to Eq. (5.78) is determined from Eqs. (5.8a), (5.15a), and (4.10) to be

$$x(\tau) = y_o H(\Omega)\sqrt{1 + (2\zeta\Omega)^2}\,\sin(\Omega\tau - \theta(\Omega) + \varphi) \tag{5.79}$$

where $\theta(\Omega)$ is given by the second of Eqs. (5.65) and

$$\varphi = \tan^{-1} 2\zeta\Omega \tag{5.80}$$

By using the appropriate trigonometric identities,[12] we can rewrite Eq. (5.79) as

$$x(\tau) = \underbrace{y_o H_{mb}(\Omega)}_{\substack{\text{Displacement}\\\text{magnitude}}}\sin(\Omega\tau - \underbrace{\psi(\Omega)}_{\text{Phase}}) \tag{5.81}$$

where the amplitude response $H_{mb}(\Omega)$ and the phase response $\psi(\Omega)$ are given by

$$H_{mb}(\Omega) = \frac{\sqrt{1 + (2\zeta\Omega)^2}}{\sqrt{(1 - \Omega^2)^2 + (2\zeta\Omega)^2}}$$

$$\psi(\Omega) = \tan^{-1}\frac{2\zeta\Omega^3}{1 + \Omega^2(4\zeta^2 - 1)} \tag{5.82}$$

The graphs of the amplitude response and the phase response are shown in Figure 5.22 for different damping factors. As seen in Figure 5.22, the curves obtained for the different damping factors all have the same amplitude value at $\Omega = \sqrt{2}$.

Velocity and Acceleration Responses

The velocity and acceleration are determined from Eq. (5.81) to be, respectively,

$$v(\tau) = y_o\Omega H_{mb}(\Omega)\cos(\Omega\tau - \psi(\Omega))$$

$$a(\tau) = -\Omega^2 x(\tau) \tag{5.83}$$

The relative displacement of the mass to the base is determined from Eqs. (5.77) and (5.81) to be

$$\begin{aligned} z = x(\tau) - y(\tau) &= y_o[H_{mb}(\Omega)\sin(\Omega\tau - \psi(\Omega)) - \sin(\Omega\tau)]\\ &= y_o[\{H_{mb}(\Omega)\cos(\psi(\Omega)) - 1\}\sin(\Omega\tau) - H_{mb}(\Omega)\sin(\psi(\Omega))\cos(\Omega\tau)]\\ &= y_o X_y(\Omega)\sin(\Omega\tau - \varphi_b(\Omega)) \end{aligned} \tag{5.84}$$

---

[12]$\tan^{-1} x \pm \tan^{-1} y = \tan^{-1}\dfrac{x \pm y}{1 \pm xy}$

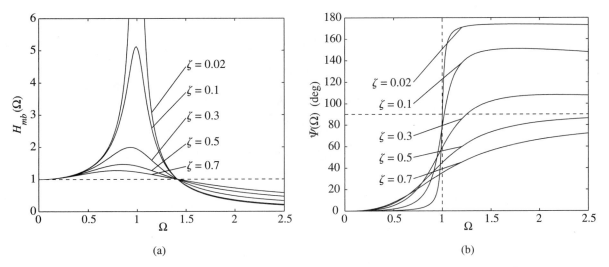

FIGURE 5.22
Excitation due to moving base: (a) amplitude response and (b) phase response.

where the amplitude function $X_y(\Omega)$ and the phase function $\varphi_b(\Omega)$ are given by

$$X_y(\Omega) = \sqrt{H_{mb}^2 - 2H_{mb}\cos(\psi(\Omega)) + 1}$$

$$\varphi_b(\Omega) = \tan^{-1}\frac{H_{mb}(\Omega)\sin(\psi(\Omega))}{H_{mb}(\Omega)\cos(\psi(\Omega)) - 1} \qquad (5.85)$$

and Eq. (4.10) has been used. A plot of $X_y(\Omega)$ versus the frequency ratio Ω is identical to that obtained for the amplitude response $H_{ub}(\Omega)$ of system with an unbalanced mass, which is shown in Figure 5.18a. However, the interpretation is different. For $\Omega < 0.3$, the mass and base move in phase and almost as a rigid body. When $\Omega > 2$, the mass moves very little; that is, it is relatively stationary, and only the base is moving. When $\Omega \approx 1$, the mass magnifies the motion of the base and there is large relative movement of the mass with respect to the base.

Response Characteristics in Different Excitation Frequency Ranges

We now examine the system response in the frequency ranges $\Omega \ll 1$ and $\Omega \gg 1$, and the frequency location $\Omega = 1$. This examination is carried out by studying $H_{mb}(\Omega)$ and $\psi(\Omega)$ in these three ranges.

$\Omega \ll 1$ In this region, we find from Eqs. (5.82) that

$$H_{mb}(\Omega) \to 1$$

$$\psi(\Omega) \to 0 \qquad (5.86)$$

and, hence, the displacement response given by Eq. (5.81) simplifies to

$$x(\tau) \cong y_o \sin(\Omega\tau) = y(\tau) \tag{5.87}$$

Hence, in this frequency region, the displacement of the mass is approximately equal to the displacement of the base. The corresponding velocity is determined from Eq. (5.87) to be

$$v(\tau) = y_o\Omega \cos(\Omega\tau) = y_o\Omega \sin(\Omega\tau + \pi/2) \tag{5.88}$$

Thus, the velocity response leads the displacement response by $\pi/2$. These results are shown in Figure 5.23.

$\Omega = 1$ In this region, we determine from Eqs. (5.82) that

$$H_{mb}(1) = \frac{\sqrt{1 + (2\zeta)^2}}{2\zeta}$$

$$\psi(1) = \tan^{-1}\frac{1}{2\zeta} \tag{5.89}$$

From Eq. (5.89), it is seen that the phase angle at $\Omega = 1$ is a function of ζ. It follows from Eq. (5.81) that the displacement response is

$$x(\tau) = y_oH_{mb}(1)\sin(\tau - \psi(1)) \tag{5.90}$$

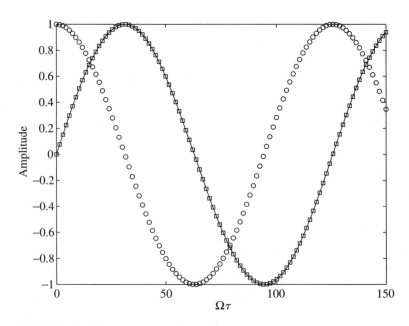

FIGURE 5.23
Phase relationships among displacement, velocity, and force of system with moving base when $\Omega \ll 1$. [$- y(\tau)/y_o$; $\square$ $x(\tau)/y_o$; $\bigcirc$ $v(\tau)/(y_o\Omega)$.]

and the corresponding velocity response is

$$v(\tau) = y_o H_{mb}(1)\cos(\tau - \psi(1)) \tag{5.91}$$

These results are shown in Figure 5.24.

$\Omega \gg 1$ In this region, Eqs. (5.82) simplify to

$$H_{mb}(\Omega) \rightarrow \frac{2\zeta}{\Omega}$$

$$\psi(\Omega) \rightarrow \psi_1(\Omega) = \tan^{-1}\frac{2\zeta\Omega}{4\zeta^2 - 1} \tag{5.92}$$

and, therefore, the displacement response given by Eq. (5.81) is approximated to

$$x(\tau) \cong \frac{2\zeta y_o}{\Omega}\sin(\Omega\tau - \psi_1(\Omega)) \tag{5.93}$$

The velocity response follows from Eq. (5.93), and it has the form

$$v(\tau) = y_o 2\zeta\cos(\Omega\tau - \psi_1(\Omega)) \tag{5.94}$$

These results are shown in Figure 5.25. Based on the discussion provided in this section and Figure 5.22, the following design guidelines are postulated.

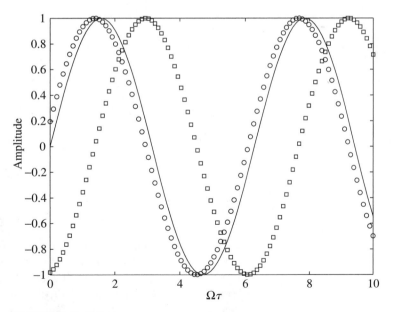

FIGURE 5.24

Phase relationships among displacement, velocity, and force of system with moving base when $\Omega = 1$. [$-$ $y(\tau)/y_o$; $\square$ $x(\tau)/(y_o H_{mb}(1))$; $\bigcirc$ $v(\tau)/(y_o H_{mb}(1)\Omega)$.]

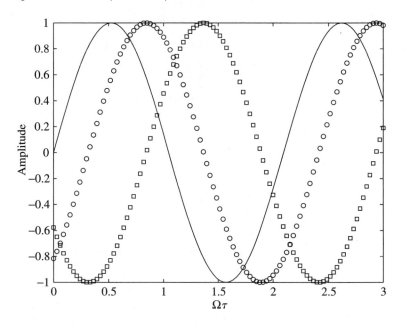

FIGURE 5.25
Phase relationships among displacement, velocity, and force of system with moving base when $\Omega \gg 1$. [$-\ y(\tau)/y_o$; $\square\ x(\tau)/(2\zeta y_o/\Omega)$; $\bigcirc\ v(\tau)/(2\zeta y_o).$]

Design Guideline: In order to reduce the displacement of the mass of a single degree-of-freedom system when the base is subjected to harmonic excitation, the natural frequency of the system should be either five times higher than the excitation frequency or the natural frequency should be at least 30% lower than the excitation frequency.

EXAMPLE 5.8   Response of an instrument subjected to base excitation

An instrument is mounted on an isolation system with a damping factor $\zeta = 0.3$. The base of the instrumentation system is subjected to harmonic motions at 120 Hz, which is three times higher than the natural frequency of the system. For the base motion amplitude of 2 cm, we shall determine the peak acceleration response of the instrument.

The magnitude of the displacement response of the instrument is given by Eqs. (5.81) and (5.82). Thus,

$$x_o = y_o H_{mb}(\Omega) \tag{a}$$

Since $\Omega = 3$ and $y_o = 0.02$ m, Eq. (5.82) is used to determine

$$H_{mb}(3) = \frac{\sqrt{1 + 4 \times 0.3^2 \times 3^2}}{\sqrt{(1 - 3^2)^2 + 4 \times 0.3^2 \times 3^2}} = 0.25 \qquad \text{(b)}$$

Hence, from Eqs. (a) and (b), we have that the displacement magnitude of the mass is

$$x_o = (0.02 \text{ m}) \times 0.25 = 5 \text{ mm} \qquad \text{(c)}$$

Since the displacement response given by Eq. (5.81) is harmonic, the acceleration response will also be harmonic and the magnitude of the peak acceleration is determined from Eq. (c) and the provided excitation frequency as

$$a_o = \omega^2 x_o = (120 \times 2\pi \text{ rad/s})^2 \times 5 \times 10^{-3} \text{ m} = 2842.5 \text{ m/s}^2 \qquad \text{(d)}$$

EXAMPLE 5.9   Frequency-response function of a tire for pavement design analysis

A major cause of pavement damage is believed to be due to the vibrations of heavy trucks, which transmit their loads through the tires to the pavement.[13] Hence, for highway pavement design work, a good model of a tire is needed. In order to determine this model, the tire is excited by a vibration shaker in the laboratory by an excitation $y(t)$ and the force $f(t)$ is measured. A representative schematic of this system is shown in Figure 5.26, where the tire is represented by a linear vibratory system with mass m, stiffness k_1, and damping c. We shall illustrate how the frequency-response function based on the displacement input $y(t)$ and the measured force output $f(t)$ can be determined.

First, the governing equation of motion in this system is determined based on force balance along the j direction, and then, based on this equation, the required transfer function for this linear system is determined. Subsequently, the required frequency-response function is obtained. The governing equation of motion is

$$m\ddot{x} + (k_1 + k_2)x + c\dot{x} = k_1 y + c\dot{y} \qquad \text{(a)}$$

The magnitude of the force $f(t)$ transmitted to the support is given by

$$f(t) = k_2 x(t) \qquad \text{(b)}$$

Assuming that the initial conditions are zero, we take the Laplace transforms of both sides of Eq. (a) and rearrange the resulting transforms to obtain the ratio

$$\frac{X(s)}{Y(s)} = \frac{cs + k_1}{ms^2 + cs + (k_1 + k_2)} \qquad \text{(c)}$$

[13]J. C. Tielking, "Conventional and wide base radial tyres," in *Proceedings of the Third International Symposium on Heavy Vehicle Weights and Dimensions,* D. Cebon and C. G. B. Mitchell, Eds., Cambridge, U.K., pp. 182–190 (28 June–2 July 1992).

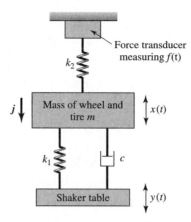

FIGURE 5.26
Model of experimental tire frequency-response measuring system.

where $X(s)$ is the Laplace transform of $x(t)$ and $Y(s)$ is the Laplace transform of $y(t)$. From Eq. (b), we find the transform

$$F(s) = k_2 X(s) \tag{d}$$

where $F(s)$ is the Laplace transform of $f(t)$. From Eqs. (c) and (d), we find the transfer function between the tire displacement and the force to be

$$\frac{F(s)}{Y(s)} = \frac{k_2(cs + k_1)}{ms^2 + cs + (k_1 + k_2)} \tag{e}$$

Based on the discussion of Section 5.3, the frequency-response function is obtained by setting $s = j\omega$ in Eq. (e) to arrive at frequency-response function

$$\frac{F(j\omega)}{Y(j\omega)} = \frac{k_2(cj\omega + k_1)}{-m\omega^2 + cj\omega + (k_1 + k_2)} \tag{f}$$

Usually, in the experiments, the magnitude of the frequency-response function is desired. Then, from Eq. (f), we obtain

$$\left| \frac{F(j\omega)}{X(j\omega)} \right| = k_2 \frac{\sqrt{k_1^2 + \omega^2 c^2}}{\sqrt{(k_1 + k_2 - \omega^2 m)^2 + \omega^2 c^2}} \tag{g}$$

Equation (g) can be used to curve fit the data to determine the tire system parameters such as the damping coefficient c. In this regard, the discussion of Figure 5.12 is applicable.

5.6 ACCELERATION MEASUREMENT: ACCELEROMETER

An accelerometer is a transducer whose electrical output is proportional to acceleration. It is a very common device used to measure the acceleration of a point on a system. Accelerometers are constructed in several ways. We shall

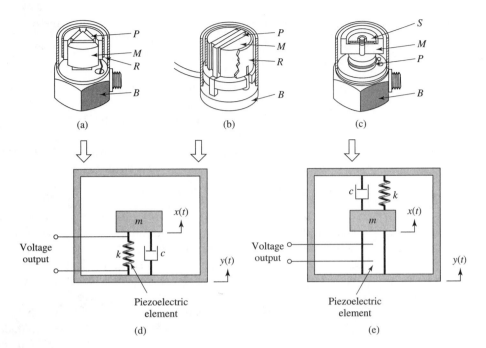

FIGURE 5.27
Piezoelectric accelerometer: (a) and (b) shear type; (c) compression type; (d) single degree-of-freedom model for shear type; and (e) single degree-of-freedom model for compression type. (B = base; M = mass; P = piezoelectric element; S = spring; R = retaining ring.) *Source:* Courtesy of Bruel & Kajer Sound and Vibration Measurement A/S.

examine one of the more common types, called a *piezoelectric accelerometer*.[14] A typical piezoelectric accelerometer is constructed in one of the three forms shown in Figure 5.27. The electrical output of the piezoelectric element is proportional to the change in its length when the piezoelectric element is in compression or it is proportional to the change in shear angle when the element is in shear. For the compression mounting, the mass is held against the element by the compression spring, so that as the mass *m* moves relative to the base by an amount z, the force on the piezoelectric element either increases or decreases. In the shear mount, the movements of the masses shear the elements as the masses move relative to the base. In this case, the compression of the elements is not a factor and the stiffness is from the piezoelectric element itself. The retaining ring ensures that all masses move as a unit. Piezoelectric elements typically have very low damping factors; that is, $0 < \zeta < 0.02$.

From Eq. (3.31), we make use of the nondimensional time variable τ to arrive at the following equation of motion of a single degree-of-freedom system with a moving base

$$\frac{d^2z}{d\tau^2} + 2\zeta \frac{dz}{d\tau} + z = -\frac{d^2y}{d\tau^2} \tag{5.95}$$

[14]In Chapter 2, a MEMS accelerometer based on the change in capacitance was discussed.

where $\tau = \omega_n t$ and $\ddot{y}$ is the acceleration of the base. If the base is subjected to a harmonically varying displacement, then

$$y(\tau) = y_o \sin(\Omega\tau) \tag{5.96}$$

Upon substituting Eq. (5.96) into Eq. (5.95), we arrive at

$$\frac{d^2z}{d\tau^2} + 2\zeta\frac{dz}{d\tau} + z = y_o\Omega^2 \sin(\Omega\tau) = a_o \sin(\Omega\tau) \tag{5.97}$$

where $a_o\omega_n^2 = y_o\omega^2$ is the acceleration of the base.

The solution to Eq. (5.97) is given by Eq. (5.17). Thus,

$$z(\tau) = a_o H(\Omega)\sin(\Omega\tau - \theta(\Omega)) \tag{5.98}$$

where $H(\Omega)$ and $\theta(\Omega)$ are given by Eqs. (5.18). We see that the relative displacement of the mass is proportional to the acceleration of the base, and note that the electrical output voltage from the piezoelectric element is proportional to the relative displacement. Therefore, if the acceleration amplitude response and the phase response are to be relatively constant over a wide frequency range, then the parameters of the accelerometer have to be chosen so that $H(\Omega)$ varies by less than $\pm d$, where $|d| \ll 1$, over that range. From Figure 5.2 and Eqs. (5.26), it is clear that for $\Omega \ll 1$ the amplitude response and the phase response are constant. Since $H(0) = 1$ and the damping ratio is very low, the frequency range is determined from

$$1 + d = \frac{1}{\sqrt{(1 - \Omega^2)^2 + (2\zeta\Omega)^2}} \approx \frac{1}{1 - \Omega^2}$$

or

$$\Omega_a = \sqrt{1 - \frac{1}{1 + d}} \tag{5.99}$$

where $\Omega_a = \omega_a/\omega_n = f_a/f_n$ and f_a is the frequency below which the amplitude response varies by less than d. Typical values of f_n for small accelerometers are between 50 kHz and 100 kHz.

EXAMPLE 5.10 Design of an accelerometer

We shall determine the working range of an accelerometer with a natural frequency of 60 kHz and whose variation in the amplitude response is less than 2%. Since we want a variation of less than 2% in the amplitude response of the accelerometer, then $d = 0.02$ and, from Eq. (5.99), $\Omega_a = 0.14$. Since the natural frequency of the accelerometer is 60 kHz, the frequency range for which the deviation in the amplitude response is less than 2% is $f_a = (0.14)(60 \text{ kHz}) = 8.4$ kHz. In addition, from Figure 5.2 it is seen that the phase response $\theta(\Omega) \approx 0$ for $0 \leq \Omega \leq \Omega_a$ when ζ is small. Hence, the working range of the accelerometer is $0 < f_a \leq 8.4$ kHz.

5.7 VIBRATION ISOLATION

As discussed in Section 5.5, there are many situations in practice that require one to either isolate a single degree-of-freedom system from transmitting vibrations to its base or to isolate a vibrating single degree-of-freedom system from vibrations of its base. This isolation may be required to reduce the magnitude of the stresses in the support structure to lessen the chances of fatigue-induced failures or to reduce annoyance from low frequency motions or to minimize interference with precision measurements and manufacturing processes. We shall show that both of these vibration isolation scenarios are equivalent and that the corresponding design guidelines are the same.

System with Direct Excitation of Inertial Element

We have shown in Eq. (5.17) that the displacement response of a single degree-of-freedom system whose mass is subjected to a harmonic force is

$$x(\tau) = \frac{F_o}{k} H(\Omega)\sin(\Omega\tau - \theta(\Omega))$$

where $H(\Omega)$ and $\theta(\Omega)$ are given by Eq. (5.18). As discussed in Example 5.7, the force transmitted to the base (ground) can be written as

$$F_T(\tau) = kx(\tau) + c\dot{x}(\tau) = k\left[x(\tau) + 2\zeta\frac{dx}{d\tau}\right] \tag{5.100}$$

Upon substituting Eq. (5.17) into Eq. (5.100), we obtain

$$\begin{aligned}F_T(\tau) &= F_o H(\Omega)[\sin(\Omega\tau - \theta(\Omega)) + 2\zeta\Omega\cos(\Omega\tau - \theta(\Omega))] \\ &= F_o H_{mb}(\Omega)\sin(\Omega\tau - \psi(\Omega))\end{aligned} \tag{5.101}$$

where we have used Eq. (4.10) and $H_{mb}(\Omega)$ and $\psi(\Omega)$ are given by Eqs. (5.82). The magnitude of the ratio of the force transmitted to the ground to that applied to the mass is

$$\left|\frac{F_T(\tau)}{F_o}\right| = H_{mb}(\Omega) \tag{5.102}$$

System with Base Excitation

Considering a system subjected to base excitation, we see from Eq. (5.81) that the ratio of the magnitude of displacement of the mass of a single degree-of-freedom system excited by a harmonic base motion to the magnitude of the applied displacement is

$$\left|\frac{x(\tau)}{y_o}\right| = H_{mb}(\Omega) \tag{5.103}$$

Transmissibility Ratio

To minimize the force transmitted to the base from the vibrations of a directly excited system or to minimize the magnitude of the base motion on a system, we require from Eqs. (5.102) and (5.103) that

$$H_{mb}(\Omega) << 1$$

We define the *transmissibility ratio TR* as a measure of either the amount of applied force to the mass that is transmitted to the ground or the amount of displacement applied to the base that gets transmitted to the mass. Thus,

$$TR = H_{mb}(\Omega) \tag{5.104}$$

To determine the region in which $TR \leq 1$, we have to determine the value of Ω that satisfies

$$H_{mb}(\Omega) \leq 1$$

which, after some algebra, results in

$$\Omega \geq \sqrt{2}$$

All the different frequency responses obtained for different ζ intersect at $\Omega = \sqrt{2}$, as shown in Figure 5.22. When $0 < \zeta \leq 0.2$ and $\Omega > 2$, there is little variation in the value of *TR*. Therefore, for these regions, we can set $\zeta = 0$ and for lightly damped systems

$$TR = \frac{1}{\Omega^2 - 1} \qquad \Omega > 2, \quad \zeta < 0.2 \tag{5.105}$$

Thus, when $\Omega = 5$, that is, when the disturbance frequency is 5 times the natural frequency of the system, $TR = 0.042$, or only 4.2% of the disturbance gets through.

The *reduction in transmissibility*, denoted *R*, is a measure of the amount of force or displacement that doesn't get through; thus,

$$R = 1 - TR$$

EXAMPLE 5.11 Modified system to limit the maximum value of *TR* due to machine start-up

The isolation that is indicated by Eq. (5.105) is valid only when the excitation frequency is much greater than the natural frequency of the system, and assumes that the system always operates at this frequency. There are many systems, however, that operate at a frequency much greater than the natural frequency, but are cycled so that they turn on and off on a regular and frequent basis. As these systems get up to operating speed (frequency) they must pass through the system's resonance region, where larger forces/displacements are temporarily transmitted to the ground/base.[15] A modification to the system that can reduce the maximum value of *TR* in the neighborhood of the resonance is shown in Figure 5.28. The modification includes an additional spring k_2 between the damper and the base, which, we recall from

[15]It is mentioned that if a vibratory system such as a rotating machine is ramped up to operate at a high frequency, then to attenuate the response while passing through resonance, one will need some amount of damping.

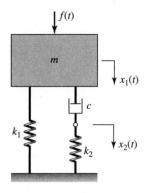

FIGURE 5.28
Single degree-of-freedom system with an additional spring added in series with the damper.

Example 4.7, results in a Maxwell element. We shall now re-derive the governing equation of this system using Lagrange's equations.

The kinetic energy and potential energy for this system are, respectively,

$$T = \frac{1}{2} m \dot{x}_1^2$$

$$V = \frac{1}{2} k_1 x_1^2 + \frac{1}{2} k_2 x_2^2 \tag{a}$$

and the dissipation function is

$$D = \frac{1}{2} c (\dot{x}_1 - \dot{x}_2)^2 \tag{b}$$

By using Eq. (3.41), the Lagrange equations, with $N = 2$, $q_1 = x_1$, $q_2 = x_2$, $Q_1 = f(t)$, and $Q_2 = 0$, we obtain

$$m\ddot{x}_1 + c(\dot{x}_1 - \dot{x}_2) + k_1 x_1 = f(t)$$

$$c(\dot{x}_1 - \dot{x}_2) - k_2 x_2 = 0 \tag{c}$$

Although the inertia element shown in Figure 5.28 is described in terms of a single coordinate—that is, x_1—here, the additional variable x_2 is also needed to determine the damper force and the spring force associated with k_2. As discussed later on, the transmitted force depends on the variable x_2.[16]

Introducing the damping factor $\zeta = c/(2m\omega_n)$ and the natural frequency $\omega_n = \sqrt{k_1/m}$, Eq. (c) is written as

$$\ddot{x}_1 + 2\zeta\omega_n(\dot{x}_1 - \dot{x}_2) + \omega_n^2 x_1 = \frac{f(t)}{m}$$

$$2\zeta\omega_n(\dot{x}_1 - \dot{x}_2) - \gamma\omega_n^2 x_2 = 0 \tag{d}$$

where $\gamma = k_2/k_1$.

We assume that the initial conditions are zero and take the Laplace transforms of Eqs. (d). This results in

$$(s^2 + 2\zeta\omega_n s + \omega_n^2)X_1(s) - 2\zeta\omega_n s X_2(s) = \frac{F(s)}{m}$$

$$2\zeta\omega_n s X_1(s) - (\gamma\omega_n^2 + 2\zeta\omega_n s)X_2(s) = 0 \tag{e}$$

where $X_1(s)$ is the Laplace transform of $x_1(t)$, $X_2(s)$ is the Laplace transform of $x_2(t)$, and $F(s)$ is the Laplace transform of $f(t)$. Solving for $X_1(s)$ and $X_2(s)$ from Eqs. (e), we obtain

$$X_1(s) = \frac{F(s)}{m} \frac{C(s)}{D_3(s)}$$

$$X_2(s) = \frac{F(s)}{m} \frac{B(s)}{D_3(s)} \tag{f}$$

[16] As discussed in Example 4.7, the system shown in Figure 5.28 is said to be a system with *one and a half degrees of freedom:* One of the governing equations is a second-order differential equation, while the other one is a first-order differential equation.

where

$$D_3(s) = \gamma\omega_n^2(s^2 + \omega_n^2) + 2\zeta\omega_n s(s^2 + \omega_n^2[1 + \gamma])$$
$$B(s) = 2\zeta\omega_n s$$
$$C(s) = \gamma\omega_n^2 + 2\zeta\omega_n s \tag{g}$$

We are interested in the dynamic force transmitted to the ground, which for the configuration shown in Figure 5.28, is

$$F_T = k_1 x_1 + k_2 x_2 \tag{h}$$

Taking the Laplace transform of Eq. (h) and using Eqs. (f) we obtain

$$\frac{F_T(s)}{F(s)} = \frac{\omega_n^2}{D_3(s)}[C(s) + \gamma B(s)] \tag{i}$$

From Section 5.3.4, we see that the transmissibility ratio is obtained by setting $s = j\omega$ in Eq. (i) and then taking its magnitude, which results in

$$TR = \left|\frac{F_T(j\omega)}{F(j\omega)}\right| = \frac{\sqrt{\gamma^2 + [2\zeta\Omega(1 + \gamma)]^2}}{\sqrt{\gamma^2(1 - \Omega^2)^2 + (2\zeta\Omega)^2[1 + \gamma - \Omega^2]^2}} \tag{j}$$

When $\gamma \to \infty$, that is, k_2 becomes rigid, Eq. (j) becomes $H_{mb}(\Omega)$ given by Eqs. (5.82).

For a fixed value of ζ and γ, the frequency ratio at which TR is an extremum, $\Omega = \Omega_{max}$, corresponds to

$$\frac{\partial(TR)}{\partial\Omega} = 0 \tag{k}$$

This leads to a cubic polynomial in Ω^2, which has one real positive root, Ω_{max}. Instead of determining another derivative of Eq. (k) and verifying if the value of TR is a minimum or a maximum at Ω_{max}, this is done numerically. From the numerical results, it is determined that TR is a maximum, and this is labeled as TR_{max}. The coefficients of the cubic equation are functions of γ and ζ. The values TR_{max} which are determined at Ω_{max} are plotted in Figure 5.29 for several values of γ and ζ. We see, for example, that TR_{max} will always be less than 2.5 when $\gamma > 2$ and $\zeta > 0.25$. In this case, if the operating frequency ratio is $\Omega > 5$, then we find from Eq. (j) that $TR < 0.1$; that is, less than 10% of the force will be transmitted to the base. In addition, on the way to reaching this operating speed, the TR will not exceed 2.5.

5.8 ENERGY DISSIPATION AND EQUIVALENT DAMPING

As discussed in the earlier chapters, viscous damping is one form of damping model. Other types of damping models include Coulomb or dry friction damping, fluid or velocity-squared damping, and structural or material damping. The viscous damping model as presented in Chapter 2 can be a

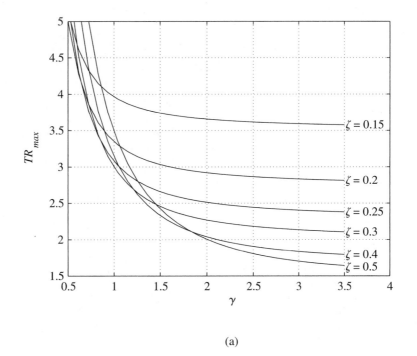

(a)

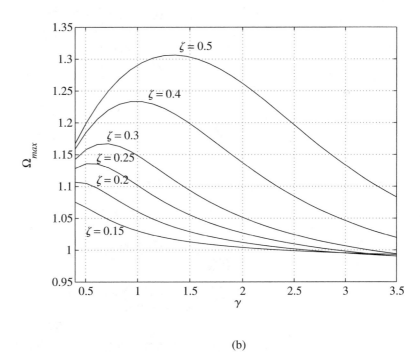

(b)

FIGURE 5.29
(a) TR_{max} as a function of ζ and γ and (b) Ω_{max} as a function of ζ and γ.

linear or nonlinear model, while the Coulomb or dry friction model and the fluid model are nonlinear. In addition, the structural damping model is a linear one. We shall now relate the three damping models to the viscous damping model through a quantity called the *equivalent viscous damping* c_{eq}, which is the value of the damping coefficient c that is required in order to dissipate the same amount of energy per period of forced harmonic oscillation. Since the energy consideration is on a per cycle basis, the following results are only applicable to a system subjected to harmonic excitation.

In a vibratory system, the spring force and inertia force are conservative and hence, the work done by each of the forces over one cycle of forced oscillation is zero.[17] Therefore, in determining the energy dissipation in a vibratory system, we only pay attention to the damper or dissipation force.

If the dissipation force is F_D, then the energy dissipation as discussed in Section 2.4 is given by the work done; that is,

$$E_d = \int F_D dx = \int F_D \frac{dx}{dt} dt = \int F_D \dot{x} dt \qquad (5.106)$$

For harmonic excitations of linear systems after the transients settle down, it can be assumed that the displacement and velocity responses have the forms

$$x(t) = X_o \sin(\omega t - \phi)$$
$$\dot{x}(t) = \omega X_o \cos(\omega t - \phi) \qquad (5.107)$$

where the displacement and the velocity responses have the period $2\pi/\omega$, where ω is the forcing frequency. It can be shown that the work done by the external force acting on the system is equal to E_d over one cycle of forcing. See Exercise 5.15.

Viscous Damping

From Eq. (2.46), we have that for a linear viscous damping model

$$F_D = c\dot{x}(t) \qquad (5.108)$$

Upon substituting Eq. (5.108) into (5.106), the energy dissipated is

$$E_{viscous} = c \int_0^{2\pi/\omega} \dot{x}^2(t) dt \qquad (5.109)$$

On substituting the velocity response from Eqs. (5.107) into Eq. (5.109), we obtain

$$E_{viscous} = c\omega^2 X_o^2 \int_0^{2\pi/\omega} \cos^2(\omega t - \phi) dt = c\pi\omega X_o^2 \qquad (5.110)$$

From Eq. (5.110) it is clear that the energy dissipated is linearly proportional to the damping coefficient c and the excitation frequency ω, and proportional to the square of the displacement amplitude X_o.

---

[17]See Exercise 2.29 for the work done by the spring force.

Coulomb (Dry Friction) Damping

From Eq. (2.52) we have that the damping force magnitude is

$$F_D = \mu mg \, \text{sgn}(\dot{x}) \tag{5.111}$$

and from Eq. (2.53) the energy dissipated is

$$E_{Coulomb} = \mu mg \int_0^{2\pi/\omega} \text{sgn}(\dot{x})\dot{x}(t)dt \tag{5.112}$$

where sgn is the signum function introduced in Section 2.4. On substituting the velocity response from Eqs. (5.107) into Eq. (5.112) and integrating, we arrive at

$$E_{Coulomb} = \mu mgX_o \int_0^{2\pi} \text{sgn}(\cos \xi)\cos \xi d\xi$$

$$= \mu mgX_o \left[\int_0^{\pi/2} \cos\xi d\xi - \int_{\pi/2}^{3\pi/2} \cos\xi d\xi + \int_{3\pi/2}^{2\pi} \cos\xi d\xi \right]$$

$$= 4\mu mgX_o \tag{5.113}$$

where the variable of integration $\xi = \omega t$.

To determine the equivalent viscous damping, we set $E_{viscous}$ equal to $E_{Coulomb}$. Thus, from Eqs. (5.113) and (5.110), we obtain

$$c_{eq}\pi\omega X_o^2 = 4 \, \mu mgX_o$$

which leads to

$$c_{eq} = \frac{4\mu mg}{\pi\omega X_o} \tag{5.114}$$

where c_{eq} is the equivalent viscous damping. Note that this equivalence of damping model is based on energy considerations only, and it should not be inferred that in this and other cases that the linear system with the equivalent viscous damping has the same stability properties as the original system. Furthermore, it is noted from Eq. (5.114) that unlike in the case of a system with viscous damping, in the case of a system with dry friction the equivalent damping coefficient is inversely proportional to the excitation frequency and the displacement response amplitude X_o.

Fluid (Velocity-Squared) Damping

From Eq. (2.54), we stated that the magnitude of the damping force is

$$F_D = c_d\dot{x}^2 \, \text{sgn}(\dot{x}) = c_d|\dot{x}|\dot{x} \tag{5.115}$$

where c_d is given by Eq. (2.55). Upon substituting Eq. (5.115) into Eq. (5.106) the energy dissipated is determined from

$$E_{fluid} = c_d \int_0^{2\pi/\omega} \text{sgn}(\dot{x})\dot{x}^3 dt \tag{5.116}$$

On substituting the velocity response from Eqs. (5.107) into Eq. (5.116), we arrive at

$$E_{fluid} = c_d\omega^2 X_o^3 \int_0^{2\pi} \text{sgn}(\cos\xi)\cos^3\xi d\xi = \frac{8}{3} c_d\omega^2 X_o^3 \tag{5.117}$$

where the integration has been carried out in a manner similar to that used to obtain Eq. (5.113).

To determine the equivalent viscous damping, we set $E_{viscous}$ from Eq. (5.110) equal to E_{fluid} given by Eq. (5.117). Thus, we arrive at

$$c_{eq}\pi\omega X_o^2 = \frac{8}{3} c_d\omega^2 X_o^3$$

which leads to

$$c_{eq} = \frac{8c_d\omega X_o}{3\pi} \tag{5.118}$$

Thus, the equivalent viscous damping is linearly proportional to the excitation frequency and the response amplitude.

Structural (Material) Damping

In Eq. (2.57), we stated that the magnitude of the damping force is given by

$$F_D = k\pi\beta \, \text{sgn}(\dot{x})|x| \tag{5.119}$$

where β is an empirically determined constant. Upon substituting Eq. (5.119) into Eq. (5.106), the energy dissipated is

$$E_{structural} = k\pi\beta \int_\pi^{2\pi/\omega} \text{sgn}(\dot{x})|x|\dot{x}dt \tag{5.120}$$

On substituting the velocity response from Eqs. (5.107) into Eq. (5.120), the resulting expression is

$$E_{structural} = k\pi\beta X_o^2 \int_0^{2\pi} \text{sgn}(\cos\xi)|\sin\xi|\cos\xi d\xi$$

$$= k\pi\beta X_o^2 \left[\int_0^{\pi/2} \sin\xi\cos\xi d\xi - \int_{\pi/2}^{\pi} \sin\xi\cos\xi d\xi \right.$$

$$\left. + \int_\pi^{3\pi/2} \sin\xi\cos\xi d\xi - \int_{3\pi/2}^{2\pi} \sin\xi\cos\xi d\xi \right]$$

$$= 2k\pi\beta X_o^2 \tag{5.121}$$

To determine the equivalent viscous damping for this case, we set $E_{viscous}$ from Eq. (5.110) equal to $E_{structural}$ in Eq. (5.121). This leads to

$$c_{eq}\pi\omega X_o^2 = 2k\pi\beta X_o^2$$

from which we obtain

$$c_{eq} = \frac{2k\beta}{\omega} \tag{5.122}$$

As in the case of dry friction, the equivalent viscous damping for structural damping is inversely proportional to the excitation frequency.

We now place the expressions for the equivalent viscous damping c_{eq} in their respective governing equations. The restriction for each of these equations is that the forcing function must be a harmonic excitation.

Viscous damping

$$m\ddot{x} + c\dot{x} + kx = F_o \sin(\omega t) \tag{5.123}$$

Coulomb (dry friction) damping

$$m\ddot{x} + \frac{4\mu mg}{\pi\omega X_o}\dot{x} + kx = F_o \sin(\omega t) \tag{5.124}$$

Fluid (velocity-squared) damping

$$m\ddot{x} + \frac{8c_d\omega X_o}{3\pi}\dot{x} + kx = F_o \sin(\omega t) \tag{5.125}$$

Structural (material) damping

$$m\ddot{x} + \frac{2k\beta}{\omega}\dot{x} + kx = F_o \sin(\omega t) \tag{5.126}$$

In what has been presented above, systems with different forms of damping have been represented by equivalent systems with equivalent viscous damping. It is repeated that these equivalent systems may not always capture the true stability properties of the original systems, in particular, in cases like Eqs. (5.124) and (5.125). We see that these equations are nonlinear equations, because the equivalent damping term is a function of X_o, the magnitude of the displacement response.

Forced Response of System with Structural Damping

The solution to Eq. (5.126) is obtained from Eq. (5.17) by replacing c with $2k\beta/\omega$ or, equivalently, $\zeta = \beta/\Omega$. Thus, in this case,

$$x_{st}(\tau) = \frac{F_o}{k} H_{st}(\Omega)\sin(\Omega\tau - \theta_{st}(\Omega)) \tag{5.127}$$

where the associated amplitude response $H_{st}(\Omega)$ and the phase response $\theta_{st}(\Omega)$ are given by

$$H_{st}(\Omega) = \frac{1}{\sqrt{(1 - \Omega^2)^2 + 4\beta^2}}$$

$$\theta_{st}(\Omega) = \tan^{-1}\frac{2\beta}{1 - \Omega^2} \tag{5.128}$$

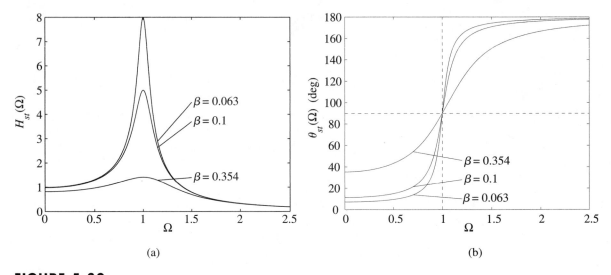

FIGURE 5.30
(a) Amplitude response and (b) phase response of system with structural damping.

Graphs of Eqs. (5.128) are given in Figure 5.30. It is observed that the value of Ω at which $H_{st}(\Omega)$ is a maximum always occurs at $\Omega = 1$.

We shall now examine $x_{st}(\tau)$ in the frequency ranges $\Omega \ll 1$ and $\Omega \gg 1$ and at $\Omega = 1$. The examination is performed by studying $H_{st}(\Omega)$ and $\theta_{st}(\Omega)$ in these three ranges.

$\boldsymbol{\Omega \ll 1}$ In this region, we approximate from Eqs. (5.128) the amplitude and phase responses as

$$H_{st}(\Omega) \to \frac{1}{\sqrt{1 + 4\beta^2}}$$

$$\theta_{st}(\Omega) \to \tan^{-1} 2\beta \qquad\qquad (5.129)$$

Thus, the displacement response in this region is determined from Eq. (5.127) to be

$$x_{st}(\tau) \cong \frac{F_o}{k\sqrt{1 + 4\beta^2}} \sin(\Omega\tau - \theta_{st}(\Omega)) \qquad\qquad (5.130)$$

Hence, the amplitude increases as β decreases.

$\boldsymbol{\Omega = 1}$ At this location, it is determined from Eq. (5.128) that

$$H_{st}(1) = \frac{1}{2\beta}$$

$$\theta_{st}(1) = \frac{\pi}{2} \qquad\qquad (5.131)$$

Then, from Eq. (5.127), we arrive at the displacement response

$$x_{st}(\tau) = -\frac{F_o}{2k\beta}\cos(\tau) \tag{5.132}$$

Hence, the magnitude of the response is proportional to $1/\beta$; that is, the peaks of $H_{st}(\Omega)$ in Figure 5.30a are inversely proportional to β. The phase angle of the displacement response is 90° out-of-phase with force, irrespective of the value of β.

$\Omega \gg 1$ In this region, we use Eq. (5.128) to arrive at the approximations

$$H_{st}(\Omega) \rightarrow \frac{1}{\Omega^2} = \frac{\omega_n^2}{\omega^2}$$

$$\theta_{st}(\Omega) \rightarrow \pi \tag{5.133}$$

In this case, Eq. (5.127) for the displacement response is simplified to

$$x_{st}(\tau) \cong \frac{F_o}{k\Omega^2}\sin(\Omega\tau - \pi) = -\frac{F_o}{k\Omega^2}\sin(\Omega\tau) = -\frac{f(\tau)}{k\Omega^2} \tag{5.134}$$

Thus, the force acts in a direction opposite to that of the displacement response.

EXAMPLE 5.12   Vibratory system with structural damping

During cyclic loading of a vibratory system, it is found that the response amplitude is 0.5 cm while the input force magnitude is 250 N. It is observed that the displacement response lags the force by 90°; that is, the system is being driven at its natural frequency. From static deflection experiments, it is found that the system stiffness is 50 kN/m. We shall determine the structural damping of this system.

Since the system is being driven at its natural frequency, $\Omega = 1$. Hence, from Eq. (5.132), the response amplitude is given by

$$X_o = \frac{F_o}{2k\beta} \tag{a}$$

On substituting the values for X_o, F_o, and k, we find from Eq. (a) that the structural damping factor is

$$\beta = \frac{F_o}{2kX_o} = \frac{250\ N}{2(50 \times 10^3\ N/m)(5 \times 10^{-2}\ m)} = 0.05 \tag{b}$$

EXAMPLE 5.13   Estimate for response amplitude of a system subjected to fluid damping

A system vibrating on a fluid medium is modeled by using fluid damping. We shall determine an estimate for the response amplitude when this system is driven at its natural frequency. From Eq. (5.118), the equivalent viscous damping coefficient is

$$c_{eq} = \frac{8c_d\omega_n X_o}{3\pi} \tag{a}$$

and the associated system damping factor is

$$\zeta = \frac{c_{eq}}{2m\omega_n} = \frac{8c_d X_o}{6\pi m} \tag{b}$$

Making use of Eqs. (5.125), (5.17), and (5.29), we find that the response amplitude for harmonic excitation at $\Omega = 1$ is given by

$$X_o = \frac{F_o H_{st}(1)}{k} = \frac{F_o}{k}\frac{1}{2\zeta} \tag{c}$$

Hence, from Eqs. (b) and (c), we find that

$$X_o = \frac{3\pi m F_o}{8k c_d X_o} = \frac{3\pi F_o}{8c_d X_o \omega_n^2} \tag{d}$$

Thus, an estimate for the response amplitude of the system is

$$X_o = \frac{1}{\omega_n}\sqrt{\frac{3\pi F_o}{8c_d}} \tag{e}$$

5.9 RESPONSE TO EXCITATION WITH HARMONIC COMPONENTS

Responses of vibratory systems subjected to periodic excitations are relevant to such widely diverse applications as internal combustion engines, propeller-driven aircraft, and weaving machinery. Here, we consider the steady-state response of a single degree-of-freedom system subjected to a forcing function that is composed of a collection of harmonic components, each at a different amplitude and frequency.

Excitation with Two Harmonic Components

Consider a harmonic excitation acting on the mass of a linear vibratory system of the form

$$f_1(t) = B\sin(\omega t) \tag{5.135a}$$

or, in the equivalent form,

$$f_1(\tau) = B\sin(\Omega\tau) \tag{5.135b}$$

where the nondimensional time $\tau = \omega_n t$ and the nondimensional frequency $\Omega = \omega/\omega_n$. Then, from Section 5.2, the steady-state displacement response is given by

$$x_1(\tau) = \frac{B}{k}H(\Omega)\sin(\Omega\tau - \theta(\Omega)) \tag{5.136}$$

where the amplitude response $H(\Omega)$ and the phase response $\theta(\Omega)$ are given by Eqs. (5.8).

When the forcing function is of the form

$$f_2(t) = A \cos(\omega t) \tag{5.137a}$$

or, in the equivalent form,

$$f_2(\tau) = A \cos(\Omega \tau) \tag{5.137b}$$

the corresponding steady-state response from Eq. (5.15) is

$$x_2(\tau) = \frac{A}{k} H(\Omega)\cos(\Omega \tau - \theta(\Omega)) \tag{5.138}$$

We now consider the linear combination of forces given by Eqs. (5.135b) and (5.137b); that is,

$$f(\tau) = A \cos(\Omega \tau) + B \sin(\Omega \tau) \tag{5.139}$$

Since the considered system is linear, the solution for the linear combination of forces given by Eq. (5.139) is the linear combination of the individual solutions[18] given by Eqs. (5.136) and (5.138). Thus,

$$\begin{aligned}
x(\tau) &= x_1(\tau) + x_2(\tau) \\
&= \frac{H(\Omega)}{k} [A \cos(\Omega \tau - \theta(\Omega)) + B \sin(\Omega \tau - \theta(\Omega))] \\
&= \frac{H(\Omega)}{k} C \sin(\Omega \tau - \theta(\Omega) + \psi)
\end{aligned} \tag{5.140}$$

where we have used Eq. (4.10) to determine that

$$C = \sqrt{A^2 + B^2}$$

$$\psi = \tan^{-1} \frac{A}{B} \tag{5.141}$$

Excitation with Multiple Harmonic Components

In general, when the forcing is of the form

$$f(t) = \sum_{i=1}^{N} [A_i \cos(\omega_i t) + B_i \sin(\omega_i t)] \tag{5.142a}$$

or, equivalently,

$$f(\tau) = \sum_{i=1}^{N} [A_i \cos(\Omega_i \tau) + B_i \sin(\Omega_i \tau)] \tag{5.142b}$$

where the ω_i are distinct, the corresponding displacement response is given by

$$\begin{aligned}
x(\tau) &= \frac{1}{k} \sum_{i=1}^{N} \{H(\Omega_i)[A_i \cos(\Omega_i \tau - \theta(\Omega_i)) + B_i \sin(\Omega_i \tau - \theta(\Omega_i))]\} \\
&= \frac{1}{k} \sum_{i=1}^{N} H(\Omega_i)C_i \sin(\Omega_i \tau - \theta(\Omega_i) + \psi_i)
\end{aligned} \tag{5.143}$$

[18]This superposition principle is an important property of linear systems.

In Eqs. (5.142b) and (5.143), the nondimensional time $\tau = \omega_n t$ and

$$\Omega_i = \frac{\omega_i}{\omega_n}$$

$$C_i = \sqrt{A_i^2 + B_i^2}$$

$$\psi_i = \tan^{-1}\frac{A_i}{B_i} \tag{5.144}$$

We see that when the input to a linear single degree-of-freedom system is a collection of harmonically varying force components each at a different frequency and amplitude, the associated displacement response is the weighted

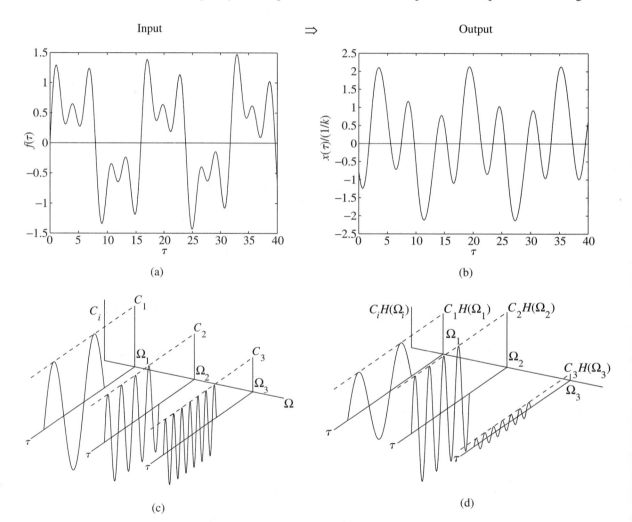

FIGURE 5.31
Time and spectral representation of the response of a system subjected to force input with three harmonic components with different frequencies and different amplitudes: (a) force input history; (b) displacement output history; (c) spectrum of input and time history of individual components; and (d) spectrum of displacement output and time history of individual components.

combination of these components comprising the input force. In the corresponding response, the amplitude of the ith input component is modified by $H(\Omega_i)$ and it is delayed an amount $\theta(\Omega_i)$. The interpretation for these results is provided in Figure 5.31. As seen from this figure, the force input consists of components at the nondimensional frequency ratios Ω_1, Ω_2, and Ω_3, with amplitudes C_1, C_2, and C_3, respectively. The time histories of the individual components comprising $f(\tau)$ are shown along with their spectral information; that is, information in the frequency domain, in Figure 5.31c. In Figure 5.31b, the displacement response of the system is shown. The individual components comprising this response are shown along with the corresponding spectral information in Figure 5.31d. These individual components are scaled and phase-shifted versions of the corresponding individual components shown in Figure 5.31c for the force input. Although in the corresponding response the components comprising $x(\tau)$ are periodic, their sum is not necessarily periodic. For the sum of the harmonic components also to be a periodic function, the different frequencies ω_i should be commensurate with respect to each other; that is, $\omega_i = p\omega_j/q$, where p and q are integers. In other words, the ratio of one frequency to another should be a rational number. This case is discussed in the following example. It is mentioned that to obtain Figure 5.31 p and q were chosen so that their ratio is not a rational number.

EXAMPLE 5.14   Response of a weaving machine

Rotating unbalanced parts in a weaving machine produce an excitation with frequency components at f_1 Hz and f_2 Hz to this machine. The vertical motions of this machine can be described by a single degree-of-freedom model

$$\ddot{x} + 2\zeta\omega_n\dot{x} + \omega_n^2 x = B_1 \sin(\omega_1 t) + B_2 \sin(\omega_2 t) \tag{a}$$

where the excitation frequency components are

$$\omega_1 = 2\pi f_1 \quad \text{and} \quad \omega_2 = 2\pi f_2 \tag{b}$$

We shall determine the response amplitude x of the weaving machine.

In this case, the vibratory system is excited by an excitation with only two distinct frequency components. Hence, Eq. (5.143) is used to obtain the response

$$x(\tau) = \frac{1}{k} \sum_{i=1}^{2} H(\Omega_i)B_i \sin(\Omega_i\tau - \theta(\Omega_i)) \tag{c}$$

where $H(\Omega_i)$ and $\theta(\Omega_i)$ are given by Eqs. (5.8) and $\Omega_i = \omega_i/\omega_n$.

It is mentioned that there are two ISO standards[19] that contain information about measurement and evaluation of machinery with unbalanced parts.

---

[19]ISO 2372, "Mechanical Vibrations of Machines with Operating Speeds from 10 to 200 rev/s for Specifying Evaluation Standards," International Standards Organization, Geneva, Switzerland (1974); and ISO 3945, "Mechanical Vibration of Large Rotating Machines with Speed Ranging from 10 to 200 rev/s—Measurement and Evaluation of Vibration Severity In Situ," International Standards Organization, Geneva, Switzerland (1985).

Next, we examine how a given periodic excitation can be broken up into harmonic components by using Fourier series, before determining the response of a vibratory system.

Fourier Series

As a special case of Eq. (5.142a), let $f(t)$ be periodic; that is, $f(t) = f(t + T)$, where $T = 2\pi/\omega_o$ and ω_o is the *fundamental frequency*. Then

$$f(t) = \frac{a_0}{2} + \sum_{i=1}^{\infty} [a_i \cos(i\omega_o t) + b_i \sin(i\omega_o t)] \tag{5.145}$$

where the quantities a_i and b_i, which are called the Fourier amplitudes, are given by

$$a_i = \frac{2}{T} \int_0^T f(t)\cos(i\omega_o t)dt \qquad i = 0, 1, 2, \ldots$$

$$b_i = \frac{2}{T} \int_0^T f(t)\sin(i\omega_o t)dt \qquad i = 1, 2, \ldots \tag{5.146}$$

Equation (5.145) is called the *Fourier-series expansion* of the signal $f(t)$. In Eq. (5.145), the frequency components $i\omega_o$ for $i > 1$ are called the higher harmonics of ω_o. For example, when $i = 2$ we have the second harmonic, when $i = 3$ we have the third harmonic, and so on.

We notice that in the definitions of a_i and b_i we have integrated over the period T. Consequently, the a_i and b_i are independent of time and these amplitudes are only a function of $i\omega_o$. In other words, we have transformed $f(t)$ into the frequency domain so that a_i and b_i represent the amplitude contributions of the sine and cosine components of the ith harmonic $i\omega_o$ comprising $f(t)$. Hence, a plot of these amplitudes as a function of $i\omega_o$ would be a plot of discrete values, since the amplitudes are zero everywhere except at the corresponding frequencies $i\omega_o$.

By using Eq. (5.143), we find that the displacement response is given by

$$x(\tau) = \frac{a_0}{2k} + \frac{1}{k} \sum_{i=1}^{\infty} \{H(\Omega_i)[a_i \cos(\Omega_i\tau - \theta(\Omega_i)) + b_i \sin(\Omega_i\tau - \theta(\Omega_i))]\}$$

$$= \frac{1}{k}\left[c_0 + \sum_{i=1}^{\infty} c_i(\Omega_i)\sin(\Omega_i\tau - \theta(\Omega_i) + \psi_i)\right] \tag{5.147}$$

where the coefficients and phases are given by

$$c_0 = \frac{a_0}{2} \quad \text{and} \quad \Omega_i = \frac{i\omega_o}{\omega_n}$$

$$c_i(\Omega_i) = H(\Omega_i)\sqrt{a_i^2 + b_i^2}$$

$$\psi_i = \tan^{-1}\frac{a_i}{b_i} \tag{5.148}$$

Fourier series expansions obtained by making use of Eqs. (5.146) for many common periodic waveforms are given in Appendix B.

Design Guidelines: When an excitation force to a linear vibratory system is composed of more than one harmonic waveform, the displacement response of the system is composed of scaled and phase shifted versions of each of these input waveforms. The displacement response of each force input waveform is modified by the amplitude response of the system at that waveform's frequency and it is delayed by system's phase response at that frequency. If the system's damping ratio is small, then in order to avoid amplifying these input waveforms, the constituent frequency components of the forcing should not be in the system's damping-dominated region. If the input waveform is composed of commensurate frequency components, then one must select the period such that neither the fundamental frequency nor any of the higher harmonics fall in the system's damping-dominated region.

EXAMPLE 5.15   Single degree-of-freedom system subjected to periodic pulse train and saw-tooth forcing

We shall consider the responses of a single degree-of-freedom system to two periodic input forcing functions: a periodic pulse train and a saw-tooth function. The pulse train and its Fourier series representation are given by waveform h of Table B in Appendix B. From this table we have that

$$f(t) = F_o\alpha\left[1 + 2\sum_{i=1}^{\infty}\frac{\sin(i\pi\alpha)}{(i\pi\alpha)}\cos(2i\pi t/T)\right]$$

or

$$f(\tau) = F_o\alpha\left[1 + 2\sum_{i=1}^{\infty}\frac{\sin(i\pi\alpha)}{(i\pi\alpha)}\cos(\Omega_i\tau)\right] \qquad (a)$$

where, $\Omega_i = i\Omega_o$, $\Omega_o = \omega_o/\omega_n$, $\alpha = t_d/T$, $T = 2\pi/\omega_o$, $\tau = \omega_n t$, and t_d is the duration of the pulse during each period. Thus, comparing Eq. (a) to Eqs. (5.145), we have

$$\frac{a_0}{2} = \alpha F_o \qquad b_i = 0 \qquad \psi_i = \tan^{-1}\frac{\sin(i\pi\alpha)/i\pi\alpha}{0}$$
$$a_i = 2\alpha F_o\frac{\sin(i\pi\alpha)}{i\pi\alpha} \qquad (b)$$

where the form of ψ_i is required in order to obtain its proper quadrant. Then, from Eqs. (5.147) and (5.148), we determine the displacement response to be

$$x(\tau) = \frac{F_o}{k}\left[\alpha + \sum_{i=1}^{\infty}c_i(\Omega_i)\sin(\Omega_i\tau - \theta(\Omega_i) + \psi_i)\right] \qquad (c)$$

where

$$c_i(\Omega_i) = \frac{2\alpha}{\sqrt{(1 - \Omega_i^2)^2 + (2\zeta\Omega_i)^2}} \left| \frac{\sin(i\pi\alpha)}{i\pi\alpha} \right| \qquad i = 1,2,...$$

$$\theta(\Omega_i) = \tan^{-1} \frac{2\zeta\Omega_i}{1 - \Omega_i^2} \tag{d}$$

Equation (c) is plotted[20] in Figure 5.32 for $\Omega_o = 0.0424$, $\alpha = 0.4$, and $\zeta = 0.1$. Thus, the non-dimensional period is $\omega_n T = 2\pi/\Omega_o = 148.2$ and $\omega_n/\omega_o = 1/\Omega_o = 23.57$, which indicates that the 23rd and 24th harmonics fall in the vicinity of ω_n. In Figure 5.32a, we see that the normalized output displacement response overshoots the input pulse's amplitude and then exhibits a decaying oscillation about the pulse's normalized height. Since $\omega_n/\omega_o = 23.57$, the period of the pulse train is 23.57 times longer than the period of the natural frequency of the system; therefore, the nondimensional period of the decaying oscillation is $148.2/23.57 \cong 2\pi$. We will show in Section 6.3 that this response is equivalent to the response of a single degree-of-freedom system subjected to a suddenly applied constant force. When damping increases substantially, these oscillations are almost eliminated as shown in Figure 5.32e.

The results plotted in Figure 5.32b are the amplitude spectrum of the pulse train before it is applied to the mass and the amplitude spectrum of the displacement of the mass in response to this force. We have also plotted the system's amplitude response function $H(\Omega)$ for reference. We see that the system's amplitude response function greatly magnifies the amplitude of those components of the force that have frequency components in the vicinity of the system's natural frequency—the damping-dominated region. It is this magnification that produces, in the time domain, the overshoot and oscillations at a frequency equal to the system's damped natural frequency. We now see why large damping eliminates these oscillations. As shown in Figure 5.32f, the amplitude components in the neighborhood of the system's natural frequency are all slightly attenuated; thus, the output response more closely follows the input.

The next case we consider is where $\Omega_o = 1/3$; that is, the natural frequency of the system is three times the fundamental frequency of the pulse train. Thus, the nondimensional period is $\omega_n T = 2\pi/\Omega_o = 18.9$ and $\omega_n/\omega_o = 1/\Omega_o = 3.0$. We see from Figure 5.32d that since the pulse train's third harmonic coincides with the natural frequency of the system, the amplitude of the third harmonic undergoes the maximum magnification. Thus, the time domain response, which is shown in Figure 5.32c, is dominated by the component whose frequency is coincident with the system's natural frequency. Consequently, the displacement response does not bear any resemblance to the input forcing function, except that it has the same period.

The responses of a linear vibratory system subjected to a saw-tooth type forcing waveform given by entry b of Table B in Appendix B are shown in Figure 5.33. Several qualitative aspects of these responses are similar to

---

[20]200 terms were used in the summation.

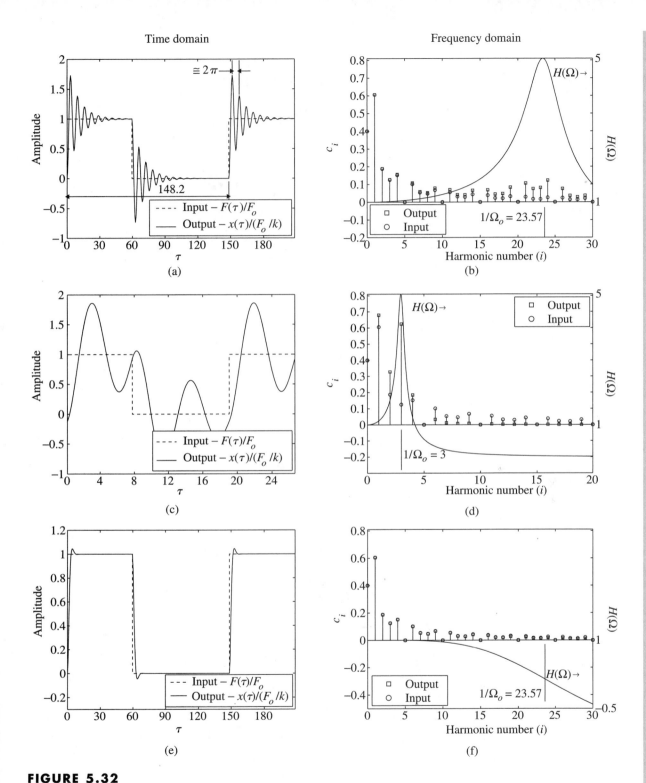

FIGURE 5.32

Comparison of responses to pulse train forcing function in the time domain and frequency domain for two different values of the system damping ratio and two different fundamental excitation frequencies: (a) and (b) $\zeta = 0.1$, $t_d/T = 0.4$, and $\Omega_o = 0.0424$; (c) and (d) $\zeta = 0.1$, $t_d/T = 0.4$, and $\Omega_o = 0.333$; and (e) and (f) $\zeta = 0.7$, $t_d/T = 0.4$, and $\Omega_o = 0.0424$. [Note: For display purposes the time axes have been shifted by $\pi\alpha/\Omega_o$.]

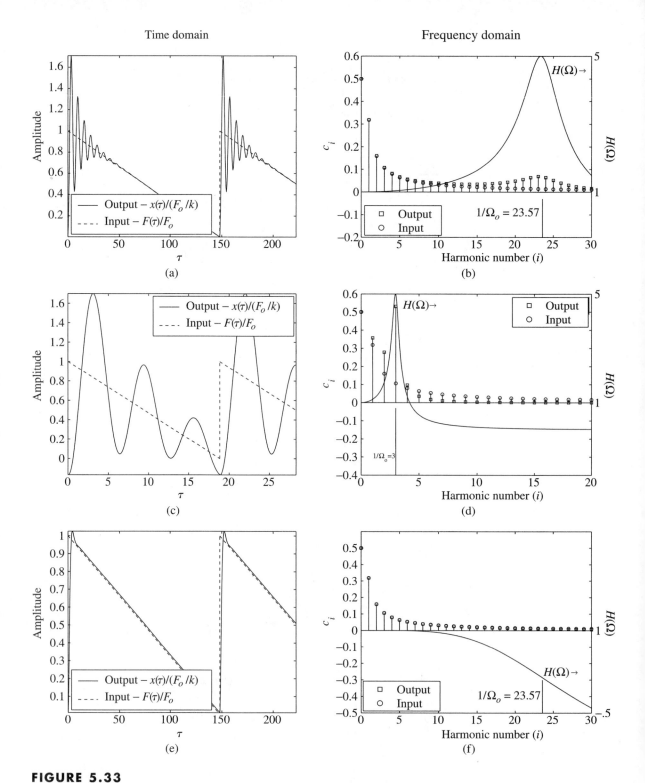

FIGURE 5.33

Comparison of the responses to a periodic saw-tooth forcing function in the time domain and frequency domain for two different values of the system damping ratio and two different fundamental excitation frequencies: (a) and (b) $\zeta = 0.1$ and $\Omega_o = 0.0424$; (c) and (d) $\zeta = 0.1$ and $\Omega_o = 0.333$; and (e) and (f) $\zeta = 0.7$ and $\Omega_o = 0.0424$.

those seen in Figure 5.32, where a periodic pulse train forcing and the corresponding response are presented. As discussed in Section 6.3, a step-like change in the input to a linear damped system always results in a "transient" component of the response. It is seen that the transient response within each period of the response is also captured.

EXAMPLE 5.16   Single degree-of-freedom system response to periodic impulses[21]

We shall consider the response of a single degree-of-freedom system that is subjected to a periodic train of force impulses of period T. We shall compare these results to those obtained for a pulse train and determine under what conditions a pulse train can be used to imitate a periodic impulse train. The periodic impulse force is given by

$$f(t) = F_o' \sum_{i=-\infty}^{\infty} \delta(t - iT) \tag{a}$$

where $\delta(t)$ is a generalized function called the *delta function*.[22] The delta function can be expressed as the following Fourier series[23]

$$f(t) = \frac{F_o'}{T}\left[1 + 2\sum_{i=1}^{\infty} \cos(2i\pi t/T)\right]$$

or, in terms of nondimensional quantities as

$$f(\tau) = \frac{F_o'}{T}\left[1 + 2\sum_{i=1}^{\infty} \cos(\Omega_i \tau)\right] \tag{b}$$

where $\Omega_i = i\Omega_o$, $\Omega_o = \omega_o/\omega_n$, $\omega_o = 2\pi/T$, and F_o' has the units N·s. Comparing Eq. (b) with Eq. (5.145), we have

$$\frac{a_0}{2} = \frac{F_o'}{T} \qquad b_i = 0$$

$$a_i = \frac{2F_o'}{T} \qquad \psi_i = \tan^{-1}\frac{2F_o'/T}{0} = \frac{\pi}{2} \tag{c}$$

From Eqs. (5.147) and (5.148), we determine the displacement response to be

$$x(\tau) = \frac{F_o'}{m\omega_n}\frac{\Omega_o}{2\pi}\left[1 + 2\sum_{i=1}^{\infty} H(\Omega_i)\cos(\Omega_i\tau - \theta(\Omega_i))\right] \tag{d}$$

---

[21]For a complete discussion of the subject, see V. I. Babitsky, *Theory of Vibro-Impact Systems and Applications,* Springer, Berlin (1998). See also S. A. Kember and V. I. Babitsky, "Excitation of vibro-impact systems by periodic excitation," *J. Sound Vibration,* Vol. 227, No. 2, pp. 427–447 (1999).

[22]See Section 6.2 for a discussion of the delta function and the units of F_o'.

[23]See, for example, A. Papoulis, *The Fourier Integral and Its Applications,* McGraw-Hill, NY, p. 44 (1962).

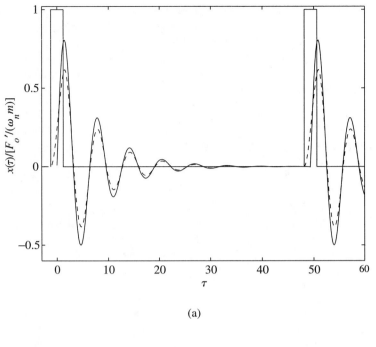

(a)

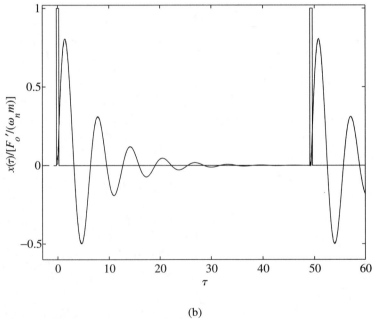

(b)

FIGURE 5.34
Responses of a single degree-of-freedom system to periodic force impulses
and periodic force pulse train: (a) $t_d/T = 0.05$ and (b) $t_d/T = 0.01$. (The
solid lines are the responses to the periodic impulses and the dashed lines
are the responses to the pulse train.)

It is noted from Eqs. (a) and (c) that the magnitude of the Fourier series coefficients are constants, independent of the fundamental frequency of the pulse train and its harmonics, and for $n \geq 1$, these constants are equal. Thus, the periodic impulse train has a discrete, uniform harmonic spectrum, with each spectral component occurring at integer multiples of ω_o.

Equation (d) is plotted in Figure 5.34 along with the results of the periodic pulse train given by Eq. (c) of Example 5.15. In order to compare the two solutions, we have to change the term multiplying Eq. (c) of Example 5.15. This change is required because the units of the magnitude of the applied force have to be the same. Thus, in Eq. (c) of Example 5.15 we have

$$\frac{\alpha F_o}{k} = \frac{t_d F_o}{kT} \rightarrow \frac{F_o'}{kT} = \frac{F_o'}{m\omega_n} \frac{\Omega_o}{2\pi} \tag{e}$$

We see from Figure 5.34b that when $t_d/T \leq 0.01$, the pulse train is a good approximation to the impulse train, since both responses overlap.

EXAMPLE 5.17   Base excitation: slider-crank mechanism

Consider the slider-crank mechanism that is attached to the base of the single degree-of-freedom system shown in Figure 5.35. We shall determine the displacement response of the mass and the corresponding response spectrum. The displacement of the base due to the slider-crank is[24]

$$\bar{y}(\tau) = \frac{R}{L} \cos(\Omega_o \tau) + \left[1 - \left(\frac{R}{L}\right)^2 \sin^2(\Omega_o \tau) \right]^{1/2} \tag{a}$$

where

$$\tau = \omega_n t \quad \bar{y} = y/L \quad \Omega_o = \omega_o/\omega_n$$

$$1 - \frac{R}{L} \leq \bar{y} \leq 1 + \frac{R}{L} \tag{b}$$

and ω_o is the rotational frequency of the crank arm of length R. The corresponding velocity is

$$\frac{d\bar{y}}{d\tau} =$$

$$-\Omega_o \frac{R}{L} \left[\sin(\Omega_o \tau) + (R/L)\sin(\Omega_o \tau)\cos(\Omega_o \tau) / \left[1 - \left(\frac{R}{L}\right)^2 \sin^2(\Omega_o \tau) \right]^{1/2} \right] \tag{c}$$

In this notation, Eq. (5.76) takes the form

$$\frac{d^2 w}{d\tau^2} + 2\zeta \frac{dw}{d\tau} + w = 2\zeta \frac{dy}{d\tau} + \bar{y} \tag{d}$$

[24]S. G. Kelly, *Fundamentals of Mechanical Vibrations*, 2nd ed., McGraw Hill, NY, pp. 171–173 (2000); E. Brusa et al., "Torsional Vibration of Crankshafts: Effects of Non-Constant Moments of Inertia," *J. Sound Vibration*, Vol. 205, No. 2, pp. 135–150 (1997).

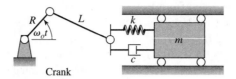

FIGURE 5.35
Slider-crank mechanism connected to the base of a system.

where the response of the slider has been scaled as

$$w = x/L \tag{e}$$

Numerical Results and Discussion

The excitation described by Eq. (a) is periodic and has the Fourier components at frequencies given by the harmonics of the (dimensionless) fundamental frequency Ω_o. Instead of using the analytical expressions presented in this section, due to the form of $\bar{y}(\tau)$ and its derivative, it is convenient to solve this equation numerically.[25] We assume the following sets of parameters: a) $R/L = 0.8$, $\zeta = 0.1$, and $\Omega_o = 0.3, .5$, and 1; and b) $R/L = 0.1$, $\zeta = 0.1$, and $\Omega_o = 0.3, .5$, and 1. The results are shown in the Figures 5.36 and 5.37, respectively. The dashed horizontal lines in these figures represent the amplitude limits of $\bar{y}$.

The presence of multiple frequency components in the excitation (the slider-crank's displacement) results in a response with more than one frequency component. By expanding Eq. (a), we can see that the second harmonic is a consequence of the second term's contribution to the displacement. Thus, for $|R/L| \ll 1$, Eq. (a) is approximated by

$$\bar{y} = 1 - \left(\frac{R}{2L}\right)^2 + \frac{R}{L}\cos(\Omega_o\tau) + \left(\frac{R}{2L}\right)^2 \cos(2\Omega_o\tau) - \ldots \tag{f}$$

While expansion given by Eq. (f) is sufficient for $R/L = 0.1$, higher order terms will be needed for $R/L = 0.8$.

We see from Figures 5.36 and 5.37 that the transient response has died out by the time $\tau = 50$. If we take the discrete Fourier transform (DFT) of the steady-state portion of the displacement response ($\tau > 50$), then we can determine the frequency content of the steady-state response. The frequency spectrum indicates that the relative magnitude of the second harmonic is a function of the rotation frequency of the slider-crank mechanism and the ratio R/L, as can be seen in Eq. (f).

The relationship of the rotation frequency of the slide-crank mechanism to the natural frequency of the system influences the relative magnitudes of the two dominant frequency components of the base's displacement. From Eq. (f), we see that the ratio of amplitude of the second harmonic to the

[25]The MATLAB function ode45 was used to obtain the solution to Eq. (d) and the MATLAB function fft was used to determine the corresponding amplitude spectrum.

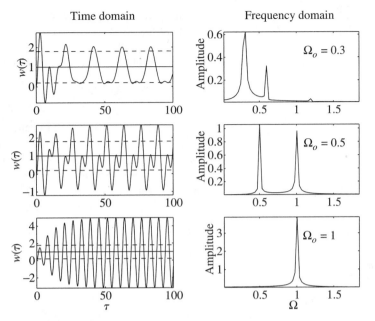

FIGURE 5.36
Slider-crank mechanism: displacement response of the mass and the steady-state amplitude spectrum of the displacement for $R/L = 0.8$ and $\zeta = 0.1$.

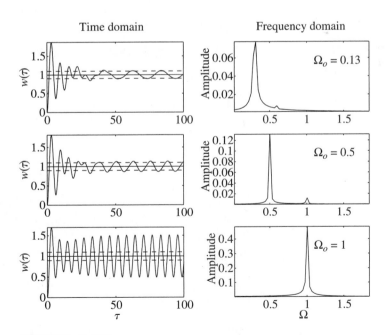

FIGURE 5.37
Slider-crank mechanism: displacement response of the mass and the steady-state amplitude spectrum of the displacement for $R/L = 0.1$ and $\zeta = 0.1$.

amplitude of fundamental frequency is $a_{21} = R/(4L)$. However, the system's amplitude response function $H_{mb}(\Omega)$, given by Eqs. (5.82), modifies each of these components at their respective frequencies. Thus, the ratio of the amplitude of the second harmonic to that of the fundamental is

$$R_{21}(\Omega_o) = \frac{R}{4L} \frac{H_{mb}(2\Omega_o)}{H_{mb}(\Omega_o)} \tag{g}$$

We are now in a position to explain the amplitude spectral plots for $\Omega_o = 0.5$. From Eq. (5.89) we have seen that the magnification of a frequency located at the natural frequency of the base-excited system is approximately $1/(2\zeta)$ for small damping factors. Since $\zeta = 0.1$, the magnification factor is 5. When $\Omega_o = 0.5$, the second harmonic coincides with the system's natural frequency, thereby magnifying it by a factor of 5. Thus, for case (1), where $R/L = 0.8$, we find that

$$R_{21}(0.5) = \frac{0.8}{4} \frac{H_{mb}(1)}{H_{mb}(0.5)} = 0.2\left(\frac{5}{1.32}\right) = 0.76$$

and when $R/L = 0.1$,

$$R_{21}(0.5) = \frac{0.1}{4} \frac{H_{mb}(1)}{H_{mb}(0.5)} = 0.025\left(\frac{5}{1.32}\right) = 0.095$$

which are comparable to the numerical values determined from the figures.

5.10   INFLUENCE OF NONLINEAR STIFFNESS ON FORCED RESPONSE

In this section, we examine the forced response of nonlinear single degree-of-freedom systems to harmonic excitation of the form

$$f(\tau) = F_o \cos(\Omega_o \tau) \tag{5.149}$$

where F_o is the excitation amplitude, $\Omega_o = \omega_o/\omega_n$ is the nondimensional frequency, $\tau = \omega_n t$ is the nondimensional time, and ω_o is the excitation frequency. The single-degree-of-freedom systems considered in this section have stiffness nonlinearity. In one case, the nonlinearity is cubic, while in another case, the nonlinearity is due to loss of contact in the system. Notions such as the backbone curve are discussed in this section and the effects of the nonlinearity on the frequency response of the system are illustrated. The dependence of the response on the excitation magnitude F_o is also studied.

Case 1: System with Cubic Nonlinearity
In this case, the system has the form

$$\frac{d^2 x}{d\tau^2} + 2\zeta \frac{dx}{d\tau} + x + \alpha x^3 = \frac{F_o}{k} \cos(\Omega_o \tau) \tag{5.150}$$

where k is the linear stiffness and α is the coefficient of the nonlinear term with units of m^{-2}.

The solution to Eq. (5.150) has to be obtained numerically[26] for the nonlinear case ($\alpha \neq 0$). For the linear and nonlinear cases, we assume that $\Omega_o = 3$, $\zeta = 0.4$, and $F_o/k = 50$ units; for the nonlinear case, we assume that $\alpha = 1.5 \ m^{-2}$. The numerical solutions to Eq. (5.150) are presented in Figure 5.38, along with their amplitude spectra. The amplitude spectra were obtained from the steady-state portions of the signal;[27] that is, for $\omega_n t > 10$. We see that in the time domain, the steady-state nonlinear response is not sinusoidal. From an examination of the amplitude response, we see that it is the sum of two components; that at the driving frequency $\Omega_o = 3$ and one additional one at the third harmonic, $3\Omega_o = 9$. The third harmonic is due to the nonlinear behavior of the spring.

For weak nonlinearity, the following approximation for the periodic amplitude response of the nonlinear system has been obtained[28] from a two-term approximate solution to Eq. (5.150)

$$X_o = \frac{S}{\sqrt{(1 \pm X_o^2 - \Omega^2)^2 + (2\zeta\Omega)^2}} \tag{5.151}$$

where $\Omega = \omega/\omega_n$ and

$$X_o = X\sqrt{\frac{3\alpha}{4}} \qquad \text{and} \qquad S = \frac{F_o}{k}\sqrt{\frac{3\alpha}{4}} \tag{5.152}$$

In Eqs. (5.151) and (5.152), X is the nondimensional magnitude of the displacement of the mass, X_o is the nondimensional response amplitude, and S is the nondimensional force amplitude. In Eq. (5.151), the plus sign is used for a hardening spring ($\alpha > 0$) and the minus sign is used for a softening spring ($\alpha < 0$). A graph of Eq. (5.151) is given in Figure 5.39.[29] Unlike the case of a linear system, the maximum value of the amplitude occurs at an excitation frequency away from the natural frequency. This type of result also appears in the design of strain gauges.[30] In Figure 5.39, the amplitude information is presented along with information about the stability of the response.

In Figure 5.39, the solid lines are used to represent the loci of stable periodic responses and the broken lines are used to represent the loci of unstable responses. The determination of stability of periodic responses is not discussed here, but broadly speaking, an unstable periodic response is not stable to disturbances.[31]

---

[26]The MATLAB function `ode45` was used.

[27]The MATLAB function `fft` was used.

[28]See, for example, L. S. Jacobsen and R. B. Ayre, *Engineering Vibrations*, McGraw-Hill, NY, pp. 286–293 (1958); or A. H. Nayfeh and D. T. Mook, *Nonlinear Oscillations*, John Wiley & Sons, NY (1979).

[29]For $\alpha < 0$, the curves would "bend" in the other direction.

[30]C. Gui et al., "Nonlinearity and Hysteresis of Resonant Strain Gauges," *J. Microelectromechanical Systems*, Vol. 7, No. 1, pp. 122–127 (March 1998).

[31]A. H. Nayfeh and B. Balachandran, *Applied Nonlinear Dynamics: Analytical, Computational, and Experimental Methods*, John Wiley & Sons, NY, Chapter 2 (1995).

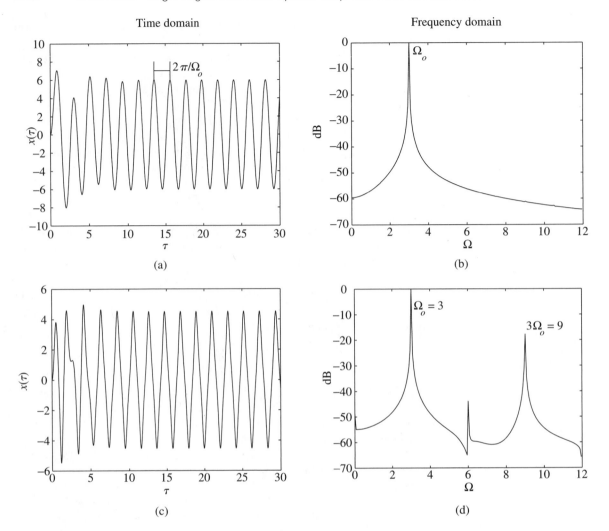

FIGURE 5.38
Displacement response of linear ($\zeta = 0.4$) and nonlinear ($\zeta = 0.4$ and $\alpha = 1.5$) systems to harmonic excitation: (a) time history of linear system; (b) amplitude spectrum of linear system; (c) time history of nonlinear system; and (d) amplitude spectrum of nonlinear system.

To examine what happens along a frequency-response curve, let us look at the curve plotted for $\zeta = 0.05$ and $S = 0.5$. In this case, as the excitation frequency is increased gradually from 0, the response amplitude follows the branch *EDA*. At the excitation frequency corresponding to point A, a jump occurs, and for further increases in the excitation frequency, the response follows the branch *BF*. On the reverse sweep, as the excitation frequency is decreased from the value corresponding to the point F, the response follows the branch *FBC* before a jump takes place at point C. For further decrease in the excitation frequency, the response follows the branch *DE*.

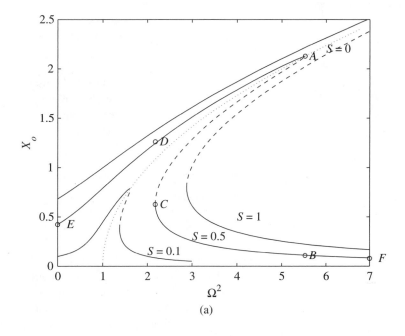

(a)

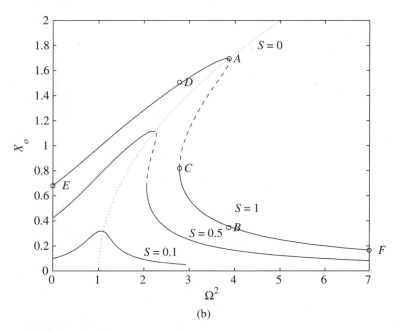

(b)

FIGURE 5.39

Representative amplitude response for system with linear and nonlinear stiffness elements with $\alpha > 0$: (a) $\zeta = 0.05$ and (b) $\zeta = 0.15$.

The locus of the peak amplitudes shown by a dotted line in Figure 5.39 is called a *backbone curve*. It is given by

$$\Omega^2 = 1 + X_o^2 \tag{5.153}$$

When the nonlinearity is of the softening type, the frequency curves bend toward the left as the excitation amplitude is increased. Again, jumps occur in the response, beyond certain excitation amplitudes. Also, in some cases, chaotic motions, which are a form of aperiodic motions with identifiable characteristics, can be obtained.[32]

Case 2: Dynamic interactions of gear teeth[33]

We shall derive the governing equation of motion for a pair of meshing gears that incorporates nonlinear stiffness characteristics, which arise in real systems due to the changing gear teeth contact regimes and the tooth separations due to tooth clearances (backlash). This governing equation is numerically solved to study the nonlinear behavior of the system.

The simplified gear system is shown in Figure 5.40a, where the gear centers are rigidly mounted and the stiffness and damping are provided by the elasticity and damping of the meshing gear teeth. The equivalent single degree-of-freedom system is shown in Figure 5.40b, where x_1 is the tangential displacement of a meshing tooth on gear 1 and x_2 is the tangential displacement of the meshing tooth on gear 2. These quantities are related to the rotations by

$$x_j = r_j \varphi_j \qquad j = 1, 2 \tag{5.154}$$

where r_j is the radius of gear j and φ_j is the angular rotation of gear j. The rotation φ_j is assumed to consist of a steady-state mesh speed ω_j, where ω_1 and ω_2 are rotational speeds of gears 1 and 2, respectively, and a variation about this steady-state value given by $\theta_j(t)$; that is,

$$\varphi_j(t) = \omega_j t + \theta_j(t) \qquad j = 1, 2 \tag{5.155}$$

Thus, Eqs. (5.154) become

$$x_j = r_j(\omega_j t + \theta_j(t)) \qquad j = 1, 2 \tag{5.156}$$

Governing Equation of Motion

The meshing gear teeth are modeled as a single degree-of-freedom system with a moving base as shown in Figure 5.40b, where the quantity m_e is de-

---

[32]A. H. Nayfeh and B. Balachandran, *ibid.*

[33]For a more complete treatment of gear systems see, for example: A. Kahraman and R. Singh, "Interactions between time-varying mesh stiffness and clearance non-linearities in a geared system," *J. Sound Vibration*, Vol. 146, No. 1, pp. 135–156 (1991); A. Kahraman and G. W. Blankenship, "Interactions between commensurate parametric and forcing excitations in a system with clearance," *J. Sound Vibration*, Vol. 194, No. 3, pp. 317–336 (1996); C. Padmanabhan and R. Singh, " Analysis of periodically forced nonlinear Hill's oscillator with application to a geared system," *J. Acoustic Soc. Amer.*, Vol. 99, No. 1, pp. 324–334 (January 1996); and S. Theodossiades and S. Natsiavas, "Non-linear dynamics of gear-pair systems with periodic stiffness and backlash," *J. Sound Vibration*, Vol. 229, No. 2, pp. 287–310 (2000).

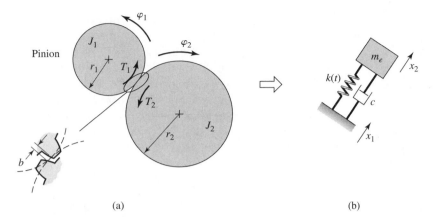

FIGURE 5.40
(a) Meshing gears and (b) equivalent single degree-of-freedom system model.

termined as follows. Since we have a system with a moving base, we form the difference between the displacements of gears 1 and 2 as

$$z = x_1 - x_2 = r_1\theta_1 - r_2\theta_2 \tag{5.157}$$

where we have used Eq. (5.156) and noted that $r_1\omega_1 = r_2\omega_2$. Next, we differentiate Eq. (5.157) twice with respect to time to obtain

$$\ddot{z} = r_1\ddot{\theta}_1 - r_2\ddot{\theta}_2 \tag{5.158}$$

However, from Figure 5.40a and Eqs. (1.17) and (1.21), we have that

$$T_1 = J_1\ddot{\varphi}_1 = J_1\ddot{\theta}_1 = r_1F$$
$$T_2 = -J_2\ddot{\varphi}_2 = -J_2\ddot{\theta}_2 = r_2F \tag{5.159}$$

where T_j is the variation of the torque and F is the corresponding force. From Eqs. (5.158) and (5.159), we obtain

$$\ddot{z} = \frac{r_1^2}{J_1}F + \frac{r_2^2}{J_2}F = \left(\frac{r_1^2}{J_1} + \frac{r_2^2}{J_2}\right)F \tag{5.160}$$

or, equivalently, as

$$F = m_e\ddot{z} \tag{5.161}$$

where the effective inertia m_e is given by

$$m_e = \left(\frac{r_1^2}{J_1} + \frac{r_2^2}{J_2}\right)^{-1} \tag{5.162}$$

We assume that the stiffness of the gear tooth is the time-varying function

$$k(t) = k_o\left(1 - \epsilon \cos \omega_M t\right) \tag{5.163}$$

where k_o is the mean stiffness, ϵ is a stiffness ratio, and $\omega_M = n_1\omega_1 = n_2\omega_2$ is the rotating speed of the gears and n_1 and n_2 are the number of teeth on gears 1 and 2, respectively. The stiffness of the gear tooth is only engaged when the mating teeth come in contact; that is, after the mating tooth traverses a small separation distance b. This is similar, in principle, to the nonlinear spring system shown in Figure 4.24. Hence, we use the following relation to express the restoring force of the spring

$$F_s = k(t)h(z) \tag{5.164}$$

where

$$
\begin{aligned}
h(z) &= 0 && |z| \leq b \\
&= z - b \, \text{sgn}(z) && |z| > b
\end{aligned}
\tag{5.165}
$$

The governing equation for the system shown in Figure 5.40 is, therefore,

$$m_e\ddot{z} + c\dot{z} + k(t)h(z) = f(t) \tag{5.166}$$

where we assume that the applied force on the teeth of the pinion is

$$f(t) = \frac{T_o}{r_1}(1 + \alpha \cos \omega_M t) \tag{5.167}$$

and T_o is the steady-state torque applied by the pinion and α is a parameter used to select the magnitude of the time-varying portion of the torque. To convert Eq. (5.166) to a nondimensional form, we introduce the following definitions

$$
\omega_n^2 = \frac{k_o}{m_e} \qquad 2\zeta = \frac{c}{m_e\omega_n} \qquad \Omega_M = \frac{\omega_M}{\omega_n}
$$

$$
\tau = \omega_n t \qquad f_o = \frac{T_o}{r_1 b k_o} \qquad p = \frac{z}{b}
\tag{5.168}
$$

Then, Eq. (5.166) becomes

$$\ddot{p} + 2\zeta\dot{p} + (1 - \epsilon \cos \Omega_M\tau)h(p) = f_o(1 + \alpha \cos \Omega_M\tau) \tag{5.169}$$

where the overdot indicates the derivative with respect to τ and

$$
\begin{aligned}
h(p) &= 0 && |p| \leq 1 \\
&= p - \text{sgn}(p) && |p| > 1
\end{aligned}
\tag{5.170}
$$

Numerical Results

The steady-state responses of the numerical evaluation[34] of Eq. (5.169) are shown in Figure 5.41. It is seen that for $\epsilon = 0$ the steady-state time histories exhibit a harmonic response. As the magnitude of ϵ increases, the steady-state response is transformed into a nonharmonic periodic solution with an increasing peak-to-peak value.

---

[34]The MATLAB function ode45 was used.

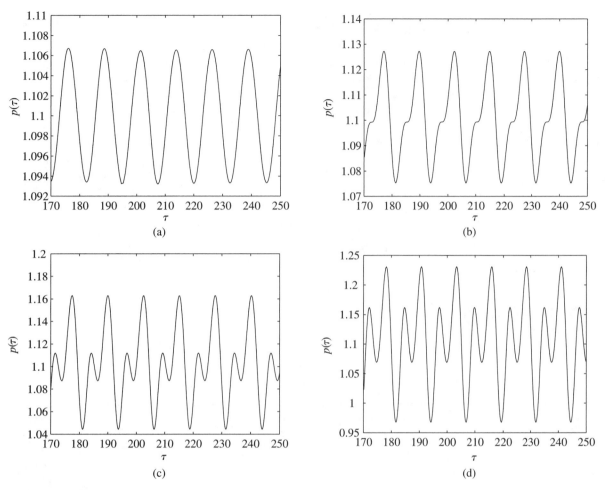

FIGURE 5.41
Steady-state time histories of gear system for $\Omega_M = 0.5$, $\zeta = 0.05$, $\alpha = 0.05$, $f_o = 0.1$: (a) $\epsilon = 0$; (b) $\epsilon = 0.1$; (c) $\epsilon = 0.2$; and (d) $\epsilon = 0.4$.

EXAMPLE 5.18   Determination of system linearity from amplitude response characteristics

Suppose that you are told that you have a single degree-of-freedom system subjected to periodically varying excitation with a known magnitude A_o. An engineer gives you a spectrum of the displacement of the mass, which is shown in Figure 5.42. The engineer wants to know what could cause the spectrum to look like this and if there is a way to determine the cause if there is more than one way that this could have occurred.

There are two possible scenarios in which this type of spectrum can occur. In one scenario, the excitation force is composed of two frequency

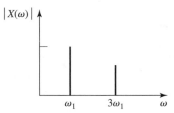

FIGURE 5.42
Spectrum of the output of single degree-of-freedom system with unknown properties.

components, one at ω_1 and the other at $3\omega_1$ and the system has a linear stiffness. The other scenario is one where the system is subjected to a force at frequency ω_1 and the system has a nonlinear cubic spring, such as the one described in this section. Both of these scenarios can produce this spectrum.

There are two independent ways to determine which of these scenarios is applicable here. One can subject the system's spring to a series of known static forces, which includes as one of its values the forcing amplitude A_o, and measure the corresponding displacements. A plot of the force levels versus the observed displacement magnitudes would reveal whether or not the spring is linear or nonlinear. (Recall Figures 2.7 and 2.12.) If the system is linear, then assuming that the inertia and damping elements are linear, the forcing excitation contains two frequency components.

Another approach to examine if the second scenario is present is to subject the mass to a periodic force at frequency ω_1 only. If the spectrum of the displacement has only one frequency component, the system is linear, whereas if it contains two components, the system is nonlinear.

5.11 SUMMARY

In this chapter, responses of single degree-of-freedom systems subjected to harmonic and other periodic excitations are studied. Different sources of forcing such as rotating unbalance and base excitation are considered. The notions of system resonance and frequency-response functions are also introduced and explained. The topic of vibration isolation is discussed at length. The underlying principle of an accelerometer is also explained. For different damping models, the notion of an equivalent viscous damping is introduced and explained. The influence of nonlinear stiffness elements on the forced response of a system is also treated.

EXERCISES

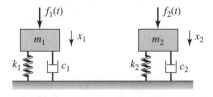

FIGURE E5.1

5.1 Consider the two independent single degree-of-freedom systems in Figure E5.1 that are each being

forced to vibrate harmonically at the same frequency ω. System 1 starts its vibration at $t = 0$ and system 2 starts its vibration at $t = t_o$; that is,

$$f_1(t) = F_1 \sin(\omega t)u(t)$$
$$f_2(t) = F_2 \sin(\omega[t - t_o])u(t - t_o)$$

a) Use Eqs. (5.1) to (5.9) to show that the steady-state responses of the two systems are

$$x_{1ss}(\tau) = \frac{F_1}{k_1} H(\Omega, \zeta_1) \sin(\Omega\tau - \theta(\Omega, \zeta_1)) u(\tau)$$

$$x_{2ss}(\tau) = \frac{F_2}{k_2} H(\Omega/\omega_r, \gamma\zeta_1) \sin(\Omega/\omega_r(\tau - \tau_o))$$
$$- \theta(\Omega/\omega_r, \gamma\zeta_1)) u(\tau - \tau_o)$$

where $\Omega = \omega/\omega_{n1}$, $\omega_r = \omega_{n2}/\omega_{n1}$, $\gamma = \zeta_2/\zeta_1$, and

$$H(a,b) = \frac{1}{\sqrt{(1 - a^2)^2 + (2ba)^2}}$$

$$\theta(a,b) = \tan^{-1}\frac{2ba}{1 - a^2}$$

b) If both systems are operating in their respective mass-dominated regions, then what is the ratio of the magnitudes of the amplitudes of system 2 to that of system 1 and their relative phase.

5.2 A vibratory system with a natural frequency of 10 Hz is suddenly excited by a harmonic excitation at 6 Hz. What should the damping factor of the system be so that the system settles down within 5% of the steady-state amplitude in 200 ms?

5.3 A 150 kg mass is suspended by a spring-damper combination with a stiffness of 30×10^3 N/m and a viscous damping constant of 1500 N·s/m. The mass is initially at rest. Calculate the steady-state displacement amplitude and phase when the mass is excited by a harmonic force amplitude of 70 N at 3 Hz.

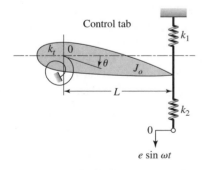

FIGURE E5.4

5.4 The control tab of an airplane elevator is shown schematically in the Figure E5.4. The mass moment

of inertia J_O of the control tab about the hinge point O is known but the torsional spring constant k_t associated with the control linkage is difficult to evaluate, and hence, the natural frequency $\omega_n = \sqrt{k_t/J_O}$ is difficult to determine. An experiment is designed to determine this natural frequency of the system. In this experiment, springs with stiffness k_1 and stiffness k_2 are attached to the control tab, and the tab is harmonically excited at an amplitude e, as illustrated in the figure. The excitation frequency ω is varied until resonance occurs at $\omega = \omega_r$, and this value is noted. Assuming that the damping in the system is negligible, determine an expression for ω_n in terms of ω_r, k_1, k_2, J_O, and L.

5.5 Consider translational motions of a vibratory system with a mass m of 200 kg and a stiffness k of 200 N/m. When a harmonic forcing of the form $F_o \sin(\omega t)$ acts on the mass of the system, where $F_o = 1.0$ N, determine the responses of the system and plot them for the following cases: a) $\omega = 0.2$ rad/s; b) $\omega = 1.0$ rad/s; and c) $\omega = 2.0$ rad/s, and discuss the results.

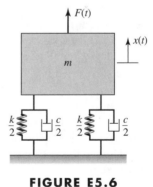

FIGURE E5.6

5.6 A machine of 25 kg mass is mounted on springs and dampers as shown in Figure E5.6. The equivalent stiffness of the spring combination is 9 kN/m and the equivalent damping of the damper combination is 150 N·s/m. An excitation force $F(t)$ is directly applied to the mass of the system, as shown in the figure. Consider the displacement $x(t)$ as the output, the forcing $F(t)$ as the input, and determine the frequency response of this system.

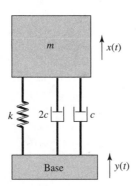

FIGURE E5.7

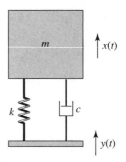

FIGURE E5.11

5.7 In the system shown in Figure E5.7, $y(t)$ is the base displacement and $x(t)$ is the displacement response of the mass m. Consider the base velocity $\dot{y}$ as the input, the acceleration response $\ddot{x}(t)$ as the output, and determine the frequency-response function of this system.

5.8 A micromechanical resonator is to be designed to have a Q factor of 1000 and a natural frequency of 2 kHz. Determine the system damping factor and the system bandwidth.

5.9 If a sensor modeled as a spring-mass-damper system is to be redesigned so that its sensitivity is increased by a factor of four, determine the corresponding percentage changes in the system natural frequency and damping ratio.

5.10 An air compressor with a total mass of 100 kg is operated at a constant speed of 2000 rpm. The unbalanced mass is 4 kg and the eccentricity is 0.12 m. The properties of the mounting are such that the damping factor $\zeta = 0.15$. Determine the following: a) the spring stiffness that the mounting should have so that only 20 percent of the unbalance force is transmitted to the foundation and (b) the amplitude of the transmitted force.

5.11 The damped single degree of freedom spring-mass system shown in Figure E5.11 has a mass $m = 20$ kg and a spring stiffness coefficient $k = 2400$ N/m. Determine the damping coefficient of the system, if it is given that the mass exhibits a response with an amplitude of 0.02 m when the support is harmonically excited at the natural frequency of the system with an amplitude $Y_0 = 0.007$ m. In addition, determine the amplitude of the dynamic force transmitted to the support.

5.12 An accelerometer is being designed to have a uniform amplitude frequency response of $\pm 1\%$ up to $f_a = 8$ kHz. What is the maximum damping ratio allowed if the phase angle is to be less than $2°$ over this frequency range?

5.13 Determine the phase shift and amplitude error in the output of an accelerometer that has a natural frequency of 25 kHz and a damping ratio of 0.1, when it is measuring vibrations at 1 kHz.

5.14 A rotating piece of machinery runs intermittently. When it is operating, it is rotating at 4200 rpm. The mass of the machine is 100 kg and it is supported by an elastic structure with an equivalent spring constant of 160 kN/m and a viscous damper of damping coefficient of 2400 N·s/m that is in parallel with the spring. The engineer would like to keep the maximum transmission ratio less than 2.3 during the period that it takes the machine to reach the operating speed. If a spring is inserted in series with the damper, then what is the best value of the stiffness of this spring in order to have a the TR < 0.08 when the machine is at its operating speed?

5.15 Show that the work done per cycle by a harmonic force acting directly on the mass of a linear spring-mass-damper system is equal to the energy dissipated by the system per forcing cycle.

5.16 A spring-mass system with $m = 20$ kg and $k = 8000$ N/m vibrates horizontally on a surface with coefficient of friction $\mu = 0.2$. When excited harmonically at 5 Hz, the steady-state displacement of the mass is 10 cm. Determine the equivalent viscous damping.

5.17 The area of the hysteresis loop of a cyclically loaded system, which is the energy dissipated per forcing cycle, is measured to be 10 N·m, and the measured maximum response X_o of the deflection is 2 cm. Calculate the equivalent viscous damping coefficient of this system if the driving force has a frequency of 30 Hz.

5.18 A compressor weighing 1000 kg operates at 1500 rpm. The compressor was originally attached to the floor of a building, but it produced undesirable vibrations to the building. To reduce these vibrations to the building, it is proposed that a concrete block be poured that is separated from the building and that the compressor then be mounted to this block. The location of the compressor will permit the block to be 1.8 m by 2.2 m. The soil on which the concrete block will rest has a compression coefficient $k_c = 20 \times 10^6$ N/m³. If the density of the concrete is 23×10^3 N/m³, then determine the height of the concrete block so that there is an 80% reduction in the force transmitted to the soil.

5.19 A motor is mounted at the end of a cantilever beam and it is found that the beam deflects 10 mm. When the motor is running at 1800 rpm, an unbalanced force of 100 N is measured. If the beam damping is negligible and its mass can be neglected, then what speed should the motor operate at so that the magnitude of the dynamic amplification is less than 15% about the static equilibrium value?

5.20 Torsional oscillations of a vibratory system are governed by the following equation

$$J_o\ddot{\theta} + c_t\dot{\theta} + k_t\theta = M_1\cos(\omega_1 t) + M_2\cos(\omega_2 t)$$

where

$$J_0 = 20 \text{ kg·m}^2, \quad k_t = 20 \text{ N·m/rad},$$
$$c_t = 20 \text{ N·m/(rad/s)}$$
$$M_1 = 10 \text{ N·m}, \quad M_2 = 20 \text{ N·m},$$
$$\omega_1 = 1.0 \text{ rad/s}, \quad \text{and} \quad \omega_2 = 2.0 \text{ rad/s}$$

Determine the steady-state response of the system.

5.21 Determine the output of an accelerometer with damping factor ζ and natural frequency ω_n when it is mounted on a system executing periodic displacement motions of the form

$$y = A_1\sin\omega_1 t + A_2\sin\omega_2 t$$

5.22 A single degree-of-freedom system is driven by the periodic triangular wave excitation as shown in case d of Appendix B. If $T = 2$ s and $f_o = 10$ N, then find the steady-state response of the system by considering the first three harmonics of the forcing. Assume that the system parameters are $k = 10$ kN/m, $c = 10$ N·s/m and $m = 1$ kg.

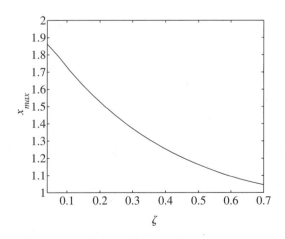

FIGURE E5.23

5.23 Consider again the results of Example 5.15, where the response of a single degree-of-freedom system to a periodic pulse train was studied. If we were to differentiate Eq. (c) of this example with respect to τ and set the result equal to zero, then the value of $\tau = \tau_{max}$ that satisfies this equation in one period is the time at which the maximum value $x_{max} = x(\tau_{max})$ occurs. Use MATLAB (or a similar program), to determine x_{max} as a function of the damping ratio for $\Omega_o = 0.042426$ and $\alpha = 0.4$. The results should look like those shown in Figure E5.23.

5.24 Repeat Exercise 5.23, except for the abscissa use the nondimensional bandwidth of the system. Explain your results.

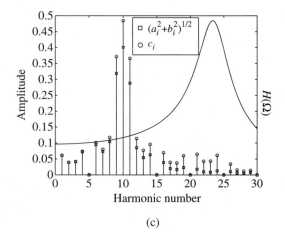

(c)

FIGURE E5.25 (continued)

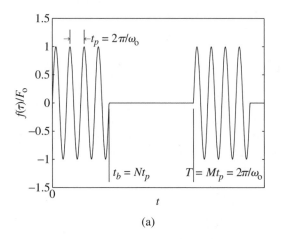

(a)

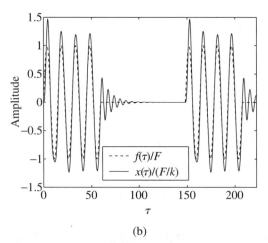

(b)

FIGURE E5.25

5.25 Consider a sine wave forcing of magnitude F_o and frequency ω_b that has a duration of N periods t_p, where $t_p = 2\pi/\omega_b$. This sine wave is repeated every period T, where $T = Mt_p = 2\pi/\omega_o$, $M \geq N$, and N and M are integers. If this periodic waveform is applied to the mass of a single degree-of-freedom system, then

$$f(t) = \sin(\omega_b t)[u(t) - u(t - t_b)]$$
$$0 \leq t \leq T, \qquad 0 < t_b < T \qquad \text{(a)}$$

where $t_b = Nt_p$ and $f(t) = f(t + T)$. See Figure E5.25a.

a) Expand Eq. (a) in a Fourier series and show that

$$a_0 = 0$$

$$a_i = \frac{1}{\pi M} \frac{1 - \cos(2\pi i N/M)}{1 - (i/M)^2} \qquad i \neq M$$

$$a_i = 0 \qquad i = M$$

$$b_i = \frac{-1}{\pi M} \frac{\sin(2\pi i N/M)}{1 - (i/M)^2} \qquad i \neq M$$

$$b_i = \frac{N}{M} \qquad i = M \qquad \text{(b)}$$

These results were obtained by noting that $\omega_b t_b = 2\pi N$, $\omega_o/\omega_b = 1/M$, and $\omega_b T = 2\pi M$.

b) Use Eq. (b) and Eqs. (5.147) and (5.148) to determine the displacement response of the mass.

c) Assume that $N = 4$, $M = 10$, $\Omega_o = 0.0424$ or 0.3333, and $\zeta = 0.1$ or 0.7. Using MATLAB, or a comparable program, generate the equivalent figures for this excitation as that shown in Figure 5.31. One pair of the results should look like that shown in Figures E5.25b and c for $\Omega_o = 0.0424$ and $\zeta = 0.1$. Explain these results.

5.26 Consider the base excitation of a single degree-of-freedom system shown in Figure 3.6 and whose motion is described by Eq. (5.76). If the displacement of the base is the periodic sawtooth waveform described by Case c in Appendix B, then obtain an expression for the displacement of the system's mass.

5.27 Consider the following nonlinear single degree of freedom system subjected to a harmonic excitation, where $f_o = 300$ N, $\omega = \omega_n/3$, $m = 100$ kg, $c = 170$ N/m/s, $k = 2$ kN/m, and $\alpha = 10$ kN/m^3. Assume that the initial conditions are $x(0) = 0.01$ m and

$v(0) = 0.1$ m/s. Compute the time response of the system and associated amplitude spectrum of steady-state motions and compare it with the corresponding quantities of the linear system; that is, when $\alpha = 0$.

$$m\ddot{x} + c\dot{x} + kx - \alpha x^3 = f_0 \cos \omega t$$

5.28 Consider the following nonlinear single degree of freedom system subjected to a harmonic excitation, with the same values of parameters as in Exercise 5.27 except that now the excitation frequency is at $\omega = \omega_n$.

$$m\ddot{x} + c\dot{x} + kx + \alpha x^3 = f_0 \cos \omega t$$

Compute the response of the system and compare it with that of the corresponding linear system. In addition, for this harmonic excitation compare the differences between the responses of the system of softening stiffness discussed in Exercise 5.27 and the system of this exercise with hardening stiffness.

Mechanical systems and civil structures must be designed to withstand dynamic environments, which are often non periodic. The supporting structure of a rock breaker has to endure the intermittent application of non-periodic impulses generated during the rock breaking process. Structures such as the Seattle Space Needle must be designed to resist intermittent wind-induced forces. (*Source:* Rockbreaker image courtesy of Kanga Loaders, Inc.; Seattle Space Needle image courtesy of Digital Vision/Getty Images.)

6

Single Degree-of-Freedom Systems
Subjected to Transient Excitations

6.1   INTRODUCTION

In Chapter 4, the general solution for the response of a single degree-of-freedom system was derived and discussed, and from this solution, responses to harmonic and other periodic excitations were addressed in Chapter 5. As illustrated in the last two chapters, a "sudden" change in the state of a system brought about by an initial condition or by a change in the profile of the forcing function results in transients in the response of the system. This transient behavior is further discussed here. Specifically, the initial conditions are zero and the responses to various types of excitations such as impulse excitations, step inputs, ramp inputs, and pulse excitations are considered at length. All of these excitations are characterized by sudden changes in their respective profiles of amplitude with time. The responses to such transient excitations can also provide a basis for determining the characteristics of a system by transforming to the frequency domain the excitation and the resulting displacement response, and several design criteria are established based on this information. In the systems considered in this chapter, both the inertia element and the base of the system are subjected to transient forcing.

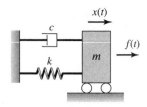

FIGURE 6.1
Spring-mass-damper system
subjected to forcing $f(t)$.

Solution for Response to Transient Excitation

When the initial displacement $X_o = 0$ and the initial velocity $V_o = 0$, the governing equation for an underdamped system $(0 < \zeta < 1)$ with a forcing acting on the mass as shown in Figure 6.1 is given by Eq. (5.2), which is repeated below.

$$x(t) = \frac{1}{m\omega_d} \int_0^t e^{-\zeta\omega_n \eta} \sin(\omega_d \eta) f(t - \eta) d\eta$$

$$= \frac{1}{m\omega_d} \int_0^t e^{-\zeta\omega_n(t-\eta)} \sin(\omega_d[t - \eta]) f(\eta) d\eta \qquad (6.1a)$$

In Eq. (6.1a), η is the variable of integration and it is recalled that the damped natural frequency ω_d is given by

$$\omega_d = \omega_n\sqrt{1 - \zeta^2}$$

In terms of nondimensional time $\tau = \omega_n t$, Eq. (6.1a) is written as

$$x(\tau) = \frac{1}{k\sqrt{1 - \zeta^2}} \int_0^\tau e^{-\zeta(\tau-\xi)} \sin((\tau - \xi)\sqrt{1 - \zeta^2}) f(\xi) d\xi \qquad (6.1b)$$

where the variable of integration $\eta = \omega_n \xi$.

Next, we use Eqs. (6.1) as a basis for determining the responses to several different forcing functions.

6.2   RESPONSE TO IMPULSE EXCITATION

An impulse excitation is described by

$$f(t) = f_o \delta(t) \qquad (6.2)$$

where f_o is the magnitude of the impulse and has the units[1] of N·s and $\delta(t)$ is a generalized function called the *delta function*, which is defined by the property[2]

$$\int_{-\infty}^{\infty} \delta(t - t_o) f(t) dt = f(t_o) \qquad (6.3)$$

---

[1]Based on Eq. (6.2) and (6.3), the magnitude f_o of the impulse acting on the system is given by

$$f_o = \int_0^t f(t) dt$$

hence, the units N·s.

[2]See, for example, A. Papoulis, *The Fourier Integral and Its Applications*, McGraw-Hill, NY, p. 270 ff (1962).

where $f(t)$ is assumed to be continuous at $t = t_o$. After substituting Eq. (6.2) into Eq. (6.1a) and evaluating the integral, one obtains

$$x(t) = \frac{f_o}{m\omega_d} \int_0^t e^{-\zeta\omega_n\eta} \sin(\omega_d\eta)\delta(t - \eta)d\eta$$

$$= \frac{f_o}{m\omega_d} e^{-\zeta\omega_n t} \sin(\omega_d t)u(t) \qquad (6.4a)$$

which is rewritten in terms of the non-dimensional time $\tau = \omega_n t$ as

$$x(\tau) = \frac{f_o}{m\omega_n} \frac{e^{-\zeta\tau}}{\sqrt{1 - \zeta^2}} \sin(\sqrt{1 - \zeta^2}\tau)u(\tau) \qquad (6.4b)$$

In Eqs. (6.4), the unit step function is included to indicate that the response is limited to those times for which $t \geq 0$. The displacement response given by Eq. (6.4b), which has the form of an exponentially decaying sinusoid for $0 < \zeta < 1$, is plotted in Figure 6.2 for a set of representative parameter values.

Similarity to Response to Initial Velocity

The graphs shown in Figure 6.2 are qualitatively similar to those shown in Section 4.3.2 for a linear vibratory system subjected to no forcing and to an initial velocity. In fact, the form of the response given by Eq. (6.4a) is identical to that given by Eq. (4.18), which is given by

$$x(t) = \frac{V_o e^{-\zeta\omega_n t}}{\omega_d} \sin(\omega_d t) \qquad (6.5)$$

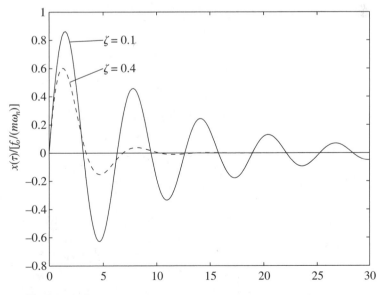

FIGURE 6.2

Response of a system to an impulse.

Comparing Eqs. (6.4a) and (6.5), we see that

$$V_o = \frac{f_o}{m}$$

which should be expected since the change in linear momentum of a system is equal to the impulse applied to a system; that is, from Eq. (1.11), we have

$$\text{change in momentum} = \int_0^t f(t)dt$$

Assuming that prior to applying the impulse the system is at rest, then V_o is the velocity of the mass after the impulse has acted on the system. In other words, the response of a system to an impulse can be determined as the response of a system with the corresponding initial velocity.

Extremum of Response to an Impulse

In view of the identical nature of the response of the system to an impulse and that due to an initial velocity given by Eqs. (6.4a) and (6.5), respectively, the maximum displacement x_{max} of the impulse response can be obtained directly from Eq. (4.23). Thus, since $V_o = f_o/m$, the first extremum is

$$x_{max} = x(\tau_m) = \frac{f_o}{m\omega_n} e^{-\varphi/\tan \varphi} \tag{6.6}$$

which, from Eq. (4.22), occurs at

$$\tau_m = \frac{\varphi}{\sqrt{1 - \zeta^2}} \tag{6.7}$$

where

$$\varphi = \tan^{-1} \frac{\sqrt{1 - \zeta^2}}{\zeta} \tag{6.8}$$

Force Transmitted to Boundary

The force transmitted to the fixed boundary of Figure 6.1 is given by Eq. (3.10). Thus,

$$F_b = kx(t) + c \frac{dx(t)}{dt} = k\left[x(t) + \frac{2\zeta}{\omega_n} \frac{dx(t)}{dt} \right] \tag{6.9}$$

After using Eq. (6.4a) and Eq. (4.10), we obtain for $t > 0$

$$F_b = \frac{f_o\omega_n}{\sqrt{1 - \zeta^2}} e^{-\zeta\omega_n t} \sin(\omega_d t + \psi) \tag{6.10}$$

where the phase angle ψ is given by

$$\psi = \tan^{-1} \frac{2\zeta\sqrt{1 - \zeta^2}}{1 - 2\zeta^2}$$

The maximum force transmitted to the fixed boundary occurs at the time t_m at which $dF_b/dt = 0$. Thus, Eq. (6.10) leads to

$$\omega_d t_m = \varphi - \psi \tag{6.11}$$

where φ is given by Eq. (6.8). Thus, the maximum normalized force[3] transmitted to the fixed boundary is given by

$$\frac{F_{b,\max}}{f_o \omega_n} = e^{-(\varphi-\psi)/\tan\varphi} \tag{6.12}$$

This result is plotted in Figure 6.3, where it is seen that $F_{b,\max}/(f_o\omega_n)$ is a minimum at $\zeta \approx 0.25$, where $F_{b,\max}/(f_o\omega_n) = 0.81$. It is noted that as the damping increases beyond $\zeta \approx 0.25$, the magnitude of the transmitted force increases. From Eqs. (6.10) and (6.12) and Figure 6.3, one can deduce the following design guideline.

Design Guideline: To obtain the maximum decrease in the amount of force transmitted to the base of a linear single degree-of-freedom system due to an impulse (shock) loading, the damping ratio should be between 0.2 and 0.35. For this range of damping values, the maximum attenuation of the force transmitted to the fixed boundary will be around 18%. For a given damping ratio and magnitude of the impulse force, decreasing the natural frequency decreases the magnitude of the force transmitted to the fixed boundary.

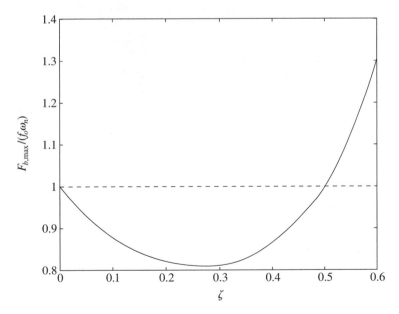

FIGURE 6.3
Maximum magnitude of impulse force transmitted to fixed boundary.

---

[3]Strictly speaking, the force transmitted has an extremum at $t = t_m$. In order to determine that this force is a maximum here, we use numerical means.

Time Domain and Frequency Domain Counterparts

We now examine the information corresponding to the displacement response given by Eq. (6.4a) in the frequency domain. To this end, we recall the following Fourier transform pair given by Eq. (5.60):

$$\frac{e^{-\zeta\omega_n t}}{\omega_d}\sin(\omega_d t)u(t) \Leftrightarrow \frac{1}{(\zeta\omega_n + j\omega)^2 + \omega_d^2} \tag{6.13}$$

where

$$\frac{1}{(\zeta\omega_n + j\omega)^2 + \omega_d^2} = \frac{1}{\omega_n^2\left[1 - (\omega/\omega_n)^2 + 2j\zeta(\omega/\omega_n)\right]}$$

$$= \frac{1}{\omega_n^2}H(\omega/\omega_n)e^{-j\theta(\omega/\omega_n)}$$

$$= mG(j\omega) \tag{6.14}$$

The amplitude response $H(\omega/\omega_n)$ and the phase response $\theta(\omega/\omega_n)$ are given by Eqs. (5.56a) and (5.56b), respectively, and $G(j\omega)$ is the frequency-response function given by Eq. (5.55). Hence, based on Eqs. (6.13), the Fourier transform of the impulse response given by Eq. (6.4a) is

$$\frac{f_o e^{-\zeta\omega_n t}}{m\omega_d}\sin(\omega_d t)u(t) \Leftrightarrow \frac{f_o}{k}H(\omega/\omega_n)e^{-j\theta(\omega/\omega_n)} = f_o G(j\omega) \tag{6.15a}$$

When the magnitude of the impulse f_o is unity, a *unit impulse* is said to be applied to the system shown in Figure 6.1. The corresponding response is called the *impulse response* of the single degree-of-freedom system and is given by

$$h(t) = \frac{e^{-\zeta\omega_n t}}{m\omega_d}\sin(\omega_d t)u(t) \tag{6.15b}$$

This response is the time domain counterpart of its frequency response function, which comprises the amplitude response $H(\omega/\omega_n)$ and the phase response $\theta(\omega/\omega_n)$; that is,

$$\frac{e^{-\zeta\omega_n t}}{m\omega_d}\sin(\omega_d t)u(t) \Leftrightarrow \frac{1}{k}H(\omega/\omega_n)e^{-j\theta(\omega/\omega_n)} = G(j\omega) \tag{6.15c}$$

Transfer Function and Frequency-Response Function

Revisiting the displacement response of the single degree-of-freedom system given by Eq. (4.4) in the Laplace domain, we find that when the initial conditions are zero, the Laplace transform of the response reduces to

$$X(s) = \frac{F(s)}{mD(s)} \tag{6.16}$$

which is rewritten as

$$X(s) = G(s)F(s) \tag{6.17}$$

where the transfer function $G(s)$ is given by

$$G(s) = \frac{1}{mD(s)} \tag{6.18}$$

In the frequency domain, Eq. (6.17) becomes

$$X(j\omega) = G(j\omega)F(j\omega) \tag{6.19}$$

When an excitation of the form given by Eq. (6.2) is applied to a system, the corresponding Laplace transform is given by transform pair 5 in Table A of Appendix A. Thus,

$$F(s) = f_o \tag{6.20}$$

Hence, the corresponding Fourier transform $F(j\omega)$ has a constant magnitude f_o over the entire frequency span. In other words, the magnitude of the amplitude spectrum of an impulse function is constant over the frequency range $0 \leq \omega \leq \infty$. In this regard, recall the fact that the Fourier coefficients for a periodic impulse were constant for all harmonics.

From a practical standpoint, an impulse function is hard to realize, since an ideal impulse has an "extremely large" magnitude that lasts an infinitesimally short time. However, a device called an impact hammer can be used in an experiment to apply a pulse of finite time duration to the mass of a system. An impact hammer usually has a built-in transducer to measure the force amplitude-time profile created by the hammer. The corresponding Fourier transform of the hammer's pulse has a fairly constant magnitude over a finite frequency bandwidth. As discussed in Section 6.6, this bandwidth becomes larger as the pulse duration is reduced.

The notion of an impulse response leads to an experimental procedure to determine the frequency-response function. This procedure is an alternative to that discussed in Chapter 5 where a harmonic excitation was used to construct this function. Thus, one would use an impact hammer or another source to apply a force (or a moment) of very short time duration to the inertia element of a system, measure and digitize the forcing and displacement response signals, use the discrete Fourier transform on these signals to convert the time domain information to the frequency domain, and then determine the frequency-response function $G(j\omega)$ based on Eq. (5.59c), which is appears in this chapter as Eq. (6.19).

Response of a Linear System to Arbitrary Inputs Based on Impulse Response

Examining Eq. (6.17), which is in the Laplace domain, one can state that for a linear vibratory system the displacement response is the product of the system's transfer function and the Laplace transform of the system input. Similarly, from Eq. (6.19), which is in the frequency domain, it can be stated that the displacement response is the product of the system's frequency-response function and the Fourier transform of the system input. This input-output relationship, which is characteristic of all linear vibratory systems, is schematically illustrated in Figure 6.4.

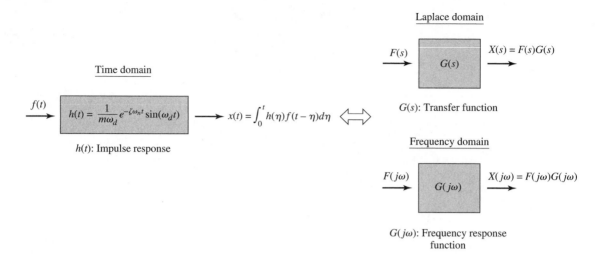

FIGURE 6.4
System input-output relationships in time and transformed domains.

From the transform pair 4 in Table A of Appendix A, we arrive at the time domain counterpart of Eq. (6.17); that is,

$$x(t) = \int_0^t h(\eta)f(t - \eta)d\eta \qquad (6.21)$$

where $f(t)$ is the forcing applied to the inertia elements and the function $h(t)$ is the impulse response of the system; that is,

$$h(t) = \frac{1}{m\omega_d} e^{-\zeta\omega_n t} \sin(\omega_d t) \qquad (6.22)$$

As noted in Section 5.1, Eq. (6.21) is the convolution integral. Thus, the applied forcing is convolved with the system impulse response function over the time interval of interest to determine the system response as a function of time.

EXAMPLE 6.1   Response of a linear vibratory system to multiple impacts[4]

An impact hammer is used to manually apply an impulse to a model of a single degree-of-freedom system as illustrated in Figure 6.5. In experiments, a single impact is preferred; however, it is difficult to realize only a single impact and multiple impacts may occur. Assume that for a system with a mass

[4]D. J. Inman, *Engineering Vibration,* Prentice Hall, Upper Saddle River, NJ (2001).

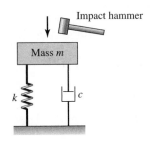

FIGURE 6.5
System inertia element struck
by an impact hammer.

of 2 kg, stiffness of 8 N/m, and damping coefficient of 2 N·s/m, a double
impact of the form[5]

$$f(t) = \delta(t) + 0.5\delta(t - 1)$$

occurs. We shall determine the system response and compare it to that obtained when the second impact at $t = 1$ s is absent.

For the given parameter values, the natural frequency ω_n and the damping factor ζ are determined, respectively, as

$$\omega_n = \sqrt{\frac{8 \text{ N/m}}{2 \text{ kg}}} = 2 \text{ rad/s}$$

$$\zeta = \frac{2 \text{ N·s/m}}{2 \times (2 \text{ kg}) \times (2 \text{ rad/s})} = 0.25$$

Since $\zeta < 1$, we have an underdamped system. Based on Eq. (6.4a), the response of the system subjected to a single impact is given by

$$x_1(t) = \frac{1}{2 \times 2\sqrt{1 - 0.25^2}} e^{-(0.25 \times 2)t} \sin(2\sqrt{1 - 0.25^2}\,t) \text{ m}$$

$$= 0.26 e^{-0.5t} \sin(1.94t)u(t) \text{ m}$$

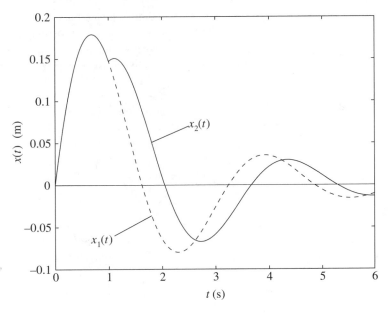

FIGURE 6.6
Displacement responses to single impulse and two impulses one second apart.

---

[5]Of course, a pulse of a finite-time duration rather than the delta function better approximates
the impact from a hammer. The responses to such pulse functions are discussed later in this
chapter.

The step function $u(t)$ is included to indicate that the first impact affects the response for $t \geq 0$. The response of the system subjected to the double impact is given by

$$x_2(t) = \underbrace{x_1(t)u(t)}_{\substack{\text{Response to} \\ \text{first impact}}} + \underbrace{0.5x_1(t-1)u(t-1)}_{\text{Response to second impact}}$$

$$= x_1(t)u(t) + (0.26 \times 0.5)e^{-0.5(t-1)}\sin(1.94(t-1))u(t-1) \text{ m}$$
$$= x_1(t)u(t) + 0.13e^{-0.5(t-1)}\sin(1.94(t-1))u(t-1) \text{ m}$$

The unit step function $u(t-1)$ is used to indicate that the second impact affects the response only for $t \geq 1$ s. Before one second, that is before the second impact occurs, the responses to the single and double impacts are identical, as seen in the graphs of the responses provided in Figure 6.6. However, after one second, the responses are different, as expected.

EXAMPLE 6.2   Stress level under impulse loading

Consider the spring-mass-damper system shown in Example 6.1, and assume that the top of the spring is welded to the mass. The welding is over an area $A_w = 4$ mm^2, and the allowed maximum stress for the weld material is $\sigma_{w,\text{max}} = 150$ MN/m^2. For an impulse of magnitude 100 N·s, we shall determine whether the stress level in the weld material will be below the maximum allowed stress. It is assumed that $\zeta = 0.25$, $m = 200$ kg, $k = 800$ N/m, and therefore, $\omega_n = 2$ rad/s.

The impulse is given by Eq. (6.2) with $f_o = 100$ N·s, and the corresponding displacement response of the system is determined from Eq. (6.4a). Then, the force acting on the weld material is determined and from the maximum value of this force, the maximum stress experienced by the weld material during the motions is determined. This value is compared to the maximum allowed stress level, $\sigma_{w,\text{max}}$. The force acting on the weld material is given by

$$f_{weld}(t) = kx(t) \tag{a}$$

The maximum force on the weld is given by Eq. (a) when $x(t) = x_{\text{max}}$, where x_{max} is given by Eq. (6.6). From Eq. (6.8), we find that

$$\varphi = \tan^{-1}\left(\frac{\sqrt{1-\zeta^2}}{\zeta}\right) = \tan^{-1}\left(\frac{\sqrt{1-.25^2}}{.25}\right) = 1.318 \text{ rad} \tag{b}$$

and, therefore,

$$\tan \varphi = \tan 1.318 = 3.873 \tag{c}$$

Then, from Eq. (6.6), x_{max} is

$$x_{\text{max}} = \frac{100}{200 \times 2}e^{-1.318/3.873} = 0.178 \text{ m} \tag{d}$$

By using Eqs. (a) and (d), the maximum force on the weld is

$$f_{weld,\max} = kx_{\max} = 800 \times 0.178 = 142.4 \text{ N} \qquad (e)$$

Hence, the maximum stress $\sigma_{w,\max}$ experienced by the weld material is given by

$$\sigma_{w,\max} = \frac{f_{weld,\max}}{A_w} = \frac{142.4 \text{ N}}{4 \times 10^{-6} \text{ m}^2} = 35.6 \text{ MN/m}^2 \qquad (f)$$

This value is well below the maximum allowed stress level in the weld material of 150 MN/m². To exceed the given value of maximum allowed stress, the impulse magnitude will have to be at least 4.2 times stronger than the present impulse magnitude.

EXAMPLE 6.3   Design of a structure subjected to sustained winds

We shall determine an estimate of the outer diameter d_o of a 10 m high steel lamppost of constant cross-section that is subjected to sustained winds so that the maximum transverse displacement of the lamps on top of the lamppost does not exceed 5 cm. The mass of the lights on top of the lamppost is 75 kg. We assume that the lamppost is a cylindrical tube whose inner diameter is 95% of d_o, that the system acts as a beam in bending, and that it can be modeled as a single degree-of-freedom system. In addition, the damping ratio of the system is assumed to be $\zeta = 0.04$. The magnitude of the turbulence-induced wind force spectrum has been experimentally determined to be

$$|F(jf)| = 400 \, fe^{-0.667f} \quad \text{N} \qquad (a)$$

where f is frequency in Hz. The spectrum given by Eq. (a) is shown in Figure 6.7.

Response in Frequency Domain

The solution is obtained by first determining from the displacement response in the frequency domain the value of the equivalent stiffness k for which the maximum amplitude $|X(jf)|$ is less than 0.05 cm over the entire frequency range of $|F(jf)|$. After the magnitude of the equivalent stiffness is known, the value of d_o is determined from the relationship for the stiffness of a cantilever beam. Noting from Eq. (6.18) that the frequency-response function is

$$G(j\omega) = \frac{1}{m[(\zeta\omega_n + j\omega)^2 + \omega_d^2]} \qquad (b)$$

the displacement response in the frequency domain is then determined from Eqs. (a) and (b), and Eq. (6.19). This response is converted from the radian frequency notation ω to frequency f in Hz resulting in

$$|X(jf)| = \frac{|F(jf)|}{k} \left[\left(1 - \left(\frac{f}{f_n}\right)^2 \right)^2 + \left(\frac{2\zeta f}{f_n}\right)^2 \right]^{-1/2} \qquad (c)$$

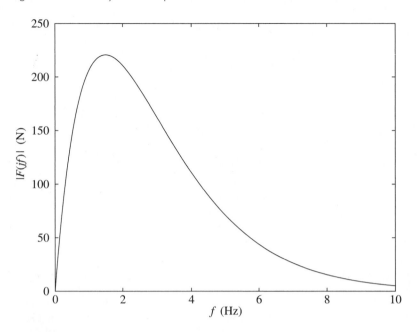

FIGURE 6.7
Assumed wind-induced force spectrum acting on the lamppost.

where the natural frequency f_n in Hz is

$$f_n = \frac{1}{2\pi} \sqrt{\frac{k}{m}} \tag{d}$$

Upon substituting Eqs. (d) and (a) into Eq. (c) and noting that $m = 75$ kg and $\zeta = 0.04$, we obtain

$$
\begin{aligned}
|X(jf)| &= \frac{400\,fe^{-0.667f}}{k}\left[\left(1 - \left(2\pi f\sqrt{\frac{m}{k}}\right)^2\right)^2 + \left(4\pi\zeta f\sqrt{\frac{m}{k}}\right)^2\right]^{-1/2} \\
&= \frac{400\,fe^{-0.667f}}{k}\left[\left(1 - \left(2\pi f\sqrt{\frac{75}{k}}\right)^2\right)^2 + \left(4\pi(0.04)f\sqrt{\frac{75}{k}}\right)^2\right]^{-1/2} \\
&= \frac{400\,fe^{-0.667f}}{k}\left[\left(1 - 2960.9\frac{f^2}{k}\right)^2 + 18.95\frac{f^2}{k}\right]^{-1/2} \tag{e}
\end{aligned}
$$

We now plot Eq. (e) for three values of k: 20,000, 30,000, and 40,000 N/m. The results are shown in Figure 6.8. It is seen that as the stiffness increases, the natural frequency increases and the maximum magnitude of $|X(jf)|$ decreases. This decrease is brought about by the fact that the spectral amplitude of $|F(jf)|$ diminishes as f increases, when f is greater than about 1.6 Hz. Since this magnitude of the forcing is decreasing in this frequency region, the magnification by the system's frequency response function $|H(jf)|$ in

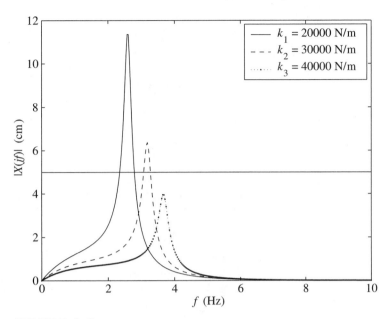

FIGURE 6.8
Displacement response spectrum of lamppost for three values of stiffness.

the damping-dominated region is magnifying a smaller quantity; hence, the peak magnitude of $|X(jf)|$ becomes smaller.

Another way to determine approximately the peak responses is to recall the response of a system in the damping-dominated region given by Eq. (5.29); that is,

$$H(1) = \frac{1}{2\zeta} \tag{f}$$

Thus, for $\zeta = 0.04$, $H(1) = 12.5$. The values of $|F(jf)|$ at the three the natural frequencies corresponding to the three stiffness values k_j are, respectively,

$$f_{n1} = \frac{1}{2\pi} \sqrt{\frac{k_1}{m}} = \frac{1}{2\pi} \sqrt{\frac{20000}{75}} = 2.6 \text{ Hz}$$

$$f_{n2} = \frac{1}{2\pi} \sqrt{\frac{k_2}{m}} = \frac{1}{2\pi} \sqrt{\frac{30000}{75}} = 3.18 \text{ Hz}$$

$$f_{n3} = \frac{1}{2\pi} \sqrt{\frac{k_3}{m}} = \frac{1}{2\pi} \sqrt{\frac{40000}{75}} = 3.68 \text{ Hz} \tag{g}$$

Then, from Eqs. (a) and (g), we have that

$$|F(jf_{n1})| = 400f_{n1}e^{-0.667\,fn1} = 400 \times 2.6e^{-0.667\times2.6} = 183.6 \text{ N}$$

$$|F(jf_{n2})| = 400f_{n2}e^{-0.667\,fn2} = 400 \times 3.18e^{-0.667\times3.18} = 152.5 \text{ N}$$

$$|F(jf_{n3})| = 400f_{n3}e^{-0.667\,fn3} = 400 \times 3.68e^{-0.667\times3.68} = 126.4 \text{ N} \tag{h}$$

These values could also have been obtained directly from Figure 6.7, but with less precision. Consequently, the peak magnitudes of the displacements at these three frequencies are, respectively,

$$|X(jf_{n1})| = \frac{1}{k_1}| F(jf_{n1}) H(1) = \frac{183.6 \times 12.5}{20000} = 0.115 \quad \text{m}$$

$$|X(jf_{n2})| = \frac{1}{k_2}| F(jf_{n2})| H(1) = \frac{152.5 \times 12.5}{30000} = 0.0635 \quad \text{m}$$

$$|X(jf_{n2})| = \frac{1}{k_3}| F(jf_{n3})| H(1) = \frac{126.4 \times 12.5}{40000} = 0.0395 \quad \text{m} \tag{i}$$

It is seen that these values closely correspond to their respective counterparts in Figure 6.8.

Lamppost Parameters

The value of k that produces a maximum magnitude of $|X(jf)|$ equal to 5 cm is determined numerically[6] to be 34,909 N/m. To determine the diameter of the cylindrical support, we note from entry 4 in Table 2.3 that

$$k = \frac{3EI}{L^3} \tag{j}$$

where

$$I = \frac{\pi}{32}(d_o^4 - d_i^4) \tag{k}$$

Upon combining Eqs. (j) and (k), and noting that $k = 34{,}909$ N/m, $d_i = 0.95d_o$, $L = 10$ m, and $E = 2.1 \times 10^{11}$ N/m^2, we obtain

$$d_o = \sqrt[4]{\frac{32kL^3}{3E\pi(1 - (0.95)^4)}}$$

$$= \sqrt[4]{\frac{32 \times 34{,}909 \times 10^3}{3\pi \times 2.1 \times 10^{11} \times (1 - (0.95)^4)}}$$

$$= 0.235 \quad \text{m} \tag{l}$$

Thus, the outer diameter of the lamppost is 23.5 cm and its inner diameter 22.3 cm, resulting in a wall thickness of 1.2 cm.

6.3 RESPONSE TO STEP INPUT

The step input to a vibratory system shown in Figure 6.1 is expressed as

$$f(t) = F_o u(t) \tag{6.23}$$

---

[6]The MATLAB function `fzero` and the MATLAB function `fminbnd` from the Optimization Toolbox were used.

where $u(t)$ is the unit step function. After substituting Eq. (6.23) into Eq. (6.1a) to determine the displacement response, we obtain

$$x(t) = \frac{F_o e^{-\zeta\omega_n t}}{m\omega_d} \int_0^t e^{\zeta\omega_n \eta} \sin(\omega_d[t - \eta])d\eta$$

$$= \frac{F_o}{k}\left[1 - \frac{e^{-\zeta\omega_n t}}{\sqrt{1 - \zeta^2}} \sin(\omega_d t + \varphi)\right]u(t) \tag{6.24}$$

which is rewritten in terms of the nondimensional time $\tau = \omega_n t$ as

$$x(\tau) = \frac{F_o}{k}\left[1 - \frac{e^{-\zeta\tau}}{\sqrt{1 - \zeta^2}} \sin(\tau\sqrt{1 - \zeta^2} + \varphi)\right]u(\tau) \tag{6.25}$$

where the phase angle φ is given by Eq. (6.8) and it has been assumed that the system is underdamped.

The maximum value x_{max} is determined from the response given by Eq. (6.25) as follows. The extremum values x_{max} are obtained by determining the times τ_m at which the time derivative of $x(\tau)$ given by Eq. (6.25) is zero; that is,

$$\frac{dx(\tau)}{d\tau} = \frac{F_o e^{-\zeta\tau}}{k}\left[\frac{\zeta}{\sqrt{1 - \zeta^2}} \sin(\tau\sqrt{1 - \zeta^2} + \varphi) - \cos(\tau\sqrt{1 - \zeta^2} + \varphi)\right]$$

$$= \frac{F_o e^{-\zeta\tau}}{k\sqrt{1 - \zeta^2}} \sin(\tau\sqrt{1 - \zeta^2}) \tag{6.26}$$

where we have made use of Eq. (4.10). Then, setting $dx(\tau)/d\tau = 0$, we find that for $\tau > 0$, the first extremum occurs at

$$\tau_m = \frac{\pi}{\sqrt{1 - \zeta^2}} \tag{6.27}$$

For the value of τ_m given by Eq. (6.27), Eq. (6.25) evaluates to[7]

$$x_{max} = x(\tau_m) = \frac{F_o}{k}\left[1 - \frac{e^{-\zeta\tau_m}}{\sqrt{1 - \zeta^2}} \sin(\tau_m\sqrt{1 - \zeta^2} + \varphi)\right]$$

$$= \frac{F_o}{k}\left[1 - \frac{e^{-\zeta\pi/\sqrt{1 - \zeta^2}}}{\sqrt{1 - \zeta^2}} \sin(\pi + \varphi)\right]$$

$$= \frac{F_o}{k}[1 + e^{-\pi/\tan\varphi}] \tag{6.28}$$

where we have made use of the first of Eq. (4.7), that is,

$$\sin\varphi = \sqrt{1 - \zeta^2}$$

The response given by Eq. (6.25) consists of a constant term F_o/k and an exponentially decaying sinusoid for $0 < \zeta < 1$. This is plotted in Figure 6.9

---

[7]Strictly speaking, to verify that this extremum is indeed a maximum, one needs to verify that the second derivative is less than zero at $\tau = \tau_m$. We have, instead, determined numerically that the extremum is a maximum.

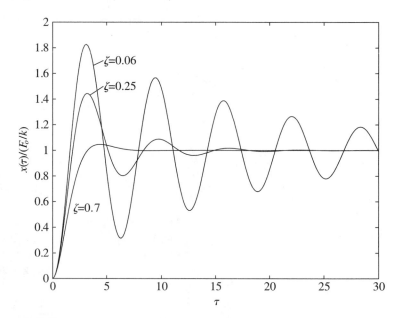

FIGURE 6.9
Responses of single degree-of-freedom systems with different damping factors to step inputs.

for three different values of ζ. From the responses shown in Figure 6.9, it is evident that the response of an underdamped system to a step input oscillates about the final settling position, which is at F_o/k, before settling down to it. This settling position, called the steady-state value, can also be determined from Eq. (6.25) by noting that

$$\lim_{\tau \to \infty} x(\tau) = \frac{F_o}{k} \tag{6.29}$$

To describe the features of the responses seen in Figure 6.9, the following definitions are introduced. A graphical illustration of these definitions is provided in Figure 6.10.

Rise Time

The *rise time* $\tau_r = \omega_n t_r$ is the nondimensional time it takes for the response of a system to a step input to go from 10% of the steady-state value to 90% of the steady-state value. This quantity is a measure of how fast one can expect a system to overcome its inertia to a sudden change in the applied force. For the normalized response considered here—that is,

$$\tilde{x}(\tau) = \frac{x(\tau)}{F_o/k}$$

—the steady-state value is 1. Then

$$\tau_r = \omega_n t_r = \tau_{0.9} - \tau_{0.1} \tag{6.30}$$

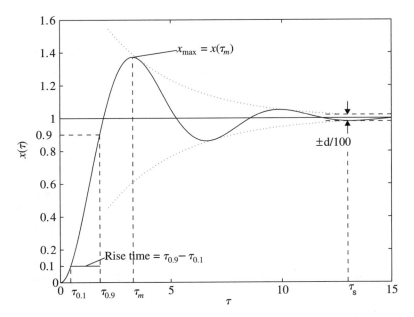

FIGURE 6.10
Definition of rise time, percentage overshoot, and settling time.

where

$$\tilde{x}(\tau_{0.9}) = 0.9$$

$$\tilde{x}(\tau_{0.1}) = 0.1$$

In general, the rise time can also be use to describe waveforms. For example, the rise time of a half-sine wave of frequency ω and duration π/ω; that is, $x(t) = \sin(\omega t)u(t - \pi/\omega)$, is

$$\omega t_r = \sin^{-1}(.9) - \sin^{-1}(.1) = 1.02$$

Percentage Overshoot

The *percentage overshoot* is determined from

$$P_o = \left(\frac{x_{\max} - x_f}{x_f}\right)100\% \tag{6.31}$$

where x_f is the final or steady-state value, which here is F_o/k, and $x_{\max}$ is the maximum value of $x(\tau)$, which is given by Eq. (6.28). Thus, after substituting Eq. (6.28) into Eq. (6.31) we obtain

$$P_o = 100e^{-\pi/\tan \varphi} \quad \% \tag{6.32}$$

The percentage overshoot is a measure of the magnification in the displacement response to a sudden change in the applied force. This quantity is important to know in the design of systems where, in a dynamic environment, interference, wear, undesirable contact between surfaces, and high stresses must be avoided.

Settling Time

Settling time t_s is the time it takes for the displacement response of a system to a step input to decay to within $\pm d$ % of the steady-state (final) value. It is a measure of a system's ability to return to an equilibrium state. Thus, from the envelope of the decay for underdamped oscillations discussed in Section 4.2 and Eq. (6.24), we have

$$\pm\, e^{-\zeta\omega_n t_s} = \frac{\pm d}{100}$$

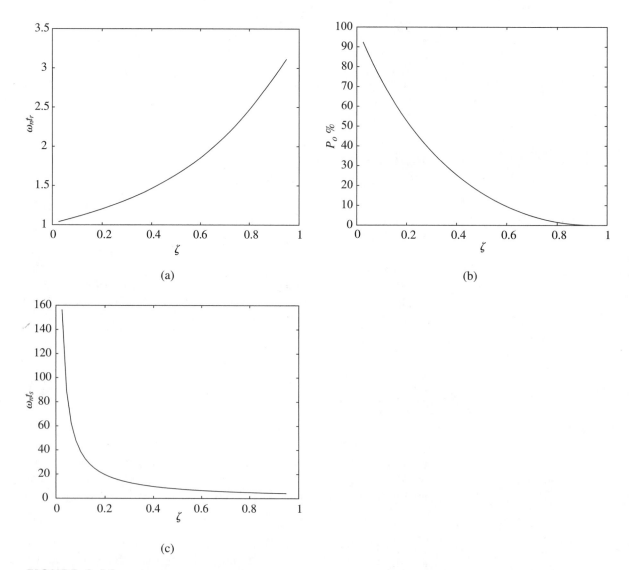

FIGURE 6.11

Underdamped oscillations: (a) rise time; (b) percentage overshoot; and (c) settling time to within $\pm 2\%$.

or, equivalently,

$$\omega_n t_s = -\frac{1}{\zeta} \ln \frac{d}{100} \tag{6.33}$$

Expressions (6.30), (6.32), and (6.33) are evaluated numerically for underdamped oscillations and the corresponding values are plotted in Figure 6.11 as a function of ζ. From these equations and Figure 6.11, one can deduce the following design limitations.

> **Design Limitations:** For a linear single degree-of-freedom system with a mass m and a natural frequency ω_n, the rise time, bandwidth, and percentage overshoot are a function of the damping ratio ζ. Selecting ζ to adjust any one characteristic affects the others. Thus, for example, decreasing ζ (decreasing c) to decrease rise time increases the percentage overshoot and the settling time.

EXAMPLE 6.4   Vehicle response to step change in road profile

Consider a car modeled as shown in Figure 6.12 with a mass of 1100 kg, a suspension stiffness of 400 kN/m, and suspension damping coefficient of 15 kN·s/m. The vehicle is traveling at a speed $v = 64$ km/h and the step change in road elevation $a = 0.03$ m is 160 m away from the vehicle at $t = 0$. We shall determine the vehicle displacement response.

The governing equation for the vehicle is given by Eq. (3.28), which is

$$\frac{d^2x}{dt^2} + 2\zeta\omega_n \frac{dx}{dt} + \omega_n^2 x = 2\zeta\omega_n \frac{dy}{dt} + \omega_n^2 y \tag{a}$$

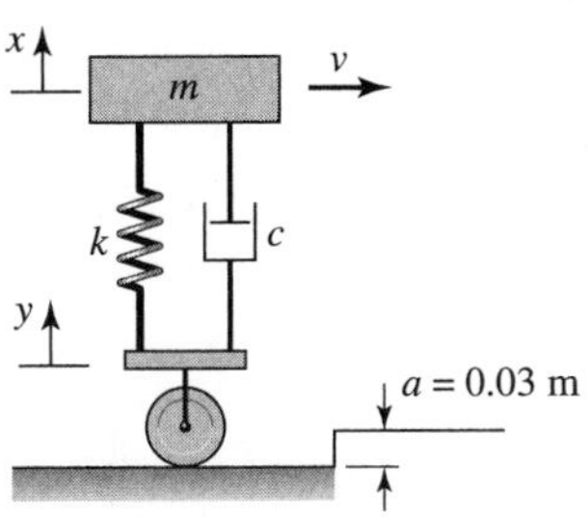

where $x(t)$ is the displacement response of the vehicle and $y(t)$ is the base motion. Noting that the vehicle encounters the step change in the road condition at $t_o = (160 \text{ m}) \times (3600 \text{ s/hr})/(64 \times 10^3 \text{ m/hr}) = 9$ s, the base motion characterized by the step change is described by

$$y(t) = au(t - t_o) \tag{b}$$

To determine the displacement response, we need to find the solution of the system

$$\frac{d^2x}{dt^2} + 2\zeta\omega_n \frac{dx}{dt} + \omega_n^2 x = 2\zeta\omega_n a\delta(t - t_o) + a\omega_n^2 u(t - t_o) \tag{c}$$

where we have made use of the fact that the derivative of a step function is the impulse function;[8] that is,

$$\frac{du(t - t_o)}{dt} = \delta(t - t_o) \tag{d}$$

FIGURE 6.12
Car suspension encountering a step change in road conditions.

(figure labels: x, m, v, k, c, y, $a = 0.03$ m)

---

[8]Papoulis, *ibid.*

For the given parameter values, the damping factor is

$$\zeta = \frac{15 \times 10^3}{2 \times \sqrt{400 \times 10^3 \times 1100}} = 0.36 \tag{e}$$

Since the system is underdamped, we can make use of Eqs. (4.1), (6.2), and (6.23) to determine that $f_o = 2a\zeta\omega_n m$ and $F_o = ak$. Then, the displacement response is given by

$$x(t) = \frac{2\zeta a}{\sqrt{1 - \zeta^2}} e^{-\zeta\omega_n(t-t_o)} \sin(\omega_d(t - t_o))u(t - t_o)$$

$$+ a\left[1 - \frac{e^{-\zeta\omega_n(t-t_o)}}{\sqrt{1 - \zeta^2}} \sin(\omega_d(t - t_o) + \varphi)\right]u(t - t_o) \tag{f}$$

where the step function $u(t - t_o)$ indicates that the step affects the response only for $t \geq t_o$. For the given parameter values, it is found that

$$\omega_n = \sqrt{\frac{4 \times 10^5 \text{ N/m}}{1100 \text{ kg}}} = 19.07 \text{ rad/s}$$

$$\zeta = \frac{15 \times 10^3 \text{ N·s/m}}{2 \times (1100 \text{ kg}) \times (19.07 \text{ rad/s})} = 0.358$$

$$\omega_d = 19.07\sqrt{1 - 0.358^2} = 17.81 \text{ rad/s}$$

$$\varphi = \tan^{-1} \frac{\sqrt{1 - 0.358^2}}{0.358} = 1.21 \text{ rad}$$

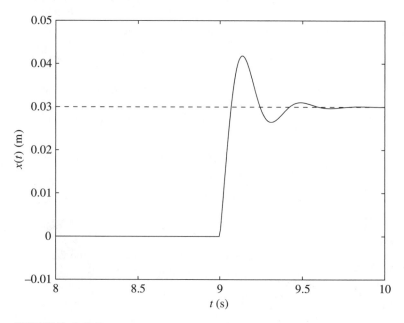

FIGURE 6.13
Displacement response of vehicle to step change in roadway.

Upon substituting these values on to Eq. (f) and noting that $a = 0.03$ m and $t_o = 9$ s, the graph of the displacement response of the vehicle shown in Figure 6.13 is obtained. The vehicle response is at the equilibrium position $x = 0$ before the sudden change due to the step is encountered. This encounter with the sudden change in road elevation produces transients in the response, which die out after some time, and the vehicle response settles down to the new position $x = a$.

EXAMPLE 6.5   Response of a foam automotive seat cushion and occupant

In this example, we study the response of a nonlinear system to a step input. The system is a car seat and its occupant, as shown in Figure 6.14. This system is subjected to a step input in the form of acceleration. The equation governing the vibrations of an open-celled polyurethane foam automotive seat cushion is given by[9]

$$M \frac{d^2z}{dt^2} + \underbrace{C_1 \frac{dz}{dt}}_{\text{Linear viscous damping}} + \underbrace{C_2 \left| \frac{dz}{dt} \right| \frac{dz}{dt}}_{\text{Quadratic fluid damping}} + \underbrace{\frac{C_3}{(cH + d_1|z|)^{p_1}} z}_{\text{Nonlinear stiffness}} = -Ma(t) \qquad \text{(a)}$$

where the coefficients in Eq. (a) are given by

$$C_1 = \frac{\mu AH}{3K\xi}, \quad C_2 = \frac{\rho A}{3K_b \xi^2}, \quad C_3 = E_f aA, \quad \text{and} \quad K = 0.012 d^2 \qquad \text{(b)}$$

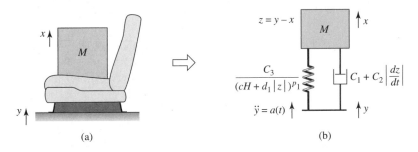

(a)   (b)

FIGURE 6.14
(a) Automotive seat cushion and (b) its single degree-of-freedom equivalent.

[9]W. N. Patten, S. Sha, and C. Mo, "A Vibration Model of Open Celled Polyurethane Foam Automotive Seat Cushions," *J. Sound Vibration,* Vol. 217, No. 1, pp. 145–161 (1998).

TABLE 6.1

Seat Parameters

Quantity	Symbol	Units	Values
Foam cell edge diameter	d	m	0.0009
Cushion height	H	m	0.067
Cushion area	A	m^2	0.095
Mass of load on car seat	M	kg	36 and 45
Young's modulus of polymer	E_s	N/m^2	6×10^7
Young's modulus of foam	E_f	N/m^2	1.1×10^4
Density of air	ρ	kg/m^3	1.22
Viscosity of air	μ	Ns/m^2	1.85×10^{-5}
Volume fraction of open cells	ξ		0.0133
Surface factor	K_b		1
Coefficients of shape function	a		0.0005
	b		0.4
	c		0.00022
	d_1		0.009
	p_1		1

Source: Reprinted from *Journal of Sound Vibration*, 217, W. N. Patten, S. Sha, and C. Mo, "A Vibration Model of Open Celled Polyurethane Foam Automotive Seat Cushions," pp. 145–161, Copyright © 1998, with permission from Elsevier Science.

The physical quantities associated with the coefficients in Eq. (b) and the numerical values of the different parameters are given in Table 6.1 for one type of seat. The second term on the left-hand side of Eq. (a) represents the linear viscous loss that is caused by the gas in the open cells of the foam being forced out of the bottom of the cushion. The third term represents the turbulent flow resistance of air escaping from the cells, and this is the fluid-damping model discussed in Section 2.4.2. The fourth term represents the nonlinear elasticity of the walls of the cells and any entrained air that doesn't escape. It is an empirically determined function. The quantity $a(t)$ is the input acceleration to the base of the seat. The input acceleration is assumed to be of the form $a(t) = a_o u(t)$, where a_o is the magnitude of the acceleration and $u(t)$ is the step function.

Equation (a) is a nonlinear, ordinary differential equation for which the solution must be obtained numerically.[10] The numerically obtained solution of Eq. (a) is used to determine the acceleration of the car seat $a_c(t)$ and the scaled acceleration $a_{cs}(t)$:

$$a_{cs}(t) = \frac{a_c(t)}{a_o} = \frac{1}{a_o} \frac{d^2 z}{dt^2} \tag{c}$$

The two different magnitudes chosen for the acceleration are $a_o = 0.15$ m/s^2 and $a_o = 0.60$ m/s^2. In Figures 6.15a and 6.15c, the acceleration responses of the car seat in the time domain for two different values of a_o and two different values of the system mass M are shown. These time domain accelera-

---

[10]The MATLAB function ode45 was used.

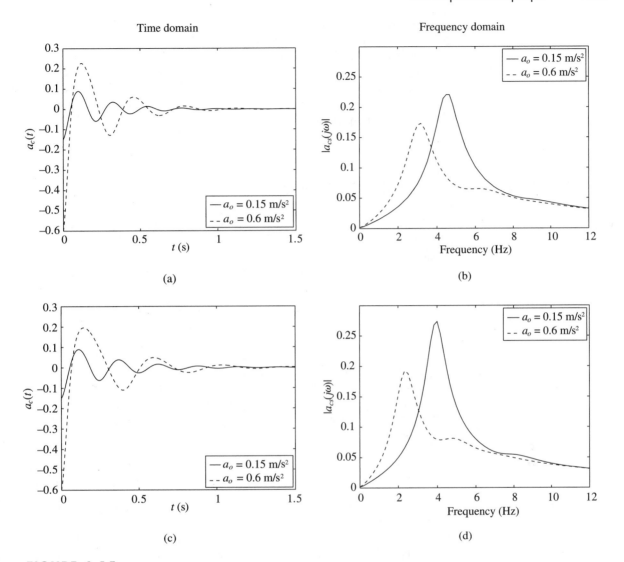

FIGURE 6.15
Time and frequency responses of car seat to step acceleration inputs: (a) and (b) $M = 36$ kg and (c) and (d) $M = 45$ kg.

tion responses are then transformed into the frequency domain by using the discrete Fourier transform[11] to obtain the results shown in Figure 6.15(b) and 6.15(d).

From the time histories, it is evident that for a given system mass, as the base acceleration magnitude is increased, the period of oscillation during the early portion of the motion increases. This characteristic is attributed to

---

[11]The MATLAB function fft was used.

the system nonlinearity. This increase in period of oscillation is reflected as a decrease in the frequency at which the peak amplitude occurs in the associated spectra. As discussed in Chapter 4, the period of oscillation of a nonlinear system can increase or decrease as the amplitude of response grows. Examining Figures 6.15(a) and 6.15(c), we note that the peak acceleration magnitude increases with an increase in the base acceleration magnitude for a given system mass.

6.4 RESPONSE TO RAMP INPUT

We now consider the response of a linear vibratory system to a ramp input and we use this response to demonstrate the effects of rise time of the input force on the rise time and overshoot of the response $x(\tau)$. The ramp waveform, which is shown in Figure 6.16, is described by

$$
\begin{aligned}
f(t) &= 0 && \text{for } t < 0 \\
&= \frac{F_o t}{t_o} && \text{for } 0 \le t \le t_o \\
&= F_o && \text{for } t \ge t_o
\end{aligned}
\tag{6.34}
$$

When $t_o \to 0$, the ramp input approaches a step input. From Eqs. (6.30) and (6.34), we find that the rise time of the ramp portion of the forcing is $t_r = 0.8t_o$. Equation (6.34) is written in terms of the unit step function as

$$
\begin{aligned}
f(t) &= F_o \left[\frac{t}{t_o} [u(t) - u(t - t_o)] + u(t - t_o) \right] \\
&= \frac{F_o}{t_o} [tu(t) - (t - t_o)u(t - t_o)]
\end{aligned}
\tag{6.35}
$$

Referring to Figure 6.16, we see that Eq. (6.35) is the sum of two ramps $f_1(t)$ and $f_2(t)$ whose slopes are the negative of each other's slope and which are time shifted with respect to each other by an amount t_o. In terms of nondimensional time τ, Eq. (6.35) is rewritten as

$$
f(\tau) = \frac{F_o}{\tau_o} [\tau u(\tau) - (\tau - \tau_o)u(\tau - \tau_o)]
\tag{6.36}
$$

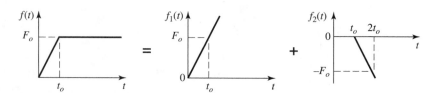

FIGURE 6.16
Ramp force composed of two ramps.

where $\tau_o = \omega_n t_o$. After substituting Eq. (6.36) into Eq. (6.1b), and defining the following quantity

$$g(\tau,\xi) = e^{-\zeta(\tau-\xi)} \sin\left((\tau - \xi)\sqrt{1 - \zeta^2}\right)$$

we determine the response of the system given in Figure 6.1 as[12]

$$x(\tau) = \frac{F_o}{\tau_o k\sqrt{1 - \zeta^2}} \left[\int_0^\tau g(\tau,\xi)\xi u(\xi)d\xi - \int_0^\tau g(\tau,\xi)(\xi - \tau_o)u(\xi - \tau_o)d\xi \right]$$

$$= \frac{F_o}{\tau_o k\sqrt{1 - \zeta^2}} \left[u(\tau)\int_0^\tau g(\tau,\xi)\xi d\xi - u(\tau - \tau_o)\int_{\tau_o}^\tau g(\tau,\xi)(\xi - \tau_o)d\xi \right]$$

Shifting the time variable in the second integral, we rewrite the above expression as

$$x(\tau) = \frac{F_o}{\tau_o k\sqrt{1 - \zeta^2}} \left[u(\tau)\int_0^\tau g(\tau,\xi)\xi d\xi - u(\tau - \tau_o)\int_0^{\tau-\tau_o} g(\tau,\xi + \tau_o)\xi d\xi \right]$$

$$= \frac{F_o}{\tau_o k\sqrt{1 - \zeta^2}} \left[u(\tau)\int_0^\tau g(\tau,\xi)\xi d\xi - u(\tau - \tau_o)\int_0^{\tau-\tau_o} g(\tau - \tau_o,\xi)\xi d\xi \right]$$

since

$$g(\tau,\xi + \tau_o) = e^{-\zeta(\tau-[\xi+\tau_o])} \sin\left((\tau - [\xi + \tau_o])\sqrt{1 - \zeta^2}\right)$$
$$= g(\tau - \tau_o,\xi)$$

After performing the integration, we obtain

$$x(\tau) = \frac{F_o}{k} [h(\tau)u(\tau) - h(\tau - \tau_o)u(\tau - \tau_o)] \tag{6.37}$$

where the first term on the right-hand side of Eq. (6.37) is the response to the ramp function $f_1(t)$ given by the first term of Eq. (6.36) and the second term on the right-hand side is the response to the ramp function $f_2(t)$ given by the second term of Eq. (6.36). The function $h(\tau)$ is given by

$$h(\tau) = \frac{1}{\tau_o}\left\{-2\zeta + \tau + e^{-\zeta\tau}\left[2\zeta \cos\left(\tau\sqrt{1 - \zeta^2}\right)\right.\right.$$

$$\left.\left. + \frac{2\zeta^2 - 1}{\sqrt{1 - \zeta^2}} \sin\left(\tau\sqrt{1 - \zeta^2}\right)\right]\right\} \tag{6.38}$$

---

[12]In these equations, the unit step function determines the lower limit of integration. In addition, we must also place the unit step function outside the integrals to explicitly indicate the regions of applicability of each integral as a function of τ.

The displacement response given by Eq. (6.37) is plotted in Figure 6.17, where it is seen that as the rise time of the ramp input decreases, the amount of overshoot increases and the displacement response deviates more from the input waveform. For comparison, the response to a step input is also included in Figure 6.17c. As the rise time of the ramp becomes shorter, the response closely resembles the response to a step input of corresponding magnitude. Also, from Eqs. (6.37) and (6.38), it is noticed that as ζ increases, the overshoot decreases.

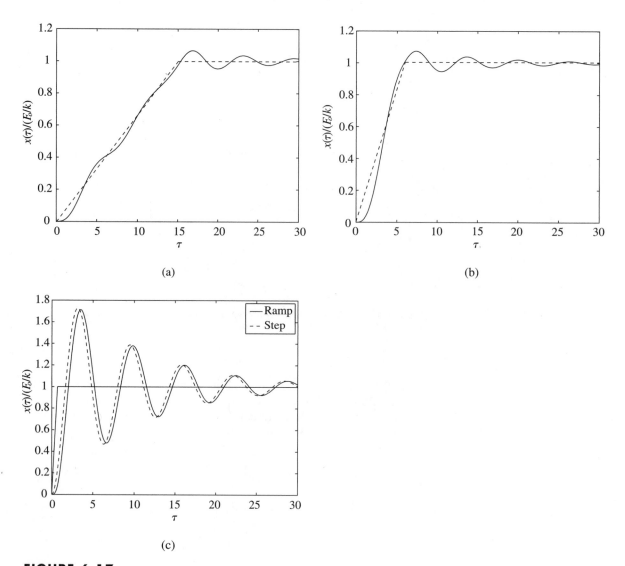

FIGURE 6.17

Response of system to the ramp force given in Figure 6.16 for $\zeta = 0.1$ and different rise times: (a) $\tau_o = \omega_n t_o = 15$; (b) $\tau_o = \omega_n t_o = 6$; and (c) $\tau_o = \omega_n t_o = 0.7$. In (a) and (b) the broken lines are used to represent the input and the solid lines are used to represent the response.

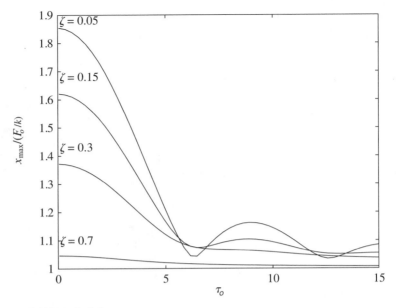

FIGURE 6.18
Maximum displacement of single degree-of-freedom system to ramp forcing shown in Figure 6.16 for several different damping factors.

The maximum displacement response x_{max}, which is a function of the ramp duration τ_o and the damping factor ζ, is determined numerically[13] based on Eqs. (6.37) and (6.38). The results are shown in Figure 6.18. Hence, we can propose the following design guideline to decrease a system's overshoot.

Design Guideline: Either increasing damping of a vibratory system or increasing the rise time of the external forcing function can decrease the amount of overshoot in the response of a single degree-of-freedom system subjected to a ramp excitation.

EXAMPLE 6.6   Response of a slab floor to transient loading

Consider the vibratory model of a slab floor shown in Figure 6.19. The slab floor has a mass of 1000 kg and the equivalent stiffness of each column supporting the system is 2×10^5 N/m. The transient loading $f(t)$ acting on the floor has the amplitude versus time profile shown in Figure 6.19. We shall

[13]The MATLAB function fminbnd from the Optimization Toolbox was used.

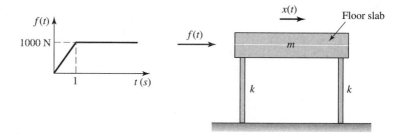

FIGURE 6.19
Floor slab subjected to ramp forcing.

determine the response $x(t)$ of the slab floor and find the earliest time at which the maximum response occurs.

The governing equation of motion of the system is given by Eq. (3.22) with $\zeta = 0$. Thus,

$$\frac{d^2x}{dt^2} + \omega_n^2 x = \frac{f(t)}{m} \tag{a}$$

where

$$\omega_n = \sqrt{\frac{2k}{m}} = \sqrt{\frac{2 \times 2 \times 10^5}{1000}} = 20 \text{ rad/s} \tag{b}$$

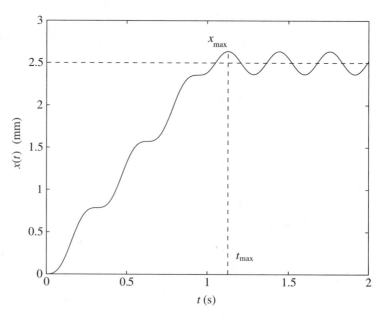

FIGURE 6.20
Response of the floor slab shown in Figure 6.19.

Making use of Eq. (6.35), the forcing function is written as

$$f(t) = 1000\{t[u(t) - u(t - 1)] + u(t - 1)\} \text{ N} \qquad \text{(c)}$$

To determine the response $x(t)$, we first recognize that in Eq. (6.38) $\tau = \omega_n t$, $\tau_o = \omega_n$, and $\zeta = 0$ because we have an undamped system. Then,

$$h(t) = \frac{1}{\omega_n}\{\omega_n t - \sin(\omega_n t)\} = t - 0.05 \sin(20t) \qquad \text{(d)}$$

Then, from Eq. (6.37), the displacement response of the slab is given by

$$x(t) = 2.5[\{t - 0.05 \sin(20t)\}u(t) - \{(t - 1) \\ - 0.05 \sin(20[t - 1])\}u(t - 1)] \text{ mm} \qquad \text{(e)}$$

where we have used the fact that $F_o/k = 1000/4 \times 10^5 \text{ m} = 2.5 \times 10^{-3} \text{ m} = 2.5$ mm. A graph of Eq. (e) is shown in Figure 6.20. The earliest time at which the maximum value $x(t)$ is found numerically[14] to be $x_{max} = 2.64$ mm at $t_m = 1.13$ s.

6.5 SPECTRAL ENERGY OF THE RESPONSE

The total energy E_T in a signal $g(t)$, which has a Laplace transform $G(s)$, is[15]

$$E_T = \int_0^\infty g^2(t)dt \qquad \text{(6.39)}$$

which, by Parseval's theorem,[16] is also given by

$$E_T = \frac{1}{\pi}\int_0^\infty |G(j\omega)|^2 \, d\omega \qquad \text{(6.40)}$$

where $|G(j\omega)|^2$ is the energy density spectrum with units $(E_u^2 \cdot s)/\text{rad/s}$, E_u has the physical or engineering unit of $g(t)$; that is, it represents N, Pa, m/s, etc., and $|G(j\omega)|$ is the amplitude density spectrum with the units $E_u/\text{rad/s}$. Hence, from either Eq. (6.39) or Eq. (6.40), the energy associated with a signal $g(t)$ can be determined. Typically, the signals of interest will be the displacement response $x(t)$ and the forcing $f(t)$.

The energy over a portion of the frequency range $0 \leq \omega \leq \omega_c$ is determined from

$$E(\omega_c) = \frac{1}{\pi}\int_0^{\omega_c} |G(j\omega)|^2 \, d\omega \qquad \text{(6.41)}$$

---

[14]The MATLAB function fminbnd from the Optimization Toolbox was used.

[15]For this equation to be valid, the signal must be bounded and its energy must be finite.

[16]See, for example, Papoulis, *ibid.*, p. 27 ff.

The fraction of total energy is this frequency band can be written as

$$\frac{E(\omega_c)}{E_T} = \frac{1}{E_T \pi} \int_0^{\omega_c} |G(j\omega)|^2 \, d\omega \qquad (6.42)$$

If ω_c is the cutoff frequency of the amplitude density spectrum, then ω_c is determined from the relation

$$|G(j\omega_c)| = \frac{1}{\sqrt{2}} |G(j\omega)|_{\omega=\omega_{\text{max}}} \qquad (6.43)$$

where $|G(j\omega)|_{\omega=\omega_{\text{max}}}$ is the maximum value of $|G(j\omega)|$, which occurs at ω_{max} and $0 \le \omega_{\text{max}} < \omega_c$.

In the next two sections, we shall develop relationships among the cutoff frequency ω_c, the rise time τ_r, the fraction of total energy in the region $0 \le \omega < \omega_c$, and the effect of $\Omega_c = \omega_c/\omega_n$ on the system's displacement response as a function of time for different pulse excitations. This information is useful for the design of components subjected to shock excitations, which are typically modeled as pulse excitations.

6.6 RESPONSE TO RECTANGULAR PULSE EXCITATION

We now consider the application of a force whose time profile is rectangular, as shown in Figure 6.21. In terms of the unit step function, this force is represented by

$$f(t) = F_o[u(t) - u(t - t_o)] \qquad (6.44)$$

Then, by using Eqs. (6.39) and (6.44), the total energy in the pulse is

$$E_T = \int_0^{t_o} F_o^2 dt = F_o^2 t_o \qquad (6.45)$$

Note that for the total energy in the pulse to remain constant, either F_o or t_o has to be adjusted so that F_o varies as $\sqrt{t_o}$. By using the Laplace transform pair 8 in Table A in Appendix A, the Laplace transform of $f(t)$ given by Eq. (6.44) is

$$F(s) = \frac{F_o}{s} (1 - e^{-st_o}) \qquad (6.46)$$

We obtain the associated amplitude spectrum and phase spectrum by making the substitution $s = j\omega$ into Eq. (6.46). This results in

$$F(j\omega) = \frac{F_o}{j\omega} (1 - e^{-j\omega t_o}) = \frac{F_o}{\omega} [\sin \omega_o t + j(\cos \omega_o t - 1)]$$
$$= G_r(\omega)e^{j\psi(\omega)} \qquad (6.47)$$

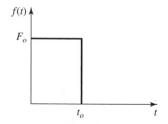

$f(t)$

F_o

t_o

t

FIGURE 6.21
Rectangular force pulse.

where the pulse amplitude function $G_r(\omega)$ and the pulse phase function $\psi(\omega)$ are given by

$$G_r(\omega) = F_o t_o \left| \frac{\sin(\omega t_o/2)}{\omega t_o/2} \right|$$

$$\psi(\omega) = \tan^{-1} \frac{\cos(\omega t_o) - 1}{\sin(\omega t_o)} = \tan^{-1}(-\tan(\omega t_o/2)) = -\omega t_o/2 \quad (6.48)$$

The quantity $G_r(\omega)/(F_o t_o)$ is plotted in Figure 6.22.

Hence, from Eqs. (6.48), the cutoff frequency ω_c associated with the input force spectrum is determined from the numerical solution to

$$\frac{G_r(\omega_c)}{F_o t_o} = \frac{1}{\sqrt{2}} = \left| \frac{\sin(\omega_c t_o/2)}{\omega_c t_o/2} \right|$$

which gives[17]

$$c_r = \omega_c t_o = 2.7831 \quad (6.49)$$

We call c_r the *pulse duration–bandwidth product;* thus, as the pulse duration t_o decreases, the bandwidth ω_c increases proportionately.

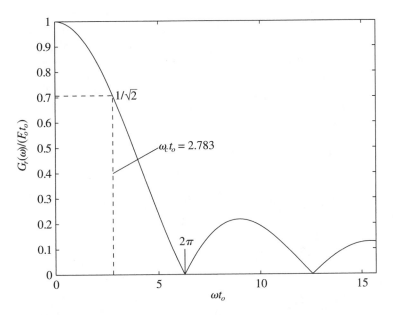

FIGURE 6.22
Normalized amplitude spectrum of rectangular pulse of duration t_o.

[17]The MATLAB function `fzero` was used.

From Eq. (6.42), the fraction of the total energy in the frequency range $0 \leq \omega \leq \omega_c$ is

$$\frac{E(\omega_c t_o)}{E_T} = \frac{1}{\pi} \int_0^{\omega_c t_o} \left| \frac{\sin(x/2)}{x/2} \right|^2 dx = 0.722 \qquad (6.50)$$

where the integral was evaluated numerically.[18] We see that 72.2% of the total energy of the rectangular pulse lies within this bandwidth. In addition, it is numerically found that $E(2\pi)/E_T = 0.903$, or about 90% of the energy of the rectangular pulse lies in the bandwidth $0 \leq \omega \leq 2\pi/t_o$.

The amplitude spectrum and the phase spectrum of the displacement response of the mass are determined from Eqs. (5.55), (6.19), and (6.47). Thus,

$$X(j\omega) = G(j\omega)F(j\omega) = X_r(\omega)e^{-j\phi(\omega)} \qquad (6.51)$$

where the response amplitude function $X(\omega)$ and the response phase function $\phi(\omega)$ are given by

$$X_r(\omega) = \frac{F_o t_o}{k\sqrt{(1 - (\omega/\omega_n)^2)^2 + (2\zeta\,\omega/\omega_n)^2}} \left| \frac{\sin(\omega t_o/2)}{\omega t_o/2} \right|$$

$$\phi(\omega) = \theta(\omega) - \psi(\omega) \qquad (6.52)$$

where $\theta(\omega)$ is given by Eq. (5.56b) and $\psi(\omega)$ is given by Eq. (6.48).

In order to be able to make comparisons for various combinations of parameters, we introduce notations in terms of E_T, the total energy of the applied force, and ω_c, the cutoff frequency. To have valid comparisons, we have chosen the total energy to be the same in all cases. We also want to be able to determine the effects of the cutoff frequency of the applied pulse in terms of the natural frequency of the system. To this end, we introduce the following notation

$$E_1 = \sqrt{\frac{c_r E_T}{\omega_n}} \qquad E_o = \sqrt{\frac{E_T \omega_n}{c_r}}$$

$$\Omega = \omega/\omega_n \qquad \Omega_c = \omega_c/\omega_n$$

$$\frac{F_o t_o}{k} = \frac{E_1}{k\sqrt{\Omega_c}} \qquad \frac{F_o}{k} = \frac{E_o\sqrt{\Omega_c}}{k} \qquad (6.53)$$

Then, Eq. (6.52) is written as

$$\frac{X_r(\omega)}{E_1/k} = \frac{1}{\sqrt{\Omega_c}\sqrt{(1 - \Omega^2)^2 + (2\zeta\Omega)^2}} \left| \frac{\sin(\Omega_c/2\Omega_c)}{\Omega_c/2\Omega_c} \right| \qquad (6.54)$$

To obtain $x(t)$, we first make use of Eqs. (6.46) and (6.17). Thus, we obtain

$$X(s) = G(s)F(s) = \frac{F_o}{m}\left[\frac{1}{sD(s)} - \frac{e^{-st_o}}{sD(s)} \right] \qquad (6.55)$$

---

[18]The MATLAB function `trapz` was used.

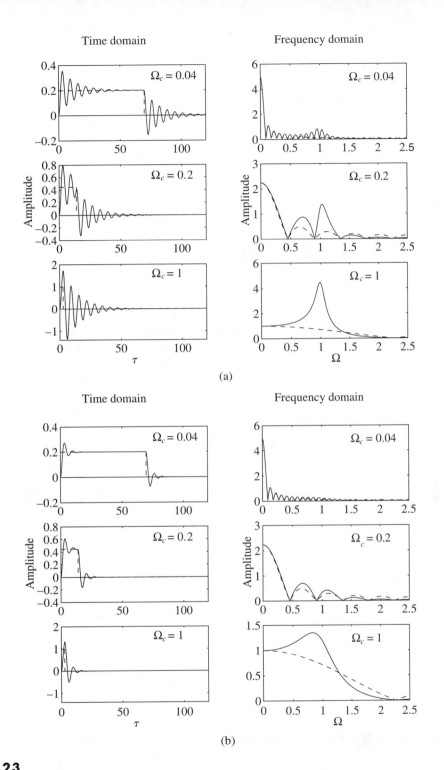

FIGURE 6.23

Comparison of responses to rectangular pulse in the time domain and the frequency domain for three different values of $\Omega_c = \omega_c/\omega_n$: (a) $\zeta = 0.08$, and (b) $\zeta = 0.3$. Dashed lines are used to represent the applied force $f(\tau)/E_o$ in the time domain and the force amplitude spectrum $G_r(\omega)/E_1$ in the frequency domain. The solid lines are used to represent the displacement response $x(\tau)/(E_o/k)$ in the time domain and the displacement amplitude spectrum $X_r(\omega)/(E_1/k)$ in the frequency domain.

From the transform pairs 15 and 3 in Table A of Appendix A, the inverse of $X(s)$ given by Eq. (6.55) is

$$\frac{x(t)}{E_o/k} = \sqrt{\Omega_c} \left[g(t)u(t) - g(t - t_o)u(t - t_o) \right] \tag{6.56}$$

where, for the system with $\zeta < 1$,

$$g(t) = 1 - \frac{e^{-\zeta\omega_n t}}{\sqrt{1 - \zeta^2}} \sin(\omega_d t + \varphi)$$

$$= 1 - \frac{e^{-\zeta(c_r/\Omega_c)(t/t_o)}}{\sqrt{1 - \zeta^2}} \sin\left[(c_r/\Omega_c)(t/t_o)\sqrt{1 - \zeta^2} + \varphi \right] \tag{6.57}$$

and φ is given by Eq. (6.8). It is noted that in this notation $\tau = \omega_n t = (c_r/\Omega_c)(t/t_o)$ and $\omega_n t_o = c_r/\Omega_c$.

Equations (6.44), (6.48), (6.54), and (6.56) are plotted in Figure 6.23 for two different values of ζ and three different values of Ω_c, which is the ratio of the cutoff frequency to the system natural frequency. The displacement response of the mass can be interpreted from the corresponding amplitude spectrum. The displacement response for $\Omega_c = 0.04$ is similar to the response of the pulse train shown in Figure 5.32 and to the response to a step change in the applied force, which is shown in Figure 6.9. In all three cases, the oscillations about the respective steady-state positions are caused by spectral components of the forcing that are in the frequency range of the single degree-of-freedom system's natural frequency. As discussed in Example 5.15, the excitation components in this range are magnified by the system's frequency response function. The period of these oscillations is equal to the period of the damped natural frequency of the system. Thus, when damping is increased, the magnitude and duration of the oscillations decrease, because the magnification of the force's spectral energy in this region is considerably less than that when the damping is small. The amount of spectral energy in the force's pulse that is in the region of the system's natural frequency is a function of the cutoff frequency; the shorter the duration of the pulse, the larger the amount of energy there is in the region of the system's natural frequency. In the limit, as $t_o \to 0$, the spectrum approaches that of the spectrum of an impulse function; that is, one whose amplitude spectrum is constant over the entire frequency range. For relatively long t_o, the displacement response of the linear single degree-of-freedom system resembles that of a step response. As t_o decreases, the displacement of the mass response looks less and less like the form of the rectangular pulse, and as t_o decreases still further, the response approaches the system's impulse response.

6.7 RESPONSE TO HALF-SINE WAVE PULSE

In the previous section, we introduced a relationship between the cutoff frequency of the spectral content of the applied force to its duration. For pulse shapes different from a rectangular shape, apart from the pulse duration,

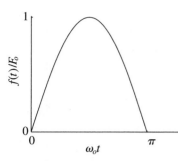

FIGURE 6.24
Half-sine force pulse.

other characteristics such as the rise time of the input can be used. Therefore, in this section, we shall develop a relationship between the rise time of the applied force and its cutoff frequency, and illustrate the effect on the system's response.

The force pulse is considered to have the form of a half-sine wave as shown in Figure 6.24. Thus,

$$f(t) = F_o \sin(\omega_o t)[u(t) - u)(t - \pi/\omega_o)] \tag{6.58}$$

where the pulse duration is $t_o = \pi/\omega_o$. From Section 6.3, the rise time of the half-sine wave is

$$\omega_o t_r = 1.02 \tag{6.59}$$

The Laplace transform of $f(t)$ given by Eq. (6.58) is obtained from transform pair 10 in Table A of Appendix A. Thus, we have

$$F(s) = \frac{F_o \omega_o}{s^2 + \omega_o^2}[1 + e^{-\pi s/\omega_o}] \tag{6.60}$$

The corresponding spectrum of the half-sine wave force pulse is obtained by setting $s = j\omega$. This leads to

$$F(j\omega) = \frac{F_o \omega_o}{\omega_o^2 - \omega^2}[1 + e^{-j\pi\omega/\omega_o}]$$

$$= \frac{F_o \omega_o}{\omega_o^2 - \omega^2}[1 + \cos(\pi\omega/\omega_o) - j\sin(\pi\omega/\omega_o)]$$

$$= G_{hs}(\omega)e^{-j\psi(\omega)} \tag{6.61}$$

where the amplitude function $G_{hs}(\omega)$ is given by

$$G_{hs}(\omega) = \frac{2F_o}{\omega_o} \left| \frac{\cos[\pi\omega/(2\omega_o)]}{1 - (\omega/\omega_o)^2} \right| \quad \text{for} \quad \omega \neq \omega_o$$

$$= \frac{\pi}{4} \frac{2F_o}{\omega_o} \quad \text{for} \quad \omega = \omega_o \tag{6.62}$$

and the phase function $\psi(\omega)$ is given by

$$\psi(\omega) = \tan^{-1} \frac{\sin(\pi\omega/\omega_o)}{1 + \cos(\pi\omega/\omega_o)} = \tan^{-1}\left(\tan\frac{\pi\omega}{2\omega_o}\right) = \frac{\pi\omega}{2\omega_o} \tag{6.63}$$

The function $G_{hs}(\omega)$ is plotted in Figure 6.25.

From Eq. (6.62), the cutoff frequency of the pulse's amplitude density spectrum is obtained from the numerical solution to

$$\frac{G_{hs}(\omega_c)}{2F_o/\omega_o} = \frac{1}{\sqrt{2}} = \left| \frac{\cos[\pi\omega_c/(2\omega_o)]}{1 - (\omega_c/\omega_o)^2} \right| \quad \text{for} \quad \omega_c \neq \omega_o \tag{6.64}$$

which leads to[19]

$$c_s = \frac{\omega_c}{\omega_o} = 1.189 \tag{6.65a}$$

---

[19]The MATLAB function `fzero` was used.

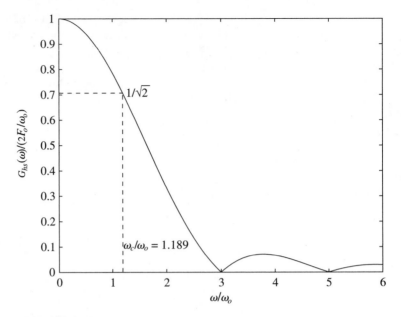

FIGURE 6.25
Amplitude density spectrum of half-sine wave pulse of duration π/ω_o.

or, upon using Eq. (6.59),

$$\omega_c t_r = 1.21 \tag{6.65b}$$

The total energy in the pulse is

$$E_T = F_o^2 \int_0^{\pi/\omega_o} \sin^2(\omega_o t)dt = \frac{F_o^2 \pi}{2\omega_o} = \frac{F_o^2 t_o}{2} \tag{6.66}$$

Comparing this result with that shown in Eq. (6.45) for a rectangular pulse, we see that the total energy of a half-sine wave whose duration is equal to that of a rectangular pulse has one-half the total energy of a rectangular pulse. Notice that for the energy of the half-sine wave pulse to remain constant, either F_o or ω_o has to be adjusted so that F_o varies as $\sqrt{\omega_o}$.

The fraction of total energy that is in the frequency range $0 \leq \omega \leq \omega_c$ is

$$\frac{E(\omega_c)}{E_T} = \frac{8}{\omega_o \pi^2} \int_0^{\omega_c} \left| \frac{\cos[\pi\omega/(2\omega_o)]}{1 - (\omega/\omega_o)^2} \right|^2 d\omega$$

$$= \frac{8}{\pi^2} \int_0^{c_s} \left| \frac{\cos[\pi x/2]}{1 - x^2} \right|^2 dx = 0.784 \tag{6.67}$$

where the integral was evaluated numerically.[20] Thus, 78.4% of the total energy of the half-sine wave pulse lies within its bandwidth. In addition, it is numerically found that $E(3)/E_T = 0.995$, or about 99.5% of the energy of the half-sine pulse lies in the bandwidth $0 \leq \omega \leq 3\omega_o$.

Proceeding along the lines of the previous section used to determine the displacement response function given by Eq. (6.54), we find that

$$\frac{X(j\omega)}{E_{s1}/k} = X_s(\Omega)e^{-j(\theta(\Omega) + \psi(\Omega))} \tag{6.68}$$

where

$$X_s(\Omega) = \frac{1}{\sqrt{\Omega_o}\sqrt{(1 - \Omega^2)^2 + (2\zeta\Omega)^2}} \left| \frac{\cos(\pi\Omega/2\Omega_o)}{1 - (\Omega/\Omega_o)^2} \right| \tag{6.69}$$

$\Omega_o = \omega_o/\omega_n$, $\psi(\Omega)$ is given by Eq. (6.63), and

$$E_{s1} = \sqrt{\frac{8E_T}{\pi\omega_n}}$$

The displacement of the mass is obtained by substituting Eq. (6.58) in Eq. (6.1b), which yields

$$\frac{x(\tau)}{E_{so}/k} = \sqrt{\Omega_o}\left[u(\tau)\int_0^\tau g(\xi,\tau)d\xi - u(\tau - \pi/\Omega_o)\int_{\pi/\Omega_o}^\tau g(\xi,\tau)d\xi \right] \tag{6.70}$$

where $u(t)$ is the unit step function, $\tau = \omega_n t$, and

$$g(\xi,\tau) = \frac{e^{-\zeta(\tau - \xi)}}{\sqrt{1 - \zeta^2}}\sin\left((\tau - \xi)\sqrt{1 - \zeta^2}\right)\sin(\Omega_o\xi)$$

$$E_{so} = \sqrt{\frac{2\omega_n E_T}{\pi}}$$

Equation (6.70) is solved numerically[21] and the results are shown in Figure 6.26. We see from Eqs. (6.65a) and (6.65b) that for a fixed ω_n, as Ω_o increases ω_c increases and, therefore, the rise time decreases. The results of Figure 6.26 are examined in the same manner as was done for the results shown in Figure 6.23 for the rectangular pulse. The introduction of E_{so} was made to ensure that the response $x(\tau)$ takes into account the differing amounts of spectral energy in the pulse as Ω_o varies. This permits us to make comparisons of the response $x(\tau)$ for different Ω_o, since the input spectral energy is constant as Ω_o varies.

It is often of interest to determine the maximum of the displacement response as a function of the pulse duration. In order to provide an idea of how this maximum varies as a function of the frequency ratio Ω_o and the

---

[20]The MATLAB function quadl was used.

[21]Although the MATLAB function trapz can be used to solve Eq. (6.70), the MATLAB function ode45 was used instead to solve Eq. (3.8) with Eq. (6.58) as the forcing.

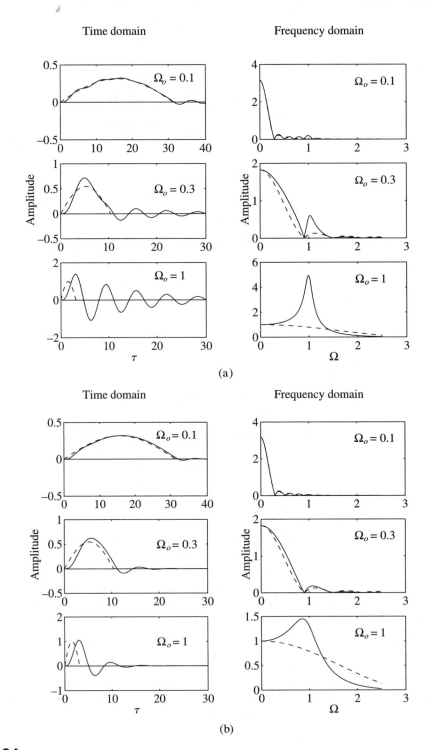

FIGURE 6.26

Comparison of responses to half-sine pulse in the time domain and the frequency domain for three different values of $\Omega_o = \omega_o/\omega_n$: (a) $\zeta = 0.08$, and (b) $\zeta = 0.3$. Dashed lines are used to represent the applied force $f(\tau)/E_{so}$ in the time domain and the force amplitude spectrum $G_{hs}(\Omega)/E_{s1}$ in the frequency domain. Solid lines are used to represent the displacement response $x(\tau)/(E_{so}/k)$ in the time domain and the displacement amplitude spectrum $X_s(\Omega)$ in the frequency domain.

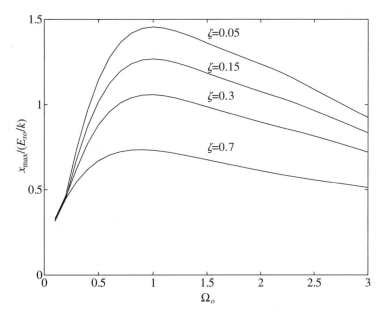

FIGURE 6.27
Maximum amplitude of response of single degree-of-freedom system to half-sine pulse for different pulse durations and different values of damping factor.

damping factor ζ, Eq. (6.70) is solved numerically[22] for the maximum displacement x_{max} and the results obtained are shown in Figure 6.27. It is seen from the graphs that the maximum of the response decreases as the damping factor increases and as the pulse duration decreases. The results are consistent with the results in Figure 6.26.

EXAMPLE 6.7   Response to half-sine pulse base excitation[23]

The governing equation describing the motion of a linear single degree-of-freedom system with a moving base is given by Eq. (3.28). Rewriting this equation in terms of the nondimensional time $\tau = \omega_n t$, we obtain

$$\frac{d^2x}{d\tau^2} + 2\zeta\frac{dx}{d\tau} + x = 2\zeta\frac{dy}{d\tau} + y \tag{a}$$

[22]Although the MATLAB functions `trapz` and `max` can be used to solve Eq. (6.70) to obtain x_{max}, the MATLAB function `ode45` is used instead to solve the Eq. (3.8) with Eq. (6.58) as the forcing and the MATLAB function `max` was used to estimate x_{max}.

[23]Extensions of these types of models to include nonlinear springs and dissipative elements can be found in the following studies: N. C. Shekhar et al., "Response of Nonlinear Dissipative Shock Isolators," *J. Sound Vibration*, Vol. 214, No. 4, pp. 589–603 (1998); and N. C. Shekhar et al., "Performance of Nonlinear Isolators and Absorbers to Shock Excitation," *J. Sound Vibration*, Vol. 227, No. 2, pp. 293–307 (1999).

We also recall from Eq. (3.29) that the relative motion of the mass with respect to the base is $z(\tau) = x(\tau) - y(\tau)$. If we assume that the motion to the base is a half-sine pulse of frequency ω_o and magnitude y_o, then we can describe the base motion as

$$y(\tau) = y_o \sin(\Omega_o \tau)[u(\tau) - u(\tau - \tau_o)] \tag{b}$$

where $\tau_o = \omega_n t_o = \pi \omega_n / \omega_o = \pi / \Omega_o$, $\Omega_o = \omega_o / \omega_n = T_n / (2 t_o)$, $\omega_o t_o = \pi$, and $T_n = 2\pi / \omega_n$ is the period of the natural frequency of the single degree-of-freedom system. After substituting Eq. (b) into the right-hand side of Eq. (a) we obtain

$$\frac{d^2 x}{d\tau^2} + 2\zeta \frac{dx}{d\tau} + x = y_o f_b(\tau) \tag{c}$$

where

$$f_b(\tau) = 2\zeta \sin(\Omega_o \tau)[\delta(\tau) - \delta(\tau - \tau_o)]$$
$$+ [\sin(\Omega_o \tau) + 2\zeta\Omega_o \cos(\Omega_o \tau)][u(\tau) - u(\tau - \tau_o)] \tag{d}$$

and we have used Eq. (d) of Example 6.4.

The solution to Eq. (c) is given by Eq. (5.2); that is, the convolution in-

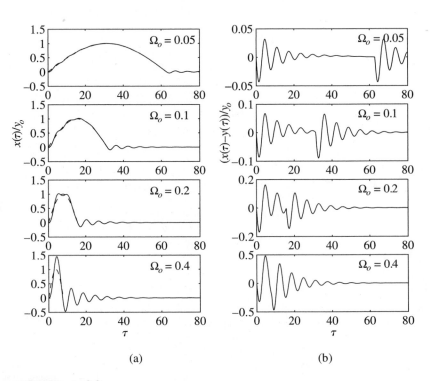

(a)   (b)

FIGURE 6.28
(a) Absolute displacement and (b) relative displacement of the mass of single degree-of-freedom system subjected to half-sine wave displacement to the base with $\zeta = 0.1$.

tegral which, when written in terms of the nondimensional time variable τ, takes the form

$$x(\tau) = \frac{y_o}{k} \int_0^\tau h(\xi,\tau)f_b(\xi)d\xi \qquad \text{(e)}$$

where

$$h(\xi,\tau) = \frac{e^{-\zeta(\tau-\xi)}}{\sqrt{1-\zeta^2}} \sin\left[\sqrt{1-\zeta^2}(\tau-\xi)\right] \qquad \text{(f)}$$

After substituting Eqs. (d) and (f) into Eq. (e) and using Eq. (6.3), we obtain

$$x(\tau) = \frac{y_o}{k}\left[\int_0^\tau g_b(\xi,\tau)d\xi - u(\tau-\tau_o)\int_{\tau_o}^\tau g_b(\xi,\tau)d\xi\right] \qquad \text{(g)}$$

where

$$g_b(\xi,\tau) = [\sin(\Omega_o\tau) + 2\zeta\Omega_o\cos(\Omega_o\tau)]h(\xi,\tau) \qquad \text{(h)}$$

The integral is solved numerically[24] and the results obtained for $x(\tau)$ and $z(\tau)$ are shown in Figure 6.28, where we see that when $\Omega_o \ll 1$, the mass follows the movement of the base as if it were rigidly connected to it. As Ω_o approaches 1, the mass amplifies the base motions and has large excursions relative to the base. This example leads to the following design guideline.

Design Guideline: In order to minimize the amplification of the motion of the base of a single degree-of-freedom, one should keep the pulse duration of the displacement applied to the base long compared to the period of the natural frequency of the system. Stated differently, for a given pulse loading it is preferable to have a system with as high a natural frequency as possible to minimize the amplification of the base motion.

EXAMPLE 6.8 Single degree-of-freedom system with moving base and nonlinear spring[25]

Consider the single degree-of-freedom system whose base is subjected to a known displacement $y(t)$ as shown in Figure 6.29. The spring is nonlinear with a force-displacement relationship given by Eq. (2.23); that is,

$$F(x) = kx + \alpha kx^3$$

---

[24]Although the MATLAB function `trapz` can be used to solve Eq. (g), the MATLAB function `ode45` was used instead to solve Eq. (c) with Eq. (d) as the forcing.

[25]N. Chandra, H. Shekhar, H. Hatwal, and A. K. Mallik, "Response of non-linear dissipative shock isolators," *J. Sound Vibration*, Vol. 214, No. 4, pp. 589–603 (1998).

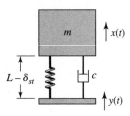

FIGURE 6.29
Base excitation of a single degree-of-freedom system with a nonlinear spring. L is the unstretched length of the spring.

Then, carrying out the force balance for vertical motions of mass m, as discussed in Sections 3.2 and 3.4, we arrive at the following equation of motion of the system

$$m\ddot{x} + c(\dot{x} - \dot{y}) + k(x - \delta_{st} - y) + k\alpha(x - \delta_{st} - y)^3 = -mg \qquad \text{(a)}$$

where x is measured from the static equilibrium position of mass m, y is the absolute displacement of the base, and δ_{st} is the static deflection. From the static equilibrium of the system, we have that

$$k\delta_{st} + k\alpha\delta_{st}^3 = mg \qquad \text{(b)}$$

Upon substituting Eq. (b) into Eq. (a), we obtain

$$m\ddot{x} + c(\dot{x} - \dot{y}) + k(x - y) + k\alpha(x - \delta_{st} - y)^3 = -k\alpha\delta_{st}^3 \qquad \text{(c)}$$

The relative displacement is

$$z(t) = x(t) - y(t) \qquad \text{(d)}$$

and, after substituting Eq. (d) into Eq. (c), we arrive at

$$m\ddot{z} + c\dot{z} + kz + k\alpha(z - \delta_{st})^3 = -k\alpha\delta_{st}^3 - m\ddot{y} \qquad \text{(e)}$$

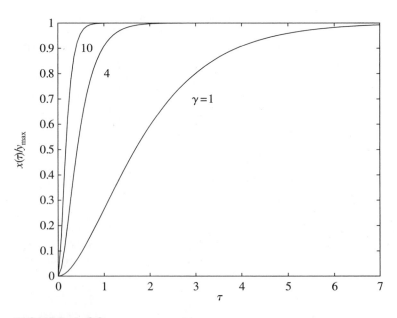

FIGURE 6.30
Base excitation waveform for several values of γ.

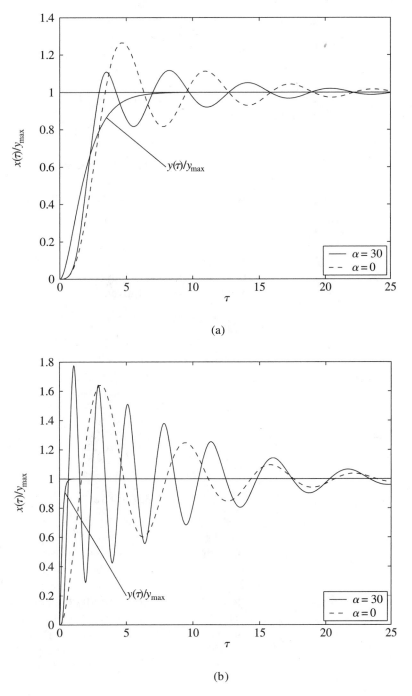

(a)

(b)

FIGURE 6.31
Normalized absolute displacement of the mass of a single degree-of-freedom system with a nonlinear spring with $\zeta = 0.15$ and subject to the base excitation shown in Figure 6.30: (a) $\gamma = 1$ and (b) $\gamma = 10$.

If we assume that the spring is very stiff—that is, k is very large—then the static displacement will be negligible and in that case Eq. (e) simplifies to

$$m\ddot{z} + c\dot{z} + kz + k\alpha z^3 = -m\ddot{y} \tag{f}$$

which is written in nondimensional form as

$$\frac{d^2z}{d\tau^2} + 2\zeta\frac{dz}{d\tau} + z + \alpha z^3 \frac{d^2y}{d\tau^2} \tag{g}$$

where the nondimensional time $\tau = \omega_n t$.

For the displacement to the base of the system, we assume a step-like disturbance that has variable rise time and a rounded shape given by

$$y(\tau) = y_{max}\left[1 - (1 + \gamma\tau)e^{-\gamma\tau}\right] \tag{h}$$

This waveform was selected instead of a unit step function, in part, because its higher order derivatives are continuous. The normalized waveform $y(\tau)/y_{max}$ is shown in Figure 6.30 for several values of the parameter γ. Taking the second derivative of Eq. (h) with respect to τ, we arrive at

$$\ddot{y}(\tau) = y_{max}\gamma^2(1 - \gamma\tau)e^{-\gamma\tau} \tag{i}$$

We now substitute Eq. (i) into Eq. (g) and introduce the nondimensional variable $z_n(\tau) = z(\tau)/y_{max}$ to obtain

$$\frac{d^2z_n}{d\tau^2} + 2\zeta\frac{dz_n}{d\tau} + z_n + \alpha z_n^3 = g(\tau) \tag{j}$$

where

$$g(\tau) = -\gamma^2(1 - \gamma\tau)e^{-\gamma\tau} \tag{k}$$

The absolute displacement $x(\tau)$ is obtained from Eq. (d) and Eq. (h); that is,

$$x(\tau) = y_{max}z_n(\tau) + y_{max}\left[1 - (1 + \gamma\tau)e^{-\gamma\tau}\right] \tag{l}$$

where $z_n(\tau)$ is a solution of Eq. (j).

The numerical evaluation[26] of Eq. (j) yields the results shown in Figure 6.31 for $\alpha = 30$ and for $\alpha = 0$; that is, when the nonlinear spring is replaced by a linear spring. We see that combination of the linear spring and the nonlinear spring is stiffer than the linear spring by itself and, therefore, the period of the oscillation is shorter than that of the linear spring by itself, especially as the rise time decreases (γ increases). In addition, as the rise time of the input displacement decreases (γ increases), the overshoot increases, which is similar to what was found for the response to a ramp input applied to a linear single degree-of-freedom system.

---

[26]The MATLAB function ode45 was used.

6.8 SUMMARY

In this chapter, responses of single degree-of-freedom systems subjected to various types of transient excitations have been studied. A common characteristic of the systems studied is that the excitation provides a "sudden" change to the state of the system. The notion of the impulse response, which is an important component for determining the response of linear systems, has been introduced, along with such notions as settling time and rise time of the response. It was also illustrated as to how the response to an input excitation with an arbitrary frequency spectrum can be determined.

EXERCISES

6.1 Plot the energy density and determine the bandwidth of the following pulses:

a)

$$f(t) = 0.5(1 - \cos \omega_o t)[u(t) - u(t - 2\pi/\omega_o)]$$

b) For $a > b$ and $a/b = 10$ and $a/b = 0.25$:

$$f(t) = (e^{-bt} - e^{-at})u(t)$$

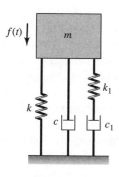

FIGURE E6.2

6.2 For the Kelvin-Voigt-Maxwell combination shown in Figure E6.2, obtain an expression for the displacement response in the Laplace transform domain of the mass when the mass is subjected to a step function force of magnitude F_o.

6.3 Determine the response of a vibratory system governed by the following equation:

$$\ddot{x}(t) + 2\dot{x}(t) + 3.5x(t) = 1.5 \sin(\omega t) + \delta(t - 1)$$

when $m = 1$ kg and the initial conditions are $x(0) = 1$ cm and $\dot{x}(0) = 0$. Plot the response.

6.4 Consider the following single-degree-of-freedom system excited by two impulses and the excitation frequency $\omega = 1.4$ rad/s, when the system is initially at rest

$$\ddot{x}(t) + 2\dot{x}(t) + 4x(t) = \delta(t) + \delta(t - 5)$$

Determine the displacement response of this vibratory system.

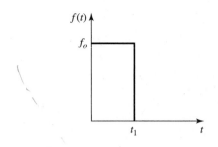

FIGURE E6.5

6.5 Determine the response of the damped second-order system described by

$$m\ddot{x}(t) + c\dot{x}(t) + kx(t) = f(t)$$

to the rectangular pulse shown in Figure E6.5. Plot the displacement response for $m = 1$ kg, $\zeta = 0.1$, $\omega_n = 4$ rad/s, $f_o = 1$ N, and $t_1 = 1$ s.

6.6 Determine the response of an underdamped single degree-of-freedom system that is subjected to the force $f(t) = F_o e^{-\alpha t}$. Assume that the system is initially at rest.

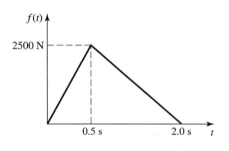

FIGURE E6.7

6.7 A machine system of mass 30 kg is mounted on an undamped foundation of stiffness 1500 N/m. During the operations, the machine is subjected to a force of the form shown in Figure E6.7, where the horizontal axis time t is in seconds and the vertical axis is the amplitude of the force in Newtons. Assume that the machine system is initially at rest and determine the displacement response of the system.

6.8 From extensive biomechanical tests, the spinal stiffness k of a person is estimated to be 50,000 N/m. Assume that the body mass is 80 kg. Let us assume that this person is driving an automobile without wearing a seat belt. On hitting an obstacle, the driver is thrown 10 cm upwards and drops in free fall onto an unpadded seat and experiences an impulse with a magnitude of 100 N·s. Determine the resulting motions if an undamped single degree-of-freedom model is used to model the vertical vibrations of this person.

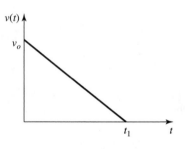

FIGURE E6.9

6.9 In Example 6.6, assume that the motions of the slab floor are damped with the damping factor being 0.2. Determine the response of this damped system for the forcing and system parameters given in Example 6.6. Also, find the earliest time at which the maximum displacement response occurs.

6.10 Determine the response of the vibratory system discussed in Example 6.3, when the forcing due to the wind spectrum is of the following form

$$|F(jf)| = 200 \, fe^{-0.5f} \, \text{N}$$

where f is the frequency in Hz. Plot the result.

6.11 Revisit Example 4.8 and determine the non-dimensional displacement response of this system when $J_1 - 0.25$. Assume that no force is applied, the initial displacement is zero, the initial velocity is V_o m/s, and the natural frequency is ω_n. Compare this response to that obtained when a unit step function is applied to the mass initially at rest and discuss.

6.12 Repeat Example 6.4 when the step change in road elevation a is 4 cm and the vehicle speed is 100 km/h.

Many systems can vibrate in multiple directions. Such systems are described by models with multiple degrees-of-freedom. The cable car and the locomotive can each simultaneously oscillate up and down as well as sway from left to right.

7

Multiple Degree-of-Freedom Systems: Governing Equations and Characteristics of Free Responses

7.1 INTRODUCTION

In Chapters 4 through 6, single degree-of-freedom systems and vibratory responses of these systems were studied. Systems with multiple degrees of freedom and their responses are studied in this chapter and the next. Systems that need to be described by more than one independent coordinate have multiple degrees of freedom. The number of degrees of freedom is determined by the inertial elements present in a system. For example, in a system with two degrees of freedom, there can be either one inertial element whose motion is described by two independent coordinates or two inertial elements whose motions are described by two independent coordinates. In general, the number of degrees of freedom of a system is not only determined by the inertial elements present in a system, but also by the constraints imposed on the system. The governing equations of motion of vibratory systems is determined by using either force-balance and moment-balance methods or Lagrange's equations. In this chapter, both of these methods will be used to develop the system equations. Furthermore, viscous damping models will be used to model dissipation in the systems.

After developing the governing equations of motion, responses of unforced multi-degree-of-freedom systems are examined at length. Forced oscillations are considered in the next chapter. To describe the responses of single degree-of-freedom systems, only time information is needed. In addition to the time information, one also needs spatial information for describing the responses of systems with more than one degree of freedom. This spatial information is expressed in terms of mode shapes, which are determined from the free-vibration solution. Each mode shape is associated with a natural frequency of the system; this shape provides information about the relative spatial positions of the inertial elements in terms of the chosen generalized coordinates. Determination of mode shapes and natural frequencies is discussed in detail in this chapter. As illustrated in the next chapter, the spatial information obtained from the free-vibration problem can also provide a basis for determining the forced response of a system with multiple degrees of freedom. It is also shown there that the properties of mode shapes can be used to construct the response of a multi-degree-of-freedom system in terms of responses of equivalent single degree-of-freedom systems. This allows the material presented in the preceding chapters to be used to determine the response of a system with multiple degrees of freedom.

The notions of stability introduced in Chapter 4 for single degree-of-freedom systems are extended in this chapter to multi-degree-of-freedom systems.

7.2 GOVERNING EQUATIONS

In this section, two approaches are presented for determining the governing equations of motion. The first approach is based on force-balance and moment-balance methods and the second approach is based on Lagrange's equations. For algebraic ease, the number of degrees of freedom for the physical systems chosen in this chapter is less than or equal to five.

7.2.1 Force-Balance and Moment-Balance Methods

The underlying principles of the force-balance and moment-balance methods are expressed by Eqs. (1.11) and (1.17), which relate the forces and moments imposed on a system to the rate of change of linear momentum and the rate of change of angular momentum, respectively.

Force-Balance Method

To illustrate the use of force-balance methods, consider the system shown in Figure 7.1. This system consists of linear springs, linear dampers, and translating inertia elements. The free-body diagrams for the point masses m_1 and m_2 are shown, along with the respective inertial forces in Figure 7.2. The generalized coordinates x_1 and x_2 are used to specify the positions of the two masses m_1 and m_2, respectively, from the fixed end on the left side. Based on

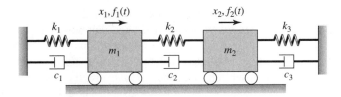

FIGURE 7.1
System with two degrees of freedom.

the free-body diagram of inertial element m_1 and carrying out the force balance along the horizontal direction i, one obtains the following equation:

$$\underbrace{-m_1\ddot{x}_1}_{\substack{\text{Inertia}\\\text{force}}} \underbrace{-k_1x_1}_{\substack{\text{Force associated}\\\text{with spring of}\\\text{stiffness } k_1}} \underbrace{+k_2(x_2 - x_1)}_{\substack{\text{Force associated}\\\text{with spring of}\\\text{stiffness } k_2}} \underbrace{-c_1\dot{x}_1}_{\substack{\text{Force associated}\\\text{with damper of}\\\text{coefficient } c_1}} \underbrace{+c_2(\dot{x}_2 - \dot{x}_1)}_{\substack{\text{Force associated}\\\text{with damper of}\\\text{coefficient } c_2}} \underbrace{+f_1(t)}_{\substack{\text{External force}\\\text{acting on}\\\text{mass } m_1}} = 0$$

This equation has been rewritten as the first of Eqs. (7.1a). Similarly, from the free-body diagram of inertial element m_2, the second of Eqs. (7.1a) is obtained.

$$m_1\ddot{x}_1 + (c_1 + c_2)\dot{x}_1 - c_2\dot{x}_2 + (k_1 + k_2)x_1 - k_2x_2 = f_1(t)$$
$$m_2\ddot{x}_2 + (c_2 + c_3)\dot{x}_2 - c_2\dot{x}_1 + (k_2 + k_3)x_2 - k_2x_1 = f_2(t) \qquad (7.1a)$$

These linear differential equations are written in matrix form[1] as

$$\begin{bmatrix} m_1 & 0 \\ 0 & m_2 \end{bmatrix}\begin{Bmatrix} \ddot{x}_1 \\ \ddot{x}_2 \end{Bmatrix} + \begin{bmatrix} c_1 + c_2 & -c_2 \\ -c_2 & c_2 + c_3 \end{bmatrix}\begin{Bmatrix} \dot{x}_1 \\ \dot{x}_2 \end{Bmatrix}$$
$$+ \begin{bmatrix} k_1 + k_2 & -k_2 \\ -k_2 & k_2 + k_3 \end{bmatrix}\begin{Bmatrix} x_1 \\ x_2 \end{Bmatrix} = \begin{Bmatrix} f_1 \\ f_2 \end{Bmatrix} \qquad (7.1b)$$

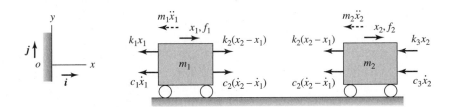

FIGURE 7.2
Free-body diagrams for masses m_1 and m_2 along with the respective inertial forces. The origin of the coordinate system is located on the fixed boundary at the left end of the spring.

---

[1]See Appendix E for a brief introduction to matrix notation.

The off-diagonal terms of the inertia matrix are zero, while the off-diagonal terms of the stiffness and damping matrices are non-zero. In addition, all of these matrices are *symmetric matrices*[2]. The equations governing the system are coupled due to these non-zero off-diagonal terms in the stiffness and damping matrices. Physically, the system is uncoupled when the damper c_2 and the spring k_2 are absent. The excitations $f_1(t)$ and $f_2(t)$ are directly applied to the inertial elements of the system as shown in the figure.

Moment-Balance Method

We now consider the system shown in Figure 7.3, where a rotor drives a system of two flywheels with rotary inertia J_{o1} and J_{o2}. The rotor inertia is considerably larger than that of each of the flywheels. Hence, the end of the shaft attached to the rotor is treated as a fixed end. The drive torque to the first flywheel is $M_o(t)$. The generalized coordinates ϕ_1 and ϕ_2 are used to describe the rotations of the flywheels about the axis k through the respective centers. The inertias of the shafts are neglected, the torsional stiffness of the shaft in between the fixed end and the flywheel closest to it is represented by k_{t1}, and the torsional stiffness of the other shaft is represented by k_{t2}. It is assumed that the flywheels are immersed in housings filled with oil and that the corresponding dissipative effect is modeled by using the damping coefficients c_{t1} and c_{t2}. In the free-body diagrams of Figure 7.3, the inertial moments $-J_{o1}\ddot{\phi}_1 k$ and $-J_{o2}\ddot{\phi}_2 k$ are also shown.

Based on the free-body diagrams shown in Figure 7.3, we apply the principle of angular momentum balance to each of the flywheels and obtain the governing equations

$$J_{o1}\ddot{\phi}_1 + c_{t1}\dot{\phi}_1 + k_{t1}\phi_1 + k_{t2}(\phi_1 - \phi_2) = M_o(t)$$
$$J_{o2}\ddot{\phi}_2 + c_{t2}\dot{\phi}_2 + k_{t2}(\phi_2 - \phi_1) = 0 \tag{7.2a}$$

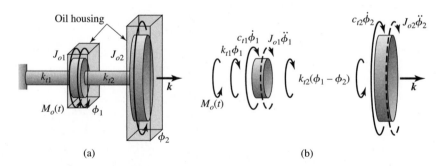

(a)   (b)

FIGURE 7.3
(a) System of two flywheels driven by a rotor and (b) free-body diagrams along with the respective inertial moments.

---

[2]As discussed in Appendix E, a matrix $[A]$ is called a symmetric matrix if the elements of the matrix $a_{ij} = a_{ji}$.

which are written in matrix form as

$$\begin{bmatrix} J_{o1} & 0 \\ 0 & J_{o2} \end{bmatrix} \begin{Bmatrix} \ddot{\phi}_1 \\ \ddot{\phi}_2 \end{Bmatrix} + \begin{bmatrix} c_{t1} & 0 \\ 0 & c_{t2} \end{bmatrix} \begin{Bmatrix} \dot{\phi}_1 \\ \dot{\phi}_2 \end{Bmatrix} + \begin{bmatrix} k_{t1} + k_{t2} & -k_{t2} \\ -k_{t2} & k_{t2} \end{bmatrix} \begin{Bmatrix} \phi_1 \\ \phi_2 \end{Bmatrix} = \begin{Bmatrix} M_o(t) \\ 0 \end{Bmatrix}$$

(7.2b)

In this case, the equations are coupled because of the non-zero off-diagonal terms in the stiffness matrix, which are due to the shaft with stiffness k_{t2}.

Both of the physical systems chosen for illustration of force-balance and moment-balance methods are described by linear models and the associated governing system of equations is written in matrix form. This is possible to do for any linear multi-degree-of-freedom system, as illustrated in Example 7.1. For a nonlinear multi-degree-of-freedom system, the governing nonlinear equations of motion are linearized to obtain a set of linear equations; the resulting linear equations are amenable to matrix form. This is illustrated in Example 7.3.

EXAMPLE 7.1   Modeling of a milling machine on a flexible floor

A milling machine and a vibratory model of this system are shown in Figure 7.4. We shall derive the governing equations of motion for this system by using the force-balance method. As shown in Figure 7.4b, the milling machine is described by using the three inertial elements m_1, m_2, and m_3 along with discrete spring elements and damper elements. All three inertial elements translate only along the i direction. The external force $f_1(t)$ in the i direction shown in the figure is a representative disturbance acting on m_1.

To obtain the governing equations of motion, we use the generalized coordinates x_1, x_2, and x_3, each measured from the system's static equilibrium position. Since the coordinates are measured from the static equilibrium position, gravity forces are not considered below. In order to apply the force-balance method to each inertial element, the free-body diagrams shown in Figure 7.4c are used. Applying the force-balance method along the i direction to each of the masses, we obtain the following equations:

$$m_1\ddot{x}_1 + k_1(x_1 - x_2) + c_1(\dot{x}_1 - \dot{x}_2) = -f_1(t)$$

$$m_2\ddot{x}_2 + (k_1 + k_2)x_2 - k_1x_1 - k_2x_3 + (c_1 + c_2)\dot{x}_2 - c_1\dot{x}_1 - c_2\dot{x}_3 = 0$$

$$m_3\ddot{x}_3 + (k_2 + k_3)x_3 - k_2x_2 + (c_2 + c_3)\dot{x}_3 - c_2\dot{x}_2 = 0 \qquad (a)$$

Equations (a) are arranged in the following matrix form:

$$\begin{bmatrix} m_1 & 0 & 0 \\ 0 & m_2 & 0 \\ 0 & 0 & m_3 \end{bmatrix} \begin{Bmatrix} \ddot{x}_1 \\ \ddot{x}_2 \\ \ddot{x}_3 \end{Bmatrix} + \begin{bmatrix} c_1 & -c_1 & 0 \\ -c_1 & c_1 + c_2 & -c_2 \\ 0 & -c_2 & c_2 + c_3 \end{bmatrix} \begin{Bmatrix} \dot{x}_1 \\ \dot{x}_2 \\ \dot{x}_3 \end{Bmatrix}$$

$$+ \begin{bmatrix} k_1 & -k_1 & 0 \\ -k_1 & k_1 + k_2 & -k_2 \\ 0 & -k_2 & k_2 + k_3 \end{bmatrix} \begin{Bmatrix} x_1 \\ x_2 \\ x_3 \end{Bmatrix} = \begin{Bmatrix} -f(t) \\ 0 \\ 0 \end{Bmatrix} \qquad (b)$$

We see that the inertia, the stiffness, and the damping matrices are symmetric matrices.

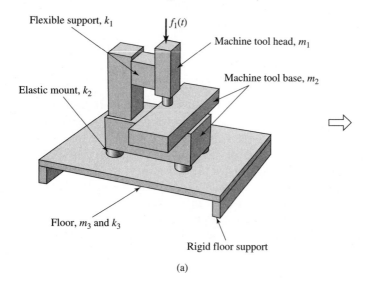

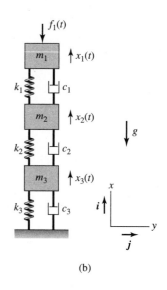

(a)

(b)

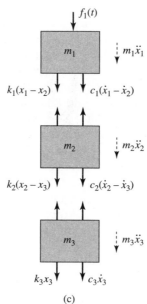

(c)

FIGURE 7.4

(a) Milling machine; (b) vibratory model for study of vertical motions; and (c) free-body diagrams of inertial elements m_1, m_2, and m_3 shown in (b) along with the inertial forces.

EXAMPLE 7.2 Conservation of linear momentum in a multi-degree-of-freedom system

We revisit Example 7.1 and discuss when the linear momentum of this multi-degree-of-freedom system is conserved along the i direction. From Eq. (1.11), it is clear that in the absence of external forces, the linear momentum of the system is conserved; that is,

$$\frac{d\mathbf{p}}{dt} = \mathbf{0} \rightarrow \mathbf{p} = \text{constant} \tag{a}$$

In the absence of the forcing $f_1(t)$ in Example 7.1, the linear momentum of this three degree-of-freedom system is conserved. To examine this, we set $f_1(t) = 0$ in Eq. (a) of Example 7.1 and arrive at the following equations:

$$m_1\ddot{x}_1 + k_1(x_1 - x_2) + c_1(\dot{x}_1 - \dot{x}_2) = 0$$
$$m_2\ddot{x}_2 + (k_1 + k_2)x_2 - k_1x_1 - k_2x_3 + (c_1 + c_2)\dot{x}_2 - c_1\dot{x}_1 - c_2\dot{x}_3 = 0 \tag{b}$$
$$m_3\ddot{x}_3 + (k_2 + k_3)x_3 - k_2x_2 + (c_2 + c_3)\dot{x}_3 - c_2\dot{x}_2 = 0$$

Each of Eqs. (b) was obtained by performing a linear momentum balance individually for each of the three inertial elements of the system. Adding all three equations of Eqs. (b), we obtain

$$m_1\ddot{x}_1 + m_2\ddot{x}_2 + m_3\ddot{x}_3 = 0 \tag{c}$$

Integrating Eq. (c) with respect to time—that is,

$$\int_0^t (m_1\ddot{x}_1 + m_2\ddot{x}_2 + m_3\ddot{x}_3)dt = 0 \tag{d}$$

—leads to

$$m_1\dot{x}_1 + m_2\dot{x}_2 + m_3\dot{x}_3 = \text{constant} \tag{e}$$

From Eq. (e), it follows that in the absence of external forces acting on the system, the total linear momentum of the system is conserved. This result could have been alternatively determined directly from Eq. (1.11); that is, Eq. (a). To examine the implication of Eq. (e) for this system of three particles, we note that the velocity of the center of mass of this three-degree-of-freedom system is given by

$$\dot{x}_{cm} = \frac{m_1\dot{x}_1 + m_2\dot{x}_2 + m_3\dot{x}_3}{m_1 + m_2 + m_3} \tag{f}$$

If the initial velocities of the three inertial elements are such that the constant on the right-hand side of Eq. (e) is zero, then it follows from Eq. (f) that

$$\dot{x}_{cm} = 0 \tag{g}$$

which, in turn, implies that the center of mass of the three degree-of-freedom system does not change during the motions of the system.

EXAMPLE 7.3   System with bounce and pitch motions

Consider the rigid bar shown in Figure 7.5a, which can rotate (pitch) about the **k** direction and translate (bounce) along the **j** direction. We locate the generalized coordinates y and θ at the center of gravity of the beam. This model provides a good representation for describing certain types of motions of motorcycles, automobiles, and other vehicles.

This particular example has been chosen to illustrate that both the force-balance and moment-balance methods are needed to obtain the governing equations. In addition, we also illustrate how the equilibrium positions are determined and how linearization of a nonlinear system is carried out.

Governing Equations of Motion

The free-body diagram shown in Figure 7.5b will be used to obtain the governing equations of motion. The inertial force and the inertial moment are also shown in Figure 7.5b. Considering force balance and moment balance with respect to the center of mass G of the rigid bar, we obtain, respectively,

$$m\ddot{y} + (c_1 + c_2)\dot{y} - (c_1L_1 - c_2L_2)\dot{\theta}\cos\theta + (k_1 + k_2)y$$
$$- (k_1L_1 - k_2L_2)\sin\theta = -mg \tag{a}$$

and

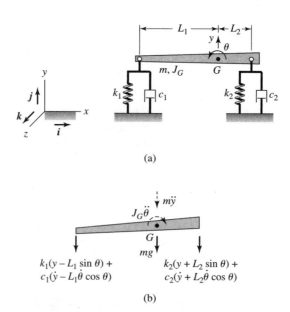

(a)

(b)

FIGURE 7.5

(a) Rigid body in the plane constrained by springs and dampers and (b) free-body diagram of the system along with the inertial force and the inertial moment.

$$J_G\ddot{\theta} - (c_1L_1 - c_2L_2)\dot{y}\cos\theta + (c_1L_1^2 + c_2L_2^2)\dot{\theta}\cos^2\theta$$

$$- (k_1L_1 - k_2L_2)y\cos\theta + (k_1L_1^2 + k_2L_2^2)\sin\theta\cos\theta = 0 \tag{b}$$

Equations (a) and (b) are nonlinear because of the $\sin\theta$ and $\cos\theta$ terms.

Static-Equilibrium Positions

The equilibrium positions y_o and θ_o of the system are obtained by setting the accelerations and velocities to zero in Eqs. (a) and (b). Thus, y_o and θ_o are solutions of

$$(k_1 + k_2)y_o - (k_1L_1 - k_2L_2)\sin\theta_o = -mg$$

$$\left\{[-(k_1L_1 - k_2L_2)y_o + (k_1L_1^2 + k_2L_2^2)\sin\theta_o\right\}\cos\theta_o = 0 \tag{c}$$

From the second of Eqs. (c), we find that

$$\cos\theta_o = 0 \quad \text{or} \quad \sin\theta_o = y_o\frac{(k_1L_1 - k_2L_2)}{(k_1L_1^2 + k_2L_2^2)} \tag{d}$$

Making use of the second of Eqs. (d) in the first of Eqs. (c), we arrive at

$$y_o = -\frac{mg(k_1L_1^2 + k_2L_2^2)}{k_1k_2(L_1 + L_2)^2} \tag{e}$$

Note that the equilibrium position $\theta_o = \pi/2$ corresponding to $\cos\theta_o = 0$ in Eqs. (d) is not considered because it is not physically meaningful. Examining Eq. (e), y_o represents a sag in the position of the bar due to the weight of the bar. From the second of Eqs. (d), θ_o represents a rotation due to the combination of the unequal stiffness at each end and the unequal mass distribution of the bar. In the absence of gravity loading or other constant loading, $y_o = 0$, and hence, $\theta_o = 0$.

Linearization and Linear System Governing "Small" Oscillations about an Equilibrium Position

We now consider "small" oscillations of the system shown in Figure 7.5 about the equilibrium position (y_o, θ_o). In order to obtain the governing equations, we substitute

$$y(t) = y_o + \hat{y}(t)$$
$$\theta(t) = \theta_o + \hat{\theta}(t) \tag{f}$$

into Eqs. (a) and (b) and carry out Taylor-series expansions[3] of $\sin\theta$ and $\cos\theta$, and retain the linear terms in y and θ. To this end, we find that

$$\ddot{y}(t) = \frac{d^2}{dt^2}(y_o + \hat{y}) = \ddot{\hat{y}}(t), \qquad \dot{y}(t) = \frac{d}{dt}(y_o + \hat{y}) = \dot{\hat{y}}(t)$$

$$\ddot{\theta}(t) = \frac{d^2}{dt^2}(\theta_o + \hat{\theta}) = \ddot{\hat{\theta}}(t), \qquad \dot{\theta}(t) = \frac{d}{dt}(\theta_o + \hat{\theta}) = \dot{\hat{\theta}}(t)$$

$$\sin\theta = \sin(\theta_o + \hat{\theta}) \approx \sin\theta_o + \hat{\theta}\cos\theta_o + \ldots$$

$$\cos\theta = \cos(\theta_o + \hat{\theta}) \approx \cos\theta_o - \hat{\theta}\sin\theta_o + \ldots \tag{g}$$

[3]T. B. Hildebrand, ibid.

On substituting Eqs. (g) into Eqs. (a) and (b), making use of Eqs. (c), and retaining only the linear terms in $\hat{y}$ and $\hat{\theta}$, we arrive at

$$
m\ddot{\hat{y}} + (c_1 + c_2)\dot{\hat{y}} - (c_1L_1 - c_2L_2)\dot{\hat{\theta}}\cos\theta_o + (k_1 + k_2)\hat{y}
$$
$$
- (k_1L_1 - k_2L_2)\hat{\theta}\cos\theta_o = 0
$$
$$
J_G\ddot{\hat{\theta}} - (c_1L_1 - c_2L_2)\dot{\hat{y}}\cos\theta_o + (c_1L_1^2 + c_2L_2^2)\dot{\hat{\theta}}\cos^2\theta_o
$$
$$
- (k_1L_1 - k_2L_2)\hat{y}\cos\theta_o + (k_1L_1^2 + k_2L_2^2)\hat{\theta}(\cos^2\theta_o - \sin^2\theta_o) = 0 \quad \text{(h)}
$$

which in matrix form reads as

$$
\begin{bmatrix} m & 0 \\ 0 & J_G \end{bmatrix}\begin{Bmatrix} \ddot{\hat{y}} \\ \ddot{\hat{\theta}} \end{Bmatrix} + \begin{bmatrix} c_1 + c_2 & -(c_1L_1 - c_2L_2)\cos\theta_o \\ -(c_1L_1 - c_2L_2)\cos\theta_o & (c_1L_1^2 + c_2L_2^2)\cos^2\theta_o \end{bmatrix}\begin{Bmatrix} \dot{\hat{y}} \\ \dot{\hat{\theta}} \end{Bmatrix}
$$
$$
+ \begin{bmatrix} k_1 + k_2 & -(k_1L_1 - k_2L_2)\cos\theta_o \\ -(k_1L_1 - k_2L_2)\cos\theta_o & (k_1L_1^2 + k_2L_2^2)(\cos^2\theta_o - \sin^2\theta_o) \end{bmatrix}\begin{Bmatrix} \hat{y} \\ \hat{\theta} \end{Bmatrix} = \begin{Bmatrix} 0 \\ 0 \end{Bmatrix} \quad \text{(i)}
$$

Equation (i) represents the linear system governing "small" oscillations of the system shown in Figure 7.5 about the static equilibrium position given by Eqs. (d) and (e). Although the gravity loading determines the equilibrium position, it does not appear explicitly in Eq. (i). Hence, when it is assumed that the generalized coordinates are measured from the static-equilibrium position, constant loading such as gravity loading is not considered for determining the equations of motion.

From the second of Eqs. (d) it is seen that if $k_1L_1 = k_2L_2$, $\theta_o = 0$, and Eq. (i) takes the form

$$
\begin{bmatrix} m & 0 \\ 0 & J_G \end{bmatrix}\begin{Bmatrix} \ddot{\hat{y}} \\ \ddot{\hat{\theta}} \end{Bmatrix} + \begin{bmatrix} c_1 + c_2 & -(c_1L_1 - c_2L_2) \\ -(c_1L_1 - c_2L_2) & (c_1L_1^2 + c_2L_2^2) \end{bmatrix}\begin{Bmatrix} \dot{\hat{y}} \\ \dot{\hat{\theta}} \end{Bmatrix}
$$
$$
+ \begin{bmatrix} k_1 + k_2 & 0 \\ 0 & (k_1L_1^2 + k_2L_2^2) \end{bmatrix}\begin{Bmatrix} \hat{y} \\ \hat{\theta} \end{Bmatrix} = \begin{Bmatrix} 0 \\ 0 \end{Bmatrix} \quad \text{(j)}
$$

If $k_1L_1 \neq k_2L_2$ and $\theta_o \approx 0$, then Eq. (i) takes the form

$$
\begin{bmatrix} m & 0 \\ 0 & J_G \end{bmatrix}\begin{Bmatrix} \ddot{\hat{y}} \\ \ddot{\hat{\theta}} \end{Bmatrix} + \begin{bmatrix} c_1 + c_2 & -(c_1L_1 - c_2L_2) \\ -(c_1L_1 - c_2L_2) & (c_1L_1^2 + c_2L_2^2) \end{bmatrix}\begin{Bmatrix} \dot{\hat{y}} \\ \dot{\hat{\theta}} \end{Bmatrix}
$$
$$
+ \begin{bmatrix} k_1 + k_2 & -(k_1L_1 - k_2L_2) \\ -(k_1L_1 - k_2L_2) & (k_1L_1^2 + k_2L_2^2) \end{bmatrix}\begin{Bmatrix} \hat{y} \\ \hat{\theta} \end{Bmatrix} = \begin{Bmatrix} 0 \\ 0 \end{Bmatrix} \quad \text{(k)}
$$

It is noted that Eqs. (k) could have been obtained directly, if it had been initially assumed that "small" oscillations about the static-equilibrium position were being considered and that the static-equilibrium position is "close" to the horizontal position. In this case, $\cos\theta$ would have been set to 1 and $\sin\theta$ would have been replaced by θ in Eqs. (a) and (b).

EXAMPLE 7.4 Governing equations of a rate gyroscope[4]

We shall obtain the governing equations of motion of a rate gyroscope, also known as a gyro-sensor. The physical system, along with the vibratory model, is shown in Figure 7.6. In the vibratory model shown in Figure 7.6b, the sensor is shown as a point mass m with two degrees of freedom and its motion is described by the coordinates x and y in the horizontal plane. The generalized coordinates are both located in a rotating reference frame. The sensor is to be designed to measure the rotational speed ω_z, which is as-

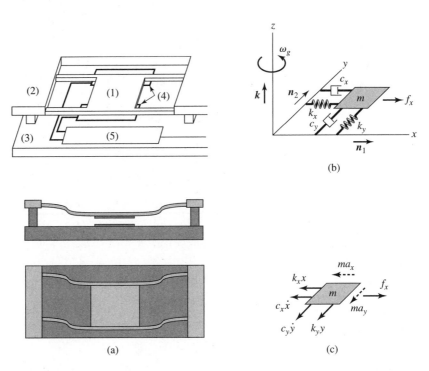

FIGURE 7.6
(a) Micromachined rate gyroscope; (b) vibratory model; and (c) free-body diagram along with the inertial forces. [(1) Suspended proof mass; (2) frame; (3) CMOS chip; (4) photodiodes; and (5) electronic circuitry.]
Source: Fig. 7.6a from O. Degani, D. J. Seter, E. Socher, S. Kaldor, and Y. Nemirovshy, "Optimal Design and Noise Consideration of Micromachined Vibrating Rate Gyroscope with Modulated Integrative Differential Optical Sensing," *Journal of Microelectromechanical Systems*, Vol. 7, No. 3, pp. 329–338 (1998). Copyright © 1998 IEEE. Reprinted with permission.

[4]O. Degani, D. J. Seter, E. Socher, S. Kaldor, and Y. Nemirovshy, "Optimal Design and Noise Consideration of Micromachined Vibrating Rate Gyroscope with Modulated Integrative Differential Optical Sensing," *J. Microelectromechanical Systems*, Vol. 7, No. 3, pp. 329–338 (1998).

sumed constant. For modeling purposes, a spring with stiffness k_x and a viscous damper with a damping coefficient c_x are used to constrain motions along the n_1 direction. Another spring-damper combination is used to constrain motions along the n_2 direction. An external force $f_x(t)$ in the n_1 direction is imposed on the system.

To obtain the governing equations, force balance is considered along the n_1 and n_2 directions. Making use of the free-body diagram shown in Figure 7.6c, we obtain the following relations

$$ma_x = -k_x x - c_x \dot{x} + f_x$$
$$ma_y = -k_y y - c_y \dot{y} \tag{a}$$

where x and y are the respective displacements along the n_1 and n_2 directions in the rotating reference frame; $\dot{x}$ and $\dot{y}$ are the respective velocities along these directions in the rotating reference frame; and a_x and a_y are the components of the absolute acceleration of the mass m along the n_1 and n_2 directions, respectively. From Exercise 1.4 and the discussion provided in Example 1.3 for a particle located in a rotating reference frame, it is found that

$$a_x = \ddot{x} - 2\omega_z \dot{y} - \omega_z^2 x$$
$$a_y = \ddot{y} + 2\omega_z \dot{x} - \omega_z^2 y \tag{b}$$

In arriving at Eqs. (b), the fact that the rotation ω_z is constant and that the corresponding angular acceleration is zero has been taken into account. From Eqs. (a) and (b), we arrive at the following set of equations

$$m(\ddot{x} - 2\omega_z \dot{y} - \omega_z^2 x) = -k_x x - c_x \dot{x} + f_x$$
$$m(\ddot{y} + 2\omega_z \dot{x} - \omega_z^2 y) = -k_y y - c_y \dot{y} \tag{c}$$

Equations (c) are written in the following matrix form:

$$[M]\begin{Bmatrix} \ddot{x} \\ \ddot{y} \end{Bmatrix} + [C]\begin{Bmatrix} \dot{x} \\ \dot{y} \end{Bmatrix} + [G]\begin{Bmatrix} \dot{x} \\ \dot{y} \end{Bmatrix} + [K]\begin{Bmatrix} x \\ y \end{Bmatrix} = \begin{Bmatrix} f_x \\ 0 \end{Bmatrix} \tag{d}$$

where the different square matrices are

$$[M] = \begin{bmatrix} m & 0 \\ 0 & m \end{bmatrix}, \quad [C] = \begin{bmatrix} c_x & 0 \\ 0 & c_y \end{bmatrix}, \quad [G] = \begin{bmatrix} 0 & -2m\omega_z \\ 2m\omega_z & 0 \end{bmatrix}$$
$$[K] = \begin{bmatrix} k_x - m\omega_z^2 & 0 \\ 0 & k_y - m\omega_z^2 \end{bmatrix} \tag{e}$$

The matrix $[G]$ is called the *gyroscopic matrix*.[5] The choice of coordinates in a rotating reference frame leads to this matrix. The gyroscopic matrix, which is a *skew-symmetric matrix*,[6] leads to coupling between the motions along the n_1 and n_2 directions. From the form of the stiffness matrix in Eqs. (e), it is clear that the effective stiffness associated with each direction of motion is reduced by the rotation.

---

[5]L. Meirovitch, *Computational Methods in Structural Dynamics*, Sijthoff and Noordhoff, The Netherlands, Chapter 2 (1980).

[6]The gyroscopic matrix $[G]$ is called a skew-symmetric matrix since its elements $g_{ij} = -g_{ji}$.

7.2.2   General Form of Equations for a Linear Multi-Degree-of-Freedom System

Based on the structure of Eqs. (7.1b) and (7.2b) and the linear systems treated in Examples 7.1, 7.3, and 7.4, the general form of the governing equations of motion for a linear N degree-of-freedom system described by the generalized coordinates q_1, q_2, ..., q_N, are put in the form

$$[M]\{\ddot{q}\} + \Big[[C] + [G]\Big]\{\dot{q}\} + \Big[[K] + [H]\Big]\{q\} = \{Q\} \tag{7.3}$$

where the different matrices and vectors in Eq. (7.3) have the following general form:

$$[M] = \begin{bmatrix} m_{11} & m_{12} & \cdots & m_{1N} \\ m_{21} & m_{22} & \cdots & m_{2N} \\ \vdots & \vdots & \ddots & \vdots \\ m_{N1} & m_{N2} & \cdots & m_{NN} \end{bmatrix}, \quad [C] = \begin{bmatrix} c_{11} & c_{12} & \cdots & c_{1N} \\ c_{21} & c_{22} & \cdots & c_{2N} \\ \vdots & \vdots & \ddots & \vdots \\ c_{N1} & c_{N2} & \cdots & c_{NN} \end{bmatrix}$$

$$[K] = \begin{bmatrix} k_{11} & k_{12} & \cdots & k_{1N} \\ k_{21} & k_{22} & \cdots & k_{2N} \\ \vdots & \vdots & \ddots & \vdots \\ k_{N1} & k_{N2} & \cdots & k_{NN} \end{bmatrix}, \quad [G] = \begin{bmatrix} g_{11} & g_{12} & \cdots & g_{1N} \\ g_{21} & g_{22} & \cdots & g_{2N} \\ \vdots & \vdots & \ddots & \vdots \\ g_{N1} & g_{N2} & \cdots & g_{NN} \end{bmatrix} \tag{7.4a}$$

$$[H] = \begin{bmatrix} h_{11} & h_{12} & \cdots & h_{1N} \\ h_{21} & h_{22} & \cdots & h_{2N} \\ \vdots & \vdots & \ddots & \vdots \\ h_{N1} & h_{N2} & \cdots & h_{NN} \end{bmatrix}$$

and

$$\{q\} = \begin{Bmatrix} q_1 \\ q_2 \\ \vdots \\ q_N \end{Bmatrix}, \quad \{\dot{q}\} = \begin{Bmatrix} \dot{q}_1 \\ \dot{q}_2 \\ \vdots \\ \dot{q}_N \end{Bmatrix}, \quad \{\ddot{q}\} = \begin{Bmatrix} \ddot{q}_1 \\ \ddot{q}_2 \\ \vdots \\ \ddot{q}_N \end{Bmatrix}, \text{ and } \{Q\} = \begin{Bmatrix} Q_1 \\ Q_2 \\ \vdots \\ Q_N \end{Bmatrix} \tag{7.4b}$$

The inertia matrix $[M]$, the stiffness matrix $[K]$, and the damping matrix $[C]$ are each an $N \times N$ matrix, and the force vector $\{Q\}$ is an $N \times 1$ vector. The $N \times N$ matrices $[G]$ and $[H]$, which are skew symmetric matrices, are called the gyroscopic matrix and the circulatory matrix, respectively.[7] The $N \times 1$ vector $\{q\}$ is called the *displacement vector,* the $N \times 1$ vector $\{\dot{q}\}$ is called the *velocity vector,* and the $N \times 1$ vector $\{\ddot{q}\}$ is called the *acceleration vector.*

---

[7]Gyroscopic forces and circulatory forces occur in rotating systems such as shafts; for example, see G. Genta, *Vibrations of Structures and Machines: Practical Aspects,* Springer-Verlag, NY (1993).

Linear Systems with N Inertial Elements, (N+1) Linear Spring Elements, and (N+1) Linear Damper Elements

As a special case of linear multi-degree-of-freedom systems, we consider the system shown in Figure 7.7. This system is an extension of the two degree-of-freedom system shown in Figure 7.1. In the figure, m_i is the mass of the ith inertial element whose motion is described by the generalized coordinate $q_i(t)$, which is measured from the point o located on the fixed boundary along the i direction. The force acting on the ith inertial element is represented by $Q_i(t)$. Carrying out a force balance based on the free-body diagram shown in Figure 7.7b, we obtain the following equation that governs the ith inertial element:

$$m_i \ddot{q}_i + (k_i + k_{i+1})q_i - k_i q_{i-1} - k_{i+1} q_{i+1}$$
$$+ (c_i + c_{i+1})\dot{q}_i - c_i \dot{q}_{i-1} - c_{i+1}\dot{q}_{i+1} = Q_i(t) \quad i = 1, 2, ..., N \quad (7.5a)$$

Assembling all of the N equations given by Eqs. (7.5a) into matrix form, we obtain Eq. (7.3) with the circulatory matrix $[H] = 0$ and the gyroscopic matrix $[G] = 0$ and with the following inertia, stiffness, and damping matrices, respectively,

$$[M] = \begin{bmatrix} m_1 & 0 & \cdots & 0 \\ 0 & m_2 & & 0 \\ \vdots & & \ddots & \vdots \\ 0 & \cdots & & m_N \end{bmatrix}$$

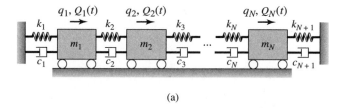

(a)

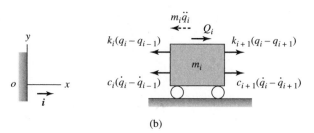

(b)

FIGURE 7.7
(a) Linear system with N inertial elements, $(N+1)$ spring elements, and $(N+1)$ damper elements and (b) free-body diagram of ith inertial element shown along with the inertial force.

$$[K] = \begin{bmatrix} k_1 + k_2 & -k_2 & 0 & 0 & \cdots & & 0 \\ -k_2 & k_2 + k_3 & -k_3 & 0 & \cdots & & 0 \\ 0 & -k_3 & & & & & \vdots \\ 0 & 0 & & \ddots & & -k_{N-1} & 0 \\ \vdots & \vdots & & & -k_{N-1} & k_{N-1} + k_N & -k_N \\ 0 & 0 & \cdots & 0 & & -k_N & k_N + k_{N+1} \end{bmatrix}$$

$$[C] = \begin{bmatrix} c_1 + c_2 & -c_2 & 0 & 0 & \cdots & & 0 \\ -c_2 & c_2 + c_3 & -c_3 & 0 & \cdots & & 0 \\ 0 & -c_3 & & & & & \vdots \\ 0 & 0 & & \ddots & & -c_{N-1} & 0 \\ \vdots & \vdots & & & -c_{N-1} & c_{N-1} + c_N & -c_N \\ 0 & 0 & \cdots & 0 & & -c_N & c_N + c_{N+1} \end{bmatrix} \tag{7.5b}$$

The inertia matrix given by Eq. (7.5b) is a diagonal matrix, while the stiffness matrix and the damping matrix given by Eq. (7.5b) are not diagonal matrices because of the presence of the off-diagonal elements. However, the stiffness and damping matrices are *banded matrices,* with each banded matrix having non-zero elements along three diagonals. All the other elements of these matrices are zero.

Conservation of Linear Momentum and Angular Momentum

In the absence of external forces—that is,

$$Q_1 = Q_2 = \cdots = Q_N = 0 \tag{7.6a}$$

—the system linear momentum is conserved. This means that

$$m_1 \dot{q}_1 + m_2 \dot{q}_2 + \cdots + m_N \dot{q}_N = \text{constant} \tag{7.6b}$$

Equation (7.6b) is obtained directly by making use of Eq. (1.11) and noting Eq. (7.6a).

For systems that experience rotational motions, Eq. (1.17) can be used to show that in the absence of external moments, the corresponding system angular momentum is conserved. This means that for the two degree-of-freedom system shown in Figure 7.3, in the absence of the external moment—that is, $M_o(t) = 0$—the angular momentum $J_{o1}\dot{\phi}_1 + J_{o2}\dot{\phi}_2 = \text{constant}$, along the $\mathbf{k}$ direction.

7.2.3 Lagrange's Equations of Motion

To use Lagrange's equations, we start from Eqs. (3.41), where N is the number of degrees of freedom and the N independent generalized coordinates used for describing the motion are $q_1, q_2, ..., q_N$. Repeating Eqs. (3.41), the system of N governing equations is given by

$$\frac{d}{dt}\left(\frac{\partial T}{\partial \dot{q}_j}\right) - \frac{\partial T}{\partial q_j} + \frac{\partial D}{\partial \dot{q}_j} + \frac{\partial V}{\partial q_j} = Q_j \qquad \text{for} \quad j = 1, 2, ..., N \tag{7.7}$$

where T is the kinetic energy of the system, V is the potential energy of the system, D is the Rayleigh dissipation function, and Q_j is the generalized force acting on the jth inertial element.

Linear Vibratory Systems

For vibratory systems with linear inertial characteristics, linear stiffness characteristics, and linear viscous damping characteristics, T, V, and D take the form given by Eqs. (3.43); that is, we have[8]

$$T = \frac{1}{2}\{\dot{q}\}^T[M]\{\dot{q}\} = \frac{1}{2}\sum_{j=1}^{N}\sum_{n=1}^{N}m_{jn}\dot{q}_j\dot{q}_n$$

$$V = \frac{1}{2}\{q\}^T[K]\{q\} = \frac{1}{2}\sum_{j=1}^{N}\sum_{n=1}^{N}k_{jn}q_jq_n$$

$$D = \frac{1}{2}\{\dot{q}\}^T[C]\{\dot{q}\} = \frac{1}{2}\sum_{j=1}^{N}\sum_{n=1}^{N}c_{jn}\dot{q}_j\dot{q}_n \tag{7.8}$$

and Q_j is given by Eq. (3.42). Since

$$\frac{\partial^2 T}{\partial \dot{q}_j \partial \dot{q}_k} = \frac{\partial^2 T}{\partial \dot{q}_k \partial \dot{q}_j}$$

$$\frac{\partial^2 V}{\partial q_j \partial q_k} = \frac{\partial^2 V}{\partial q_k \partial q_j}$$

$$\frac{\partial^2 D}{\partial \dot{q}_j \partial \dot{q}_k} = \frac{\partial^2 D}{\partial \dot{q}_k \partial \dot{q}_j} \tag{7.9a}$$

the inertia coefficients m_{jk}, the stiffness coefficients k_{jk}, and the damping coefficients c_{jk} are symmetric; that is,

$$m_{jk} = m_{kj}$$

$$k_{jk} = k_{kj} \tag{7.9b}$$

$$c_{jk} = c_{kj}$$

In light of Eqs. (7.9b), for a two degree-of-freedom system with linear characteristics, Eq. (7.8) is written in expanded form as

$$T = \frac{1}{2}\sum_{j=1}^{2}\sum_{n=1}^{2}m_{jn}\dot{q}_j\dot{q}_n = \frac{1}{2}m_{11}\dot{q}_1^2 + m_{12}\dot{q}_1\dot{q}_2 + \frac{1}{2}m_{22}\dot{q}_2^2$$

$$V = \frac{1}{2}\sum_{j=1}^{2}\sum_{n=1}^{2}k_{jn}q_jq_n = \frac{1}{2}k_{11}q_1^2 + k_{12}q_1q_2 + \frac{1}{2}k_{22}q_2^2$$

$$D = \frac{1}{2}\sum_{j=1}^{2}\sum_{n=1}^{2}c_{jn}\dot{q}_j\dot{q}_n = \frac{1}{2}c_{11}\dot{q}_1^2 + c_{12}\dot{q}_1\dot{q}_2 + \frac{1}{2}c_{22}\dot{q}_2^2 \tag{7.10}$$

[8]The quadratic forms shown for T, V, and D in Eqs. (7.8) are strictly valid for systems with linear characteristics. However, the kinetic energy T is not always a function of only velocities as shown here. Systems in which the kinetic energy has the quadratic form shown in Eqs. (7.8) are called *natural systems*. As discussed later in this chapter, for systems such as that given in Example 7.4, where rotation effects are present, the form of the kinetic energy T is different from that shown in Eqs. (7.8). As noted in Chapter 2, gravity forces can also contribute to the potential energy of the system. For a general system with holonomic constraints, Eqs. (7.7) will be used directly in this book to obtain the governing equations.

On substituting Eqs. (7.10) into Eqs. (7.7) and performing the indicated operations, we obtain the following system of two coupled ordinary differential equations

$$m_{11}\ddot{q}_1 + m_{12}\ddot{q}_2 + c_{11}\dot{q}_1 + c_{12}\dot{q}_2 + k_{11}q_1 + k_{12}q_2 = Q_1$$
$$m_{12}\ddot{q}_1 + m_{22}\ddot{q}_2 + c_{12}\dot{q}_1 + c_{22}\dot{q}_2 + k_{12}q_1 + k_{22}q_2 = Q_2 \tag{7.11}$$

which are written in the matrix form

$$[M]\{\ddot{q}\} + [C]\{\dot{q}\} + [K]\{q\} = \{Q\} \tag{7.12}$$

where the square matrix $[M]$ is the *inertia matrix,* the square matrix $[K]$ is the *stiffness matrix,* and square matrix $[C]$ is the *damping matrix.* They are given by, respectively,

$$[M] = \begin{bmatrix} m_{11} & m_{12} \\ m_{12} & m_{22} \end{bmatrix}, \quad [K] = \begin{bmatrix} k_{11} & k_{12} \\ k_{12} & k_{22} \end{bmatrix}, \quad \text{and} \quad [C] = \begin{bmatrix} c_{11} & c_{12} \\ c_{12} & c_{22} \end{bmatrix} \tag{7.13}$$

The displacement vector $\{q\}$, the velocity vector $\{\dot{q}\}$, the acceleration vector $\{\ddot{q}\}$, and the vector of generalized forces $\{Q\}$ are given by the column vectors

$$\{\ddot{q}\} = \begin{Bmatrix} \ddot{q}_1 \\ \ddot{q}_2 \end{Bmatrix}, \quad \{\dot{q}\} = \begin{Bmatrix} \dot{q}_1 \\ \dot{q}_2 \end{Bmatrix}, \quad \{q\} = \begin{Bmatrix} q_1 \\ q_2 \end{Bmatrix}, \quad \text{and} \quad \{Q\} = \begin{Bmatrix} Q_1 \\ Q_2 \end{Bmatrix} \tag{7.14}$$

Illustration of Derivation of Governing Equations for a Two Degree-of-Freedom System

Consider again the two degree-of-freedom system shown in Figure 7.1, with linear springs and linear dampers. The kinetic energy and potential energy are given by

$$T = \frac{1}{2} m_1\dot{x}_1^2 + \frac{1}{2} m_2\dot{x}_2^2$$

$$V = \frac{1}{2} k_1 x_1^2 + \frac{1}{2} k_2(x_1 - x_2)^2 + \frac{1}{2} k_3 x_2^2 \tag{7.15}$$

$$= \frac{1}{2} (k_1 + k_2)x_1^2 - k_2 x_1 x_2 + \frac{1}{2} (k_2 + k_3)x_2^2$$

the dissipation function D takes the form

$$D = \frac{1}{2} c_1\dot{x}_1^2 + \frac{1}{2} c_2(\dot{x}_1 - \dot{x}_2)^2 + \frac{1}{2} c_3\dot{x}_2^2$$

$$= \frac{1}{2} (c_1 + c_2)\dot{x}_1^2 - c_2\dot{x}_1\dot{x}_2 + \frac{1}{2} (c_2 + c_3)\dot{x}_2^2 \tag{7.16}$$

and $Q_1 = f_1$ and $Q_2 = f_2$.

Comparing Eqs. (7.15) and (7.16) with Eqs. (7.10), the corresponding inertia coefficients m_{jk}, stiffness coefficients k_{jk}, and damping coefficients c_{jk} are identified. Upon using Eqs. (7.4a), this leads to the governing system

$$[M]\{\ddot{x}\} + [C]\{\dot{x}\} + [K]\{x\} = \{F\} \tag{7.17}$$

where the different square matrices are

$$[M] = \begin{bmatrix} m_1 & 0 \\ 0 & m_2 \end{bmatrix}, \quad [K] = \begin{bmatrix} k_1 + k_2 & -k_2 \\ -k_2 & k_2 + k_3 \end{bmatrix}, \quad \text{and}$$

$$[C] = \begin{bmatrix} c_1 + c_2 & -c_2 \\ -c_2 & c_2 + c_3 \end{bmatrix} \tag{7.18}$$

and the different column vectors are

$$\{\ddot{x}\} = \begin{Bmatrix} \ddot{x}_1 \\ \ddot{x}_2 \end{Bmatrix}, \quad \{\dot{x}\} = \begin{Bmatrix} \dot{x}_1 \\ \dot{x}_2 \end{Bmatrix}, \quad \{x\} = \begin{Bmatrix} x_1 \\ x_2 \end{Bmatrix}, \quad \text{and} \quad \{F\} = \begin{Bmatrix} f_1 \\ f_2 \end{Bmatrix} \tag{7.19}$$

In certain situations, it is necessary to use Eqs. (7.7) directly. To illustrate this procedure, we evaluate the following derivatives based on Eqs. (7.15) and (7.16):

$$\frac{d}{dt}\left(\frac{\partial T}{\partial \dot{x}_1}\right) = m_1\ddot{x}_1, \qquad \frac{d}{dt}\left(\frac{\partial T}{\partial \dot{x}_2}\right) = m_2\ddot{x}_2$$

$$\frac{\partial T}{\partial x_1} = 0, \qquad \frac{\partial T}{\partial x_2} = 0$$

$$\frac{\partial V}{\partial x_1} = k_1 x_1 + k_2(x_1 - x_2) = (k_1 + k_2)x_1 - k_2 x_2$$

$$\frac{\partial V}{\partial x_2} = -k_2(x_1 - x_2) + k_3 x_2 = (k_2 + k_3)x_2 - k_2 x_1$$

$$\frac{\partial D}{\partial \dot{x}_1} = c_1\dot{x}_1 + c_2(\dot{x}_1 - \dot{x}_2) = (c_1 + c_2)\dot{x}_1 - c_2\dot{x}_2$$

$$\frac{\partial D}{\partial \dot{x}_2} = -c_2(\dot{x}_1 - \dot{x}_2) + c_3\dot{x}_2 = (c_2 + c_3)\dot{x}_2 - c_2\dot{x}_1 \tag{7.20}$$

On substituting Eqs. (7.20) into Eqs. (7.7); that is,

$$\frac{d}{dt}\left(\frac{\partial T}{\partial \dot{x}_1}\right) - \frac{\partial T}{\partial x_1} + \frac{\partial D}{\partial \dot{x}_1} + \frac{\partial V}{\partial x_1} = Q_1$$

$$\frac{d}{dt}\left(\frac{\partial T}{\partial \dot{x}_2}\right) - \frac{\partial T}{\partial x_2} + \frac{\partial D}{\partial \dot{x}_2} + \frac{\partial V}{\partial x_2} = Q_2 \tag{7.21a}$$

we obtain

$$m_1\ddot{x}_1 + (c_1 + c_2)\dot{x}_1 - c_2\dot{x}_2 + (k_1 + k_2)x_1 - k_2 x_2 = f_1(t)$$

$$m_2\ddot{x}_2 + (c_2 + c_3)\dot{x}_2 - c_2\dot{x}_1 + (k_2 + k_3)x_2 - k_2 x_1 = f_2(t) \tag{7.21b}$$

which yields the same matrix form as Eqs. (7.1b).

In light of Eqs. (7.9a) and (7.9b), the equations of motion obtained by using Lagrange's equations for linear systems always have symmetric inertia and stiffness matrices. For systems in which the dissipation is modeled by the Rayleigh dissipation function, the damping matrix is also symmetric. These symmetry properties are not necessarily explicit when the governing equations are obtained by using force-balance and moment-balance methods. However, the form of the governing equations obtained by using force-balance and moment-balance methods can be manipulated to obtain the same form as that obtained through Lagrange's equations.

Next, different examples are provided to illustrate the use of Lagrange's equations for deriving the equations of motion of multi-degree-of-freedom systems. When using Lagrange's equations, in cases where the expressions for kinetic energy, potential energy, and dissipation function are in the form of Eqs. (7.8), these equations are directly used to identify the coefficients in the inertia, stiffness, and damping matrices in the governing equations.

EXAMPLE 7.5 System with a translating mass attached to an oscillating disk

Consider the system shown in Figure 7.8 with linear springs and linear dampers and a rotating element. A massless rigid bar connects the rotating element to the base of the combination of the spring k_2 and the damper c_2. The inertial element m_1 is treated as a translating point mass and the other inertial element is treated as a rigid body rotating about the point o. The position x of mass m_1 is the displacement measured from the fixed end and the other generalized coordinate θ is the angular position of the disk measured from the vertical. We shall use Lagrange's equations to determine the governing equations of motion of the system.

The kinetic energy of the system is given by

$$T = \underbrace{\frac{1}{2} m_1 \dot{x}^2}_{\substack{\text{Kinetic energy} \\ \text{of mass } m_1}} + \underbrace{\frac{1}{2} J_o \dot{\theta}^2}_{\substack{\text{Kinetic energy of} \\ \text{rigid body } J_o}} \tag{a}$$

We assume that the mass center of the disk coincides with the point o and that the angular oscillations in θ are "small." The "small" angle assumption allows us to express the horizontal translation of point A as being equal to $r\theta$. The potential energy is then given by

$$V = \underbrace{\frac{1}{2} k_1 x^2}_{\substack{\text{Potential energy} \\ \text{associated with} \\ \text{spring } k_1}} + \underbrace{\frac{1}{2} k_2 (x - r\theta)^2}_{\substack{\text{Potential energy} \\ \text{associated with} \\ \text{spring } k_2}}$$

$$= \frac{1}{2} k_1 x^2 + \frac{1}{2} k_2 r^2 \theta^2 - k_2 r \theta x + \frac{1}{2} k_2 x^2$$

$$= \frac{1}{2} (k_1 + k_2) x^2 - k_2 r \theta x + \frac{1}{2} k_2 r^2 \theta^2 \tag{b}$$

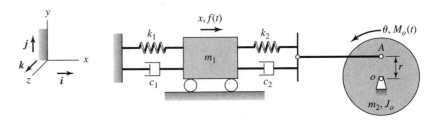

FIGURE 7.8
Translating and rotating system with two degrees of freedom.

The dissipation function takes the form

$$D = \underbrace{\frac{1}{2} c_1 \dot{x}^2}_{\substack{\text{Dissipation} \\ \text{associated with} \\ \text{damper } c_1}} + \underbrace{\frac{1}{2} c_2 (\dot{x} - r\dot{\theta})^2}_{\substack{\text{Dissipation} \\ \text{associated with} \\ \text{damper } c_2}}$$

$$= \frac{1}{2} c_1 \dot{x}^2 + \frac{1}{2} c_2 r^2 \dot{\theta}^2 - c_2 r \dot{\theta} \dot{x} + \frac{1}{2} c_2 \dot{x}^2$$

$$= \frac{1}{2} (c_1 + c_2) \dot{x}^2 - c_2 r \dot{\theta} \dot{x} + \frac{1}{2} c_2 r^2 \dot{\theta}^2 \qquad (c)$$

Since Eqs. (a) through (c) are in the standard form of Eqs. (7.8), we identify the inertial, stiffness, and damping coefficients by comparing Eqs. (a) through (c) to Eqs. (7.10) and associating x with q_1 and θ with q_2. In addition, we make use of Eq. (3.42) to recognize the generalized forces as $Q_1 = f(t)$ and $Q_2 = M_o(t)$. Then, the governing equations become

$$\begin{bmatrix} m & 0 \\ 0 & J_O \end{bmatrix} \begin{Bmatrix} \ddot{x} \\ \ddot{\theta} \end{Bmatrix} + \begin{bmatrix} c_1 + c_2 & -rc_2 \\ -rc_2 & r^2 c_2 \end{bmatrix} \begin{Bmatrix} \dot{x} \\ \dot{\theta} \end{Bmatrix} + \begin{bmatrix} k_1 + k_2 & -rk_2 \\ -rk_2 & r^2 k_2 \end{bmatrix} \begin{Bmatrix} x \\ \theta \end{Bmatrix} = \begin{Bmatrix} f(t) \\ M_o(t) \end{Bmatrix} \quad (d)$$

Since the damping and stiffness matrices have non-zero off-diagonal terms, Eqs. (d) are coupled.

EXAMPLE 7.6   System with bounce and pitch motions revisited

Consider the rigid bar shown in Figure 7.5a. The generalized coordinates are y and θ, which are located at the center of gravity of the beam. We shall use Lagrange's equations to obtain the governing equations of motion of this two degree-of-freedom system.

For a rigid bar undergoing planar motions, we use Eq. (1.24) to find that the kinetic energy is given by

$$T = \underbrace{\frac{1}{2} m \dot{y}^2}_{\substack{\text{Kinetic energy} \\ \text{associated with} \\ \text{translation of center} \\ \text{of mass } m}} + \underbrace{\frac{1}{2} J_G \dot{\theta}^2}_{\substack{\text{Kinetic energy} \\ \text{associated with} \\ \text{rotation about} \\ \text{center of mass}}} \qquad (a)$$

Taking into account the potential energy due to the spring displacement and the work done by the gravity loading, the potential energy is

$$V = \underbrace{\frac{1}{2} k_1 (y - L_1 \sin \theta)^2}_{\substack{\text{Potential energy associated} \\ \text{with spring } k_1}} + \underbrace{\frac{1}{2} k_2 (y + L_2 \sin \theta)^2}_{\substack{\text{Potential energy associated} \\ \text{with spring } k_2}} + \underbrace{mgy}_{\substack{\text{Potential energy} \\ \text{associated with} \\ \text{gravity loading}}}$$

$$= \frac{1}{2} (k_1 + k_2) y^2 - (k_1 L_1 - k_2 L_2) y \sin \theta$$

$$+ \frac{1}{2} (k_1 L_1^2 + k_2 L_2^2) \sin^2 \theta + mgy \qquad (b)$$

The dissipation function is of the form

$$D = \frac{1}{2} c_1(\dot{y} - L_1\dot{\theta} \cos \theta)^2 + \frac{1}{2} c_2(\dot{y} + L_2\dot{\theta} \cos \theta)^2$$

$$\underbrace{\phantom{\frac{1}{2} c_1(\dot{y} - L_1\dot{\theta} \cos \theta)^2}}_{\substack{\text{Dissipation associated} \\ \text{with damper } c_1}} \underbrace{\phantom{\frac{1}{2} c_2(\dot{y} + L_2\dot{\theta} \cos \theta)^2}}_{\substack{\text{Dissipation associated} \\ \text{with damper } c_2}}$$

$$= \frac{1}{2} (c_1 + c_2)\dot{y}^2 - (c_1L_1 - c_2L_2)\dot{\theta}\dot{y} \cos \theta$$

$$+ \frac{1}{2} (c_1L_1^2 + c_2L_2^2)\dot{\theta}^2 \cos^2 \theta \tag{c}$$

Since Eqs. (b) and (c) are not in the standard form of Eqs. (7.8), the equations of motion are obtained directly by using Eqs. (7.7). Thus, recognizing that $q_1 = y$, the first equation of Eqs. (7.7) takes the form

$$\frac{d}{dt}\left(\frac{\partial T}{\partial \dot{y}}\right) - \frac{\partial T}{\partial y} + \frac{\partial D}{\partial \dot{y}} + \frac{\partial V}{\partial y} = Q_y \tag{d}$$

Noting that $Q_y = 0$, we obtain from Eqs. (a) through (d) that the first equation of motion is

$$m\ddot{y} + (c_1 + c_2)\dot{y} - (c_1L_1 - c_2L_2)\dot{\theta} \cos \theta + (k_1 + k_2)y$$
$$- (k_1L_1 - k_2L_2)\sin \theta = -mg \tag{e}$$

Recognizing that $q_2 = \theta$, the second equation of Eqs. (7.7) assumes the form

$$\frac{d}{dt}\left(\frac{\partial T}{\partial \dot{\theta}}\right) - \frac{\partial T}{\partial \theta} + \frac{\partial D}{\partial \dot{\theta}} + \frac{\partial V}{\partial \theta} = Q_\theta \tag{f}$$

Recognizing that $Q_\theta = 0$, we arrive from Eqs. (a), (b), (c), and (f) at the second equation of motion

$$J_G\ddot{\theta} - (c_1L_1 - c_2L_2)\dot{y} \cos \theta + (c_1L_1^2 + c_2L_2^2)\dot{\theta} \cos^2 \theta$$
$$- (k_1L_1 - k_2L_2)y \cos \theta + (k_1L_1^2 + k_2L_2^2)\sin \theta \cos \theta = 0 \tag{g}$$

Equations (e) and (g) are identical to Eqs. (a) and (b) of Example 7.3, which were obtained by using force-balance and moment-balance methods.

EXAMPLE 7.7 Pendulum absorber

Consider the two degree-of-freedom system shown in Figure 7.9, in which the generalized coordinate x is used to locate the mass m_1 and the other generalized coordinate θ is used to specify the angular position of the pendulum. This type of system can model a pendulum absorber,[9] which is used in many

[9]J. J. Hollkamp, R. L. Bagley, and R. W. Gordon, "A Centrifugal Pendulum Absorber for Rotating, Hollow Engine Blades," *J. Sound Vibration,* Vol. 219, No. 3, pp. 539–549 (1999); Z.-M. Ge and T.-N. Lin, "Regular and Chaotic Dynamic Analysis and Control of Chaos of an Elliptical Pendulum on a Vibrating Basement," *J. Sound Vibration,* Vol. 230, No. 5, pp. 1045–1068 (2000); and A. Ertas et al., "Performance of Pendulum Absorber for a Nonlinear System of Varying Orientation," *J. Sound Vibration,* Vol. 229, No. 4, pp. 913–933 (2000).

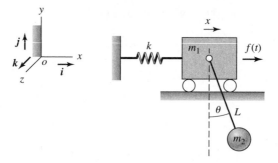

FIGURE 7.9
Pendulum absorber.

applications. The mass of the rod of length L is assumed negligible. In this example, we obtain the nonlinear equations of motion and then linearize the governing nonlinear equations about an equilibrium position of the system.

Governing Equations

We see from Figure 7.9 that the position of mass m_2 is

$$\boldsymbol{r}_m = (x + L \sin \theta)\boldsymbol{i} - L \cos \theta\boldsymbol{j}$$

Therefore, the velocity of mass m_2 is given by

$$\boldsymbol{V}_m = \frac{d\boldsymbol{r}_m}{dt} = (\dot{x} + L\dot{\theta} \cos \theta)\boldsymbol{i} + L\dot{\theta} \sin \theta\boldsymbol{j} \tag{a}$$

Making use of Eq. (1.22) and Eq. (a), the kinetic energy for the system is

$$T = \underbrace{\frac{1}{2} m_1\dot{x}^2}_{\substack{\text{Kinetic energy} \\ \text{of mass } m_1}} + \underbrace{\frac{1}{2} m_2(\boldsymbol{V}_m{\cdot}\boldsymbol{V}_m)}_{\substack{\text{Kinetic energy of} \\ \text{pendulum of mass } m_2}}$$

$$= \frac{1}{2} m_1\dot{x}^2 \frac{1}{2} m_2[(\dot{x} + L\dot{\theta} \cos \theta)^2 + (L\dot{\theta} \sin \theta)^2]$$

$$= \frac{1}{2} (m_1 + m_2)\dot{x}^2 + m_2L \cos \theta\dot{x}\dot{\theta} + \frac{1}{2} m_2L^2\dot{\theta}^2 \tag{b}$$

The potential energy of the system is

$$V = \underbrace{\frac{1}{2} kx^2}_{\substack{\text{Potential energy} \\ \text{associated with} \\ \text{spring } k}} + \underbrace{m_2gL(1 - \cos \theta)}_{\substack{\text{Potential energy} \\ \text{associated with} \\ \text{gravity loading}}} \tag{c}$$

where the datum has been chosen at the bottom position of the pendulum. Since Eqs. (b) and (c) are not in the form of Eqs. (7.8), we use Eqs. (7.7) directly to obtain the equations of motion. On using Eqs. (7.7) with $q_1 = x$, $q_2 = \theta$, and recognizing that the generalized forces are $Q_1 = f(t)$ and $Q_2 = 0$ and $D = 0$, Eqs. (7.7) becomes

$$\frac{d}{dt}\left(\frac{\partial T}{\partial \dot{x}}\right) - \frac{\partial T}{\partial x} + \frac{\partial V}{\partial x} = f(t)$$

$$\frac{d}{dt}\left(\frac{\partial T}{\partial \dot{\theta}}\right) - \frac{\partial T}{\partial \theta} + \frac{\partial V}{\partial \theta} = 0 \tag{d}$$

Upon substituting Eqs. (b) and (c) into Eq. (d), we obtain

$$(m_1 + m_2)\ddot{x} + m_2 L \ddot{\theta} \cos\theta - m_2 L \dot{\theta}^2 \sin\theta + kx = f(t)$$

$$m_2 L^2 \ddot{\theta} + m_2 L \ddot{x} \cos\theta + m_2 g L \sin\theta = 0 \tag{e}$$

Static-Equilibrium Positions

Setting the accelerations and velocities and the time-dependent forcing $f(t)$ to zero in Eqs. (e), the equations governing the equilibrium positions x_o and θ_o of the system are

$$kx_o = 0$$

$$m_2 g L \sin\theta_o = 0 \tag{f}$$

From Eqs. (f), the equilibrium positions of the system are obtained as

$$(x_o, \theta_o) = (0, 0) \quad \text{and} \quad (x_o, \theta_o) = (0, \pi) \tag{g}$$

where the first of Eqs. (g) corresponds to the bottom position of the pendulum and the second of Eqs. (g) corresponds to the pendulum being rotated by 180°.

Linearization

Considering "small" oscillations about the equilibrium position (0,0) and linearizing the equations of motion given by Eqs. (e) along the lines of what was illustrated with the use of Eqs. (g) of Example 7.3, we obtain

$$(m_1 + m_2)\ddot{x} + m_2 L \ddot{\theta} + kx = f(t)$$

$$m_2 L^2 \ddot{\theta} + m_2 L \ddot{x} + m_2 g L \theta = 0 \tag{h}$$

which in matrix form reads as

$$\begin{bmatrix} m_1 + m_2 & m_2 L \\ m_2 L & m_2 L^2 \end{bmatrix}\begin{Bmatrix} \ddot{x} \\ \ddot{\theta} \end{Bmatrix} + \begin{bmatrix} k & 0 \\ 0 & m_2 g L \end{bmatrix}\begin{Bmatrix} x \\ \theta \end{Bmatrix} = \begin{Bmatrix} f(t) \\ 0 \end{Bmatrix} \tag{i}$$

In Eq. (i), the gravity loading appears explicitly in the equations governing "small" oscillations about the static-equilibrium position. In this case, the governing equations are coupled due to the non-zero off-diagonal terms in the inertia matrix.

EXAMPLE 7.8   Governing equations of a rate gyroscope

We revisit the gyro-sensor presented in Example 7.4 and obtain the equations of motion by using Lagrange's equations. Therefore, we construct the kinetic energy T, the potential energy V, and the dissipation function D as follows:

$$T = \frac{1}{2} m(\dot{x} - \omega_z y)^2 + \frac{1}{2} m(\dot{y} + \omega_z x)^2$$

$$= \frac{1}{2} m(\dot{x}^2 + \dot{y}^2) + m\omega_z(-y\dot{x} + x\dot{y}) + \frac{1}{2} m\omega_z^2(y^2 + x^2) \qquad \text{(a)}$$

$$V = \frac{1}{2} k_x x^2 + \frac{1}{2} k_y y^2$$

$$D = \frac{1}{2} c_x \dot{x}^2 + \frac{1}{2} c_y \dot{y}^2$$

In arriving at the kinetic energy, we have made use of Eq. (b) of Example 1.3, which provides the velocity of a particle located in a rotating reference frame. Because of the rotation ω_z, the kinetic energy is not in the standard form shown in Eqs. (7.8) and we have an example of a *nonnatural system*.[10]

Upon using Eqs. (a) and Eqs. (7.7), and recognizing that the generalized coordinates are $q_1 = x$ and $q_1 = y$ and that the associated generalized forces are

$$Q_1 = f_x \quad \text{and} \quad Q_2 = 0 \qquad \text{(b)}$$

we carry out the indicated operations in Eqs. (7.7) to obtain equations that are identical to those given by Eqs. (c) and (d) in Example 7.4.

EXAMPLE 7.9   Governing equations of hand-arm vibrations[11]

In Chapter 3, we considered the development of the governing equation of motion of a hand-arm system modeled as a rigid body with one degree of freedom. In this example, a multi-degree-of-freedom system model is used to study hand-arm vibrations. The model shown in Figure 7.10 has five degrees of freedom and the associated generalized coordinates are x_1, x_2, x_3, y_3, and $\tilde{\theta}_3$. The angle $\tilde{\theta}$ shows the nominal position of the system about which we wish to study the vibrations of the system. It is assumed that $\tilde{\theta} = 0$ throughout this discussion. The coordinates x_1, x_2, and x_3 describe the longitudinal motions of the hand, the forearm, and the elbow, respectively. The coordinate y_3 describes the vertical motion of the elbow and the coordinate θ_3 describes the rotation of the elbow joint about the nominal position. In Figure 7.10, x_o represents the external disturbance acting on the hand.

---

[10]L. Meirovitch, *ibid.*

[11]C. Thomas, S. Rakheja, R. B. Bhat, and I. Stiharu, "A Study of the Modal Behavior of the Human Hand-Arm System," *J. Sound Vibration,* Vol. 191, No. 1, pp. 171–176 (1996).

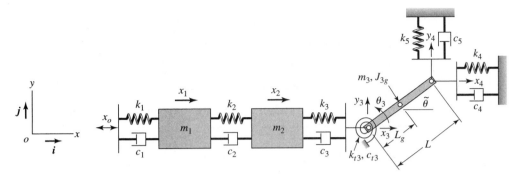

FIGURE 7.10

A five degree-of-freedom model used to study planar vibrations of the human hand-arm system. The nominal position of the upper arm is specified by $\tilde{\theta}$.

Source: Reprinted from *Journal of Sound Vibration,* 191, C. Thomas, S. Rakheja, R. B. Bhat, and I. Stiharu, "A Study of the Modal Behavior of the Human Hand-Arm System," pp. 171–176, Copyright © 1996, with permission from Elsevier Science.

The hand is treated as a point mass of mass m_1, the forearm is treated as a point mass of mass m_2, and the upper arm is treated as a rigid body of mass m_3 and rotary inertia J_{3g}. In addition, translation springs with stiffness of k_1 through k_5 and a torsion spring of stiffness k_{t3} are used to represent the flexibility of the various members. Similarly, translation dampers with damping coefficients of c_1 through c_5 and a torsion damper with damping coefficient c_{t3} are also used.

To construct the different quantities, we first note that the position vector of the center of mass of the rigid mass m_3 from the origin o is given by

$$\boldsymbol{r}_g = (x_3 + L_g \cos(\tilde{\theta} + \theta_3))\boldsymbol{i} + (y_3 + L_g \sin(\tilde{\theta} + \theta_3))\boldsymbol{j} \tag{a}$$

and, therefore, the velocity of the center of mass is

$$\boldsymbol{V}_g = (\dot{x}_3 + L_g \dot{\theta}_3 \sin(\tilde{\theta} + \theta_3))\boldsymbol{i} + (\dot{y}_3 + L_g \dot{\theta}_3 \cos(\tilde{\theta} + \theta_3))\boldsymbol{j} \tag{b}$$

In addition, the displacements x_4 and y_4 are related to x_3, y_3, and θ_3 as follows:

$$x_4 = x_3 + L \cos(\tilde{\theta} + \theta_3)$$
$$y_4 = y_3 + L \sin(\tilde{\theta} + \theta_3) \tag{c}$$

Then, making use of Eq. (1.22) and Eqs. (a) through (c), the kinetic energy of the system is

$$T = \underbrace{\frac{1}{2} m_1 \dot{x}_1^2}_{\substack{\text{Kinetic energy} \\ \text{of the hand}}} + \underbrace{\frac{1}{2} m_2 \dot{x}_2^2}_{\substack{\text{Kinetic energy} \\ \text{of the forearm}}} + \underbrace{\frac{1}{2} m_3 (\boldsymbol{V}_g \cdot \boldsymbol{V}_g) + \frac{1}{2} J_{3g} \dot{\theta}_3^2}_{\text{Kinetic energy of the upper arm}}$$

$$= \frac{1}{2} m_1 \dot{x}_1^2 + \frac{1}{2} m_2 \dot{x}_2^2 + \frac{1}{2} m_3 \left[(\dot{x}_3 - L_g \dot{\theta}_3 \sin(\tilde{\theta} + \theta_3))^2 \right. $$
$$\left. + (\dot{y}_3 + L_g \dot{\theta}_3 \cos(\tilde{\theta} + \theta_3))^2 \right] + \frac{1}{2} J_{3g} \dot{\theta}_3^2 \tag{d}$$

and the potential energy of the system is given by

$$V = \underbrace{\frac{1}{2} k_1(x_1 - x_o)^2}_{\substack{\text{Potential energy} \\ \text{associated with spring } k_1}} + \underbrace{\frac{1}{2} k_2(x_2 - x_1)^2}_{\substack{\text{Potential energy} \\ \text{associated with spring } k_2}} + \underbrace{\frac{1}{2} k_3(x_3 - x_2)^2}_{\substack{\text{Potential energy} \\ \text{associated with spring } k_3}} + \underbrace{\frac{1}{2} k_{t3}(\tilde{\theta} + \theta_3)^2}_{\substack{\text{Potential energy} \\ \text{associated with spring } k_{t3}}}$$

$$+ \underbrace{\frac{1}{2} k_4(x_3 + L\cos(\tilde{\theta} + \theta_3))^2}_{\text{Potential energy associated with spring } k_4} + \underbrace{\frac{1}{2} k_5(y_3 + L\sin(\tilde{\theta} + \theta_3))^2}_{\text{Potential energy associated with spring } k_5} \tag{e}$$

$$+ \underbrace{m_3 g(y_3 + L_g \sin(\tilde{\theta} + \theta_3))}_{\text{Potential energy associated with gravity loading}}$$

The dissipation function D is given by

$$D = \underbrace{\frac{1}{2} c_1(\dot{x}_1 - \dot{x}_o)^2}_{\substack{\text{Dissipation associated} \\ \text{with damper } c_1}} + \underbrace{\frac{1}{2} c_2(\dot{x}_2 - \dot{x}_1)^2}_{\substack{\text{Dissipation associated} \\ \text{with damper } c_2}} + \underbrace{\frac{1}{2} c_3(\dot{x}_3 - \dot{x}_2)^2}_{\substack{\text{Dissipation associated} \\ \text{with damper } c_3}} + \underbrace{\frac{1}{2} c_{t3}\dot{\theta}_3^2}_{\substack{\text{Dissipation} \\ \text{associated} \\ \text{with damper } c_{t3}}}$$

$$+ \underbrace{\frac{1}{2} c_4(\dot{x}_3 - L\dot{\theta}_3 \sin(\tilde{\theta} + \theta_3))^2}_{\text{Dissipation associated with damper } c_4} + \underbrace{\frac{1}{2} c_5(\dot{y}_3 + L\dot{\theta}_3 \cos(\tilde{\theta} + \theta_3))^2}_{\text{Dissipation associated with damper } c_5} \tag{f}$$

From the form of Eqs. (d) through (f), it is clear that we need to make use of Eqs. (7.7) directly to obtain the governing equations of motion. In terms of the chosen coordinates for this five degree-of-freedom system, we have

$$\frac{d}{dt}\left(\frac{\partial T}{\partial \dot{x}_1}\right) - \frac{\partial T}{\partial x_1} + \frac{\partial D}{\partial \dot{x}_1} + \frac{\partial V}{\partial x_1} = Q_1$$

$$\frac{d}{dt}\left(\frac{\partial T}{\partial \dot{x}_2}\right) - \frac{\partial T}{\partial x_2} + \frac{\partial D}{\partial \dot{x}_2} + \frac{\partial V}{\partial x_2} = Q_2$$

$$\frac{d}{dt}\left(\frac{\partial T}{\partial \dot{x}_3}\right) - \frac{\partial T}{\partial x_3} + \frac{\partial D}{\partial \dot{x}_3} + \frac{\partial V}{\partial x_3} = Q_3$$

$$\frac{d}{dt}\left(\frac{\partial T}{\partial \dot{y}_3}\right) - \frac{\partial T}{\partial y_3} + \frac{\partial D}{\partial \dot{y}_3} + \frac{\partial V}{\partial y_3} = Q_4$$

$$\frac{d}{dt}\left(\frac{\partial T}{\partial \dot{\theta}_3}\right) - \frac{\partial T}{\partial \theta_3} + \frac{\partial D}{\partial \dot{\theta}_3} + \frac{\partial V}{\partial \theta_3} = Q_5 \tag{g}$$

where the generalized forces are given by

$$Q_i = 0 \quad i = 1, 2, \dots, 5 \tag{h}$$

since the external disturbance acting on the hand has already been taken into account in determining the kinetic energy, the potential energy, and the dissipation function. In addition, gravity loading has been accounted for in the system potential energy. After substituting Eqs. (d) through (f) and Eq. (h) into Eq. (g) and carrying out the different differentiations in Eqs. (g), we arrive at the following governing equations of motion of the hand-arm system:

$$m_1\ddot{x}_1 + (k_1 + k_2)x_1 - k_2x_2 + (c_1 + c_2)\,\dot{x}_1 - c_2\dot{x}_2 = k_1x_o + c_1\dot{x}_o$$

$$m_2\ddot{x}_2 + (k_2 + k_3)x_2 - k_2x_1 - k_3x_3 + (c_2 + c_3)\dot{x}_2 - c_2\dot{x}_1 - c_3\dot{x}_3 = 0$$

$$m_3\left[\ddot{x}_3 - L_g\ddot{\theta}_3\sin(\tilde{\theta} + \theta_3) - L_g\dot{\theta}_3^2\cos(\tilde{\theta} + \theta_3)\right] + (k_3 + k_4)x_3 - k_3x_2$$
$$+ k_4L\cos(\tilde{\theta} + \theta_3) + (c_3 + c_4)\dot{x}_3 - c_3\dot{x}_2 - c_4L\dot{\theta}_3\sin(\tilde{\theta} + \theta_3) = 0$$

$$m_3\left[\ddot{y}_3 + L_g\ddot{\theta}_3\cos(\tilde{\theta} + \theta_3) - L_g\dot{\theta}_3^2\sin(\tilde{\theta} + \theta_3)\right] + k_5\left[y_3 + L\sin(\tilde{\theta} + \theta_3)\right]$$
$$+ c_5\left[\dot{y}_3 + L\dot{\theta}_3\cos(\tilde{\theta} + \theta_3)\right] = -m_3g$$

$$J_{3g} + m_3L_g^2)\ddot{\theta}_3 + m_3L_g\left[-\ddot{x}_3\sin(\tilde{\theta} + \theta_3) + \ddot{y}_3\cos(\tilde{\theta} + \theta_3)\right] + k_{t3}(\tilde{\theta} + \theta_3)$$
$$- k_4L\sin(\tilde{\theta} + \theta_3)[x_3 + L\cos(\tilde{\theta} + \theta_3)]$$
$$+ k_5L\cos(\tilde{\theta} + \theta_3)[y_3 + L\sin(\tilde{\theta} + \theta_3)]$$

$$c_{t3}\dot{\theta}_3 - c_4L\sin(\tilde{\theta} + \theta_3)[\dot{x}_3 - L\dot{\theta}_3\sin(\tilde{\theta} + \theta_3)]$$
$$+ c_5L\cos(\tilde{\theta} + \theta_3)[\dot{y}_3 + L\dot{\theta}_3\cos(\tilde{\theta} + \theta_3)]$$
$$+ m_3gL_g\cos(\tilde{\theta} + \theta_3) = 0 \tag{i}$$

Linearization

We now linearize the nonlinear system of equations given by Eqs. (i) about the nominal position $\tilde{\theta}$ to describe "small" oscillations about this position. To this end, we use Eqs. (g) of Example 7.3 to determine that

$$\sin(\tilde{\theta} + \theta_3) \approx \sin\tilde{\theta} + \theta_3\cos\tilde{\theta}$$
$$\cos(\tilde{\theta} + \theta_3) \approx \cos\tilde{\theta} - \theta_3\sin\tilde{\theta} \tag{j}$$

Making use of Eqs. (j) in Eqs. (i) and retaining only linear terms, we obtain the following system of linear equations:

$$m_1\ddot{x}_1 + (k_1 + k_2)x_1 - k_2x_2 + (c_1 + c_2)\,\dot{x}_1 - c_2\dot{x}_2 = k_1x_o + c_1\dot{x}_o$$

$$m_2\ddot{x}_2 + (k_2 + k_3)x_2 - k_2x_1 - k_3x_3 + (c_2 + c_3)\dot{x}_2 - c_2\dot{x}_1 - c_3\dot{x}_3 = 0$$

$$m_3\left[\ddot{x}_3 - L_g\ddot{\theta}_3\sin\tilde{\theta}\right] + (k_3 + k_4)x_3 - k_3x_2 - k_4L\theta_3\sin\tilde{\theta}$$
$$+ (c_3 + c_4)\dot{x}_3 - c_3\dot{x}_2 - c_4L\dot{\theta}_3\sin\tilde{\theta} = -k_4L\cos\tilde{\theta}$$

$$m_3\left[\ddot{y}_3 + L_g\ddot{\theta}_3\cos\tilde{\theta}\right] + k_5\left[y_3 + L\theta_3\cos\tilde{\theta}\right]$$
$$+ c_5\left[\dot{y}_3 + L\dot{\theta}_3\cos\tilde{\theta}\right] = -m_3g - k_5L\sin\tilde{\theta}$$

$$(J_{3g} + m_3L_g^2)\ddot{\theta}_3 + m_3L_g\left[-\ddot{x}_3\sin\tilde{\theta} + \ddot{y}_3\cos\tilde{\theta}\right] + k_{t3}\theta_3$$
$$- k_4L\sin\tilde{\theta}[x_3 - L\theta_3\sin\tilde{\theta}]$$
$$- k_4L^2\theta_3\cos^2\tilde{\theta} + k_5L\cos\tilde{\theta}[y_3 + L\theta_3\cos\tilde{\theta}] - k_5L^2\theta_3\sin^2\tilde{\theta} + c_{t3}\dot{\theta}_3$$
$$- c_4L\sin\tilde{\theta}[\dot{x}_3 - L\dot{\theta}_3\sin\tilde{\theta}] + c_5L\cos\tilde{\theta}[\dot{y}_3 + L\dot{\theta}_3\cos\tilde{\theta}]$$
$$- m_3gL_g\theta_3\sin\tilde{\theta} =$$
$$- m_3gL_g\cos\tilde{\theta} - k_{t3}\tilde{\theta} + (k_4 - k_5)L^2\sin\tilde{\theta}\cos\tilde{\theta} \tag{k}$$

Equation (k) is assembled in the following matrix form

$$\begin{bmatrix} m_1 & 0 & 0 & 0 & 0 \\ 0 & m_2 & 0 & 0 & 0 \\ 0 & 0 & m_3 & 0 & -m_3L_g \\ 0 & 0 & 0 & m_3 & m_3L_g \\ 0 & 0 & -m_3L_g & m_3L_g & J_{3g} + m_3L_g^2 \end{bmatrix} \begin{Bmatrix} \ddot{x}_1 \\ \ddot{x}_2 \\ \ddot{x}_3 \\ \ddot{y}_3 \\ \ddot{\theta}_3 \end{Bmatrix}$$

$$
+
\begin{bmatrix}
c_1 + c_2 & -c_2 & 0 & 0 & 0 \\
-c_2 & c_2 + c_3 & -c_3 & 0 & 0 \\
0 & -c_3 & c_3 + c_4 & 0 & -c_4 L \sin \tilde{\theta} \\
0 & 0 & 0 & c_5 & c_5 L \cos \tilde{\theta} \\
0 & 0 & -c_4 L \sin \tilde{\theta} & c_5 L \cos \tilde{\theta} & c_{55}
\end{bmatrix}
\begin{Bmatrix}
\dot{x}_1 \\
\dot{x}_2 \\
\dot{x}_3 \\
\dot{y}_3 \\
\dot{\theta}_3
\end{Bmatrix}
$$

$$
+
\begin{bmatrix}
k_1 + k_2 & -k_2 & 0 & 0 & 0 \\
-k_2 & k_2 + k_3 & -k_3 & 0 & 0 \\
0 & -k_3 & k_3 + k_4 & 0 & -k_4 L \sin \tilde{\theta} \\
0 & 0 & 0 & k_5 & k_5 L \cos \tilde{\theta} \\
0 & 0 & -k_4 L \sin \tilde{\theta} & k_5 L \cos \tilde{\theta} & k_{55}
\end{bmatrix}
\begin{Bmatrix}
x_1 \\
x_2 \\
x_3 \\
y_3 \\
\theta_3
\end{Bmatrix}
\quad (1)
$$

$$
=
\begin{Bmatrix}
k_1 x_o + c_1 \dot{x}_o \\
0 \\
-k_4 L \cos \tilde{\theta} \\
-m_3 g - k_5 L \sin \tilde{\theta} \\
\hat{Q}_5
\end{Bmatrix}
$$

where

$$
c_{55} = c_{t3} + c_4 L^2 \sin \tilde{\theta} + c_5 L^2 \cos \tilde{\theta}
$$
$$
k_{55} = k_{t3} + k_4 L^2 (\sin^2 \tilde{\theta} - \cos^2 \tilde{\theta}) + k_5 L^2 (\cos^2 \tilde{\theta} - \sin^2 \tilde{\theta}) - m_3 g L_g \sin \tilde{\theta}
$$
$$
\hat{Q}_5 = -m_3 g L_g \cos \tilde{\theta} - k_{t3} \tilde{\theta} + (k_4 - k_5) L^2 \sin \tilde{\theta} \cos \tilde{\theta}
$$

It is clear from Eq. (1) that the system inertia, damping, and stiffness matrices are symmetric matrices. Equation (1) is a system of five second-order ordinary differential equations that is used to study "small" motions about the nominal position $\tilde{\theta}$. It is noted that this position is not an equilibrium position, but rather an operating point where $\ddot{\theta} = 0$. It is noted that depending on the application, linearization of a nonlinear system may need to be carried out about a reference position different from the equilibrium position.

7.3 FREE RESPONSES

In this section, we examine free responses of undamped and damped systems. As in the case of single degree-of-freedom systems, when the forcing is absent, the responses exhibited by a multi-degree-of-freedom system are called free responses. In Sections 7.3.1 and 7.3.2, undamped systems are considered and characteristics such as natural frequencies and mode shapes are examined. Following this discussion, modes of damped systems are examined in Section 7.3.3, and the notion of proportional damping is introduced. Subsequently, conservation of energy during free oscillations of a multi-degree-of-freedom system is examined. Throughout Section 7.3, characteristics of free responses of multi-degree-of-freedom systems are examined without explicitly determining the solution for the response. In Section 7.4, we briefly discuss how the stability of a multi-degree-of-freedom system is assessed. The explicit determination of the solution for free responses is carried out in Section 8.2.

7.3.1 Undamped Systems: Natural Frequencies and Mode Shapes

For illustration, consider the system of equations given by Eq. (7.3), which govern the motion of a linear multi-degree-of-freedom system. Setting the damping, the circulatory and gyroscopic terms, and the external forces to zero, and replacing q_i by x_i, we obtain

$$[M]\{\ddot{x}\} + [K]\{x\} = \{0\} \tag{7.22}$$

Since the system given by Eq. (7.22) is a linear system of ordinary differential equations with constant coefficients, the solution for Eq. (7.22) is assumed to be of the form[12]

$$\{x(t)\} = \{X\}e^{\lambda t} \tag{7.23a}$$

where t is time, the exponent λ can be complex valued, and the displacement vector $\{x\}$ and the constant vector $\{X\}$ are given by

$$\{x\} = \begin{Bmatrix} x_1(t) \\ x_2(t) \\ \vdots \\ x_N(t) \end{Bmatrix} \quad \text{and} \quad \{X\} = \begin{Bmatrix} X_1 \\ X_2 \\ \vdots \\ X_N \end{Bmatrix} \tag{7.23b}$$

On substituting Eq. (7.23a) into Eq. (7.22), we obtain

$$[[K] + \lambda^2[M]]\{X\}e^{\lambda t} = \{0\} \tag{7.24}$$

Eigenvalue Problem

Since Eq. (7.24) should be satisfied for all time, we find that

$$[[K] + \lambda^2[M]]\{X\} = \{0\} \tag{7.25}$$

The system of Eq. (7.25) is a system of N linear algebraic equations. This system is satisfied for $X_1 = X_2 = ... = X_N = 0$, which is a trivial solution. Since we seek nontrivial solutions for $X_1, X_2, ..., X_N$, Eq. (7.25) represents an *eigenvalue problem*.[13] The unknowns are $\lambda^2, X_1, X_2, ..., X_N$, and since there are only N equations to solve for these $(N + 1)$ unknowns, at best, what one could do is solve for λ^2, and the ratios $X_2/X_1, X_3/X_1, ..., X_N/X_1$. The quantity λ^2 is referred to as the *eigenvalue* and the vector $\{X_1 \ X_2 \ ... \ X_N\}^T$ is called the *eigenvector*. The eigenvalues λ^2 are determined by finding the roots of *the characteristic equation*

$$\det[[K] + \lambda^2[M]] = 0 \tag{7.26}$$

---

[12]For a solution of the form of Eq. (7.25a), we note that the ratio of any two elements $x_j(t)/x_k(t) = X_j/X_k$ is always time independent. This type of motion is called *synchronous motion* because both generalized coordinates have the same time dependence.

[13]In general, for a system $[A]\{X\}=\lambda'[B]\{X\}$, the problem of finding the constants λ' for which the vector $\{X\}$ is nontrivial, is called an *eigenvalue* or *characteristic value problem*. A scalar version of this problem was discussed for single degree-of-freedom systems in Section 4.4.

Since the stiffness and mass matrices are $N \times N$ matrices, the expansion of Eq. (7.26) is a polynomial of degree $2N$ in λ for an N degree-of-freedom system. Alternatively, one can view this polynomial as an Nth order polynomial in λ^2 with N roots or eigenvalues $\lambda_1^2, \lambda_2^2, ..., \lambda_N^2$. The associated eigenvectors are solutions of the equations

$$\left[[K] + \lambda_j^2 [M] \right] \{X\}_j = \{0\}$$

(7.27)

where $\{X\}_j$ is the eigenvector associated with the eigenvalue λ_j^2 and we have a total of N eigenvectors.

To solve for the eigenvectors associated with the first eigenvalue λ_1^2, the second eigenvalue λ_2^2, and so forth, we construct the eigenvectors

$$\{X\}_1 = \begin{Bmatrix} X_{11} \\ X_{21} \\ \vdots \\ X_{N1} \end{Bmatrix}, \quad \{X\}_2 = \begin{Bmatrix} X_{12} \\ X_{22} \\ \vdots \\ X_{N2} \end{Bmatrix}, \quad ... \quad \{X\}_N = \begin{Bmatrix} X_{1N} \\ X_{2N} \\ \vdots \\ X_{NN} \end{Bmatrix}$$

(7.28a)

which can be written as

$$\{X\}_1 = X_{11} \begin{Bmatrix} 1 \\ X_{21}/X_{11} \\ \vdots \\ X_{N1}/X_{11} \end{Bmatrix}, \quad \{X\}_2 = X_{12} \begin{Bmatrix} 1 \\ X_{22}/X_{12} \\ \vdots \\ X_{N2}/X_{12} \end{Bmatrix}, \quad ...$$

$$\{X\}_N = X_{1N} \begin{Bmatrix} 1 \\ X_{2N}/X_{1N} \\ \vdots \\ X_{NN}/X_{1N} \end{Bmatrix}$$

(7.28b)

In Eqs. (7.28), $\{X\}_1$, which is associated with the eigenvalue λ_1^2, is called the *first eigenvector, first eigenmode,* or *first mode shape,* and $\{X\}_2$, which is associated with the eigenvalue λ_2^2, is called the *second eigenvector, second eigenmode,* or *second mode shape,* and so on.

Due to the nature of the eigenvalue problem, the eigenvectors $\{X\}_j$ are arbitrary to a scaling constant. A convenient normalization that is often used to remove this arbitrariness is

$$X_{11} = 1, \quad X_{12} = 1, \quad ..., \quad X_{1N} = 1$$

(7.29)

Another choice for normalization is

$$X_{21} = 1, \quad X_{22} = 1, \quad ..., \quad X_{2N} = 1$$

(7.30)

and so forth. Physically, the eigenvectors provide information about the relative spatial positions of the different inertial elements in terms of the generalized coordinates. The process of normalizing the mode shapes (eigenvectors) of a system is called *normalization,* and the resulting modes are called *normal modes.*

The mode shapes are placed in a *modal matrix* $[\Phi]$, which is

$$[\Phi] = \left\{ \{X\}_1 \quad \{X\}_2 \quad \cdots \quad \{X\}_N \right\} = \begin{bmatrix} X_{11} & X_{12} & \cdots & X_{1N} \\ X_{21} & X_{22} & \cdots & X_{2N} \\ \vdots & \vdots & \ddots & \vdots \\ X_{N1} & X_{N2} & \cdots & X_{NN} \end{bmatrix} \quad (7.31)$$

where $\{X\}_j$ is the mode shape associated with the jth eigenvalue λ_j^2. When the normalization of the mode shapes is carried out according to Eqs. (7.29), then the modal matrix takes the form

$$[\Phi] = \begin{bmatrix} 1 & 1 & \cdots & 1 \\ X_{21}/X_{11} & X_{22}/X_{12} & \cdots & X_{2N}/X_{1N} \\ \vdots & \vdots & \cdots & \vdots \\ X_{N1}/X_{11} & X_{N2}/X_{12} & \cdots & X_{NN}/X_{1N} \end{bmatrix} \quad (7.32)$$

When the normalization is carried out according to Eqs. (7.30), then the modal matrix takes the form

$$[\Phi] = \begin{bmatrix} X_{11}/X_{21} & X_{12}/X_{22} & \cdots & X_{1N}/X_{2N} \\ 1 & 1 & \cdots & 1 \\ \vdots & \vdots & \cdots & \vdots \\ X_{N1}/X_{21} & X_{N2}/X_{22} & \cdots & X_{NN}/X_{2N} \end{bmatrix} \quad (7.33)$$

and so forth. When the eigenvectors are normalized so that their magnitude is one, the corresponding normalization equation is

$$\|\{X\}_j\| = \sqrt{X_{1j}^2 + X_{2j}^2 + \cdots + X_{Nj}^2} = 1 \quad (7.34)$$

However, regardless of the choice of the normalization, the ratios of the different components in an eigenvector are always preserved.

For real and symmetric matrices $[M]$ and $[K]$, the eigenvalues λ^2 of Eq. (7.25) are real and the associated eigenvectors $\{X\}_j$ are also real.[14] Hence, it is common to write

$$\lambda^2 = (j\omega)^2 = -\omega^2 \quad (7.35)$$

where ω will be shown later to be one of the N natural frequencies of the N degree-of-freedom system. On substituting for λ^2 from Eq. (7.35) into Eqs. (7.26) and (7.27), we find that the natural frequencies are determined by solving the characteristic equation

$$\det\left[[K] - \omega^2[M] \right] = 0 \quad (7.36)$$

which is an Nth order polynomial in ω^2 and that the eigenvectors $\{X\}_j$ associated with the natural frequencies ω_j are determined from

$$\left[[K] - \omega_j^2[M] \right]\{X\}_j = \{0\} \quad (7.37)$$

---

[14]For a comprehensive discussion of eigenvalue problems associated with structural and mechanical systems, see L. Meirovitch, *ibid.*

For a system with N degrees of freedom, Eq. (7.36) provides the N natural frequencies ω_1, ω_2, ..., ω_N and Eq. (7.37) provides the associated eigenvectors $\{X\}_1$, $\{X\}_2$, ..., $\{X\}_N$. The natural frequencies are ordered so that

$$\omega_1 \leq \omega_2 \leq ... \leq \omega_N$$

Hence, the first natural frequency is lower than or equal to the second natural frequency, and so forth. It is noted that this ordering should not be expected when software such as MATLAB is used to solve Eq. (7.36).

To illustrate how the eigenvalues and eigenvectors is determined for a multi-degree-of-freedom system, we use two degree-of-freedom systems. However, the discussion provided below is valid for any linear multi-degree-of-freedom system.

Free Oscillations of Two Degree-of-Freedom Systems

Setting $N = 2$ in Eq. (7.37), and using the definitions of $[K]$ and $[M]$ from Eqs. (7.5b), we obtain

$$\left[-\omega^2 \begin{bmatrix} m_1 & 0 \\ 0 & m_2 \end{bmatrix} + \begin{bmatrix} k_1 + k_2 & -k_2 \\ -k_2 & k_2 + k_3 \end{bmatrix} \right] \begin{Bmatrix} X_1 \\ X_2 \end{Bmatrix} = \begin{Bmatrix} 0 \\ 0 \end{Bmatrix} \tag{7.38a}$$

which is rewritten as

$$\begin{bmatrix} k_1 + k_2 - \omega^2 m_1 & -k_2 \\ -k_2 & k_2 + k_3 - \omega^2 m_2 \end{bmatrix} \begin{Bmatrix} X_1 \\ X_2 \end{Bmatrix} = \begin{Bmatrix} 0 \\ 0 \end{Bmatrix} \tag{7.38b}$$

In this case, the characteristic equation given by Eq. (7.36) translates to

$$\det \begin{bmatrix} k_1 + k_2 - \omega^2 m_1 & -k_2 \\ -k_2 & k_2 + k_3 - \omega^2 m_2 \end{bmatrix} = 0 \tag{7.39a}$$

which, when expanded, takes the form

$$(k_1 + k_2 - \omega^2 m_1)(k_2 + k_3 - \omega^2 m_2) - k_2^2 = 0 \tag{7.39b}$$

Equation (7.39b) is rewritten as

$$m_1 m_2 \omega^4 - [(k_1 + k_2)m_2 + (k_2 + k_3)m_1]\omega^2 + (k_1 + k_2)(k_2 + k_3) - k_2^2 = 0 \tag{7.39c}$$

which is a fourth-order polynomial in ω. Due to the form of this equation, one can treat it as a quadratic polynomial in ω^2. From Eq. (7.37), the eigenvectors associated with the natural frequencies ω_1 and ω_2 are determined from the following system of equations:

$$\begin{bmatrix} k_1 + k_2 - \omega_j^2 m_1 & -k_2 \\ -k_2 & k_2 + k_3 - \omega_j^2 m_2 \end{bmatrix} \begin{Bmatrix} X_{1j} \\ X_{2j} \end{Bmatrix} = \begin{Bmatrix} 0 \\ 0 \end{Bmatrix} \quad j = 1, 2 \tag{7.40}$$

Next, we present an example to show the explicit details of determining the natural frequencies and mode shapes of a two degree-of-freedom system before examining a nondimensional form of the system given by Eq. (7.38a). The nondimensional form is better suited for examining the influences of the different parameters on the system natural frequencies and mode shapes.

EXAMPLE 7.10 Natural frequencies and mode shapes of a two degree-of-freedom system

We shall illustrate how the algebraic system given by Eq. (7.38b) is solved to determine the natural frequencies and mode shapes associated with a specific two degree-of-freedom system. The modal matrix $[\Phi]$ of the system is also constructed. We choose the stiffness parameters k_1, k_2, and k_3 and the mass parameters m_1 and m_2 so that

$$k_1 = k_2 = k_3 = k \quad \text{and} \quad m_1 = m_2 = m \tag{a}$$

Thus, making use of Eqs. (a) in Eq. (7.38b), we obtain

$$\begin{bmatrix} 2k - \omega^2 m & -k \\ -k & 2k - \omega^2 m \end{bmatrix} \begin{Bmatrix} X_1 \\ X_2 \end{Bmatrix} = \begin{Bmatrix} 0 \\ 0 \end{Bmatrix} \tag{b}$$

To determine the natural frequencies of the system, we make use of Eq. (7.36) and Eqs. (a) to arrive at

$$\det \begin{bmatrix} 2k - \omega^2 m & -k \\ -k & 2k - \omega^2 m \end{bmatrix} = 0 \tag{c}$$

On expanding Eq. (c), the result is the characteristic equation

$$(2k - \omega^2 m)(2k - \omega^2 m) - k^2 = 0 \tag{d}$$

which is written as

$$m^2 \omega^4 - 4km\omega^2 + 3k^2 = 0 \tag{e}$$

Equation (e) could have also been obtained by using Eq. (7.39c) and Eqs. (a). This equation is a quadratic equation in ω^2 whose roots are given by

$$\omega_{1,2}^2 = \frac{1}{2m^2}\left[4km \mp \sqrt{16k^2 m^2 - 4(m^2)(3k^2)} \right]$$

which simplifies to

$$\omega_1^2 = \frac{k}{m} \quad \text{and} \quad \omega_2^2 = \frac{3k}{m} \tag{f}$$

or

$$\omega_1 = \sqrt{\frac{k}{m}} \quad \text{and} \quad \omega_2 = \sqrt{\frac{3k}{m}}$$

which are the two natural frequencies of the system and they have been ordered so that $\omega_1 < \omega_2$.

To determine the associated mode shapes, we make use of Eqs. (7.40), (b), and (f). Therefore, to determine the mode shape associated with ω_1, we set $\omega = \omega_1$ in Eq. (b) to obtain

$$\begin{bmatrix} 2k - \omega_1^2 m & -k \\ -k & 2k - \omega_1^2 m \end{bmatrix} \begin{Bmatrix} X_{11} \\ X_{21} \end{Bmatrix} = \begin{Bmatrix} 0 \\ 0 \end{Bmatrix}$$

which, upon using the first of Eqs. (f), reduces to

$$\begin{bmatrix} k & -k \\ -k & k \end{bmatrix} \begin{Bmatrix} X_{11} \\ X_{21} \end{Bmatrix} = \begin{Bmatrix} 0 \\ 0 \end{Bmatrix} \tag{g}$$

From the first of Eq. (g) we find that

$$kX_{11} - kX_{21} = 0$$

or

$$\frac{X_{21}}{X_{11}} = 1 \tag{h}$$

The second of Eq. (g) also provides the same ratio of modal amplitudes, as expected. Then, choosing the normalization given by Eq. (7.29), Eqs. (7.28b) and (h) lead to

$$\{X\}_1 = X_{11} \begin{Bmatrix} 1 \\ 1 \end{Bmatrix} \tag{i}$$

To determine the second mode shape associated with ω_2, we set $\omega = \omega_2$ in Eqs. (b) to obtain

$$\begin{bmatrix} 2k - \omega_2^2 m & -k \\ -k & 2k - \omega_2^2 m \end{bmatrix} \begin{Bmatrix} X_{12} \\ X_{22} \end{Bmatrix} = \begin{Bmatrix} 0 \\ 0 \end{Bmatrix}$$

which, upon using the second of Eqs. (f), reduces to

$$\begin{bmatrix} -k & -k \\ -k & -k \end{bmatrix} \begin{Bmatrix} X_{12} \\ X_{22} \end{Bmatrix} = \begin{Bmatrix} 0 \\ 0 \end{Bmatrix} \tag{j}$$

From the first of Eq. (j) we find that

$$\frac{X_{22}}{X_{12}} = -1 \tag{k}$$

The ratio of the modal amplitudes shown in Eq. (k) could have also been determined from the second of Eq. (j). Again choosing the normalization given by Eq. (7.29), Eqs. (7.28b) and (k) lead to

$$\{X\}_2 = X_{12} \begin{Bmatrix} 1 \\ -1 \end{Bmatrix} \tag{l}$$

For the normalization chosen, the modal matrix is obtained from Eqs. (7.31), (i), and (l) as

$$[\Phi] = \begin{bmatrix} 1 & 1 \\ 1 & -1 \end{bmatrix} \tag{m}$$

In Example 7.10, the natural frequencies and mode shapes were determined for a two degree-of-freedom system with a specific set of parameters. In order to explore the natural frequencies and mode shapes associated with arbitrary system parameters, we introduce many nondimensional parameters and then solve the system given by Eq. (7.38b) in terms of these nondimensional parameters.

Eigenvalue Problem in Terms of Nondimensional Parameters

The different parameters to be used are given by

$$\omega_r = \frac{\omega_{n2}}{\omega_{n1}} = \frac{1}{\sqrt{m_r}} \sqrt{\frac{k_2}{k_1}}, \quad m_r = \frac{m_2}{m_1}, \quad \Omega = \frac{\omega}{\omega_{n1}}$$

$$k_{32} = \frac{k_3}{k_2}, \quad \text{and} \quad \omega_{nj}^2 = \frac{k_j}{m_j} \quad \text{for} \quad j = 1, 2 \tag{7.41}$$

On substituting the different quantities from Eqs. (7.41) into Eq. (7.38b), we find that the resulting system is

$$[1 + \omega_r^2 m_r - \Omega^2]X_1 - m_r \omega_r^2 X_2 = 0$$
$$-\omega_r^2 X_1 + [\omega_r^2(1 + k_{32}) - \Omega^2]X_2 = 0 \tag{7.42a}$$

which, in matrix form, is

$$\begin{bmatrix} 1 + \omega_r^2 m_r - \Omega^2 & -m_r \omega_r^2 \\ -\omega_r^2 & \omega_r^2(1 + k_{32}) - \Omega^2 \end{bmatrix} \begin{Bmatrix} X_1 \\ X_2 \end{Bmatrix} = \begin{Bmatrix} 0 \\ 0 \end{Bmatrix} \tag{7.42b}$$

For the eigenvalue formulation given by Eq. (7.42a) or (7.42b), the eigenvalue is Ω^2 and the corresponding eigenvector is $\{X\}$. This system of equations has a nontrivial solution for $\{X\}$ only when the determinant of the coefficient matrix from Eq. (7.42b) is zero. Thus, from Eqs. (7.42b), we arrive at

$$\det \begin{bmatrix} 1 + \omega_r^2 m_r - \Omega^2 & -m_r \omega_r^2 \\ -\omega_r^2 & \omega_r^2(1 + k_{32}) - \Omega^2 \end{bmatrix} = 0 \tag{7.43}$$

which gives the characteristic equation

$$[1 + \omega_r^2 m_r - \Omega^2][\omega_r^2(1 + k_{32}) - \Omega^2] - m_r \omega_r^4 = 0 \tag{7.44}$$

Equation (7.44) is Eq. (7.39c) rewritten in terms of the nondimensional quantities given by Eqs. (7.41). It is important to note that the nondimensionalization introduced in Eqs. (7.41) led to the compact form of the characteristic equation, Eq. (7.44), which enables one to readily identify the parameters on which the natural frequencies depend.

Expanding Eq. (7.44) leads to

$$\Omega^4 - a_1 \Omega^2 + a_2 = 0 \tag{7.45}$$

where

$$a_1 = 1 + \omega_r^2(1 + m_r + k_{32})$$
$$a_2 = \omega_r^2[1 + k_{32}(1 + \omega_r^2 m_r)] \tag{7.46}$$

The two positive roots of this *characteristic equation* given by Eq. (7.45) are

$$\Omega_{1,2} = \sqrt{\frac{1}{2}[a_1 \mp \sqrt{a_1^2 - 4a_2}]} \tag{7.47}$$

and the frequency ratios Ω_1 and Ω_2 are ordered such that $\Omega_1 < \Omega_2$.[15] The frequency ω_1 associated with Ω_1 is called the *first natural frequency* and the frequency ω_2 associated with Ω_2 is called the *second natural frequency*.

We see that when the interconnecting spring k_2 is absent from the system shown in Figure 7.1, we can set $k_2 = 0$ ($\omega_r = 0$) in Eq. (7.39b) and, as expected, the resulting system is uncoupled. The two natural frequencies are, respectively, the natural frequencies of two independent single degree-of-freedom systems. One natural frequency is $\omega_{n1} = \sqrt{k_1/m_1}$ and the other natural frequency is $\omega_{n2} = \sqrt{k_3/m_2}$. When $k_1 \neq 0$, $k_2 \neq 0$, and $k_3 = 0$, appropriate expressions are similarly obtained.[16]

Based on Eqs. (7.45), (7.46), and (7.41), we see that, in general, the natural frequencies of this system are functions of the three ratios: m_r, ω_r, and k_{32}. If we assume that $k_{32} = 0$ (i.e., $k_3 = 0$), then we can graph the first and second natural frequency ratios as functions of m_r and ω_r. The results are shown in Figures 7.11 and 7.12, and they lead to the following design guideline.

Design Guideline. For a two degree-of-freedom system with two springs and two masses, as the mass ratio $m_r = m_2/m_1$ increases, the first nondimensional natural frequency Ω_1 decreases and the second nondimensional natural frequency Ω_2 increases. Thus, increasing the ratio of the two masses tends to drive the two natural frequencies away from each other. For a constant mass ratio, we see that an increase in the stiffness ratio k_2/k_1 increases both Ω_1 and Ω_2.

We return to Eqs. (7.42a) and determine the eigenvectors associated with Ω_1 and Ω_2. For $\Omega = \Omega_j$, we have

$$[1 + \omega_r^2 m_r - \Omega_j^2]X_{1j} - m_r\omega_r^2 X_{2j} = 0$$
$$-\omega_r^2 X_{1j} + [\omega_r^2(1 + k_{32}) - \Omega_j^2]X_{2j} = 0 \tag{7.48}$$

where X_{1j} and X_{2j} are the respective displacements of the two masses oscillating at the frequency $\omega_{n1}\Omega_j$. Since we can solve only for the ratio of X_{1j}/X_{2j} or X_{2j}/X_{1j}, from the first of Eqs. (7.48), we arrive at

$$\frac{X_{1j}}{X_{2j}} = \frac{m_r\omega_r^2}{1 + \omega_r^2 m_r - \Omega_j^2} \text{ for } j = 1, 2 \tag{7.49}$$

---

[15]In practice, when software such as MATLAB is used to determine the eigenvalues of a system of the form, the eigenvalues obtained are not ordered in terms of magnitude from the lowest to the highest. See Chapter 9 of E. B. Magrab et al., *An Engineer's Guide to MATLAB*, Prentice Hall, Upper Saddle River NJ (2000).

[16]We shall continue to include k_3 and c_3 in determining the necessary equations, but when we numerically evaluate related expressions, these coefficients are frequently set to zero.

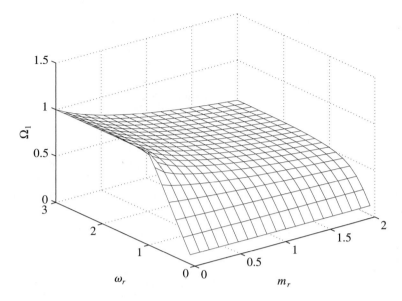

FIGURE 7.11
Variation of the first nondimensional natural frequency of two degree-of-freedom system shown in Figure 7.1 as a function of m_r and ω_r when $k_3 = 0$.

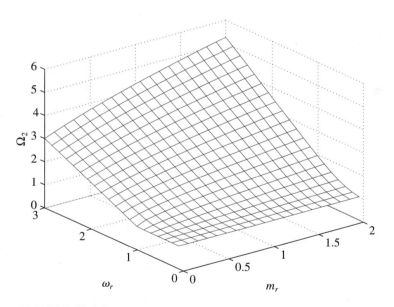

FIGURE 7.12
Variation of the second nondimensional natural frequency of the two degree-of-freedom system shown in Figure 7.1 as a function of m_r and ω_r when $k_3 = 0$.

and from the second of Eqs. (7.48) we arrive at

$$\frac{X_{1j}}{X_{2j}} = \frac{\omega_r^2(1 + k_{32}) - \Omega_j^2}{\omega_r^2} \quad \text{for} \quad j = 1, 2 \tag{7.50}$$

Although Eqs. (7.49) and (7.50) appear to have algebraically different forms, they can be shown to be identical by making use of Eqs. (7.47). It is remarked again that the nondimensionalization introduced in Eqs. (7.41) enables us to determine the dependence of the mode shapes on the various system parameters, as seen from the compact forms of Eqs. (7.45) through (7.50).

For a special case of interest, we let $k_{32} = 0$ and $m_r \ll 1$ for the system shown in Figure 7.1. Then, Eqs. (7.46) lead to

$$a_1 \to 1 + \omega_r^2$$
$$a_2 \to \omega_r^2 \tag{7.51}$$

and the associated natural frequency ratios Ω_j are determined from Eqs. (7.47) to be

$$\left.\begin{array}{l} \Omega_1^2 \to \omega_r^2 \\ \Omega_2^2 \to 1 \end{array}\right\} \quad \text{for} \quad \omega_r \leq 1 \tag{7.52}$$

and

$$\left.\begin{array}{l} \Omega_1^2 \to 1 \\ \Omega_2^2 \to \omega_r^2 \end{array}\right\} \quad \text{for} \quad \omega_r > 1 \tag{7.53}$$

Substituting these limiting values into Eq. (7.50), we find that the modal matrices in these two regions are as follows. For $\omega_r \leq 1$

$$[\Phi] = \begin{bmatrix} 0 & 1 - 1/\omega_r^2 \\ 1 & 1 \end{bmatrix} \tag{7.54a}$$

and for $\omega_r > 1$

$$[\Phi] = \begin{bmatrix} 1 - 1/\omega_r^2 & 0 \\ 1 & 1 \end{bmatrix} \tag{7.54b}$$

The modal matrix plays a key role in determining the response of vibratory systems. In the next section, properties of mode shapes are examined. In the remainder of this section, we illustrate the determination and interpretation of natural frequencies and mode shapes through different examples.

EXAMPLE 7.11 Rigid-body mode of a railway car system

A special case of interest is when

$$k_1 = k_3 = c_1 = c_2 = c_3 = 0 \tag{a}$$

for the system shown in Figure 7.1; that is, we have two masses connected by a spring as shown in Figure 7.13. This system is used to model two inter-

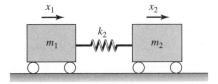

FIGURE 7.13
Two carts with a spring interconnection.

connected railway cars, a truck towing a car, and other such systems. Through this example, we illustrate what is meant by a *rigid-body mode of oscillation*. In this case, Eq. (7.38b) reduces to

$$\begin{bmatrix} k_2 - \omega^2 m_1 & -k_2 \\ -k_2 & k_2 - \omega^2 m_2 \end{bmatrix} \begin{Bmatrix} X_1 \\ X_2 \end{Bmatrix} = \begin{Bmatrix} 0 \\ 0 \end{Bmatrix} \tag{b}$$

and the characteristic equation given by Eq. (7.39c) simplifies to

$$(k_2 - \omega^2 m_1)(k_2 - \omega^2 m_2) - k_2^2 = 0 \tag{c}$$

or

$$\omega^2 \left(m_1 m_2 \omega^2 - k_2(m_1 + m_2) \right) = 0$$

The roots of this equation are

$$\omega_1 = 0$$
$$\omega_2 = \omega_{n2} \sqrt{1 + m_r} \tag{d}$$

where ω_{n2} and m_r are defined in Eqs. (7.41). From Eqs. (b), the corresponding mode shape ratios are

$$\frac{X_{11}}{X_{21}} = \frac{k_2}{k_2 - \omega_1^2 m_1} = 1$$

$$\frac{X_{12}}{X_{22}} = \frac{k_2}{k_2 - \omega_2^2 m_1} = \frac{k_2}{k_2 - \omega_{n2}^2 m_1(1 + m_r)} = \frac{1}{1 - (1 + m_r)/m_r} = -m_r \tag{e}$$

The first mode shape, which corresponds to $\omega_1 = 0$, is a *rigid-body mode;* that is, one in which there is no relative displacement between the two masses. The second mode shape, which corresponds to ω_2, indicates that the displacements of the two masses are always out of phase; that is, when m_1 moves in one direction, m_2 moves in the opposite direction.

In general, the presence of a rigid-body mode is determined by examining the stiffness matrix $[K]$. If the size of the square matrix is n and the rank of the matrix is m, then there are $(n - m)$ zero eigenvalues and, correspondingly, $(n - m)$ rigid-body modes.[17]

---

[17]E. B. Magrab et al., *ibid.,* Chapter 9.

EXAMPLE 7.12   Natural frequencies and mode shapes of a two-mass-three-spring system

For the system shown in Figure 7.1, let $m_1 = 1.2$ kg, $m_2 = 2.7$ kg, $k_1 = 10$ N/m, $k_2 = 20$ N/m, and $k_3 = 15$ N/m. This example is identical in spirit to Example 7.10, except that we will use the nondimensional quantities in carrying out the computations. We shall find the natural frequencies and mode shapes of this system and illustrate how the mode shapes are graphically illustrated. The modal matrix is also constructed.

First, we compute the quantities

$$\omega_{n1} = \sqrt{\frac{10}{1.2}} = 2.887 \text{ rad/s}, \quad \omega_{n2} = \sqrt{\frac{20}{2.7}} = 2.722 \text{ rad/s}$$

$$\omega_r = \frac{2.722}{2.887} = 0.943, \quad m_r = \frac{2.7}{1.2} = 2.25, \quad \text{and} \quad k_{32} = \frac{15}{20} = 0.75 \quad \text{(a)}$$

From Eqs. (a) and (7.46), we find that

$$a_1 = 1 + (1 + 2.25 + 0.75) \times 0.943^2 = 4.556$$
$$a_2 = [1 + 0.75 \times (1 + 2.25 \times 0.943^2)] \times 0.943^2 = 2.889 \quad \text{(b)}$$

Making use of Eqs. (b) and Eqs. (7.47), we obtain the nondimensional natural frequency ratios

$$\Omega_1 = \sqrt{0.5 \times (4.556 - \sqrt{4.556^2 - 4 \times 2.889})} = 0.873$$
$$\Omega_2 = \sqrt{0.5 \times (4.556 + \sqrt{4.556^2 - 4 \times 2.889})} = 1.948 \quad \text{(c)}$$

Noting that the natural frequencies $\omega_j = \omega_{n1}\Omega_j$, we find from Eqs. (a) and (c) that

$$\omega_1 = 2.887 \times 0.873 = 2.519 \text{ rad/s}$$
$$\omega_2 = 2.887 \times 1.948 = 5.623 \text{ rad/s} \quad \text{(d)}$$

Next, the mode shape ratios are computed from Eqs. (7.49), (a), and (c) to be

$$\frac{X_{11}}{X_{21}} = \frac{2.25 \times 0.943^2}{1 + 2.25 \times 0.943^2 - 0.873^2} = 0.893$$

$$\frac{X_{12}}{X_{22}} = \frac{2.25 \times 0.943^2}{1 + 2.25 \times 0.943^2 - 1.948^2} = -2.518 \quad \text{(e)}$$

After choosing the normalization given by Eqs. (7.30), we set $X_{2j} = 1$ and construct the modal matrix as

$$[\Phi] = \begin{bmatrix} 0.893 & -2.518 \\ 1 & 1 \end{bmatrix} \quad \text{(f)}$$

We see that for oscillations in the first mode, both masses move in the same direction, with m_1 moving an amount that is 0.893 times that of m_2. For oscillation in the second mode, the masses move in opposite directions, with m_1 moving 2.518 times as far in one direction as m_2 moves in the opposite direction. It is common, as seen in this example, that the mode shape ratios are positive in the first mode, and there is a sign change when we go to the second mode. The modes $\{X\}_1$ and $\{X\}_2$ are illustrated in Figure 7.14.

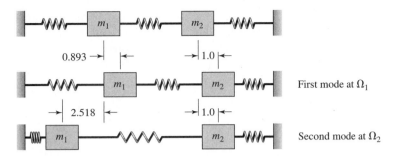

FIGURE 7.14
Mode shapes for system shown in Figure 7.1.

EXAMPLE 7.13 Natural frequencies and mode shapes of a pendulum attached to a translating mass

Consider the system shown in Figure 7.15. We shall derive the governing equations of motion for small $|\theta|$ and then use these equations to determine the natural frequencies and mode shapes associated with this system. The equations of motion are obtained by using Lagrange's equations. We will also illustrate how the expressions for the kinetic energy and the potential energy are appropriately truncated to obtain the linear equations of motion.

Considering the translation of the point mass m, and the rotation of the rigid bar M, the system kinetic energy is

$$T = \frac{1}{2} m\dot{x}^2 + \frac{1}{2} J_o \dot{\theta}^2 \tag{a}$$

and the system potential energy is

$$V = \frac{1}{2} k(x - L\sin\theta)^2 + Mga(1 - \cos\theta) \tag{b}$$

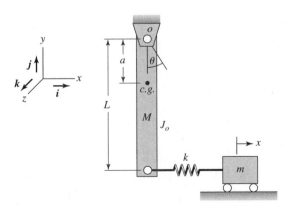

FIGURE 7.15
System with two degrees of freedom.

where the datum for computing the potential energy due to gravity loading has been chosen at the center of mass of the pendulum. For "small" oscillations about $\theta = 0$, Taylor-series expansions lead to

$$\sin \theta \approx \theta$$
$$\cos \theta \approx 1 - \frac{\theta^2}{2} \tag{c}$$

and the potential energy given by Eq. (b) is approximated as[18]

$$V = \frac{1}{2} k(x - L\theta)^2 + \frac{1}{2} Mga\theta^2 \tag{d}$$

Making use of Eqs. (a) and (c) in Eqs. (7.7) for $q_1 = x$ and $q_2 = \theta$ and recognizing that $Q_x = 0$ and $Q_\theta = 0$, we obtain

$$\begin{bmatrix} m & 0 \\ 0 & J_o \end{bmatrix} \begin{Bmatrix} \ddot{x} \\ \ddot{\theta} \end{Bmatrix} + \begin{bmatrix} k & -kL \\ -kL & kL^2 + Mga \end{bmatrix} \begin{Bmatrix} x \\ \theta \end{Bmatrix} = \begin{Bmatrix} 0 \\ 0 \end{Bmatrix} \tag{e}$$

Comparing Eq. (e) with Eq. (7.22), we substitute a solution of the form

$$\begin{Bmatrix} x \\ \theta \end{Bmatrix} = \begin{Bmatrix} X_o \\ \Theta_o \end{Bmatrix} e^{\lambda t} \tag{f}$$

into Eq. (e), choose $\lambda^2 = -\omega^2$, and arrive at the following eigenvalue formulation from Eq. (7.38b)

$$-\omega^2 \begin{bmatrix} m & 0 \\ 0 & J_o \end{bmatrix} \begin{Bmatrix} X_o \\ \Theta_o \end{Bmatrix} + \begin{bmatrix} k & -kL \\ -kL & kL^2 + Mga \end{bmatrix} \begin{Bmatrix} X_o \\ \Theta_o \end{Bmatrix} = \begin{Bmatrix} 0 \\ 0 \end{Bmatrix} \tag{g}$$

where ω^2 is the eigenvalue and the eigenvector is given by $\{X_o \ \Theta_o\}^T$.

Introducing the notation

$$\omega_{n1}^2 = \frac{k}{m}, \quad \omega_{n2}^2 = \frac{K_o}{J_o}, \quad K_o = kL^2 + Mga$$

$$J_r = \frac{mL^2}{J_o}, \quad \Omega = \frac{\omega}{\omega_{n1}}, \quad \text{and} \quad \omega_r = \frac{\omega_{n2}}{\omega_{n1}} \tag{h}$$

Eq. (g) is written as

$$\begin{bmatrix} (1 - \Omega^2) & -1 \\ -J_r & (\omega_r^2 - \Omega^2) \end{bmatrix} \begin{Bmatrix} X_o \\ L\Theta_o \end{Bmatrix} = \begin{Bmatrix} 0 \\ 0 \end{Bmatrix} \tag{i}$$

where $\{X_o \ L\Theta_o\}^T$ is the eigenvector and Ω^2 is the eigenvalue.

To determine the nontrivial solution of Eq. (i), we obtain a solution to

$$\det \begin{bmatrix} (1 - \Omega^2) & -1 \\ -J_r & (\omega_r^2 - \Omega^2) \end{bmatrix} = 0 \tag{j}$$

Thus, we obtain

$$(1 - \Omega^2)(\omega_r^2 - \Omega^2) - J_r = 0 \tag{k}$$

Expanding Eq. (k) results in the characteristic equation

$$\Omega^4 - (1 + \omega_r^2)\Omega^2 + \omega_r^2 - J_r = 0 \tag{l}$$

---

[18]For obtaining the linear equations of motion, retaining up to quadratic terms in the functions T and V is sufficient, as illustrated in this example.

whose roots are

$$\Omega_{2,1} = \sqrt{1 + \omega_r^2 \pm \sqrt{(1 - \omega_r^2)^2 + 4J_r}} / \sqrt{2} \qquad (m)$$

where the nondimensional natural frequencies have been ordered so that $\Omega_1 < \Omega_2$. The corresponding mode shapes are obtained from the expanded form of Eq. (i); that is,

$$(1 - \Omega_j^2)X_{oj} - L\Theta_{oj} = 0$$
$$-J_r X_{oj} + (\omega_r^2 - \Omega_j^2)L\Theta_{oj} = 0 \quad \text{for} \quad j = 1, 2 \qquad (n)$$

From the first of Eqs. (n), the mode shape ratio is

$$\frac{X_{oj}}{L\Theta_{oj}} = \frac{1}{1 - \Omega_j^2} \quad \text{for} \quad j = 1, 2 \qquad (o)$$

If, instead, we chose the second of Eqs. (n), then the mode shape ratio will take the form

$$\frac{X_{oj}}{L\Theta_{oj}} = \frac{\omega_r^2 - \Omega_j^2}{J_r} \quad \text{for} \quad j = 1, 2 \qquad (p)$$

Again, the introduction of nondimensional quantities has enabled us to express the nondimensional frequencies given by Eqs. (m) and the mode shapes given by Eqs. (o) or (p) in compact form to show the dependence on the various system parameters.

When $\omega_{n1} = \omega_{n2}$, we see from Eqs. (h) that $\omega_r = 1$ and Eq. (m) becomes

$$\Omega_{2,1} = \sqrt{1 + 1 \pm \sqrt{(1 - 1)^2 + 4J_r}} / \sqrt{2} = \sqrt{1 \pm \sqrt{J_r}} \qquad (q)$$

and the corresponding mode shape ratios given by Eqs. (o) become

$$\frac{X_{oj}}{L\Theta_{oj}} = \frac{1}{1 - (1 \pm \sqrt{J_r})} = \mp \frac{1}{\sqrt{J_r}} \quad \text{for} \quad j = 1, 2 \qquad (r)$$

Alternatively, the mode shape ratios can also be obtained from Eqs. (p); the result is

$$\frac{X_{oj}}{L\Theta_{oj}} = \frac{1 - (1 \pm \sqrt{J_r})}{J_r} = \mp \frac{1}{\sqrt{J_r}} \quad \text{for} \quad j = 1, 2 \qquad (s)$$

which is identical to Eqs. (r), as expected.

Based on the convention that in the first mode, the mode shape ratios are positive, the second mode corresponds to the negative sign and the first mode

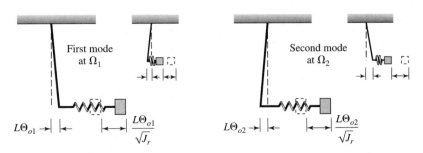

FIGURE 7.16
Mode shapes for the system shown in Figure 7.15 when $\omega_r = 1$.

corresponds to the positive sign. Plots of the two mode shapes are given in Figure 7.16. We see that for free oscillation in the first mode, the masses move in the same direction, but with relatively different displacements. In the second mode, the masses move in opposite directions. Based on Eqs. (r) and (s), it is remarked that, although the relative magnitudes of the mode shape displacement components may change as the physical characteristics of a system change, the directions of the relative motions do not.

EXAMPLE 7.14 Natural frequencies and mode shapes of a system with bounce and pitch motions

We now return to Example 7.3, where we determined the equations governing the system shown in Figure 7.5. To determine the natural frequencies and mode shapes of the undamped system, we set $c_1 = c_2 = 0$ in Eq. (k) of Example 7.3. In addition, we shall introduce the notion of a *node* of a mode shape. We first introduce the quantities

$$X = L_1\Theta, \qquad \omega_r^2 = \frac{\omega_{n2}^2}{\omega_{n1}^2}, \qquad \omega_{n1}^2 = \frac{k_1}{m}, \qquad \omega_{n2}^2 = \frac{k_2 L_2^2}{J_G}$$

$$k_{21} = \frac{k_2}{k_1}, \qquad L_{21} = \frac{L_2}{L_1}, \quad \text{and} \quad \Omega = \frac{\omega}{\omega_{n1}} \tag{a}$$

Then the eigenvalue formulation obtained from Eq. (k) of Example 7.3 in expanded form is

$$(-\Omega^2 + 1 + k_{21})Y - (1 - k_{21}L_{21})X = 0$$

$$-(1 - k_{21}L_{21})Y + \left(-\Omega^2 \frac{k_{21}L_{21}^2}{\omega_r^2} + 1 + k_{21}L_{21}^2\right)X = 0 \tag{b}$$

where Ω^2 is the eigenvalue and $\{Y \, X\}^T$ is the associated mode shape. Note that $X = L_1\Theta$ is the displacement of the left end of the bar associated with the rotation Θ.

Setting the determinant of the coefficient matrix in Eqs. (b) to zero gives the following characteristic equation

$$b_1\Omega^4 - b_2\Omega^2 + b_3 = 0 \tag{c}$$

where the coefficients in the quartic polynomial are

$$b_1 = \frac{k_{21}L_{21}^2}{\omega_r^2}$$

$$b_2 = 1 + b_1(1 + k_{21} + \omega_r^2)$$

$$b_3 = k_{21}(1 + L_{21})^2 \tag{d}$$

The solutions of Eq. (c) provide the nondimensional natural frequencies

$$\Omega_{1,2} = \sqrt{\frac{b_2 \mp \sqrt{b_2^2 - 4b_1b_3}}{2b_1}} \tag{e}$$

where $\Omega_1 < \Omega_2$. The mode shapes is obtained from either the first of Eqs. (b) as

$$\frac{Y_j}{X_j} = \frac{1 - k_{21}L_{21}}{-\Omega_j^2 + 1 + k_{21}} \quad \text{for} \quad j = 1, 2 \tag{f}$$

or from the second of Eqs. (b) as

$$\frac{Y_j}{X_j} = \frac{1}{1 - k_{21}L_{21}} \left(-\Omega^2 \frac{k_{21}L_{21}^2}{\omega_r^2} + 1 + k_{21}L_{21}^2 \right) \quad \text{for} \quad j = 1, 2 \tag{g}$$

We notice that when the ratios $k_{21} = L_{21} = 1$; that is, when the spring constants are equal and the center of gravity is midway between both ends, the system equations are uncoupled as seen from Eqs. (b). In this case, from Eqs. (b), we obtain

$$(-\Omega^2 + 2)Y = 0$$

$$\left(-\frac{\Omega^2}{\omega_r^2} + 2 \right)X = 0$$

indicating that the rotation is independent of the translation. Thus, in this special case, if the bar is subjected to a force at its center of gravity, then the system can only translate. The translation occurs at the nondimensional frequency $\Omega_1^2 = 2$, or equivalently, at the dimensional frequency $\omega_{nt} = \sqrt{2}\omega_{n1} = \sqrt{2k_1/m}$.

In order to illustrate the mode shapes, a numerical case is considered next. Thus, for the choice, $k_{21} = 0.6$, $L_{21} = 1.1$, and $\omega_r = 1.32$, we find from Eqs. (d) that $b_1 = 0.417$, $b_2 = 2.393$, and $b_3 = 2.646$. Upon substituting these values into Eqs. (e), we find that $\Omega_1 = 1.223$ and $\Omega_2 = 2.061$. From Eqs. (f), the respective mode shape ratios are

$$\frac{Y_1}{X_1} = \frac{1 - 0.6 \times 1.1}{-1.223^2 + 1 + 0.6} = 3.261$$

$$\frac{Y_2}{X_2} = \frac{1 - 0.6 \times 1.1}{-2.061^2 + 1 + 0.6} = -0.128 \tag{h}$$

Thus, for oscillation in the first mode, we see from Eqs. (h) that if we assume that there is a displacement d in the negative y-direction at the left end of the bar, the amount of positive rotation of the bar is $\Theta = d/L_1$ about the center of gravity. In addition, the center of gravity of the bar moves a distance that is $3.261d$ in the opposite (positive, in this case) direction. The mode shapes given by Eqs. (h) are shown in Figure 7.17. It is pointed out that there is a point in each mode shape at which the bar (or an extension of it) intersects the reference (equilibrium) position. This point does not undergo any motion. Such a point is called a *node point*. For the first mode shape, the node point does not physically lie on the bar. Even so, one can consider this node point as being the fulcrum for the motion of the rigid bar at its first natural frequency. In other words, the bar oscillates through an arc as if it were pivoted at that point. For the second mode shape, the node point lies in the span of the bar. When the bar is vibrating in the second natural frequency, the bar is pivoting about this stationary point.

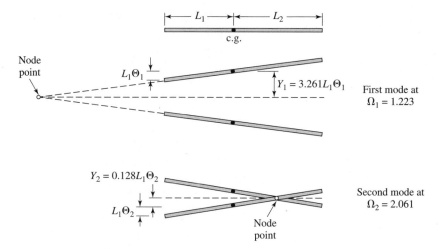

FIGURE 7.17
Modes shapes and node points for the system shown in Figure 7.5: $k_{21} = 0.6$, $L_{21} = 1.1$, and $\omega_r = 1.32$. [Note: Figure not to scale.]

Design Guideline. For a system with two or more degrees of freedom, if a sensor needs to be located so as not to pick up vibrations of a certain mode, one should locate the sensor at the node point of this mode. Alternatively, if a sensor should sense vibrations in a certain mode, one should not locate it at a node of the mode of interest.

Nodes in vibratory systems play a significant role in determining the placement of vibration sensors and vibration actuators. For instance, a sensor placed at the node of a certain mode will not detect that particular mode. Similarly, an actuator placed at the node of a certain mode cannot excite that mode. The observation that when oscillating in a certain mode shape, the displacement is zero at the node point of a mode leads to the following design guideline.

EXAMPLE 7.15 Determination of system parameters

Consider the system shown in Figure 7.1, which is described by the following inertia matrix $[M]$ and stiffness matrix $[K]$

$$[M] = \begin{bmatrix} 2 & 0 \\ 0 & 3 \end{bmatrix} \text{kg} \quad \text{and} \quad [K] = \begin{bmatrix} 25{,}000 & k_{12} \\ k_{21} & k_{22} \end{bmatrix} \text{N/m} \tag{a}$$

The modes of the system are given by

$$\{X\}_1 = \begin{Bmatrix} -3 \\ 1 \end{Bmatrix} \quad \text{and} \quad \{X\}_2 = \begin{Bmatrix} 1/2 \\ 1 \end{Bmatrix} \tag{b}$$

and the natural frequency ω_1 associated with mode $\{X\}_1$ is 100 rad/s. We shall illustrate how the unknown elements in the stiffness matrix and the other natural frequency ω_2 is determined.

To determine the needed quantities, we make use of Eq. (7.38b). Comparing the elements in the given stiffness matrix with that used in Eq. (7.38b), we determine that

$$k_{11} = k_1 + k_2 = 25{,}000$$
$$k_{12} = k_{21} = -k_2$$
$$k_{22} = k_2 + k_3 \tag{c}$$

Therefore, from the first of Eq. (7.40) and Eqs. (b), the mode shape ratio is

$$\frac{X_{21}}{X_{11}} = \frac{k_1 + k_2 - m_1\omega_1^2}{k_2} = -\frac{k_{11} - m_1\omega_1^2}{k_{12}} = -\frac{1}{3} \tag{d}$$

Upon substituting the known numerical values into Eq. (d), we obtain

$$k_{12} = 3 \times (25{,}000 - 2 \times 100^2) = 15{,}000 \text{ N/m} \tag{e}$$

From the second of Eq. (7.40), we see that

$$\frac{X_{21}}{X_{11}} = \frac{k_2}{k_2 + k_3 - m_2\omega_1^2} = -\frac{k_{12}}{k_{22} - m_2\omega_1^2} = -\frac{1}{3} \tag{f}$$

From Eq. (f), we find that

$$k_{22} = 3k_{12} + \omega_1^2 m_2 = 3 \times 15{,}000 + 3 \times 100^2 = 75{,}000 \text{ N/m} \tag{g}$$

The other natural frequency is found by making use of the eigenvector $\{X\}_2$ and the first of Eq. (7.40) with $j = 2$. Thus,

$$\frac{X_{22}}{X_{12}} = -\frac{k_{11} - m_1\omega_2^2}{k_{12}} = \frac{1}{1/2} = 2 \tag{h}$$

Solving Eq. (h) for ω_2, we obtain

$$\omega_2 = \sqrt{\frac{1}{m_1}\left(k_{12}\frac{X_{22}}{X_{12}} + k_{11}\right)} = \sqrt{\frac{1}{2}(15{,}000 \times 2 + 25{,}000)}$$
$$= 165.83 \text{ rad/s}$$

7.3.2 Undamped Systems: Properties of Mode Shapes

In this section, we examine the properties of the eigenvectors of undamped systems described by Eq. (7.25), which, after setting $\lambda^2 = -\omega^2$, is

$$\left[[K] - \omega^2 [M]\right]\{X\} = \{0\} \tag{7.55}$$

It is shown that the eigenvectors $\{X\}_j$, and hence, the modal matrix $[\Phi]$, have a special property called *orthogonality*, which will be described shortly. This property follows from the properties of eigenvectors associated with real, symmetric matrices. We shall take advantage of this orthogonality to solve the coupled equations of the form given by Eq. (7.3) and other systems in the next chapter. Another important property is that the eigenvectors $\{X\}_j$ form a linearly independent set. We address the orthogonality of the modes next, and explain the linear independence of the modes at the end of this section.

Orthogonality of Modes

We see from Eq. (7.55) that

$$\omega_j^2[M]\{X\}_j = [K]\{X\}_j \quad \text{for} \quad j = 1, 2, ..., N \tag{7.56}$$

where the matrices $[M]$ and $[K]$ are symmetric and ω_j^2 are the different eigenvalues and $\{X\}_j$ are the associated eigenvectors. We consider Eq. (7.56) for two distinct frequencies ω_l and ω_m; thus, we have

$$\omega_l^2[M]\{X\}_l = [K]\{X\}_l$$
$$\omega_m^2[M]\{X\}_m = [K]\{X\}_m \tag{7.57}$$

Pre-multiplying the first equation of Eqs. (7.57) by $\{X\}_m^T$ and the second equation of Eqs. (7.57) by $\{X\}_l^T$ leads to

$$\omega_l^2\{X\}_m^T[M]\{X\}_l = \{X\}_m^T[K]\{X\}_l$$
$$\omega_m^2\{X\}_l^T[M]\{X\}_m = \{X\}_l^T[K]\{X\}_m \tag{7.58}$$

Taking the transpose[19] of the second equation of Eqs. (7.58) we find that

$$\omega_m^2[\{X\}_l^T[M]\{X\}_m]^T = [\{X\}_l^T[K]\{X\}_m]^T$$
$$\omega_m^2\{X\}_m^T[M]^T\{X\}_l = \{X\}_m^T[K]^T\{X\}_l$$
$$\omega_m^2\{X\}_m^T[M]\{X\}_l = \{X\}_m^T[K]\{X\}_l \tag{7.59}$$

since we have assumed that $[M]$ and $[K]$ are symmetric matrices. Then, Eqs. (7.58) and (7.59) lead to

$$\omega_l^2\{X\}_m^T[M]\{X\}_l = \{X\}_m^T[K]\{X\}_l$$
$$\omega_m^2\{X\}_m^T[M]\{X\}_l = \{X\}_m^T[K]\{X\}_l \tag{7.60}$$

On subtracting one equation from the other in Eqs. (7.60), we arrive at

$$(\omega_l^2 - \omega_m^2)\{X\}_m^T[M]\{X\}_l = 0 \tag{7.61}$$

Since $\omega_l \neq \omega_m$, Eq. (7.61) implies that

$$\{X\}_m^T[M]\{X\}_l = 0 \quad \text{for} \quad \omega_l \neq \omega_m \tag{7.62}$$

Equation (7.62) provides a definition of *orthogonality* for the eigenvectors (or eigenmodes or modes) of a system. Also, from Eqs. (7.60), it follows that

$$\{X\}_m^T[K]\{X\}_l = 0 \quad \text{for} \quad \omega_l \neq \omega_m \tag{7.63}$$

From Eqs. (7.62) and (7.63), we see that the modes are orthogonal with respect to both the mass matrix $[M]$ and the stiffness matrix $[K]$, and this important property is one that we will make use of in the normal-mode approach for determining responses of multi-degree-of-freedom systems in Section 8.2.

Equations (7.62) and (7.63) can be shown to be true for cases where the eigenvalues are not distinct;[20] that is, when $\omega_l = \omega_m$. Thus, in general,

---

[19]From linear algebra, recall that $([A][B])^T = [B]^T[A]^T$. See Appendix E.

[20]D. C. Murdoch, *Linear Algebra,* John Wiley & Sons, NY, Chapter 6 (1970).

$$\{X\}_l^T[M]\{X\}_m = 0$$
$$\{X\}_l^T[K]\{X\}_m = 0 \quad l \neq m \tag{7.64}$$

where ω_j and the corresponding $\{X\}_j$ are obtained from the solutions to, respectively,

$$\det[-\omega^2[M] + [K]] = 0$$
$$-\omega_j^2[M]\{X\}_j + [K]\{X\}_j = 0 \quad \text{for} \quad j = 1, 2, ..., N \tag{7.65}$$

Modal Mass, Modal Stiffness, and Modal Matrix

After pre-multiplying each side of Eq. (7.56) with $\{X\}_j^T$, we arrive at

$$\omega_j^2\{X\}_j^T[M]\{X\}_j = \{X\}_j^T[K]\{X\}_j$$
$$\omega_j^2\hat{M}_{jj} = \hat{K}_{jj} \qquad j = 1, 2, ..., N$$
$$\omega_j^2 = \hat{K}_{jj}/\hat{M}_{jj} \tag{7.66}$$

where the *modal mass* $\hat{M}_{jj}$ of the jth mode and the *modal stiffness* $\hat{K}_{jj}$ of the jth mode are given by, respectively,

$$\hat{M}_{jj} = \{X\}_j^T[M]\{X\}_j \quad j = 1, 2, ..., N$$
$$\hat{K}_{jj} = \{X\}_j^T[K]\{X\}_j \quad j = 1, 2, ..., N \tag{7.67}$$

By using Eqs. (7.31) for the modal matrix $[\Phi]$ and taking advantage of Eqs. (7.62) and (7.67), we see that

$$[\Phi]^T[M][\Phi] = \begin{Bmatrix} \{X\}_1^T \\ \{X\}_2^T \\ \vdots \\ \{X\}_N^T \end{Bmatrix} [M] \begin{Bmatrix} \{X\}_1 & \{X\}_2 & \cdots & \{X\}_N \end{Bmatrix}$$

$$= \begin{Bmatrix} \{X\}_1^T \\ \{X\}_2^T \\ \vdots \\ \{X\}_N^T \end{Bmatrix} \begin{Bmatrix} [M]\{X\}_1 & [M]\{X\}_2 & \cdots & [M]\{X\}_N \end{Bmatrix}$$

$$= \begin{bmatrix} \{X\}_1^T[M]\{X\}_1 & \{X\}_1^T[M]\{X\}_2 & \cdots & \{X\}_1^T[M]\{X\}_N \\ \{X\}_2^T[M]\{X\}_1 & \{X\}_2^T[M]\{X\}_2 & \cdots & \{X\}_2^T[M]\{X\}_N \\ \vdots & \vdots & \ddots & \\ \{X\}_N^T[M]\{X\}_1 & \{X\}_N^T[M]\{X\}_2 & \cdots & \{X\}_N^T[M]\{X\}_N \end{bmatrix}$$

$$= \begin{bmatrix} \hat{M}_{11} & 0 & \cdots & 0 \\ 0 & \hat{M}_{22} & \cdots & 0 \\ \vdots & \vdots & \ddots & \vdots \\ 0 & 0 & \cdots & \hat{M}_{NN} \end{bmatrix} = [M_D] \tag{7.68a}$$

and in a similar manner, by making use of Eqs. (7.63) and (7.67), we obtain the diagonal matrix

$$[\Phi]^T [K][\Phi] = \begin{bmatrix} \hat{K}_{11} & 0 & \cdots & 0 \\ 0 & \hat{K}_{22} & \cdots & 0 \\ \vdots & \vdots & \ddots & \vdots \\ 0 & 0 & \cdots & \hat{K}_{NN} \end{bmatrix} = [K_D] \tag{7.68b}$$

Therefore, Eqs. (7.66) are written in matrix form as

$$[\omega_D^2][M_D] = [K_D]$$
$$[\omega_D^2] = [M_D]^{-1} [K_D] \tag{7.69a}$$

which means that

$$\begin{bmatrix} \omega_1^2 & 0 & \cdots & 0 \\ 0 & \omega_2^2 & \cdots & 0 \\ \vdots & \vdots & \ddots & \vdots \\ 0 & 0 & \cdots & \omega_N^2 \end{bmatrix} = \begin{bmatrix} 1/\hat{M}_{11} & 0 & \cdots & 0 \\ 0 & 1/\hat{M}_{22} & \cdots & 0 \\ \vdots & \vdots & \ddots & \vdots \\ 0 & 0 & \cdots & 1/\hat{M}_{NN} \end{bmatrix} \begin{bmatrix} \hat{K}_{11} & 0 & \cdots & 0 \\ 0 & \hat{K}_{22} & \cdots & 0 \\ \vdots & \vdots & \ddots & \vdots \\ 0 & 0 & \cdots & \hat{K}_{NN} \end{bmatrix}$$

$$= \begin{bmatrix} \hat{K}_{11}/\hat{M}_{11} & 0 & \cdots & 0 \\ 0 & \hat{K}_{22}/\hat{M}_{22} & \cdots & 0 \\ \vdots & \vdots & \ddots & \vdots \\ 0 & 0 & \cdots & \hat{K}_{NN}/\hat{M}_{NN} \end{bmatrix} \tag{7.69b}$$

in agreement with Eqs. (7.66).

Mass Normalized Modes

Equation (7.68a) can also be used for normalizing the mode shapes in lieu of Eqs. (7.29) and (7.30) or Eq. (7.34). If one normalizes the mode shapes so that for the first mode

$$\{X_{11} \quad X_{12} \quad \cdots \quad X_{1N}\}[M] \begin{Bmatrix} X_{11} \\ X_{12} \\ \vdots \\ X_{1N} \end{Bmatrix} = 1 \tag{7.70a}$$

and for the second mode

$$\{X_{21} \quad X_{22} \quad \cdots \quad X_{2N}\}[M] \begin{Bmatrix} X_{21} \\ X_{22} \\ \vdots \\ X_{2N} \end{Bmatrix} = 1 \tag{7.70b}$$

and so forth, then the mode shapes are said to be *mass-normalized,* and these normalized mode shapes are also referred to as *ortho-normal modes.* In general, for a system with N degrees of freedom, Eq. (7.62) together with Eqs. (7.70a) and (7.70b) are expressed as

$$[\Phi]^T [M][\Phi] = [I] \tag{7.71a}$$

where $[I]$ is the identity matrix. It follows from Eqs. (7.60) and (7.63) that

$$[\Phi]^T [K][\Phi] = [\omega_D^2] \tag{7.71b}$$

We take advantage of the orthogonality of the modes in developing a solution for the response of multi-degree-of-freedom systems in Section 8.2.

Linear Independence of Eigenvectors

We now consider the fact that the eigenvectors $\{X\}_j$ form a linearly independent set. This means that for a system with N degrees of freedom with N modes $\{X\}_j$, any N-dimensional vector is constructed as a linear combination of these eigenvectors. In physical terms, the implication is that any vibratory motion of a system is viewed as a weighted sum of oscillations in the individual modes. This observation, along with the orthogonality of the modes, forms the basis of the normal-mode approach discussed in Section 8.2. Mathematically, linear dependence of the eigenvectors means that

$$c_1\{X\}_1 + c_2\{X\}_2 + \cdots + c_n\{X\}_n = \{0\} \tag{7.72}$$

for non-zero constants c_j. The orthogonality properties given by Eqs. (7.64) is used to show that Eq. (7.72) is true only if the c_j are all zero, thus verifying that the eigenvectors are not linearly dependent, and hence, they form a linearly independent set.

EXAMPLE 7.16 Orthogonality of modes, modal masses, and modal stiffness of a spring-mass system

We now illustrate how the orthogonality of the modes of a vibratory system is examined, and how the modal masses and modal stiffness are computed. We consider the two degree-of-freedom system treated in Example 7.12 where $m_1 = 1.2$ kg, $m_2 = 2.7$ kg, $k_1 = 10$ N/m, $k_2 = 20$ N/m, and $k_3 = 15$ N/m. Then, from Eq. (7.38a)

$$[M] = \begin{bmatrix} 1.2 & 0 \\ 0 & 2.7 \end{bmatrix} \text{kg} \quad \text{and} \quad [K] = \begin{bmatrix} 30 & -20 \\ -20 & 35 \end{bmatrix} \text{N/m} \tag{a}$$

Since the mass and stiffness matrices are real and symmetric matrices, the modes will be orthogonal to the mass and stiffness matrices. We shall now show this through numerical calculations. To this end, we recall the modal matrix $[\Phi]$ from Eq. (f) of Example 7.12, which is

$$[\Phi] = \begin{bmatrix} 0.893 & -2.518 \\ 1 & 1 \end{bmatrix} \tag{b}$$

To check the orthogonality of the mode shapes, we note that

$$\{X\}_2^T[M]\{X\}_1 = \frac{\{-2.518 \quad 1\}}{}\begin{bmatrix} 1.2 & 0 \\ 0 & 2.7 \end{bmatrix}\begin{Bmatrix} 0.893 \\ 1 \end{Bmatrix} = 0$$

$$\{X\}_2^T[K]\{X\}_1 = \frac{\{-2.518 \quad 1\}}{}\begin{bmatrix} 30 & -20 \\ -20 & 35 \end{bmatrix}\begin{Bmatrix} 0.893 \\ 1 \end{Bmatrix} = 0 \quad \text{(c)}$$

To determine the modal masses, we find that

$$\hat{M}_{11} = \{X\}_1^T[M]\{X\}_1 = \frac{\{0.893 \quad 1\}}{}\begin{bmatrix} 1.2 & 0 \\ 0 & 2.7 \end{bmatrix}\begin{Bmatrix} 0.893 \\ 1 \end{Bmatrix} = 3.658$$

$$\hat{M}_{22} = \{X\}_2^T[M]\{X\}_2 = \frac{\{-2.518 \quad 1\}}{}\begin{bmatrix} 1.2 & 0 \\ 0 & 2.7 \end{bmatrix}\begin{Bmatrix} -2.518 \\ 1 \end{Bmatrix} = 10.311 \quad \text{(d)}$$

and to determine the modal stiffness associated with each mode, we compute

$$\hat{K}_{11} = \{X\}_1^T[K]\{X\}_1 = \frac{\{0.893 \quad 1\}}{}\begin{bmatrix} 30 & -20 \\ -20 & 35 \end{bmatrix}\begin{Bmatrix} 0.893 \\ 1 \end{Bmatrix} = 23.209$$

$$\hat{K}_{22} = \{X\}_2^T[K]\{X\}_2 = \frac{\{-2.51 \quad 1\}}{}\begin{bmatrix} 30 & -20 \\ -20 & 35 \end{bmatrix}\begin{Bmatrix} -2.518 \\ 1 \end{Bmatrix} = 326.01 \quad \text{(e)}$$

In addition, to compare with the results provided in Example 7.12 for the natural frequencies, we carry out the following. From Eq. (7.69b) and the computed quantities above, we find that

$$[\omega_D^2] = \begin{bmatrix} \hat{K}_{11}/\hat{M}_{11} & 0 \\ 0 & \hat{K}_{22}/\hat{M}_{22} \end{bmatrix} = \begin{bmatrix} 23.21/3.658 & 0 \\ 0 & 326.01/10.311 \end{bmatrix}$$

$$\begin{bmatrix} \omega_1^2 & 0 \\ 0 & \omega_2^2 \end{bmatrix} = \begin{bmatrix} 6.345 & 0 \\ 0 & 31.618 \end{bmatrix} \quad \text{(f)}$$

and

$$\begin{bmatrix} \omega_1 & 0 \\ 0 & \omega_2 \end{bmatrix} = \begin{bmatrix} 2.519 & 0 \\ 0 & 5.623 \end{bmatrix} \quad \text{(g)}$$

which agree with the values presented in Eq. (d) of Example 7.12.

7.3.3 Characteristics of Damped Systems

In Sections 7.3.1 and 7.3.2, we treated the natural frequencies and mode shapes of undamped multi-degree-of-freedom systems with symmetric inertia and symmetric stiffness matrices. The eigenvalues and eigenvectors of such systems are real-valued quantities with a physical interpretation. Here, we examine the eigenvalues and eigenvectors for damped multi-degree-of-freedom systems.

We revisit Eq. (7.3), set the external forces to zero, replace q_i by x_i, and arrive at

$$[M]\{\ddot{x}\} + \Big[[C] + [G]\Big]\{\dot{x}\} + \Big[[K] + [H]\Big]\{x\} = \{0\} \quad \text{(7.73)}$$

Since we have a linear system of ordinary differential equations with constant coefficients, we can assume a solution of the form given by Eq. (7.23a). On substituting Eq. (7.23a) into Eq. (7.73), the result is

$$\left[\lambda^2[M] + \lambda\left([C] + [G]\right) + [K] + [H]\right]\{X\}e^{\lambda t} = \{0\} \tag{7.74}$$

Eigenvalue Problem

Noting that Eq. (7.74) should be satisfied for all time t, we arrive at

$$\left[\lambda^2[M] + \lambda\left([C] + [G]\right) + [K] + [H]\right]\{X\} = \{0\} \tag{7.75}$$

Although the trivial solution

$$\{X\} = \{0\} \tag{7.76}$$

satisfies Eq. (7.75), we are looking for nontrivial solutions of this system, which has N algebraic equations in the $(N + 1)$ unknowns λ, X_1, X_2, ..., X_N. The special values of λ for which we have a nontrivial solution are called eigenvalues.

The eigenvalues of Eq. (7.75) are given by the roots of the characteristic equation

$$\det\left[\lambda^2[M] + \lambda\left([C] + [G]\right) + [K] + [H]\right] = 0 \tag{7.77}$$

which is a polynomial in λ of order $2N$. Since all of the matrices in Eq. (7.77) are real-valued, we end up with $2N$ roots for this characteristic equation of the N degree-of-freedom system. For mechanical systems known as *lightly damped systems,* these roots are in the form of N complex conjugates pairs, and they are given by[21,22]

$$\lambda_k = \delta_k \pm j\omega_k \quad k = 1, 2, ..., N \tag{7.78}$$

For systems that do not fall under the category of lightly damped systems, one or more of the eigenvalues is real. The associated eigenvectors are determined by solving the algebraic system

$$\left[\lambda_k^2[M] + \lambda_k\left([C] + [G]\right) + [K] + [H]\right]\{X\}_k = \{0\} \quad k = 1, 2, ..., N \tag{7.79}$$

In Section 8.3, we shall see why the state-space form of Eq. (7.73) is convenient to use and to interpret the eigenvalues given by Eqs. (7.78) and the eigenvectors determined from Eqs. (7.79).

Damped Systems Without Gyroscopic and Circulatory Forces

For the cases where the gyroscopic and circulatory forces are absent, Eq. (7.73) is of the form

$$[M]\{\ddot{x}\} + [C]\{\dot{x}\} + [K]\{x\} = \{0\} \tag{7.80}$$

---

[21]L. Meirovitch, 1980, *ibid.*

[22]P. C. Müller and W. O. Schiehlen, *Linear Vibrations: A Theoretical Treatment of Multi-Degree-of-Freedom Vibrating Systems,* Martinus Nijhoff Publishers, Dordrecht, The Netherlands, Chapters 4 and 6 (1985).

Following the steps that were used to obtain Eqs. (7.77) and (7.79), the eigenvalues associated with Eq. (7.80) are given by the roots of

$$\det [\lambda^2[M] + \lambda[C] + [K]] = 0 \qquad (7.81)$$

and they have the form of Eqs. (7.78). The corresponding eigenvectors are determined from

$$[\lambda_k^2[M] + \lambda_k[C] + [K]]\{X\}_k = \{0\} \quad k = 1, 2, ..., N \qquad (7.82)$$

In Sections 8.2 and 8.3, we shall use the normal-mode formulation and the state-space formulation to interpret the eigenvalues given by Eqs. (7.78) and the eigenvectors determined from Eqs. (7.82). While the state-space formulation is applicable to arbitrary forms of the damping matrix $[C]$, the normal mode approach is applicable to only certain forms of damping, called *proportional damping*.

Proportional Damping

For the case of proportional damping, the damping matrix $[C]$ is given by

$$[C] = \alpha[M] + \beta[K] \qquad (7.83)$$

where α and β are real-valued constants. Since, in Eq. (7.83), the damping matrix $[C]$ is a combination of a matrix proportional to the mass matrix $[M]$ and a matrix proportional to a stiffness matrix $[K]$, we use the designation proportional damping.

On substituting Eq. (7.83) into Eq. (7.81), we arrive at the eigenvalue problem

$$[\lambda^2[M] + \lambda\alpha[M] + \beta\lambda[K] + [K]]\{X\} = \{0\} \qquad (7.84)$$

which is rewritten in the form

$$[(1 + \beta\lambda)[K] + \lambda(\lambda + \alpha)[M]]\{X\} = \{0\} \qquad (7.85)$$

Next, we compare the eigenvalues λ_{dk} of the proportionally damped system with the eigenvalues λ_k of the undamped system determined from Eq. (7.25). These eigenvalues are determined by the following characteristic equations determined from Eq. (7.26) and Eq. (7.85), respectively.

Undamped system.

$$\det[[K] + \lambda^2[M]] = 0 \qquad (7.86)$$

Damped system.

$$\det[(1 + \beta\lambda)[K] + \lambda(\lambda + \alpha)[M]] = 0 \qquad (7.87)$$

When the two characteristic polynomials given by Eqs. (7.86) and (7.87) are compared, it is clear that $\lambda_{dk} \neq \lambda_k$; that is, the eigenvalues for the proportionally damped case are not the same as those for the undamped case. Since $[K]$ and $[M]$ are real and symmetric matrices, the eigenvalues of the undamped system λ_k^2 are real and $\lambda_k = \pm j\omega_k$, where ω_k are the system natural frequencies. By contrast, the eigenvalues λ_{dk}^2 of the proportionally damped system are complex-valued quantities.

To carry out a proper comparison of the eigenvectors of a damped system with those of the corresponding undamped system, the state-space formulation discussed in Section 8.3 is needed. From such a formulation, it can be established that the eigenvectors of the proportionally damped system and the eigenvectors of the associated undamped system have similar structure; in particular, the ratios of the modal components corresponding to the displacement states are the same in the undamped and damped cases.[23] This information will now be used to determine the relationship between λ_{dk} and λ_k.

Let $\{X\}_k$ be the eigenvector associated with the eigenvalue λ_k^2 of the undamped system described by Eq. (7.25). Then, setting $\lambda = \lambda_{dk}$ and $\{X\} = \{X\}_k$ in Eq. (7.85) we obtain

$$\left[(1 + \beta\lambda_{dk})[K] + \lambda_{dk}(\lambda_{dk} + \alpha)[M]\right]\{X\}_k = \{0\} \quad k = 1, 2, ..., N \quad (7.88)$$

Pre-multiplying Eqs. (7.88) by $\{X\}_k^T$, we arrive at

$$\{X\}_k^T\left[(1 + \beta\lambda_{dk})[K] + \lambda_{dk}(\lambda_{dk} + \alpha)[M]\right]\{X\}_k = \{X\}_k^T\{0\} \quad k = 1, 2, ..., N \quad (7.89)$$

which, upon expanding the different terms, leads to

$$(1 + \beta\lambda_{dk})\{X\}_k^T[K]\{X\}_k + \lambda_{dk}(\lambda_{dk} + \alpha)\{X\}_k^T[M]\{X\}_k = 0 \quad k = 1, 2, ..., N \quad (7.90)$$

Making use of Eqs. (7.67) for the modal mass $\hat{M}_{kk}$ and the modal stiffness $\hat{K}_{kk}$ in Eqs. (7.90), we find that

$$(1 + \beta\lambda_{dk})\hat{K}_{kk} + \lambda_{dk}(\lambda_{dk} + \alpha)\hat{M}_{kk} = 0 \quad k = 1, 2, ..., N \quad (7.91)$$

The natural frequency associated with the kth mode is given by Eqs. (7.66); that is,

$$\omega_k^2 = \frac{\hat{K}_{kk}}{\hat{M}_{kk}} \quad k = 1, 2, ..., N \tag{7.92}$$

Then, Eqs. (7.91) become the quadratic equation

$$\lambda_{dk}^2 + (\alpha + \beta\omega_k^2)\lambda_{dk} + \omega_k^2 = 0 \quad k = 1, 2, ..., N \tag{7.93a}$$

whose roots are given by

$$\lambda_{dk_{1,2}} = \frac{1}{2}\left[-(\alpha + \beta\omega_k^2) \mp \sqrt{(\alpha + \beta\omega_k^2)^2 - 4\omega_k^2}\right] \quad k = 1, 2, ..., N \quad (7.93b)$$

Thus, Eqs. (7.93b) establish how the eigenvalues λ_{dk} of the proportionally damped system are related to the eigenvalues $\lambda_k = \pm j\omega_k$ of the undamped system.

Equations (7.93a), which are associated with the free oscillation of the kth mode of the proportionally damped system, has the same form of the

---

[23]P. C. Müller and W. O. Schiehlen, *ibid.*

characteristic equation obtained for a single degree-of-freedom system; that is, Eq. (4.53). Comparing these two equations, we introduce the *modal damping factor* ζ_k associated with the kth mode as

$$\zeta_k = \frac{1}{2}\left(\frac{\alpha}{\omega_k} + \beta\omega_k\right) \quad k = 1, 2, ..., N \tag{7.94}$$

From Eq. (7.94), we see that if $\beta = 0$, then $\zeta_k = \alpha/\omega_k$ and ζ_k decreases as ω_k increases. On the other hand, when $\alpha = 0$, $\zeta_k = \beta\omega_k$ and ζ_k increases as ω_k increases.

In terms of the modal damping factor, one can rewrite Eqs. (7.93a) as

$$\lambda_{dk}^2 + 2\zeta_k\omega_k\lambda_{dk} + \omega_k^2 = 0 \quad k = 1, 2, ..., N \tag{7.95}$$

Hence, if the damping factor ζ_k of the kth mode is such that $0 \le \zeta_k < 1$, then one can define the corresponding damped natural frequency of the system as

$$\omega_{dk} = \omega_k\sqrt{1 - \zeta_k^2} \tag{7.96}$$

The roots of Eqs. (7.93a), given by Eqs. (7.93b), are expressed in terms of the damping factor ζ_k, the natural frequency ω_k, and the damped natural frequency ω_{dk} as

$$\lambda_{dk_{1,2}} = -\zeta_k\omega_k \mp j\omega_{dk} \tag{7.97}$$

Examining Eqs. (7.97), it is clear that the real parts of the complex-conjugate pair of eigenvalues associated with a particular mode contain information about the associated damping factor and undamped natural frequency and that the imaginary parts of these eigenvalues contain information about the associated damped natural frequency.

From Eqs. (7.93), it is also clear that the characteristic polynomial associated with the proportionally damped system is of the form

$$(\lambda_{d1}^2 + (\alpha + \beta\omega_1^2)\lambda_{d1} + \omega_1^2)(\lambda_{d2}^2 + (\alpha + \beta\omega_2^2)\lambda_{d2} + \omega_2^2)$$
$$...(\lambda_{dN}^2 + (\alpha + \beta\omega_N^2)\lambda_{dN} + \omega_N^2) = 0 \tag{7.98}$$

In other words, Eq. (7.98) is the result of expanding the determinant given in Eq. (7.87) in terms of N quadratic polynomials, each being associated with an equivalent single degree-of-freedom system in the considered mode of free oscillation. The interpretation of the modal damping factors of the different modes and the associated damped natural frequencies will be further clarified in Section 8.2, when presenting the normal-mode approach.

In Section 7.3.2, we saw that the modal matrix $[\Phi]$, which consisted of the N eigenvectors determined for the undamped system, is used to establish the diagonal inertia matrix $[M_D]$ and the diagonal stiffness matrix $[K_D]$ by making use of the orthogonality of the eigenvectors. Similarly, for a proportionally damped system, the damping matrix $[C]$ given by Eq. (7.83) can be transformed to a diagonal matrix $[C_D]$. To examine this, let us start from

$$[C_D] = [\Phi]^T [C][\Phi] \tag{7.99}$$

Then substituting from Eq. (7.83) into Eq. (7.99), we arrive at

$$[C_D] = [\Phi]^T[\alpha[M] + \beta[K]][\Phi] = \alpha[\Phi]^T[M][\Phi] + \beta[\Phi]^T[K][\Phi]$$

$$= \alpha[M_D] + \beta[K_D] = \alpha[M_D] + \beta[\omega_D^2][M_D]$$
$$= [\alpha[I] + \beta[\omega_D^2]][M_D] \tag{7.100}$$

where we have made use of Eqs. (7.68) and (7.69). On expanding the right-hand side of Eq. (7.100), we have that

$$[C_D] = \begin{bmatrix} (\alpha + \beta\omega_1^2)\hat{M}_{11} & 0 & \cdots & 0 \\ 0 & (\alpha + \beta\omega_2^2)\hat{M}_{22} & \cdots & 0 \\ \vdots & \vdots & \ddots & \vdots \\ 0 & 0 & \cdots & (\alpha + \beta\omega_N^2)\hat{M}_{NN} \end{bmatrix} \tag{7.101}$$

from which it is clear that we have a diagonal damping matrix $[C_D]$ for a proportionally damped system. The matrix $[C_D]$ is rewritten in terms of the modal damping factors given by Eqs. (7.94) as

$$[C_D] = \begin{bmatrix} 2\zeta_1\omega_1\hat{M}_{11} & 0 & \cdots & 0 \\ 0 & 2\zeta_2\omega_2\hat{M}_{22} & \cdots & 0 \\ \vdots & \vdots & \ddots & \vdots \\ 0 & 0 & \cdots & 2\zeta_N\omega_N\hat{M}_{NN} \end{bmatrix} \tag{7.102}$$

On examining the form of Eq. (7.102), it is clear that each of the diagonal terms is an equivalent damping coefficient associated with a certain damping mode. The transformation given by Eq. (7.99) can also be used to determine if the damping matrix for a system can be labeled as a proportional damping matrix. In other words, if the resulting transformed damping matrix is a diagonal matrix, then the system is proportionally damped; in all other cases, the transformed matrix is not a diagonal matrix.

Lightly Damped Systems and Other Cases

There are lightly damped systems ($0 < \zeta < 0.1$) in which $[\Phi]^T[C][\Phi]$ does not result in a diagonal matrix.[24] In this case, the off-diagonal terms are neglected, and only the diagonal terms $([\Phi]^T[C][\Phi])_{kk}$ are retained. In this case, the damping factor ζ_k takes the form

$$\zeta_k = \frac{1}{2\omega_k\hat{M}_{kk}}([\Phi]^T[C][\Phi])_{kk} \tag{7.103}$$

Another case of interest is one where the damping factor is constant and equal for each mode. In this case, the modal damping factor is assumed to be

$$\zeta_k = \zeta \quad k = 1, 2, ..., N \tag{7.104}$$

The last case that we shall consider corresponds to the physical system shown in Figure 7.18, which is used as a vibratory model of such diverse

[24]J. H. Ginsberg, *Mechanical and Structural Vibrations,* John Wiley & Sons, NY, Chapter 4 (2001).

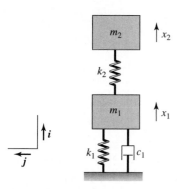

FIGURE 7.18
Vibratory model of a two
degree-of-freedom system with
one damper.

systems as an animal paw[25] or a system with a vibration absorber, which is discussed in Section 8.6. For this two degree-of-freedom system, the damping matrix $[C]$ and the modal damping $[\Phi]$ are given by, respectively,

$$[C] = \begin{bmatrix} c_1 & 0 \\ 0 & 0 \end{bmatrix}$$

$$[\Phi] = \begin{bmatrix} X_{11} & X_{12} \\ X_{21} & X_{22} \end{bmatrix} \tag{7.105}$$

Making use of Eqs. (7.105), the matrix $[\Phi]^T[C][\Phi]$ is determined as

$$
\begin{aligned}
[\Phi]^T[C][\Phi] &= \begin{bmatrix} X_{11} & X_{21} \\ X_{12} & X_{22} \end{bmatrix} \begin{bmatrix} c_1 & 0 \\ 0 & 0 \end{bmatrix} \begin{bmatrix} X_{11} & X_{12} \\ X_{21} & X_{22} \end{bmatrix} \\
&= \begin{bmatrix} X_{11} & X_{21} \\ X_{12} & X_{22} \end{bmatrix} \begin{bmatrix} c_1 X_{11} & c_1 X_{12} \\ 0 & 0 \end{bmatrix} \\
&= \begin{bmatrix} c_1 X_{11}^2 & c_1 X_{12} X_{21} \\ c_1 X_{11} X_{12} & c_1 X_{12}^2 \end{bmatrix}
\end{aligned}
\tag{7.106}
$$

Thus, we see that for a two degree-of-freedom system where one damper is connected to only one inertial element, the resulting transformed matrix is a nondiagonal matrix. This information is useful for deciding which approach to use in determining a solution for the response of a multi-degree-of-freedom system, as discussed in Chapter 8.

EXAMPLE 7.17   Character of the damping matrix

For the following mass, stiffness, and damping matrices, we shall determine whether the system has proportional damping.

$$[M] = \begin{bmatrix} 2 & 0 \\ 0 & 1 \end{bmatrix}, \quad [K] = \begin{bmatrix} 2 & -1 \\ -1 & 1 \end{bmatrix}, \quad [C] = \begin{bmatrix} 0.8 & -0.2 \\ -0.2 & 0.4 \end{bmatrix} \tag{a}$$

To determine whether the damping matrix is proportional, we consider Eq. (7.83) and determine if there are constants α and β for which the damping matrix $[C]$ in Eq. (a) is written as

$$
\begin{aligned}
[C] &= \alpha[M] + \beta[K] \\
&= \alpha \begin{bmatrix} 2 & 0 \\ 0 & 1 \end{bmatrix} + \beta \begin{bmatrix} 2 & -1 \\ -1 & 1 \end{bmatrix} \\
&= \begin{bmatrix} 2\alpha + 2\beta & -\beta \\ -\beta & \alpha + \beta \end{bmatrix}
\end{aligned}
\tag{b}
$$

Comparing the elements of matrix $[C]$ from the third equation of Eqs. (a) with those of Eq (b), we obtain

[25]R. M. Alexander, *Elastic Mechanisms in Animal Movement,* Cambridge University Press, Cambridge, Great Britain Chapter 7 (1988).

$$2\alpha + 2\beta = 0.8$$
$$-\beta = -0.2$$
$$\alpha + \beta = 0.4 \tag{c}$$

from which we find that $\alpha = 0.2$ and $\beta = 0.2$. Therefore, it is possible to express the damping matrix in the form of Eq. (7.83), and hence, the given damping matrix represents proportional damping.

EXAMPLE 7.18 Free oscillation characteristics of a proportionally damped system

We revisit Example 7.17 to determine the characteristic equation associated with this system, and from this equation, we solve for the eigenvalues associated with the damped system. We shall illustrate how the modal damping factors are calculated and the associated damped natural frequencies are determined.

Undamped System

To determine the natural frequencies of the undamped system, we make use of the stiffness and inertia matrices from Eqs. (a) of Example 7.17 and Eq. (7.26) and obtain the following characteristic equation:

$$\det \left[\lambda^2 \begin{bmatrix} 2 & 0 \\ 0 & 1 \end{bmatrix} + \begin{bmatrix} 2 & -1 \\ -1 & 1 \end{bmatrix} \right] = 0$$

or

$$\det \begin{bmatrix} 2(\lambda^2 + 1) & -1 \\ -1 & (\lambda^2 + 1) \end{bmatrix} = 0 \tag{a}$$

Equation (a) leads to the quartic equation

$$2\lambda^4 + 4\lambda^2 + 1 = 0 \tag{b}$$

whose roots are

$$\lambda^2 = -\left(1 \mp \frac{1}{\sqrt{2}} \right) \tag{c}$$

Since $[M]$ and $[K]$ are real symmetric matrices, the eigenvalues are real. To determine the associated natural frequencies, one can use Eq. (7.35)—that is, $\lambda^2 = -\omega^2$—and obtain

$$\omega_1 = \sqrt{1 - \frac{1}{\sqrt{2}}} = 0.541 \text{ rad/s}$$

$$\omega_2 = \sqrt{1 + \frac{1}{\sqrt{2}}} = 1.307 \text{ rad/s} \tag{d}$$

Damped System

From Eq. (7.81) and Eqs. (a) of Example 7.17, we find that the characteristic equation in the proportionally damped case is given by

$$\det\left[\lambda^2\begin{bmatrix}2 & 0\\0 & 1\end{bmatrix} + \lambda\begin{bmatrix}0.8 & -0.2\\-0.2 & 0.4\end{bmatrix} + \begin{bmatrix}2 & -1\\-1 & 1\end{bmatrix}\right] = 0$$

which is rewritten as

$$\det\begin{bmatrix}2\lambda^2 + 0.8\lambda + 2 & -0.2\lambda - 1\\-0.2\lambda - 1 & \lambda^2 + 0.4\lambda + 1\end{bmatrix} = 0 \tag{e}$$

Expanding the determinant in Eq. (e), we arrive at the quartic polynomial

$$2\lambda^4 + 1.6\lambda^3 + 4.28\lambda^2 + 1.2\lambda + 1 = 0 \tag{f}$$

Solving[26] Eq. (f), we find that the eigenvalues in the proportionally damped case are given by

$$\lambda_{d1_{1,2}} = \lambda_{1,2} = -0.129 \mp j0.526$$
$$\lambda_{d2_{1,2}} = \lambda_{3,4} = -0.271 \mp j1.278 \tag{g}$$

Equations (g) could have been obtained directly from Eqs. (7.93b) after making use of the undamped natural frequencies given in Eqs. (d) and the values of α and β determined in Example 7.17; that is,

$$\lambda_{d1_{1,2}} = \frac{1}{2}[-(0.2 + 0.2 \times 0.541^2) \mp \sqrt{(0.2 + 0.2 \times 0.541^2)^2 - 4 \times 0.541^2}]$$
$$= -0.129 \mp j0.526 \tag{h}$$

and

$$\lambda_{d2_{1,2}} = \frac{1}{2}[-(0.2 + 0.2 \times 0.1.307^2) \mp \sqrt{(0.2 + 0.2 \times 1.307^2)^2 - 4 \times 1.307^2}]$$
$$= -0.271 \mp j1.278 \tag{i}$$

From Eqs. (7.94), we find that the modal damping factors are given by

$$\zeta_1 = \frac{1}{2}\left(\frac{0.2}{0.541} + 0.2 \times 0.541\right) = 0.239$$
$$\zeta_2 = \frac{1}{2}\left(\frac{0.2}{1.307} + 0.2 \times 1.307\right) = 0.207 \tag{j}$$

Since both damping factors are less than 1, the corresponding damped natural frequencies are determined by making use of Eqs. (7.96), (d), and (j). The calculations lead to

$$\omega_{d1} = 0.541 \times \sqrt{1 - 0.239^2} = 0.525 \text{ rad/s}$$
$$\omega_{d2} = 1.307 \times \sqrt{1 - 0.207^2} = 1.279 \text{ rad/s} \tag{k}$$

---

[26]The MATLAB function `roots` was used.

From the numerical values provided in Eqs. (d), (g), (j), and (k), it is clear that the eigenvalues of the proportionally damped system are in the form of Eq. (7.97); that is,

$$\lambda_{d1_{1,2}} = -\zeta_1\omega_1 \mp j\omega_{d1}$$
$$\lambda_{d2_{1,2}} = -\zeta_2\omega_2 \mp j\omega_{d2} \tag{l}$$

EXAMPLE 7.19 Free oscillation characteristics of a system with gyroscopic forces

We revisit Example 7.4, where we discussed a gyro-sensor, and illustrate the effects that gyroscopic forces have on the eigenvalues. The characteristic polynomial is determined for the damped case and the eigenvalues are explicitly determined only for the undamped case.

Setting the external force f_x to zero in Eq. (d) of Example 7.4, we obtain the following system of equations:

$$[M]\begin{Bmatrix} \ddot{x} \\ \ddot{y} \end{Bmatrix} + [C]\begin{Bmatrix} \dot{x} \\ \dot{y} \end{Bmatrix} + [G]\begin{Bmatrix} \dot{x} \\ \dot{y} \end{Bmatrix} + [K]\begin{Bmatrix} x \\ y \end{Bmatrix} = \begin{Bmatrix} 0 \\ 0 \end{Bmatrix} \tag{a}$$

Considering a special case where $c_x = c_y = c$ and $k_x = k_y = k$ in Figure 7.6; that is, the stiffness-damper combinations are identical in both directions, we find that the different matrices in Eqs. (a) are determined from Eqs. (e) of Example 7.4 as

$$[M] = \begin{bmatrix} m & 0 \\ 0 & m \end{bmatrix}, \qquad [K] = \begin{bmatrix} k - m\omega_z^2 & 0 \\ 0 & k - m\omega_z^2 \end{bmatrix}$$

$$[C] = \begin{bmatrix} c & 0 \\ 0 & c \end{bmatrix}, \qquad [G] = \begin{bmatrix} 0 & -2m\omega_z \\ 2m\omega_z & 0 \end{bmatrix} \tag{b}$$

To determine the eigenvalues associated with this system, we set the circulatory terms $[H] = [0]$ and substitute Eqs. (b) into Eq. (7.77) to obtain the characteristic equation

$$\det\left[\lambda^2 \begin{bmatrix} m & 0 \\ 0 & m \end{bmatrix} + \lambda\left(\begin{bmatrix} c & 0 \\ 0 & c \end{bmatrix} + \begin{bmatrix} 0 & -2m\omega_z \\ 2m\omega_z & 0 \end{bmatrix} \right) \right.$$

$$\left. + \begin{bmatrix} k - m\omega_z^2 & 0 \\ 0 & k - m\omega_z^2 \end{bmatrix} \right] = 0 \tag{c}$$

Equation (c) is rearranged to give

$$\det\begin{bmatrix} m\lambda^2 + c\lambda + k - m\omega_z^2 & -2m\omega_z\lambda \\ 2m\omega_z\lambda & m\lambda^2 + c\lambda + k - m\omega_z^2 \end{bmatrix} = 0 \tag{d}$$

On expanding this determinant, the result is the quartic polynomial

$$\lambda^4 + 2\frac{c}{m}\lambda^3 + \left[\left(\frac{c}{m} \right)^2 + 2\frac{k}{m} + 2\omega_z^2 \right]\lambda^2$$

$$+ 2\left(\frac{k}{m} - \omega_z^2 \right)\frac{c}{m}\lambda + \left(\frac{k}{m} - \omega_z^2 \right)^2 = 0 \tag{e}$$

Equation (e) is the characteristic equation for the damped system, whose four roots are determined numerically for given values of k, c, m, and ω_z.

In order to determine the eigenvalues of the undamped system explicitly, we set $c = 0$ in Eq. (e) and obtain

$$\lambda^4 + 2\left(\frac{k}{m} + \omega_z^2\right)\lambda^2 + \left(\frac{k}{m} - \omega_z^2\right)^2 = 0 \tag{f}$$

The roots of this quadratic polynomial in λ^2 are given by

$$\lambda_{1,2}^2 = -\left(\frac{k}{m} + \omega_z^2\right) \pm 2\omega_z\sqrt{\frac{k}{m}} \tag{g}$$

When the gyroscopic force is zero—that is, $\omega_z = 0$—these results reduce to the natural frequencies of two uncoupled single degree-of-freedom systems, each of which has the same mass and the same stiffness. If

$$2\omega_z\sqrt{\frac{k}{m}} < \left(\frac{k}{m} + \omega_z^2\right)$$

then all of the eigenvalues are imaginary, as in the case of an undamped system free of gyroscopic forces; that is, when $\omega_z = 0$.

7.3.4 Conservation of Energy

In Section 7.2.1, we discussed how the linear momentum and the angular momentum of a multi-degree-of-freedom system is conserved when the external forces and external moments are absent. Here, we examine when the energy of a multi-degree-of-freedom system is conserved during free oscillations. To this end, we start from Eq. (7.73), which are the equations for a damped system with gyroscopic and circulatory terms; that is

$$[M]\{\ddot{x}\} + [[C] + [G]]\{\dot{x}\} + [[K] + [H]]\{x\} = \{0\}$$

Pre-multiplying this equation by $\{\dot{x}\}^T$, we arrive at

$$\{\dot{x}\}^T[M]\{\ddot{x}\} + \{\dot{x}\}^T[[C] + [G]]\{\dot{x}\} + \{\dot{x}\}^T[[K] + [H]]\{x\} = \{0\} \tag{7.107}$$

Noting that $\{\dot{x}\}^T[G]\{x\} = 0$ because the gyroscopic matrix is a skew symmetric matrix, and expanding and rearranging Eq. (7.107) leads to

$$\{\dot{x}\}^T[M]\{\ddot{x}\} + \{\dot{x}\}^T[K]\{\dot{x}\} = -\{\dot{x}\}^T[C]\{\dot{x}\} - \{\dot{x}\}^T[H]\{x\} \tag{7.108}$$

Since $[M]$ and $[K]$ are symmetric matrices, the left-hand side of Eq. (7.108) is expressed in terms of a time derivative as

$$\frac{d}{dt}\left(\frac{1}{2}\{\dot{x}\}^T[M]\{\dot{x}\} + \frac{1}{2}\{x\}^T[K]\{x\}\right) = -\{\dot{x}\}^T[C]\{\dot{x}\} - \{\dot{x}\}^T[H]\{x\} \tag{7.109}$$

Examining Eq. (7.109), we find that if the system is undamped and free of circulatory forces, the right-hand side is zero and Eq. (7.109) becomes

$$\frac{d}{dt}\left(\frac{1}{2}\{\dot{x}\}^T[M]\{\dot{x}\} + \frac{1}{2}\{x\}^T[K]\{x\}\right) = 0 \tag{7.110}$$

which means that

$$\frac{1}{2}\{\dot{x}\}^T[M]\{\dot{x}\} + \frac{1}{2}\{x\}^T[K]\{x\} = \text{constant} \tag{7.111}$$

Making use of Eqs. (7.8) for a natural system, we recognize that the left-hand side of Eq. (7.111) is the sum of the system kinetic energy and the system potential energy. In other words, Eq. (7.111) means that

$$E = T + V = \text{constant} \tag{7.112}$$

where E is the total energy of the natural system. Thus, the energy of a multi-degree-of-freedom system is only conserved in the absence of damping and circulatory forces. For nonnatural systems, the left-hand side of Eq. (7.111) is referred to as the *Hamiltonian* of the system.[27]

EXAMPLE 7.20   Conservation of energy in a three degree-of-freedom system

Consider the three degree-of-freedom system used to model the milling machine shown in Figure 7.4. We set $f_1(t)$ to zero so that we can examine free oscillations of this system. Then Eqs. (b) of Example 7.1 becomes

$$
\begin{bmatrix} m_1 & 0 & 0 \\ 0 & m_2 & 0 \\ 0 & 0 & m_3 \end{bmatrix} \begin{Bmatrix} \ddot{x}_1 \\ \ddot{x}_2 \\ \ddot{x}_3 \end{Bmatrix} + \begin{bmatrix} c_1 & -c_1 & 0 \\ -c_1 & c_1 + c_2 & -c_2 \\ 0 & -c_2 & c_2 + c_3 \end{bmatrix} \begin{Bmatrix} \dot{x}_1 \\ \dot{x}_2 \\ \dot{x}_3 \end{Bmatrix}
$$
$$
+ \begin{bmatrix} k_1 & -k_1 & 0 \\ -k_1 & k_1 + k_2 & -k_2 \\ 0 & -k_2 & k_2 + k_3 \end{bmatrix} \begin{Bmatrix} x_1 \\ x_2 \\ x_3 \end{Bmatrix} = \begin{Bmatrix} 0 \\ 0 \\ 0 \end{Bmatrix} \tag{a}
$$

From the form of Eqs. (a) it is clear that the system is free of gyroscopic and circulatory forces. However, due to the presence of damping, we see from Eq. (7.109) that the sum of the kinetic energy of the system and the potential energy of the system is not conserved; in other words,

$$\underbrace{\frac{1}{2} m_1 \dot{x}_1^2 + \frac{1}{2} m_2 \dot{x}_2^2 + \frac{1}{2} m_3 \dot{x}_3^2}_{\text{System kinetic energy}} + \underbrace{\frac{1}{2} k_1 (x_1 - x_2)^2 + k_2 (x_2 - x_3)^2 + \frac{1}{2} k_3 x_3^2}_{\text{System potential energy}}$$

$$\neq \text{constant} \tag{b}$$

Making use of $E = T + V$ to represent the total energy of the system and using Eq. (7.109), we see that

$$\frac{dE}{dt} = -\{\dot{x}\}^T [C]\{\dot{x}\} = -\left[\frac{1}{2} c_1 (\dot{x}_1 - \dot{x}_2)^2 + c_2 (\dot{x}_2 - \dot{x}_3)^2 + \frac{1}{2} c_3 \dot{x}_3^2 \right] \tag{c}$$

Since the value of the right-hand side of Eq. (c) is always negative, except when $\dot{x}_1 = \dot{x}_2 = \dot{x}_3 = 0$ (i.e., at the system's equilibrium position), the total energy of the system continues to decrease and eventually comes to zero at the system's equilibrium position. In the absence of damping, Eq. (c) becomes

$$\frac{dE}{dt} = 0 \tag{d}$$

and hence, energy is conserved.

[27]L. Meirovitch, *ibid.*

7.4 STABILITY

Extending the notion of bounded stability presented in Section 4.4, a linear multi-degree-of-freedom system described by the generalized coordinates x_1, x_2, ..., x_N and subjected to finite initial conditions and finite forcing functions is said to be *stable* if

$$\|\{x\}\| = \sqrt{x_1^2 + x_2^2 + \cdots + x_N^2} \leq A \text{ for all time} \tag{7.113}$$

where A has a finite value. This is a boundedness condition, which requires the system response $\{x\}$ to be bounded for bounded system inputs. If this condition is not satisfied, then the system is said to be *unstable*.

EXAMPLE 7.21 Stability of an undamped system with gyroscopic forces

We revisit Examples of 7.4 and 7.19 and determine under what conditions the gyro-sensor is unstable during free oscillations at angular speeds ω_z. Based on Eq. (7.113), we need to determine if

$$\sqrt{x^2 + y^2} \leq A \text{ for all time} \tag{a}$$

is true. For free oscillations of the gyro-sensor based on the discussion of Section 7.3.3, the motion is described by

$$\begin{Bmatrix} x(t) \\ y(t) \end{Bmatrix} = \begin{Bmatrix} X \\ Y \end{Bmatrix} e^{\lambda t} \tag{b}$$

where λ are the eigenvalues. If one or more of the eigenvalues have a positive real part, then Eq. (a) will not be true for all time. We can use this as a basis to determine the angular speeds ω_z for which the system will be either unstable or stable.

For the special case treated in Example 7.19, the eigenvalues are given by Eqs. (g). From these eigenvalues, we arrive at the following angular speed range for instability:

$$2\omega_z\sqrt{\frac{k}{m}} > \left(\frac{k}{m} + \omega_z^2\right) \tag{c}$$

Thus, if the angular speed is selected, one should select ω_z based on Eq. (c) to avoid instability.

7.5 SUMMARY

In this chapter, the derivation of the governing equations of a multi-degree-of-freedom system was illustrated by making use of force-balance and moment-balance methods and Lagrange's equations. Linearization of equations governing a nonlinear system was also addressed. Free-response characteristics of undamped and damped systems were considered and notions of orthogonality of modes, modal mass, modal stiffness, proportional damping, modal damping factor, node of a mode, and rigid-body mode were introduced. The conditions under which conservation of energy, conservation of linear momentum, and conservation of angular momentum hold during free-oscillations of a multi-degree-of-freedom system were determined. Finally, the notion of stability of a linear multi-degree-of-freedom system was introduced.

EXERCISES

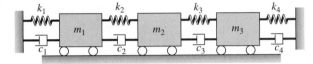

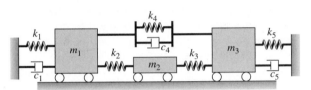

FIGURE E7.1

7.1 Derive the equations of motion of the systems shown in Figures E7.1a and E7.1b and present the resulting equations in each case in matrix form.

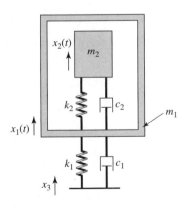

FIGURE E7.2

7.2 Derive the equations of motion for the model of an electronic system m_2 contained in a package m_1, as shown in Figure E7.2 and present them in matrix form.

7.3 Consider the "small" amplitude motions of the pendulum-absorber system shown in Figure 7.9 and derive the equations of motion by using force-balance and moment-balance methods.

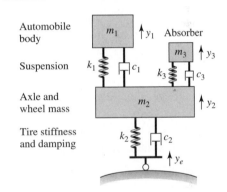

FIGURE E7.4

7.4 Derive the equations of motion of the vehicle model shown in Figure E7.4.

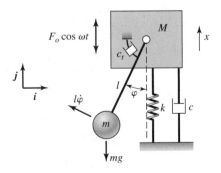

FIGURE E7.5

7.5 Derive the equations of motion of the pendulum absorber shown in Figure E7.5 for large oscillations and then linearize these equations about the static equilibrium position corresponding to the bottom position of the pendulum. Present the final equations in matrix form.

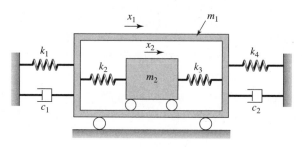

FIGURE E7.6

7.6 Derive the equations of motion of the system shown in Figure E7.6 by using Lagrange's equations.

7.7 Replace each of the linear springs k_2 and k_3 shown in Figure E7.6 by a nonlinear spring whose force-displacement characteristic is given by

$$F(x) = k(x + \alpha x^3)$$

and determine the resulting equations of motion.

7.8 Derive the equations of the milling model shown in Figure 7.4b by using Lagrange's equations.

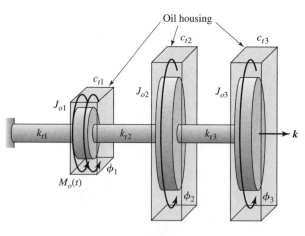

FIGURE E7.9

7.9 Derive the equations of motion of the system shown in Figure E7.9, which is an extended version of a two-degree-of-freedeom system discussed in Section 7.2. Let $M_o(t)$ be the external torque that acts on the disc whose motion is described by the angular variable φ_1.

7.10 Derive the equations of the hand-arm system treated in Example 7.9 by using force-balance and moment-balance methods for "large" and "small" oscillations about the nominal position.

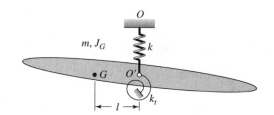

FIGURE E7.11

7.11 The experimental arrangement for an airfoil mounted in a wind tunnel is described by the model shown in Figure E7.11. Determine the equations of motion governing this system when the stiffness of the translation spring is k, the stiffness of the torsion spring is k_t, G is the center of the mass of the airfoil located a distance l from the attachment point O', m is the mass of the airfoil, and J_G is the mass moment of inertia of the airfoil about the center of mass. Use the generalized coordinates x and θ discussed in Exercise 1.19.

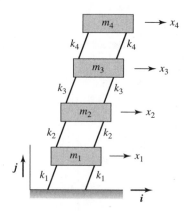

FIGURE E7.12

7.12 A multistory building is described by the model shown in Figure E7.12. Derive the equations of motion of this system and present them in matrix form. Are the mass and stiffness matrices symmetric?

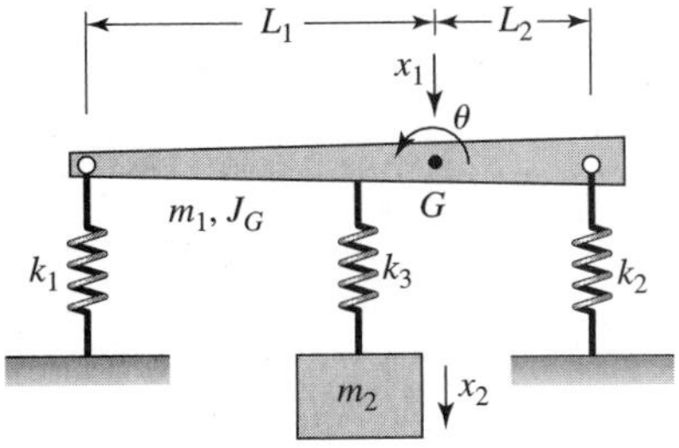

FIGURE E7.13

7.13 Obtain the governing equations of motion for large oscillations of the system shown in Figure E7.13 in terms of the generalized coordinates x_1, x_2, and θ. The spring k_3 is attached at the midpoint of the bar. Linearize the resulting system of equations for "small" motions about the system equilibrium position and present the resulting equations in matrix form. Point G is the center of mass of the bar.

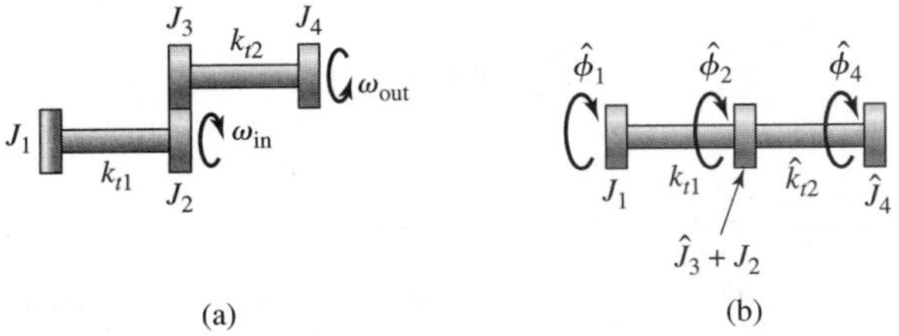

FIGURE E7.14

7.14 A pair of shafts are linked by a set of gears as shown in Figure E7.14a. An equivalent system for the system shown in Figure E7.14a is determined as shown in Figure E7.14b, where $\hat{J}_i = J_i T_R^2$, $\hat{k}_{t2} = k_{t2} T_R^2$, and the transmission ratio $T_R = \omega_{out}/\omega_{in}$. Determine the governing equations of motion of the system shown in Figure E7.14b in terms of the generalized coordinates $\phi_1, \hat{\phi}_2$, and $\hat{\phi}_4$ wheres $\hat{\phi}_i = \phi_i/T_R$.

7.15 A two-degree-of-freedom system with a nonlinear spring element is described by the following system of equations:

$$m_1\ddot{x}_1 + k_1x_1 + \alpha x_1^3 + k_2(x_1 - x_2) + k_2\alpha(x_1 - x_2)^3 = 0$$
$$m_2\ddot{x}_2 + k_2(x_2 - x_1) + k_2\alpha(x_2 - x_1)^3 = 0$$

Determine the system equilibrium positions. Assume that m_1, m_2, k_1, and k_2 are positive and α is negative.

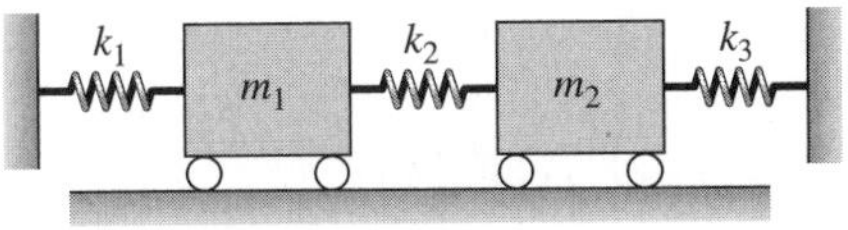

FIGURE E7.16

7.16 Determine the natural frequencies and mode shapes associated with the system shown in Figure E7.16 for $m_1 = 10$ kg, $m_2 = 20$ kg, $k_1 = 100$ N/m, $k_2 = 100$ N/m, and $k_3 = 50$ N/m.

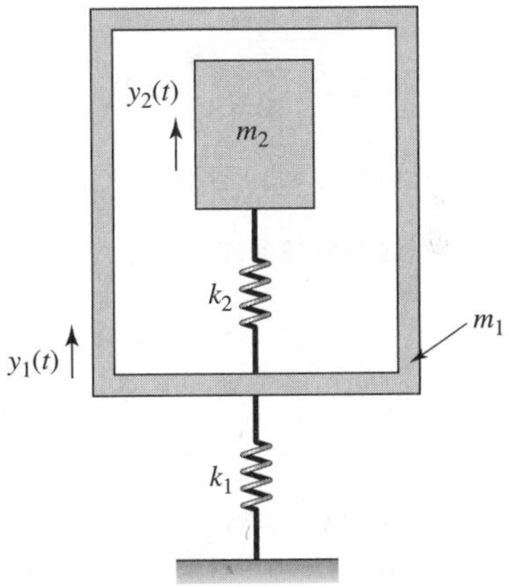

FIGURE E7.17

7.17 Determine the natural frequencies and mode shapes associated with the system shown in Figure E7.17 for $m_1 = 10^{-3}$ kg, $m_2 = 0.01$ kg, and $k_1 = k_2 = 2$ kN/m. Include plots of the mode shapes.

7.18 For the system considered in Exercise 7.7, remove the nonlinear spring k_3, set the damping coefficients $c_1 = c_4 = 0$ and consider the free oscillations of this system. Choose the values of the parameters as follows: $m_1 = 10$ kg, $m_2 = 2$ kg, $k_1 = k_4 = k = 10$ N/m, and $\alpha = 2$ m^{-2}. Assume that the nonlinear spring k_2 is initially compressed by 0.05 m. Determine the natural frequencies and mode shapes associated with free oscillations about the equilibrium position.

7.19 Consider a system with the following inertia and stiffness matrices:

$$[M] = \begin{bmatrix} 2 & 0 \\ 0 & 2 \end{bmatrix} \text{kg}; \quad [K] = \begin{bmatrix} 10{,}000 & -k \\ -k & k_{22} \end{bmatrix} \text{N/m}$$

If the modes of the system are given by

$$\{X\}_1 = \begin{Bmatrix} -2 \\ 1 \end{Bmatrix} \quad \text{and} \quad \{X\}_2 = \begin{Bmatrix} 1/2 \\ 1 \end{Bmatrix}$$

and the natural frequency associated with $\{X\}_1$ is 100 rad/s, then determine the unknown coefficients in the stiffness matrix and the other natural frequency of the system.

whose mass matrix, stiffness matrix, and damping matrix are given by the following:

$$[M] = \begin{bmatrix} 1 & 0 \\ 0 & 2 \end{bmatrix}, \quad [K] = \begin{bmatrix} 1 & -1 \\ -1 & 2 \end{bmatrix},$$

$$[C] = \begin{bmatrix} 3 & -2 \\ -2 & 6 \end{bmatrix}$$

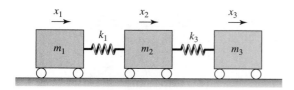

FIGURE E7.20

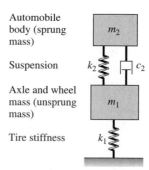

FIGURE E7.26

7.20 Determine if the system shown in Figure E7.20 has any rigid-body modes.

7.21 Let the system shown in Figure E7.20 represent a system of three railroad cars with masses $m_1 = m_2 = m_3 = 1200$ kg and interconnections $k_1 = k_3 = 4800$ kN/m. Determine the natural frequencies and mode shapes of this system and plot the corresponding mode shapes.

7.22 Show that the linear independence of eigenvectors given by Eq. (7.72) is true by making use of the orthogonality of eigenvectors.

7.23 Is it possible for a three degree-of-freedom system to have the following eigenvectors?

$$\{X\}_1 = \begin{Bmatrix} 1 \\ 1 \\ 1/2 \end{Bmatrix}, \quad \{X\}_2 = \begin{Bmatrix} -1 \\ 1 \\ 1/2 \end{Bmatrix}, \quad \{X\}_3 = \begin{Bmatrix} 0 \\ 2 \\ 1 \end{Bmatrix}$$

7.24 Show that the system treated in Example 7.17 is proportionally damped by making use of Eq. (7.99).

7.25 Determine the modal mass, modal stiffness, and modal damping factors associated with a system

7.26 To describe the vertical motions of an automobile, the two degree-of-freedom system shown in Figure E7.26 is used. This model is known as a *quarter-car model*. If the parameters of the system are $m_1 = 80$ kg, $m_2 = 1100$ kg, $k_2 = 30$ kN/m, $k_1 = 300$ kN/m, and $c_2 = 5000$ N/(m/s), determine if the system is proportionally damped.

7.27 If the modal matrix and the damping matrix for a two degree-of-freedom system are, respectively,

$$[\Phi] = \begin{bmatrix} 1 & -1 \\ 1 & 1 \end{bmatrix} \quad \text{and} \quad [C] = \begin{bmatrix} 5 & 0 \\ 0 & 0 \end{bmatrix}$$

Can this system be proportionally damped?

7.28 Consider free oscillations of the gyro-sensor treated in Example 7.4 and examine if energy is conserved in the system.

7.29 Consider the three degree-of-freedom system shown in Figure E7.20 and determine if the linear momentum of this system and the total energy of this system are conserved.

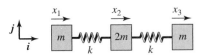

FIGURE E7.30

7.30 A flexible structural system is represented by the model shown in Figure E7.30. Determine the governing equations of motion of this system and from these three equations determine the eigenvalues and eigenvectors associated with free oscillations of this system. Find the locations of the nodes for the different mode shapes. The stiffness of the structural members is $k = 3EI/L^3$.

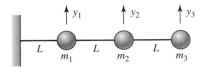

FIGURE E7.31

7.31 Consider the system shown in Figure E7.31 in which the three masses m_1, m_2, and m_3 are located on a uniform cantilever beam with flexural rigidity EI. The inverse of the stiffness matrix for this system $[K]$, which is called the *flexibility matrix*, is given by

$$[K]^{-1} = \frac{L^3}{3EI} \begin{bmatrix} 27 & 14 & 4 \\ 14 & 8 & 2.5 \\ 4 & 2.5 & 1 \end{bmatrix}$$

Determine the following:
  a) the stiffness matrix of the system
  b) the governing equations of motion
  c) when $m_1 = m_2 = m_3 = m$, determine the natural frequencies and mode shapes of the system.

For part c), let $\alpha = 3EI/mL^3$ and express the natural frequencies in terms of α.

7.32 If the stiffness matrix for the system shown in Figure E7.12 is given by the following matrix

$$[K] = \frac{EI}{L^3} \begin{bmatrix} 24 & -12 & 0 & 0 \\ -12 & 24 & -12 & 0 \\ 0 & -12 & 24 & -12 \\ 0 & 0 & -12 & 12 \end{bmatrix}$$

and if all the masses of the floors of the four-story building are equal—that is, $m_1 = m_2 = m_3 = m_4 = m$—then determine the system eigenvalues and eigenvectors and plot the eigenvectors of this system. Let $\alpha = EI/L^3$ and express the natural frequencies in terms of α.

7.33 For the system shown in Figure E7.9, assume that the drive torque $M_o(t) = 0$. Let $J_{o1} = 1$ kg·m², $J_{o2} = 4$ kg·m², $J_{o3} = 1$ kg·m² and let the torsional stiffness of each shaft be as follows: $k_{t1} = k_{t2} = 10$ Nm/rad and $k_{t3} = 5$ Nm/rad. In addition, let the damping coefficients be such that $c_{t1} = 0.5$ Nm/rad/s, $c_{t2} = 2$ Nm/rad/s, and $c_{t3} = 0.5$ Nm/rad/s. Determine the following:
  a) is the system proportionally damped?
  b) if the damping associated with the second flywheel changes from 2 Nm/rad/s to 1.8 Nm/rad/s can the system be approximated as a system with modal damping?

7.34 Repeat Example 7.14 for the following values of the nondimensional ratios: $k_{21} = 0.5$, $L_{21} = 2.0$, and $\omega_r = 1.5$.

7.35 Repeat Example 7.13 when $\omega_r = 2.0$ and $J_r = 1.0$.

7.36 The eigenvalues associated with a damped three degree-of-freedom system are given by the following:

$$\lambda_{d1_{1,2}} = -0.1 \mp j0.995$$
$$\lambda_{d2_{1,2}} = -0.75 \mp j1.299$$
$$\lambda_{d3_{1,2}} = -0.4 \mp j1.960$$

Determine the system natural frequencies, modal damping factors and damped natural frequencies.

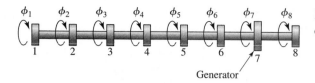

ϕ_1 ϕ_2 ϕ_3 ϕ_4 ϕ_5 ϕ_6 ϕ_7 ϕ_8

Generator

FIGURE E7.37

7.37 A six-cylinder, four-cycle engine driving a generator is modeled by the eight degree-of-freedom system shown in Figure E7.37.[28] Free oscillations of this system are described by the following system

$$[J]\{\ddot{\phi}\} + [K_t]\{\phi\} = \{0\}$$

where

$$[J] = \begin{bmatrix} 21 & 0 & 0 & 0 & 0 & 0 & 0 & 0 \\ 0 & 21 & 0 & 0 & 0 & 0 & 0 & 0 \\ 0 & 0 & 21 & 0 & 0 & 0 & 0 & 0 \\ 0 & 0 & 0 & 21 & 0 & 0 & 0 & 0 \\ 0 & 0 & 0 & 0 & 21 & 0 & 0 & 0 \\ 0 & 0 & 0 & 0 & 0 & 21 & 0 & 0 \\ 0 & 0 & 0 & 0 & 0 & 0 & 98 & 0 \\ 0 & 0 & 0 & 0 & 0 & 0 & 0 & 49 \end{bmatrix} \text{kg·m}^2$$

and

$$[K_t] =$$

$$\begin{bmatrix} 51 & -51 & 0 & 0 & 0 & 0 & 0 & 0 \\ -51 & 102 & -51 & 0 & 0 & 0 & 0 & 0 \\ 0 & -51 & 102 & -51 & 0 & 0 & 0 & 0 \\ 0 & 0 & -51 & 102 & -51 & 0 & 0 \\ 0 & 0 & 0 & -51 & 102 & -51 & 0 & 0 \\ 0 & 0 & 0 & 0 & -51 & 117 & -66 & 0 \\ 0 & 0 & 0 & 0 & 0 & -66 & 81 & -15 \\ 0 & 0 & 0 & 0 & 0 & 0 & -15 & 15 \end{bmatrix}$$

$$\times 10^6 \text{ Nm/rad}$$

Determine the natural frequencies and mode shapes associated with the system and plot the mode shapes. Does the system have any rigid-body modes?

[28]C. Genta, *ibid.*

7.38 The eigenvalues determined for the two different three degree-of-freedom systems are as follows:
a)

$$\lambda_{d1_{1,2}} = -a_1 \mp jb_1; \qquad a_1 > 0, b_1 > 0$$
$$\lambda_{d2_{1,2}} = -a_2 \mp jb_2; \qquad a_2 > 0, b_2 > 0$$
$$\lambda_{d3_{1,2}} = -a_3 \mp jb_3; \qquad a_3 > 0, b_3 > 0$$

b)

$$\lambda_{d1_{1,2}} = -a_1 \mp jb_1; \qquad a_1 > 0, b_1 > 0$$
$$\lambda_{d2_{1,2}} = -a_2 \mp jb_2; \qquad a_2 > 0, b_2 > 0$$
$$\lambda_{d3_{1,2}} = a_3, -b_3; \qquad a_3 > 0, b_3 > 0$$

Determine which of these systems is stable.

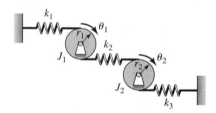

FIGURE E7.39

7.39
a) Determine the equations of motion for the system shown in Figure E7.39 and put them in matrix form.
b) From the results in a) determine the determinant form of the characteristic equation.

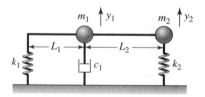

FIGURE E7.40

7.40 Consider the vibratory system shown in Figure E7.40, which consists of two masses that are connected by rigid massless rods. Determine the equations of motion of the system.

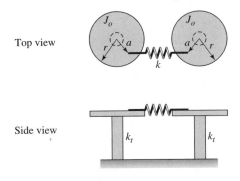

FIGURE E7.41

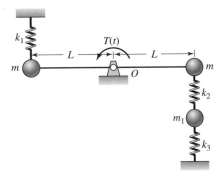

FIGURE E7.43

7.41 Two identical discs of rotary inertia J_O and radius r are mounted on identical shafts and undergo torsional oscillations. Each shaft has a torsional spring constant k_t. At a radial distance a from their respective centers, the rotors are connected via a translation spring of constant k, as shown in Figure E7.41. If $J_O = 10$ kg·m^2, $r = 0.25$ m, $k_t = 600$ N/m, and $a = 0.15$ m, then determine the natural frequencies of the system.

each end of the rod a mass m is attached. To one of the end masses another spring-mass system is attached. A torque $T(t)$ is applied to the rod at the pivot as shown in the figure. Obtain the equations of motion for this system.

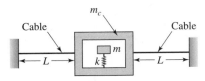

FIGURE E7.42

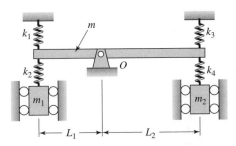

FIGURE E7.44

7.44 Consider the vibratory system shown in Figure E7.44.

a) If the connecting rod is rigid and uniform with a total mass m, then determine the equations of motion for the system.

b) If the connecting rod is rigid and weightless, then determine the equations of motion for the system.

c) If the connecting rod is rigid and weightless and the pivot is removed—that is, the rod can translate vertically and rotate in the plane of the page—then determine the equations of motion for the system.

7.42 A container of mass m_c is suspended by two taut cables of length L as shown in Figure E7.42. The tension in the cables is T_o. Inside the container, a mass m is elastically supported by a spring k.

a) Determine the equivalent spring constant for the cable-mass system and sketch the equivalent vibratory system.

b) For the equivalent system determined in part a), determine the equations governing the motion of this system.

7.43 A rigid weightless rod of length $2L$ is attached to a pivot at its center as shown in Figure E7.43. At

7.45 Consider the autoparametric vibration absorber shown in Figure E7.45. The system composed

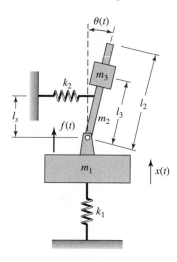

$$\gamma_2 = \frac{1}{2} \frac{l_c^2(m_2 + 2m_3l_3/l_2)}{m_2l_2^2/3 + m_3l_3^2}$$

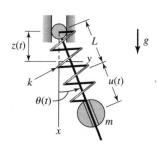

FIGURE E7.46

FIGURE E7.45

of mass m_1 and spring k_1 is externally excited by a harmonically oscillating force

$$f(t) = f_o \cos \omega t$$

The system's oscillations are to be attenuated by attaching to the system another system composed of a mass m_3 that is attached to a rigid rod of mass m_2 and length l_2. The base of the rod is pivoted on mass m_1. Attached to the rod at a distance l_s from the pivot is a spring k_2.

a) Determine expressions for the kinetic energy and the potential energy of the system.

b) Show that the nonlinear equations of motions for this system are

$$\frac{d^2X}{d\tau^2} + X - \gamma_1\left[\left(\frac{d\theta}{d\tau}\right)^2 \cos\theta + \frac{d^2\theta}{d\tau^2}\sin\theta\right]$$
$$= F_o \cos\Omega\tau$$

$$\frac{d^2\theta}{d\tau^2} - \gamma_2\frac{d^2X}{d\tau^2}\sin\theta + \frac{\omega_2^2}{\omega_1^2}\sin\theta\cos\theta = 0$$

where

$$\tau = \omega_1 t, \quad \Omega = \frac{\omega}{\omega_1}, \quad \omega_1^2 = \frac{k_1}{m_1 + m_2 + m_3},$$

$$\omega_2^2 = \frac{k_2 l_s^2}{m_2 l_2^2/3 + m_3 l_3^2} \quad X = x/l_2, \quad F_o = \frac{f_o}{k_1 l_2},$$

$$\gamma_1 = \frac{1}{2}\frac{m_2 + 2m_3l_3/l_2}{m_1 + m_2 + m_3}, \quad \text{and}$$

7.46 The mass m shown in Figure E7.46 slides in a gravity field along a massless rod. Wrapped around the rod is a spring of constant k that has an unstretched length L. One end of the rod is attached to a vertically moving pivot that oscillates harmonically as

$$z(t) = z_o \cos \omega t$$

The position of the mass along the rod is $u(t)$, which is measured from the unstretched position of the spring. Use Lagrange's equations to obtain the nonlinear equations of motion.

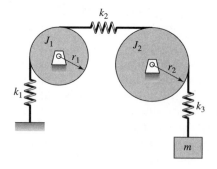

FIGURE E7.47

7.47 A cable fixed at one end and carrying a mass m at the other end is stretched over two pulleys as shown in Figure E7.47. The pulleys have rotary inertia J_1 and J_2 and corresponding radii are r_1 and r_2, respectively. The stiffness of the various sections of the cables are as indicated in the figure. Assume that there is sufficient friction so that the cable does not slip on the pulleys.

a) Determine the equations governing this vibratory system and place them in matrix form.

b) If $m = 12$ kg, $J_1 = 0.2$ kg m^2, $J_2 = 0.3$ kg m^2, $r_1 = 120$ mm, $r_2 = 160$ mm, $k_1 = k_3 = 25 \times 10^3$ N/m, and $k_2 = 40 \times 10^3$ N/m, then determine the natural frequencies and mode shapes.

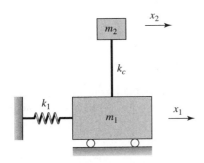

FIGURE E7.48

7.48 An electrical motor and pump system operates at 1800 rpm and is elastically mounted to a support structure. The mass of the system is $m_1 = 25$ kg and the effective viscous damping factor of the mount is 0.15. Unfortunately, it is found that 1800 rpm coincides with the natural frequency ω_{n1} of the system and the horizontal amplitude of the system is excessive. To decrease the magnitude of the horizontal amplitude, it is decided that rather than change the stiffness of the support, a second mass m_2 will be added to the system by attaching $m_2 = 0.25$ kg to the end of a cantilever beam as shown in Figure E7.48. The cantilever beam is a solid circular rod 6 mm in diameter. Use $E = 1.96 \times 10^{11}$ N/m^2 for the Young's modulus of elasticity of the beam. What should be the length of the rod so that the natural frequencies of the modified system are not in the range $\omega_{n1} \pm 15\%$?

7.49 Consider the pulley-cable-mass system shown in Figure E7.49. Use Lagrange's equations to deter-

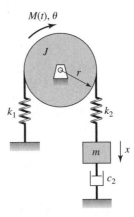

FIGURE E7.49

mine the equation of motion of this system and place these equations in matrix form.

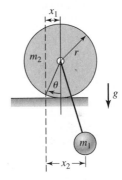

FIGURE E7.50

7.50 For the system shown in Figure E7.50, determine the equations of motion. The length of the pendulum is L.

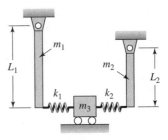

FIGURE E7.51

7.51

a) Use Lagrange's equations to derive the equations of motion of the coupled pendulum system shown in Figure E7.51. Include the effects of gravity. Put the results in matrix form.

b) For $m_1 = 10$ kg, $m_2 = 15$ kg, $m_3 = 5$ kg, $k_1 = k_2 = 100$ N/m, $L_1 = 0.5$ m, and $L_2 = 0.4$ m, determine the natural frequencies and mode shapes of the system.

7.52 Consider the system shown in Figure 7.1 undergoing forced harmonic vibrations of mass m_1. Set the viscous damping to zero and let the three springs be represented by the structural damping model given in Eq. (5.135); that is, $k_m = k_m(1 + j\beta_m)$, $m = 1, 2, 3$. Obtain a matrix expression from which the real and imaginary parts of the displacements of the two inertial elements can be determined.

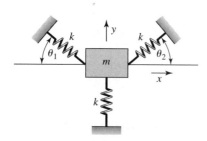

FIGURE E7.53

7.53 Determine the natural frequencies and mode shapes for the system shown in Figure E7.53, when $\theta_1 = 30°$ and $\theta_2 = 45°$ at the equilibrium position. Let L be the length of each spring at the equilibrium position and assume that the deflections in the springs at an angle are small. Gravity acts normal to the plane of the system.

7.54 An elastically supported machine tool with a total mass of 4000 kg has a resonant frequency of 80 Hz. An 800 kg absorber system with a natural frequency of 80 Hz is attached to the machine tool.

Determine the natural frequencies and mode shapes of this system.

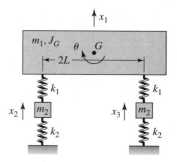

FIGURE E7.55

7.55 One model that has been used to study the vibratory motion of motor vehicles is shown in Figure E7.55. The body of the vehicle has a mass m_1 and a rotary inertia J_G about an axis through the center. The elasticity of the tires is represented by springs k_2 and the elasticity of the suspension by springs k_1. The mass of the tire assemblies is m_2.

a) Determine the matrix form for the governing equations of the system.

b) Obtain the natural frequencies and mode shapes for the case where $m_1 = 800$ kg, $m_2 = 25$ kg, $k_1 = 60$ kN/m, $k_2 = 20$ kN/m, $L = 1.4$ m, and $J_G = 180$ kg·m².

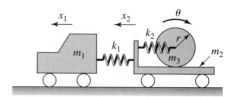

FIGURE E7.56

7.56 A tractor-trailer is hauling a large cylindrical drum that is elastically supported by spring k_2 as shown in Figure E7.56. The drum rolls on the floor of the trailer without slipping. The trailer is attached to the tractor by a system that has an equivalent spring k_1. The mass of the tractor is m_1, that of the trailer is m_2, and that of the drum is m_3.

a) Obtain the equations of motion for this system.

b) If $k_1 = k_2 = k$, $m_1 = m_2 = m$, and $m_3 = 2m/3$, then obtain the natural frequencies in terms of k/m and the corresponding mode shapes.

The transmission of undesired vibrations can be attenuated by various techniques, one of which is based on the principle of the vibration absorber. The attachment of an elastically supported mass creates a tuned mass absorber that reduces vibration levels in systems excited at a constant frequency. An externally energized elastically mounted mass whose motion is controlled in a specific manner is a more sophisticated version of the tuned mass absorber.

8

Multiple Degree-of-Freedom Systems: General Solution for Response and Forced Oscillations

8.1   INTRODUCTION

In the previous chapter, we studied systems with multiple degrees of freedom and discussed the ways in which one can obtain the governing equations of motion. In addition, free-response characteristics of damped and undamped systems were examined. In this chapter, we build on this material, and as laid out in Table 8.1,

385

TABLE 8.1

Application Domains
of Different Solution
Methods for Multiple
Degree-of-Freedom Systems

System	Methods	Typically Used to Determine the Following	Comments	Section
Linear systems	Normal-mode approach	Natural frequencies, mode shapes, and responses of inertial elements to initial conditions and external forcing	For damped systems with given initial conditions and forcing; requires special form of linear viscous damping matrix—analytical and numerical solutions	8.2
	Laplace transform approach	Natural frequencies, transfer functions, and responses of inertial elements to initial conditions and external forcing for general form of damping matrix	Cumbersome for systems with high degrees of freedom—analytical and numerical solutions	8.4
	State-space formulation	Responses of inertial elements to initial conditions and external forcing for general form of damping matrix	Analytical and numerical solutions	8.3
Nonlinear systems	Weakly nonlinear analysis and numerical approach	Responses of inertial elements to initial conditions and external forcing for arbitrary damping models	For "strongly" nonlinear systems, usually, only possible to get a numerical solution	

employ different approaches to obtain the solution for the responses of multi-degree-of-freedom systems.

As with the case of a single degree-of-freedom system, there are different approaches to determine the response of a multi-degree-of-freedom-system. These approaches include the a) normal-mode approach, b) Laplace transform approach, and c) state-space formulation. The first approach is unique to multi-degree-of-freedom-systems, while the second and third approaches are similar in spirit to those used for single degree-of-freedom systems. It is remarked that unlike the other two approaches, the Laplace transform approach is not based in the time domain but, instead, this approach is based in the Laplace domain or the s-domain.

With the objective of providing a summary of the domain of applicability of the above-mentioned approaches, Table 8.1 has been constructed. This summary is organized around a) whether the system is linear or nonlinear and b) the nature of the damping properties of the system.

As presented in Table 8.1, all three approaches can be used to study systems subjected to periodic and other excitations. However, the form of the damping matrix places a restriction on the domain of applicability of the normal-mode approach. As discussed in Chapter 7 and elaborated further in this chapter, the damping matrix needs to have a special structure in order to use the normal-mode approach. However, whenever this approach is applicable, the solution obtained through this approach provides insight into the spatial responses of the inertial elements. This type of information is indispensable for system designs where one is minimizing or maximizing the response to external excitations and where one is determining where to place an actuator and/or a sensor.

The Laplace transform approach is convenient to use since one can obtain transfer functions and frequency-response functions, which can be used to design systems such as vibration absorbers and mechanical filters. In addition, an attractive feature of this approach is that there is no restriction needed on the form of damping. However, for systems with more than two degrees of freedom, the algebra is cumbersome. Since the Laplace transform approach is not based in the time domain, it is limited to linear vibratory systems.

The state-space formulation, where the governing equations are put in first-order form, lends itself to an analytical solution in linear cases and to numerical solutions in both linear and nonlinear cases. This formulation is applicable to systems with all forms of damping and forcing. For the nonlinear systems presented in this book, only numerical solutions are pursued, although methods to determine analytical approximations exist for "weakly" nonlinear systems.[1] Whenever one is seeking a solution of a nonlinear system, it is important to note that the approach is based in the time domain.

After presenting the different approaches to determine the response of a multi-degree-of-freedom system, we present transfer functions and the relationship between the frequency-response function and the transfer function in the same manner as was done for single degree-of-freedom systems. In the latter part of the chapter, we introduce vibration absorbers, the notion of transmissibility ratio, and examine systems subjected to base excitation.

8.2 NORMAL-MODE APPROACH

8.2.1 General Solution

In Section 7.3, we discussed the characteristics associated with free oscillations of multi-degree-of-freedom systems such as natural frequencies and mode shapes without explicitly determining them. In this section, we determine the general solution for the response of a multi-degree-of-freedom system and present explicit forms for the responses due to both forcing and initial conditions. This development follows along the lines of Chapter 4, except that one now needs to consider spatial information apart from time information.

In this section, we determine the response of the system given by Eq. (7.3), which is repeated below after replacing $q_i(t)$ by $x_i(t)$, $\{Q\}$ by $\{F\}$, and dropping the gyroscopic and circulatory force terms.

$$[M]\{\ddot{x}\} + [C]\{\dot{x}\} + [K]\{x\} = \{F\} \tag{8.1}$$

In general, Eq. (8.1) is a coupled system of equations. One way to solve this system is to uncouple them by using an appropriate coordinate transformation.

[1]A. H. Nayfeh and D. T. Mook, *Nonlinear Oscillations,* John Wiley & Sons, NY (1979).

One candidate transformation is the modal matrix $[\Phi]$ given by Eq. (7.31), which has the desirable property of being orthogonal to the matrices $[M]$ and $[K]$. Thus, we assume a solution to Eq. (8.1) of the form

$$\{x(t)\} = \underbrace{[\Phi]}_{\substack{\text{Modal} \\ \text{matrix}}} \underbrace{\{\eta(t)\}}_{\substack{\text{Modal} \\ \text{amplitudes}}} \tag{8.2}$$

where $\{\eta(t)\}$ is a column vector of generalized (modal) coordinates that are to be determined. On substituting Eq. (8.2) into Eq. (8.1), we obtain

$$[M][\Phi]\{\ddot{\eta}\} + [C][\Phi]\{\dot{\eta}\} + [K][\Phi]\{\eta\} = \{F\} \tag{8.3}$$

Pre-multiplying Eq. (8.3) by $[\Phi]^T$ results in

$$[\Phi]^T[M][\Phi]\{\ddot{\eta}\} + [\Phi]^T[C][\Phi]\{\dot{\eta}\} + [\Phi]^T[K][\Phi]\{\eta\} = [\Phi]^T\{F\} \tag{8.4}$$

Upon using Eqs. (7.68), we obtain

$$[M_D]\{\ddot{\eta}\} + [\Phi]^T[C][\Phi]\{\dot{\eta}\} + [K_D]\{\eta\} = [\Phi]^T\{F\} \tag{8.5}$$

We now pre-multiply Eq. (8.5) by $[M_D]^{-1}$, the inverse of $[M_D]$, and use Eq. (7.69) to arrive at

$$\{\ddot{\eta}\} + [M_D]^{-1}[\Phi]^T[C][\Phi]\{\dot{\eta}\} + [\omega_D^2]\{\eta\} = \{Q\} \tag{8.6}$$

where we have assumed that $[M_D]^{-1}$ exists and reintroduced the force vector $\{Q\}$ so that

$$\{Q\} = [M_D]^{-1}[\Phi]^T\{F\} \tag{8.7}$$

Equations (8.6) and (8.7) are written in expanded form as

$$\begin{bmatrix} 1 & 0 & \cdots & 0 \\ 0 & 1 & \cdots & 0 \\ \vdots & \vdots & \ddots & \vdots \\ 0 & 0 & \cdots & 1 \end{bmatrix} \begin{Bmatrix} \ddot{\eta}_1(t) \\ \ddot{\eta}_2(t) \\ \vdots \\ \ddot{\eta}_N(t) \end{Bmatrix} + [M_D]^{-1}[\Phi]^T[C][\Phi] \begin{Bmatrix} \dot{\eta}_1(t) \\ \dot{\eta}_2(t) \\ \vdots \\ \dot{\eta}_N(t) \end{Bmatrix}$$

$$+ \begin{bmatrix} \omega_1^2 & 0 & \cdots & 0 \\ 0 & \omega_2^2 & \cdots & 0 \\ \vdots & \vdots & \ddots & \vdots \\ 0 & 0 & \cdots & \omega_N^2 \end{bmatrix} \begin{Bmatrix} \eta_1(t) \\ \eta_2(t) \\ \vdots \\ \eta_N(t) \end{Bmatrix} = \begin{Bmatrix} Q_1(t) \\ Q_2(t) \\ \vdots \\ Q_N(t) \end{Bmatrix} \tag{8.8}$$

From the form of Eq. (8.8), it is clear that these equations in the transformed coordinates can be uncoupled into N individual second-order differential equations if the matrix

$$[M_D]^{-1}[\Phi]^T[C][\Phi]$$

is a diagonal matrix. As discussed in Section 7.3.3, this is possible when the system is proportionally damped. For this case, we obtain from Eqs. (7.99) and (7.102) that

$$[C_D] = [\Phi]^T [C][\Phi]$$

$$= \begin{bmatrix} 2\zeta_1\omega_1\hat{M}_{11} & 0 & \cdots & 0 \\ 0 & 2\zeta_2\omega_2\hat{M}_{22} & \cdots & 0 \\ \vdots & \vdots & \ddots & \vdots \\ 0 & 0 & \cdots & 2\zeta_N\omega_N\hat{M}_{NN} \end{bmatrix} \tag{8.9}$$

where $\hat{M}_{kk}$ is the modal mass of the kth mode. From Eqs. (7.68a) and (8.9), it follows that

$$[M_D]^{-1}[C_D] = \begin{bmatrix} 1/\hat{M}_{11} & 0 & \cdots & 0 \\ 0 & 1/\hat{M}_{22} & \cdots & 0 \\ \vdots & \vdots & \ddots & \vdots \\ 0 & 0 & \cdots & 1/\hat{M}_{NN} \end{bmatrix}$$

$$\times \begin{bmatrix} 2\zeta_1\omega_1\hat{M}_{11} & 0 & \cdots & 0 \\ 0 & 2\zeta_2\omega_2\hat{M}_{22} & \cdots & 0 \\ \vdots & \vdots & \ddots & \vdots \\ 0 & 0 & \cdots & 2\zeta_N\omega_N\hat{M}_{NN} \end{bmatrix}$$

$$= \begin{bmatrix} 2\zeta_1\omega_1 & 0 & \cdots & 0 \\ 0 & 2\zeta_2\omega_2 & \cdots & 0 \\ \vdots & \vdots & \ddots & \vdots \\ 0 & 0 & \cdots & 2\zeta_N\omega_N \end{bmatrix} = [(2\zeta\omega)_D] \tag{8.10}$$

Hence, for a system with proportional damping, from Eqs. (8.6) and (8.10), we arrive at the uncoupled set of equations

$$\{\ddot{\eta}\} + [(2\zeta\omega)_D]\{\dot{\eta}\} + [\omega_D^2]\{\eta\} = \{Q\} \tag{8.11}$$

in which the jth equation has the form

$$\ddot{\eta}_j(t) + 2\zeta_j\omega_j\dot{\eta}_j(t) + \omega_j^2\eta_j(t) = Q_j(t) \quad j = 1, 2, \ldots, N \tag{8.12}$$

In Eqs. (8.12), ζ_j is the damping factor associated with the jth mode and ω_j is the natural frequency associated with the jth mode. As discussed in Section 7.3.3, since the damping matrix in Eq. (8.9) is diagonal, the system is said to have *modal damping*. The coordinates η_j, in which the governing equations of motion are uncoupled, are referred to as the *principal coordinates* or *modal coordinates*.

The equation describing the oscillation in the jth mode has the form of the governing equation of a single degree-of-freedom system, namely, Eq. (4.1). Hence the solution for the response of the jth mode is given by Eqs. (4.8) and (4.9) if $0 \le \zeta_j < 1$. Thus,

$$\eta_j(t) = \underbrace{A_j e^{-\zeta_j\omega_j t} \sin(\omega_{dj}t + \varphi_{dj})}_{\text{Response to initial conditions}} + \underbrace{\frac{1}{\omega_{dj}} \int_0^t e^{-\zeta_j\omega_j(t-\xi)} \sin(\omega_{dj}[t - \xi])Q_j(\xi)d\xi}_{\text{Response to forcing}}$$

$$\tag{8.13}$$

where the constants A_j and φ_{dj} are determined by the initial conditions,

$$A_j = \sqrt{\eta_j^2(0) + \left(\frac{\dot{\eta}_j(0) + \zeta_j\omega_j\eta_j(0)}{\omega_{dj}}\right)^2}$$

$$\varphi_{dj} = \tan^{-1}\frac{\omega_{dj}\eta_j(0)}{\dot{\eta}_j(0) + \zeta_j\omega_j\eta_j(0)} \tag{8.14a}$$

and the damped natural frequency ω_{dj} of the jth mode is given by

$$\omega_{dj} = \omega_j\sqrt{1 - \zeta_j^2} \tag{8.14b}$$

From Eqs. (7.94), the modal damping factor ζ_j is

$$\zeta_j = \frac{1}{2}\left(\frac{\alpha}{\omega_j} + \beta\omega_j\right) \tag{8.15}$$

where α and β are such that

$$[C] = \alpha[M] + \beta[K]$$

In Section 8.2.2, we show how to obtain the initial modal displacements $\eta_j(0)$ and initial modal velocities $\dot{\eta}_j(0)$ in terms of the system displacements $x_j(0)$ and velocities $\dot{x}_j(0)$.

Having determined the responses of the individual modes—that is, $\eta_j(t)$—we return to Eq. (8.2) and find that the system response can be constructed by making use of Eq. (7.31) as

$$\begin{Bmatrix} x_1(t) \\ x_2(t) \\ \vdots \\ x_N(t) \end{Bmatrix} = [\Phi]\{\eta(t)\} = \underbrace{\begin{bmatrix} X_{11} & X_{12} & \cdots & X_{1N} \\ X_{21} & X_{22} & \cdots & X_{2N} \\ \vdots & \vdots & \ddots & \vdots \\ X_{N1} & X_{N2} & \cdots & X_{NN} \end{bmatrix}}_{\text{Modal matrix}} \underbrace{\begin{Bmatrix} \eta_1(t) \\ \eta_2(t) \\ \vdots \\ \eta_N(t) \end{Bmatrix}}_{\substack{\text{Modal} \\ \text{Amplitudes}}} \tag{8.16}$$

where

$$x_i(t) = \sum_{j=1}^{N} X_{ij}\eta_j(t)$$

$$= \underbrace{X_{i1}\eta_1(t)}_{\substack{\text{Component of response} \\ \text{in the first mode}}} + \underbrace{X_{i2}\eta_2(t)}_{\substack{\text{Component of response} \\ \text{in the second mode}}} + \cdots + \underbrace{X_{iN}\eta_N(t)}_{\substack{\text{Component of response} \\ \text{in the Nth mode}}}$$

$$i = 1, 2, \ldots, N \tag{8.17}$$

Equation (8.16) is rewritten in the compact form as

$$\{x(t)\} = \sum_{j=1}^{N} \underbrace{\{X\}_j\eta_j(t)}_{\substack{\text{Oscillation in the} \\ \text{jth mode}}} \tag{8.18}$$

We see from Eq. (8.18) that the displacement of each mass $x_i(t)$ is composed of response components in the different modes of the vibratory system. Thus, the response of a multi-degree-of-freedom system is a weighted combination of the individual modes of the system. The modal amplitude provides the weighting for each mode.

Response of Damped System

On substituting Eq. (8.13) into Eq. (8.18), we see that the response of the linear multi-degree-of-freedom system given by Eq. (8.1) takes the form

$$\{x(t)\} = \sum_{j=1}^{N} \{X\}_j \left[A_j e^{-\zeta_j \omega_j t} \sin(\omega_{dj} t + \varphi_{dj}) \right.$$

$$\left. + \frac{1}{\omega_{dj}} \int_0^t e^{-\zeta_j \omega_j (t-\xi)} \sin(\omega_{dj}[t - \xi]) Q_j(\xi) d\xi \right] \quad (8.19)$$

where

$$x_i(t) = \sum_{j=1}^{N} X_{ij} \left[A_j e^{-\zeta_j \omega_j t} \sin(\omega_{dj} t + \varphi_{dj}) \right.$$

$$\left. + \frac{1}{\omega_{dj}} \int_0^t e^{-\zeta_j \omega_j (t-\xi)} \sin(\omega_{dj}[t - \xi]) Q_j(\xi) d\xi \right]$$

Response of Undamped System

When the system is undamped, we set $\zeta_j = 0$ in Eq. (8.19), note that $\omega_{dj} = \omega_j$ from Eq. (8.14b), and obtain

$$\{x(t)\} = \sum_{j=1}^{N} \{X\}_j \left[A_j \sin(\omega_j t + \varphi_j) + \frac{1}{\omega_j} \int_0^t \sin(\omega_j[t - \xi]) Q_j(\xi) d\xi \right] \quad (8.20)$$

where from Eqs. (8.14a), we find that

$$A_j = \sqrt{\eta_j^2(0) + \left(\frac{\dot{\eta}_j(0)}{\omega_j} \right)^2}$$

$$\varphi_j = \tan^{-1} \frac{\omega_j \eta_j(0)}{\dot{\eta}_j(0)} \quad (8.21)$$

8.2.2 Response to Initial Conditions

When the forcing is zero—that is, $\{F\} = \{0\}$ in Eq. (8.1)—then the modal forced vector $\{Q\} = \{0\}$ in Eq. (8.7), and the response of the system follows from Eq. (8.19) as

$$\{x(t)\} = \sum_{j=1}^{N} \underbrace{\{X\}_j A_j e^{-\zeta_j \omega_j t} \sin(\omega_{dj} t + \varphi_{dj})}_{\text{Free oscillation component in the } j\text{th mode}} \quad (8.22)$$

where A_j and φ_{dj} are determined from Eqs. (8.14a) in terms of the initial displacement and the initial velocity. For $0 \leq \zeta_j < 1$, Eq. (8.22) consists of a sum of damped sinusoids associated with the individual modes. The damped natural frequency ω_{dj} of the jth mode is given by Eq. (8.14b).

For the determination of φ_{dj} in Eqs. (8.14a), the proper quadrant as determined by the sign of the numerator and the denominator must be used. For example, when $\dot{\eta}_j(0) = 0$, we see that the quadrant is determined by the sign of $\eta_j(0)$; when $\eta_j(0) > 0$, φ_{dj} is in the first quadrant and when $\eta_j(0) < 0$, φ_{dj} is in the third quadrant.

The initial conditions and $\dot{\eta}_j(0)$ and $\eta_j(0)$ can now be determined from Eq. (8.2) in the following manner. We note from Eq. (8.18) that

$$\{x(0)\} = \sum_{j=1}^{N} \{X\}_j \eta_j(0)$$

$$\{\dot{x}(0)\} = \sum_{j=1}^{N} \{X\}_j \dot{\eta}_j(0) \qquad (8.23)$$

If we pre-multiply Eqs. (8.23) by $\{X\}_i^T[M]$; that is,

$$\{X\}_i^T[M]\{x(0)\} = \sum_{j=1}^{N} \{X\}_i^T[M]\{X\}_j \eta_j(0)$$

$$\{X\}_i^T[M]\{\dot{x}(0)\} = \sum_{j=1}^{N} \{X\}_i^T[M]\{X\}_j \dot{\eta}_j(0) \qquad (8.24)$$

and make use of the orthogonality of the modes given by Eq. (7.62) and the modal mass given by Eqs. (7.67), we obtain

$$\eta_j(0) = \{X\}_j^T[M]\{x(0)\}/\hat{M}_{jj}$$
$$\dot{\eta}_j(0) = \{X\}_j^T[M]\{\dot{x}(0)\}/\hat{M}_{jj} \qquad (8.25)$$

If the modes are orthonormal, then $\hat{M}_{jj} = 1$.

Free Oscillations for a Special Case

Now we consider the special case where the initial velocity $\{\dot{x}(0)\} = \{0\}$, and the system is provided an initial displacement in its nth mode shape; that is,

$$\{x(0)\} = a_o\{X\}_n \qquad (8.26)$$

From Eqs. (7.62) and (7.67), we find that

$$\frac{1}{\hat{M}_{jj}}\{X\}_j^T[M]\{X\}_n = \delta_{nj} \qquad (8.27)$$

where the *Kronecker delta function* $\delta_{nj} = 0$, $n \neq j$ and $\delta_{nj} = 1$, $n = j$. Then, from Eqs. (8.25), we find that the initial modal displacement and initial modal velocity are given by, respectively,

$$\eta_j(0) = a_o\delta_{nj}$$
$$\dot{\eta}_j(0) = 0 \qquad (8.28)$$

Therefore, the only mode for which the associated initial condition is not equal to zero is the one corresponding to the nth mode shape, and it follows that the only mode that has a non-zero response is $\eta_n(t)$. Then, from Eq. (8.18)

$$\{x(t)\} = \{X\}_n \, \eta_n(t) \tag{8.29}$$

where η_n is found from Eq. (8.13) as

$$\eta_n(t) = A_n e^{-\zeta_n \omega_n t} \sin(\omega_{dn} t + \varphi_{dn}) \tag{8.30}$$

and from Eqs. (8.14a), (8.14b), and (8.28) we obtain

$$A_n = \sqrt{\eta_n^2(0) + \left(\frac{\zeta_n \omega_n \eta_n(0)}{\omega_{dn}}\right)^2} = \frac{\eta_n(0)}{\sqrt{1 - \zeta_n^2}} = \frac{a_o}{\sqrt{1 - \zeta_n^2}}$$

$$\varphi_{dn} = \tan^{-1} \frac{\omega_{dn} \eta_n(0)}{\zeta_n \omega_n \eta_n(0)} = \tan^{-1} \frac{\sqrt{1 - \zeta_n^2}}{\zeta_n} \tag{8.31}$$

In other words, the system only oscillates in its nth mode shape; none of the other modes is excited.

EXAMPLE 8.1 Undamped free oscillations of a two degree-of-freedom system

We return to the system shown in Figure 7.1 and consider the case when the forcing and the damping are absent. We shall find the response of the system when the initial conditions are as follows

$$\{x(0)\} = \begin{Bmatrix} x_{10} \\ x_{20} \end{Bmatrix} \quad \text{and} \quad \{\dot{x}(0)\} = \begin{Bmatrix} \dot{x}_{10} \\ \dot{x}_{20} \end{Bmatrix} \tag{a}$$

The response in this case is determined by making use of the general solution given by Eqs. (8.20) and (8.21) for the undamped system; that is,

$$\{x(t)\} = \sum_{j=1}^{2} \{X\}_j A_j \sin(\omega_j t + \varphi_{dj}) \tag{b}$$

where

$$A_j = \sqrt{\eta_j^2(0) + \frac{\dot{\eta}_j^2(0)}{\omega_j^2}}$$

$$\varphi_{dj} = \tan^{-1} \frac{\omega_j \eta_j(0)}{\dot{\eta}_j(0)} \tag{c}$$

The initial conditions from the modes $\eta_j(0)$ and $\dot{\eta}(0)$ are determined from Eqs. (8.25) as

$$\eta_j(0) = \frac{1}{\hat{M}_{jj}} \{X\}_j^T [M] \begin{Bmatrix} x_{10} \\ x_{20} \end{Bmatrix}$$

$$\dot{\eta}_j(0) = \frac{1}{\hat{M}_{jj}} \{X\}_j^T [M] \begin{Bmatrix} \dot{x}_{10} \\ \dot{x}_{20} \end{Bmatrix} \tag{d}$$

Thus, from Eq. (b) the undamped oscillations of the system are described as the weighted sum of the free oscillations of the first and second modes of the system.

EXAMPLE 8.2 Damped and undamped free oscillations of a two degree-of-freedom system

We now continue with Examples 7.12 and 7.16 and assume that damping is present in the form of Eq. (7.104) with $\zeta = 0.05$. The damped and undamped free responses of the system are determined for the initial conditions

$$\begin{Bmatrix} x_{10} \\ x_{20} \end{Bmatrix} = \begin{Bmatrix} 0.1 \\ -0.1 \end{Bmatrix} \text{ m} \quad \text{and} \quad \begin{Bmatrix} \dot{x}_{10} \\ \dot{x}_{20} \end{Bmatrix} = \begin{Bmatrix} 0 \\ 0 \end{Bmatrix} \text{ m/s} \tag{a}$$

The response of the system has the form of Eq. (8.20) for the undamped case and Eq. (8.19) for the damped case. Based on the information in Examples 7.6 and 7.10 and from Eqs. (a) and (d) of Example 8.1, we find that the initial modal velocities $\dot{\eta}_j(0) = 0$ and the initial modal displacements $\eta_j(t)$ are

$$\eta_1(0) = \{X\}_1^T[M]\{x(0)\}/\hat{M}_{11} = \frac{1}{3.658}\left(\{0.893 \quad 1\}\begin{bmatrix} 1.2 & 0 \\ 0 & 2.7 \end{bmatrix}\begin{Bmatrix} 0.1 \\ -0.1 \end{Bmatrix}\right)$$

$$= -0.045$$

$$\eta_2(0) = \{X\}_2^T[M]\{x(0)\}/\hat{M}_{22} = \frac{1}{10.311}\left(\{-2.518 \quad 1\}\begin{bmatrix} 1.2 & 0 \\ 0 & 2.7 \end{bmatrix}\begin{Bmatrix} 0.1 \\ -0.1 \end{Bmatrix}\right)$$

$$= -0.056 \tag{b}$$

where from Eq. (b) of Example 7.12 we determined that

$$[\Phi] = \begin{bmatrix} 0.893 & -2.518 \\ 1 & 1 \end{bmatrix}$$

Response for Damped Case

For $\zeta = 0.05$ and the previously determined natural frequencies, we obtain

$$\zeta\omega_1 = 0.05 \times 2.519 = 0.126$$
$$\zeta\omega_2 = 0.05 \times 5.623 = 0.281 \tag{c}$$

From Eqs. (8.14), we compute the damped natural frequencies ω_{dj} and the phases φ_{dj} as

$$\omega_{d1} = \omega_1\sqrt{1 - \zeta^2} = 2.519\sqrt{1 - 0.05^2} = 2.516$$
$$\omega_{d2} = \omega_2\sqrt{1 - \zeta^2} = 5.623\sqrt{1 - 0.05^2} = 5.616$$
$$\varphi_{d1} = \tan^{-1}\frac{\omega_{d1}\eta_1(0)}{\zeta\omega_1\eta_1(0)} = \tan^{-1}\frac{\eta_1(0)\sqrt{1 - \zeta^2}}{\zeta\eta_1(0)}$$
$$= \tan^{-1}\frac{-\sqrt{1 - 0.05^2}}{-0.05} = -1.621$$
$$\varphi_{d2} = \tan^{-1}\frac{\omega_{d2}\eta_2(0)}{\zeta\omega_2\eta_2(0)} = \tan^{-1}\frac{\eta_2(0)\sqrt{1 - \zeta^2}}{\zeta\eta_2(0)}$$
$$= \tan^{-1}\frac{-\sqrt{1 - 0.05^2}}{-0.05} = -1.621 \tag{d}$$

and the amplitudes A_j as

$$A_1 = \sqrt{\eta_1^2(0) + \left(\frac{\zeta\omega_1\eta_1(0)}{\omega_{d1}}\right)^2} = \sqrt{(-0.045)^2 + \left(\frac{-0.045 \times 0.126}{2.516}\right)^2}$$

$$= 0.045$$

$$A_2 = \sqrt{\eta_2^2(0) + \left(\frac{\zeta\omega_2\eta_2(0)}{\omega_{d2}}\right)^2} = \sqrt{(-0.056)^2 + \left(\frac{-0.056 \times 0.281}{5.616}\right)^2}$$

$$= 0.056 \tag{e}$$

Then, from Eq. (8.13), in the absence of forcing (i.e., $Q_j = 0$), we obtain the modal responses and compute them based on Eqs. (c), (d), and (e), as follows.

$$\eta_1(t) = A_1 e^{-\zeta\omega_1 t} \sin(\omega_{d1}t + \varphi_{d1}) = 0.045 e^{-0.126t} \sin(2.516t - 1.621)$$

$$\eta_2(t) = A_2 e^{-\zeta\omega_2 t} \sin(\omega_{d2}t + \varphi_{d2}) = 0.056 e^{-0.281t} \sin(5.616t - 1.621) \tag{f}$$

The free response of the damped system is determined from Eq. (8.22) and Eqs. (f). Thus,

$$\begin{aligned} x_1(t) &= \eta_1(t)X_{11} + \eta_2(t)X_{12} \\ &= A_1 X_{11} e^{-\zeta\omega_1 t} \sin(\omega_{d1}t + \varphi_{d1}) + A_2 X_{12} e^{-\zeta\omega_2 t} \sin(\omega_{d2}t + \varphi_{d2}) \\ &= 0.045 \times 0.893 \times e^{-0.126t} \sin(2.516t - 1.621) \\ &\quad - 0.056 \times 2.518 \times e^{-0.281t} \sin(5.616t - 1.621) \\ &= 0.040 e^{-0.126t} \sin(2.516t - 1.621) \\ &\quad - 0.140 e^{-0.281t} \sin(5.616t - 1.621) \quad \text{m} \end{aligned} \tag{g}$$

and

$$\begin{aligned} x_2(t) &= \eta_1(t)X_{21} + \eta_2(t)X_{22} \\ &= A_1 X_{21} e^{-\zeta\omega_1 t} \sin(\omega_{d1}t + \varphi_{d1}) + A_2 X_{22} e^{-\zeta\omega_2 t} \sin(\omega_{d2}t + \varphi_{d2}) \\ &= 0.045 \times 1 \times e^{-0.126t} \sin(2.516t - 1.621) \\ &\quad - 0.056 \times 1 \times e^{-0.281t} \sin(5.616t - 1.621) \\ &= 0.045 e^{-0.126t} \sin(2.516t - 1.621) \\ &\quad - 0.056 e^{-0.281t} \sin(5.616t - 1.621) \quad \text{m} \end{aligned} \tag{h}$$

Response for Undamped Case

The results for the undamped case are computed as follows. From Eqs. (b) and (e), with
$\zeta = 0$, we have

$$A_1 = 0.045$$

$$A_2 = 0.056 \tag{i}$$

and from Eqs. (d) we obtain

$$\omega_{d1} = \omega_1 = 2.519$$

$$\omega_{d2} = \omega_2 = 5.623$$

$$\varphi_{d1} = -\pi/2$$

$$\varphi_{d2} = -\pi/2 \tag{j}$$

FIGURE 8.1
Displacements of a two
degree-of-freedom system with
$\zeta = 0.05$ when both masses
are subjected to equal, but
opposite, initial displacements. [Solid line $x_1(t)$;
dashed line $x_2(t)$.]

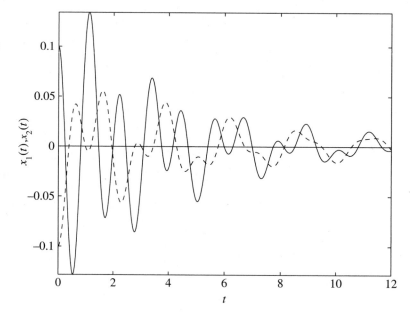

Then, from Eq. (8.22) with $\zeta_j = 0$ and Eqs. (i) and (j), we obtain

$$x_1(t) = A_1 X_{11} \sin(\omega_1 t - \pi/2) + A_2 X_{12} \sin(\omega_2 t - \pi/2)$$
$$= -0.045 \times 0.893 \cos(2.516t) + 0.056 \times 2.518 \cos(5.623t)$$
$$= -0.040 \cos(2.516t) + 0.140 \cos(5.623t) \quad \text{m} \tag{k}$$

and

FIGURE 8.2
Displacements of a two
degree-of-freedom system with
$\zeta = 0$ when both masses are
subjected to equal, but
opposite, initial displacements. [Solid line $x_1(t)$;
dashed line $x_2(t)$.]

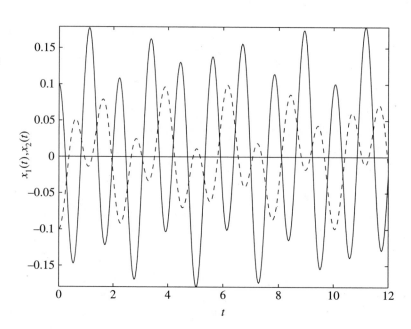

$$x_2(t) = A_1 X_{21} \sin(\omega_1 t - \pi/2) + A_2 X_{22} \sin(\omega_2 t - \pi/2)$$
$$= -0.045 \times 1 \times \cos(2.516t) - 0.056 \times 1 \times \cos(5.623t)$$
$$= -0.045 \cos(2.516t) - 0.056 \cos(5.623t) \quad \text{m} \qquad (1)$$

In the damped case, the response given by Eqs. (g) and (h) consists of exponentially decaying sinusoidal oscillations in each mode. In the undamped case, the response given by Eqs. (k) and (l) consists of harmonic oscillations in each mode.

The time histories given by Eqs. (g) and (h) for the damped case are plotted in Figure 8.1, and the time histories given by Eqs. (k) and (l) for the undamped case are plotted in Figure 8.2. Comparing Figures 8.1 and 8.2, we see that in the damped case, the displacements $x_1(t)$ and $x_2(t)$ approach zero as $t \to \infty$; that is, the system eventually settles down to the equilibrium position. This is not true in the undamped case.

8.2.3 Response to Harmonic Forcing and the Frequency-Response Function

In this section, we determine the response of the physical systems described by Eq. (8.1) to harmonic forcing. First, we consider the response of undamped systems and illustrate how the frequency-response function can be constructed from this response. The notion of resonance in a multi-degree-of-freedom system is also discussed. A direct approach that can be used to obtain the forced response of a multi-degree-of-freedom system is presented and it is followed by a presentation of the normal-mode approach. Proportionally damped systems are treated at the end of the chapter.

Undamped Systems: Direct Approach

For illustration, we consider the two degree-of-freedom system shown in Figure 7.1 without damping; that is, $c_1 = c_2 = c_3 = 0$ in Eq. (7.1b), and let the forcing be of the form

$$f_1(t) = F_1 \cos \omega t$$
$$f_2(t) = F_2 \cos \omega t \qquad (8.32)$$

where F_j are constants. Then, the governing equations of motion given by Eq. (7.1b) reduce to

$$\begin{bmatrix} m_1 & 0 \\ 0 & m_2 \end{bmatrix} \begin{Bmatrix} \ddot{x}_1 \\ \ddot{x}_2 \end{Bmatrix} + \begin{bmatrix} k_1 + k_2 & -k_2 \\ -k_2 & k_2 + k_3 \end{bmatrix} \begin{Bmatrix} x_1 \\ x_2 \end{Bmatrix} = \begin{Bmatrix} F_1 \\ F_2 \end{Bmatrix} \cos \omega t \qquad (8.33)$$

To determine a response of the two degree-of-freedom system to the assumed forcing, a solution of the form

$$\begin{Bmatrix} x_1(t) \\ x_2(t) \end{Bmatrix} = \begin{Bmatrix} X_1 \\ X_2 \end{Bmatrix} \cos \omega t \qquad (8.34)$$

is assumed where X_1 and X_2 are the unknowns to be determined. Equation (8.34) means that the masses m_1 and m_2 in Figure 7.1 are assumed to respond at the same frequency as the excitation frequency ω. Furthermore, this solution form is consistent with the discussion presented in Section 5.2 where harmonically forced single degree-of-freedom systems were considered. On substituting Eq. (8.34) into Eq. (8.33) and canceling the common factor cos ωt on both sides of the equation, we arrive at

$$\begin{bmatrix} k_1 + k_2 - \omega^2 m_1 & -k_2 \\ -k_2 & k_2 + k_3 - \omega^2 m_2 \end{bmatrix} \begin{Bmatrix} X_1 \\ X_2 \end{Bmatrix} = \begin{Bmatrix} F_1 \\ F_2 \end{Bmatrix} \tag{8.35}$$

After using the nondimensional quantities given by Eqs. (7.41), Eq. (8.35) is recast into the form

$$[1 + \omega_r^2 m_r - \Omega^2]X_1 - m_r\omega_r^2 X_2 = F_1/k_1$$
$$-\omega_r^2 X_1 + [\omega_r^2(1 + k_{32}) - \Omega^2]X_2 = \omega_r^2 F_2/k_2 \tag{8.36}$$

Solving for X_j from Eqs. (8.36), we obtain

$$X_1 = \frac{1}{D_o}[(F_1/k_1)[\omega_r^2(1 + k_{32}) - \Omega^2] + (F_2/k_2)m_r\omega_r^4]$$

$$X_2 = \frac{\omega_r^2}{D_o}[(F_1/k_1) + (F_2/k_2)[1 + \omega_r^2 m_r - \Omega^2]] \tag{8.37}$$

where the term in the denominator has the form

$$D_o = \Omega^4 - a_1\Omega^2 + a_2 \tag{8.38}$$

and a_1 and a_2 are given by Eqs. (7.46). When $D_o = 0$, we have the characteristic equation given by Eq. (7.45).

From Eqs. (8.37), it is evident that whenever the excitation frequency ω is equal to a natural frequency of the system, that is, when $\omega = \omega_1$ or $\omega = \omega_2$, the displacement responses X_j become infinite; hence, each of these frequency relationships between the excitation frequency and a system natural frequency is called a *resonance relation* and at these excitation frequencies, the system is said to be in *resonance*. In Chapter 5, it was seen that for a linear single degree-of-freedom system there is one excitation frequency at which we have a resonance. Since a linear two degree-of-freedom system has two natural frequencies, there are two excitation frequencies at which we have a resonance. By extension, since a linear system with N degrees of freedom has N natural frequencies, there are N excitation frequencies at which we can have a resonance. It is also noted from the stability discussion in Section 7.4 that an undamped multi-degree-of-freedom system is unbounded when excited at one of its resonances.

In a manner similar to that for a single degree-of-freedom system, we construct the system frequency-response functions. To this end, we set $F_2 = 0$ in Eqs. (8.36) and obtain

$$\frac{X_1}{F_1/k_1} = H_{11}(\Omega) = \frac{\omega_r^2(1 + k_{32}) - \Omega^2}{D_o}$$

$$\frac{X_2}{F_1/k_1} = H_{21}(\Omega) = \frac{\omega_r^2}{D_o} \tag{8.39}$$

Similarly, setting $F_1 = 0$ in Eqs. (8.36), we find that

$$\frac{X_1}{F_2/k_2} = H_{12}(\Omega) = \frac{m_r\omega_r^4}{D_o}$$

$$\frac{X_2}{F_2/k_2} = H_{22}(\Omega) = \frac{\omega_r^2(1 + \omega_r^2 m_r - \Omega^2)}{D_o} \qquad (8.40)$$

In Eqs. (8.39) and (8.40), the frequency-response functions $H_{ij}(\Omega)$ are nondimensional, the subscript i is associated with the response of inertial element m_i and the subscript j is associated with the force input to the inertial element m_j. Transfer functions and associated frequency-response functions are obtained for more general cases in Section 8.5.

From Eqs. (8.39), it is clear that for a given excitation frequency one can choose the ratio ω_r and k_{32} so that

$$\Omega^2 = \omega_r^2(1 + k_{32}) \qquad (8.41)$$

For this choice of system parameters, then, the displacement of the first inertial element is zero at the excitation frequency that satisfies Eq. (8.41). This observation is used to design an undamped vibration absorber, which is the subject of Example 8.3. In general, the excitation frequencies at which the response amplitude of an inertial element is zero are referred to as the frequencies at which the *zeros of the forced response* occur for that inertial element.

The frequency-response functions given by Eqs. (8.39) are plotted as a function of the nondimensional excitation frequency Ω for the following parameter values: $m_r = 1$, $\omega_r = 1$, and $k_{32} = 1$. In this case, $a_1 = 4$ and $a_2 = 3$ and from Eqs. (7.41), (8.38), and (8.39), the frequency-response functions have the following forms:

$$H_{11}(\Omega) = \frac{\omega_r^2(1 + k_{32}) - \Omega^2}{\Omega^4 - a_1\Omega^2 + a_2} = \frac{2 - \Omega^2}{\Omega^4 - 4\Omega^2 + 3}$$

$$H_{21}(\Omega) = \frac{\omega_r^2}{\Omega^4 - a_1\Omega^2 + a_2} = \frac{1}{\Omega^4 - 4\Omega^2 + 3} \qquad (8.42)$$

Graphs of $H_{11}(\Omega)$ and $H_{21}(\Omega)$ are plotted as functions of Ω in Figure 8.3. From Eqs. (7.47), it is found that $\Omega = 1$ and $\Omega = \sqrt{3}$. Thus, it is seen that $H_{11}(\Omega)$ and $H_{21}(\Omega)$ become infinite-valued for $\Omega = 1$ and $\Omega = \sqrt{3}$. This is reflected in Figure 8.3. Furthermore, from this figure it is seen that for $\Omega < 1$, the ratio of $H_{11}(\Omega)/H_{21}(\Omega)$ is positive and that for $\Omega > \sqrt{3}$, this ratio is negative. This is explained in terms of participation of first and second modes, as discussed in Example 8.5. On examining Figure 8.3, it is clear that there is a sign change for $H_{ik}(\Omega)$ whenever a resonance location is crossed. This sign change means that the response phase goes for $0°$ to $180°$ or from $180°$ to $0°$ as we go from a frequency on one side of a resonance to a frequency on the other side of the resonance.

FIGURE 8.3

Frequency-response functions of an undamped two degree-of-freedom system when harmonic forcing is applied to mass m_1. [Solid lines: $H_{11}(\Omega)$; dashed lines: $H_{21}(\Omega)$.]

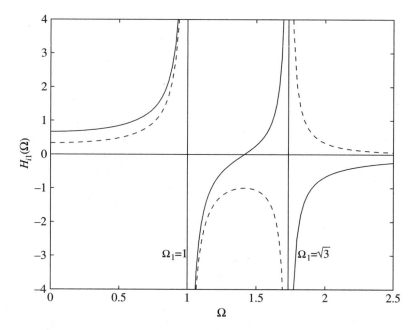

Undamped Systems: Normal-Mode Approach

In this case, the system of equations given by Eq. (8.33) is uncoupled by making use of the transformation given by Eq. (8.2); that is,

$$\begin{Bmatrix} x_1(t) \\ x_2(t) \end{Bmatrix} = [\Phi] \begin{Bmatrix} \eta_1(t) \\ \eta_2(t) \end{Bmatrix} = \begin{bmatrix} X_{11} & X_{12} \\ X_{21} & X_{22} \end{bmatrix} \begin{Bmatrix} \eta_1(t) \\ \eta_2(t) \end{Bmatrix} \tag{8.43}$$

where the modal matrix $[\Phi]$ is first determined from the unforced problem given by Eq. (7.42) and the unknowns to be determined are the modal amplitudes $\eta_1(t)$ and $\eta_2(t)$. On substituting Eq. (8.43) into Eq. (8.33) and following the steps used for obtaining Eqs. (8.3) to (8.8), we obtain the following system of equations:

$$\begin{Bmatrix} \ddot{\eta}_1(t) \\ \ddot{\eta}_2(t) \end{Bmatrix} + \begin{bmatrix} \omega_1^2 & 0 \\ 0 & \omega_2^2 \end{bmatrix} \begin{Bmatrix} \eta_1(t) \\ \eta_2(t) \end{Bmatrix} = \begin{bmatrix} 1/\hat{M}_{11} & 0 \\ 0 & 1/\hat{M}_{22} \end{bmatrix} \begin{bmatrix} X_{11} & X_{21} \\ X_{12} & X_{22} \end{bmatrix} \begin{Bmatrix} F_1 \\ F_2 \end{Bmatrix} \cos \omega t$$

$$= \begin{Bmatrix} (X_{11}F_1 + X_{21}F_2)/\hat{M}_{11} \\ (X_{12}F_1 + X_{22}F_2)/\hat{M}_{22} \end{Bmatrix} \cos \omega t \tag{8.44}$$

In Eq. (8.44), the system natural frequencies ω_1 and ω_2 are determined from the eigenvalue problem associated with free oscillations. The modal masses $\hat{M}_{kk}$ are determined from Eqs. (7.67) to be

$$\hat{M}_{11} = \{ X_{11} \quad X_{21} \} \begin{bmatrix} m_1 & 0 \\ 0 & m_2 \end{bmatrix} \begin{Bmatrix} X_{11} \\ X_{21} \end{Bmatrix} = m_1 X_{11}^2 + m_2 X_{21}^2$$

$$\hat{M}_{22} = \{ X_{12} \quad X_{22} \} \begin{bmatrix} m_1 & 0 \\ 0 & m_2 \end{bmatrix} \begin{Bmatrix} X_{12} \\ X_{22} \end{Bmatrix} = m_1 X_{12}^2 + m_2 X_{22}^2 \tag{8.45}$$

From Eqs. (8.44) and (8.45), it is seen that the system given by Eq. (8.33) has been transformed into two uncoupled single degree-of-freedom systems, each of which is treated as in Chapter 5. After solving for the modal amplitudes $\eta_1(t)$ and $\eta_2(t)$, one can use Eq. (8.2) to determine the system responses.

Proportionally Damped Systems: Normal-Mode Approach

In this case, the governing equations of motion given by Eq. (7.1b) and for the forcing described by Eqs. (8.32) take the form

$$
\begin{bmatrix} m_1 & 0 \\ 0 & m_2 \end{bmatrix} \begin{Bmatrix} \ddot{x}_1 \\ \ddot{x}_2 \end{Bmatrix} + \begin{bmatrix} c_1 + c_2 & -c_2 \\ -c_2 & c_2 + c_3 \end{bmatrix} \begin{Bmatrix} \dot{x}_1 \\ \dot{x}_2 \end{Bmatrix} + \begin{bmatrix} k_1 + k_2 & -k_2 \\ -k_2 & k_2 + k_3 \end{bmatrix} \begin{Bmatrix} x_1 \\ x_2 \end{Bmatrix}
$$

$$
= \begin{Bmatrix} F_1 \\ F_2 \end{Bmatrix} \cos \omega t \tag{8.46}
$$

where the damping matrix $[C]$ is assumed to have the form of proportional damping as given by Eq. (7.83); that is,

$$
\begin{bmatrix} c_1 + c_2 & -c_2 \\ -c_2 & c_2 + c_3 \end{bmatrix} = \alpha \begin{bmatrix} m_1 & 0 \\ 0 & m_2 \end{bmatrix} + \beta \begin{bmatrix} k_1 + k_2 & -k_2 \\ -k_2 & k_2 + k_3 \end{bmatrix} \tag{8.47}
$$

where α and β are real-valued quantities. Equation (8.47) means that the damping coefficients need to satisfy the relationships

$$
\begin{aligned}
c_1 + c_2 &= \alpha m_1 + \beta(k_1 + k_2) \\
c_2 &= \beta k_2 \\
c_2 + c_3 &= \alpha m_2 + \beta(k_2 + k_3)
\end{aligned} \tag{8.48a}
$$

from which we find that

$$
\begin{aligned}
c_1 &= \alpha m_1 + \beta k_1 \\
c_2 &= \beta k_2 \\
c_3 &= \alpha m_2 + \beta k_3
\end{aligned} \tag{8.48b}
$$

Recall that Eqs. (8.48b) were used in Example 7.17 to determine if the given system was a proportionally damped system.

To determine the response of the proportionally damped system, we again uncouple Eq. (8.46) by making use of the transformation given by Eq. (8.43). Noting that the unknowns to be determined are the modal amplitudes, we substitute Eq. (8.43) into Eq. (8.46) and follow the steps used to obtain Eqs. (8.3) to (8.8) to obtain the following system of equations:

$$
\begin{aligned}
\begin{Bmatrix} \ddot{\eta}_1(t) \\ \ddot{\eta}_2(t) \end{Bmatrix} &+ \begin{bmatrix} 2\zeta_1\omega_1 & 0 \\ 0 & 2\zeta_2\omega_2 \end{bmatrix} \begin{Bmatrix} \dot{\eta}_1(t) \\ \dot{\eta}_2(t) \end{Bmatrix} + \begin{bmatrix} \omega_1^2 & 0 \\ 0 & \omega_2^2 \end{bmatrix} \begin{Bmatrix} \eta_1(t) \\ \eta_2(t) \end{Bmatrix} \\
&= \begin{bmatrix} 1/\hat{M}_{11} & 0 \\ 0 & 1/\hat{M}_{22} \end{bmatrix} \begin{bmatrix} X_{11} & X_{21} \\ X_{12} & X_{22} \end{bmatrix} \begin{Bmatrix} F_1 \\ F_2 \end{Bmatrix} \cos \omega t \\
&= \begin{Bmatrix} (X_{11}F_1 + X_{21}F_2)/\hat{M}_{11} \\ (X_{12}F_1 + X_{22}F_2)/\hat{M}_{22} \end{Bmatrix} \cos \omega t
\end{aligned} \tag{8.49}
$$

In Eq. (8.49), the system natural frequencies ω_1 and ω_2 and the system damping factors ζ_1 and ζ_2 are determined from the eigenvalue problem associated with free oscillations. The modal masses $\hat{M}_{kk}$ are given by Eqs. (8.45). From Eqs. (8.46) and (8.49), it is seen that the system of equations given by Eq. (8.46) has been transformed into two uncoupled single-degree-of-freedom systems, each of which is treated as in Chapter 5. After solving for the modal amplitudes $\eta_1(t)$ and $\eta_2(t)$, one uses Eq. (8.2) to determine the system responses. For underdamped systems, the solutions to Eq. (8.49) are given by Eq. (8.19).

For arbitrarily damped systems, one cannot use the normal-mode approach to determine the solution for the response. However, one can use either the state-space formulation discussed in Section 8.3 or the Laplace transform approach discussed in Section 8.4 to determine the response of a multi-degree-of-freedom system.

EXAMPLE 8.3   Undamped vibration absorber system

Consider the model of a physical system shown in Figure 7.18 with $c_1 = 0$. This model is an idealization of a primary undamped mechanical system composed of m_1 and k_1 to which a secondary mechanical system composed of m_2 and k_2 is attached. The secondary mechanical system is called a *vibration absorber* as illustrated in Figure 8.4. The generalized coordinates x_1 and x_2 represent the displacements measured from the static-equilibrium position of the system.

Single Degree-of-Freedom System

When the secondary system is absent and a harmonic force

$$f_1(t) = F_1 \sin \omega t \tag{a}$$

is applied to m_1, the governing equation of motion is that of a single degree-of-freedom system given by

$$m_1\ddot{x}_1 + k_1 x_1 = F_1 \sin \omega t \tag{b}$$

For the harmonically forced single degree-of-freedom system given by Eq. (b), the resulting forced response is given by Eqs. (5.17) and (5.18) with $\zeta = 0$. Thus

FIGURE 8.4
Undamped vibration absorber system.

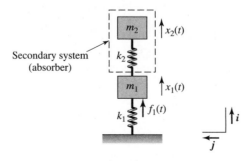

$$x_1(t) = \frac{F_1}{k_1} \frac{\sin \omega t}{1 - \Omega^2} \tag{c}$$

It is seen from Eq. (c) that if the excitation frequency is in the vicinity of the natural frequency ω_{n1} of the primary system—that is, when $\Omega \approx 1$—then the resulting response is undesirable. However, if we are required to operate the system at $\Omega = 1$ ($\omega = \omega_{n1}$) or close to it, then a secondary system can be introduced in order to have a finite amplitude response at the resonance of the single degree-of-freedom system as illustrated below.

Two Degree-of-Freedom System

The governing equations of the two degree-of-freedom systems are obtained from Eqs. (7.1a) with $c_1 = c_2 = c_3 = k_3 = f_2(t) = 0$. Thus,

$$m_1\ddot{x}_1 + (k_1 + k_2)x_1 - k_2 x_2 = F_1 \sin \omega t$$
$$m_2\ddot{x}_2 + k_2 x_2 - k_2 x_1 = 0 \tag{d}$$

Assuming a solution of the form,

$$\begin{Bmatrix} x_1(t) \\ x_2(t) \end{Bmatrix} = \begin{Bmatrix} X_1 \\ X_2 \end{Bmatrix} \sin \omega t \tag{e}$$

and substituting Eq. (e) into Eqs. (d), we obtain Eq. (8.35) with $F_2 = 0$. Then the response amplitudes X_1 and X_2 are obtained from Eqs. (8.39) with $k_{32} = 0$. Thus, the forced response is given by

$$X_1 = \frac{F_1}{k_1} \frac{\omega_r^2 - \Omega^2}{D_o}$$
$$X_2 = \frac{F_1}{k_1} \frac{\omega_r^2}{D_o} \tag{f}$$

where upon making use of Eqs. (8.38) and (7.46), we find that

$$D_o = \Omega^4 - [1 + \omega_r^2(1 + m_r)]\Omega^2 + \omega_r^2 \tag{g}$$

It is recalled from Eqs. (7.41) that $\omega_r = \omega_{n2}/\omega_{n1}$ is the ratio of the system's uncoupled natural frequencies and $\Omega = \omega/\omega_{n1}$ is the excitation frequency ratio.

Since we are forcing the primary system at $\Omega = 1$, we see from the first of Eqs. (f) that $X_1 = 0$ when

$$\omega_r = \Omega = 1 \tag{h}$$

or, equivalently, from Eq. (7.41) that

$$\omega_{n1} = \omega_{n2} \quad \text{and} \quad \omega = \omega_{n1} \tag{i}$$

Evaluating D_o at $\Omega = 1$ from Eq. (g) we find that $D_o = -m_r\omega_r^2 = -k_2/k_1$. Then, Eqs. (f) lead to

$$X_1 = 0$$
$$X_2 = \frac{F_1}{k_1} \frac{1}{D_o} = \frac{-F_1}{k_1(k_2/k_1)} = \frac{-F_1}{k_2} \tag{j}$$

Although the harmonic excitation of the primary mass is at a frequency equal to the primary system's natural frequency, the choice of the secondary system's

FIGURE 8.5
Free-body diagram of mass m_1 for $\omega_r = 1$ and $\Omega = 1$.

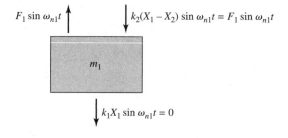

$F_1 \sin \omega_{n1}t$ $k_2(X_1 - X_2) \sin \omega_{n1}t = F_1 \sin \omega_{n1}t$

m_1

$k_1 X_1 \sin \omega_{n1}t = 0$

parameters are such that $\omega_{n1} = \omega_{n2}$, which makes the response of the primary system zero. To understand why the response of the primary mass m_1 is zero when the excitation frequency is at the natural frequency of the single degree-of-freedom system, we consider the free-body diagram shown in Figure 8.5. In this diagram, the spring forces have been evaluated from Eqs. (j). From this figure, it is seen that the spring force generated by the secondary system, the absorber, is equal and opposite to the excitation force at $\Omega = 1$; that is, the primary mass m_1 does not experience any effective excitation at $\Omega = 1$ when the secondary system is designed so that $\omega_r = \omega_{n2}/\omega_{n1} = 1$. If one considers the two degree-of-freedom system from an input energy-output energy perspective, all of the energy input to the system at the excitation frequency $\Omega = 1$ goes into the secondary system, the absorber. In other words, the secondary system absorbs all of the input energy.

The natural frequencies of the two degree-of-freedom system are given by the roots of $D_o = 0$. We find from Eq. (g) that when $\omega_r = 1$, the natural frequencies are solutions of

$$\Omega^4 - (2 + m_r)\Omega^2 + 1 = 0 \tag{k}$$

Although the presence of the absorber is good for the system in terms of attenuating the response of the primary mass at $\omega = \omega_{n1}$, there are still two resonances given by the solution to Eq. (k). At these two resonances, the system response is unbounded since we have an undamped system. This unbounded response can be eliminated by the inclusion of damping, which is considered in Section 8.6.

EXAMPLE 8.4   Absorber for a diesel engine[2]

An engine of mass 300 kg is found to experience undesirable vibrations at an operating speed of 6000 rpm. If the magnitude of the excitation force is 240 N, design a vibration absorber for this system so that the maximum amplitude of the absorber mass does not exceed 3 mm.

To design the vibration absorber for this system, we make use of the analysis of Example 8.3 and the parameters given above to determine the ab-

[2]S. S. Rao, *Mechanical Vibrations,* Addison-Wesley, Reading, MA, Chapter 9 (1995).

sorber stiffness k_2 and the absorber mass m_2. From the information provided, the excitation frequency is given by

$$\omega = \frac{(2\pi \text{ rad/rev})(6000 \text{ rev/min})}{60 \text{ s/min}} = 628.32 \text{ rad/s} \tag{a}$$

and the excitation amplitude is

$$F_1 = 240 \text{ N} \tag{b}$$

The amplitude X_2 of the absorber mass has to be such that

$$|X_2| \leq 3 \times 10^{-3} \text{ m} \tag{c}$$

It is assumed that the system is operating at the natural frequency of the engine—that is, $\omega = \omega_{n1}$—and that the absorber is designed so that $\omega_{n1} = \omega_{n2}$ (the absorber's natural frequency is the same as the engine's natural frequency). From Eq. (a) and Eqs. (7.41), we arrive at

$$\omega_{n2}^2 = \frac{k_2}{m_2} = \omega_{n1}^2 = \omega^2 = (628.32)^2 \text{ rad}^2/\text{s}^2 \tag{d}$$

Equation (d) provides us one of the equations needed to determine the parameters k_2 and m_2 of the absorber. To determine another equation, we make use of Eq. (b) and Eq. (j) from Example 8.3 to arrive at

$$|X_2| = \frac{F_1}{k_2} = \frac{240}{k_2} \tag{e}$$

From Eqs. (c) and (e), we find that the absorber stiffness should be such that

$$|X_2| = \frac{F_1}{k_2} = \frac{240}{k_2} \leq 0.003 \text{ m}$$

or

$$k_2 \geq 80 \times 10^3 \text{ N/m} \tag{f}$$

Making use of Eq. (d), we determine the absorber mass to be

$$m_2 = \frac{k_2}{(628.32)^2} \tag{g}$$

Choosing $k_2 = 100 \times 10^3$ N/m so that Eq. (f) is satisfied leads to $m_2 = 0.253$ kg.

EXAMPLE 8.5 Forced response of a system with bounce and pitch motions

We build on Example 7.14 and construct the frequency-response functions and then graph them. Since the system is undamped, we can use the direct approach presented in Section 8.2.3. Based on Eqs. (a) and (b) of Example 7.14 and Eq. (8.34), we obtain the following set of equations when the harmonic forcing of amplitude F_1 is applied to the center of gravity of the system; that is, point G in Figure 7.5.

$$(-\Omega^2 + 1 + k_{21})Y + (1 - k_{21}L_{21})X = F_1/k_1$$

$$(1 - k_{21}L_{21})Y + \left(-\Omega^2 \frac{k_{21}L_{21}^2}{\omega_r^2} + 1 + k_{21}L_{21}^2\right)X = 0 \tag{a}$$

where $X = L_1\Theta$. Upon solving Eqs. (a) for the forced response amplitudes, we obtain

$$\frac{Y}{F_1/k_1} = \frac{1}{D'_o}\left(-\Omega^2\frac{k_{21}L_{21}^2}{\omega_r^2} + 1 + k_{21}L_{21}^2\right)$$

$$\frac{X}{F_1/k_1} = -\frac{(1 - k_{21}L_{21})}{D'_o} \tag{b}$$

where the term D'_o has the form

$$D'_o = b_1\Omega^4 - b_2\Omega^2 + b_3 \tag{c}$$

The coefficients b_j in Eqs. (c) are given in Eqs. (d) of Example 7.14. The frequency response functions given by Eqs. (b) are plotted in Figure 8.6 for the parameters used in Example 7.14. We see that the relative motion of the beam at different frequencies is

$$\frac{Y}{X} = -\left(-\Omega^2\frac{k_{21}L_{21}^2}{\omega_r^2} + 1 + k_{21}L_{21}^2\right)\Big/(1 - k_{21}L_{21}) \tag{d}$$

provided that $k_{21}L_{21} \neq 1$ and $D'_o \neq 0$; that is, when $\Omega \neq \Omega_j$. When $D'_o = 0$—that is, when $\Omega = \Omega_j$—the responses are unbounded and the solution given by Eq. (8.34) is not valid.

Equation (d) is used to plot the relative motion of the beam at frequencies other than those at Ω_j in Figure 8.6.

From this figure, it is seen that the displacement amplitude Y and the rotation Θ of the bar in Figure 8.6 become unbounded when the excitation fre-

FIGURE 8.6
Amplitudes of frequency-response functions for the system in Figure 7.5 and the envelopes of motion of the bar at selected frequencies. The solid dot shown on the envelopes is the point G in Figure 7.5.

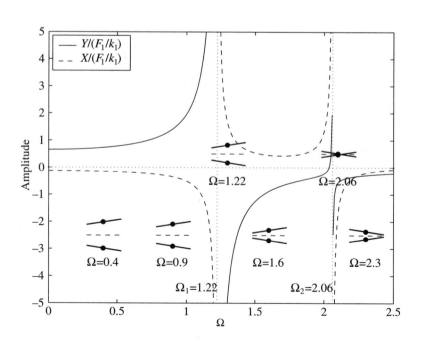

quency is equal to either one of the natural frequencies; that is, when $\omega = \omega_j$ ($\Omega = \Omega_j$). When the excitation frequency is below the first natural frequency ($\Omega < \Omega_1$), it is clear from the displacement pattern and the discussion of Example 7.14 that the first mode dominates the response. On the other hand, when the excitation frequency is above the second natural frequency ($\Omega > \Omega_2$), it is clear from the displacement pattern that the second mode dominates the response. In the intermediate response region, $\Omega_1 < \Omega < \Omega_2$, neither the first nor the second mode is dominant.

8.3 STATE-SPACE FORMULATION

In Section 8.2, we discussed how the response of a damped system can be determined for proportionally damped systems. In this section, we discuss how the response of a multi-degree-of-freedom system can be determined for a system with an arbitrary damping matrix.

The approach is based on a standard solution procedure from the theory of ordinary differential equations, where the governing equations are rewritten as a set of first-order equations called the *state-space* form. Although one can obtain this form for both linear and nonlinear systems, here, we restrict our discussion to linear multi-degree-of-freedom systems. To this end, we start with the governing equations given by Eq. (7.3) for a system with damping, circulatory forces, and gyroscopic forces. These equations are repeated below after replacing the force vector $\{Q\}$ on the right-hand side by $\{F\}$. Thus,

$$[M]\{\ddot{q}\} + [[C] + [G]]\{\dot{q}\} + [[K] + [H]]\{q\} = \{F\} \qquad (8.50)$$

where

$$\{\ddot{q}\} = \begin{Bmatrix} \ddot{q}_1 \\ \ddot{q}_2 \\ \vdots \\ \ddot{q}_N \end{Bmatrix}, \quad \{\dot{q}\} = \begin{Bmatrix} \dot{q}_1 \\ \dot{q}_2 \\ \vdots \\ \dot{q}_N \end{Bmatrix}, \quad \{q\} = \begin{Bmatrix} q_1 \\ q_2 \\ \vdots \\ q_N \end{Bmatrix}, \quad \text{and} \quad \{F\} = \begin{Bmatrix} f_1(t) \\ f_2(t) \\ \vdots \\ f_N(t) \end{Bmatrix} \qquad (8.51)$$

We now introduce the new vectors $\{Y_1\}$ and $\{Y_2\}$, which are defined as

$$\{Y_1\} = \{q\} = \begin{Bmatrix} q_1 \\ q_2 \\ \vdots \\ q_N \end{Bmatrix}, \quad \{Y_2\} = \{\dot{Y}_1\} = \{\dot{q}\} = \begin{Bmatrix} \dot{q}_1 \\ \dot{q}_2 \\ \vdots \\ \dot{q}_N \end{Bmatrix},$$

$$\text{and} \quad \{\dot{Y}_2\} = \{\ddot{q}\} = \begin{Bmatrix} \ddot{q}_1 \\ \ddot{q}_2 \\ \vdots \\ \ddot{q}_N \end{Bmatrix} \qquad (8.52)$$

From Eqs. (8.52), it is clear that $\{Y_1\}$ is the displacement vector containing the N displacements q_i and $\{Y_2\}$ is the velocity vector containing the N velocities $\dot{q}_i$. On substituting Eqs. (8.52) into Eq. (8.50), we arrive at the following system of N first-order equations

$$[M]\{\dot{Y}_2\} + [[C] + [G]]\{Y_2\} + [[K] + [H]]\{Y_1\} = \{F\} \tag{8.53}$$

We also note from Eqs. (8.52) that we have the second set of N first-order equations

$$\{\dot{Y}_1\} = [I]\{Y_2\} \tag{8.54}$$

where $[I]$ is an $N \times N$ identity matrix given by

$$[I] = \begin{bmatrix} 1 & 0 & \cdots & 0 \\ 0 & 1 & \cdots & 0 \\ \vdots & \vdots & \ddots & \vdots \\ 0 & 0 & \cdots & 1 \end{bmatrix}$$

Equation (8.53) is written as

$$\{\dot{Y}_2\} + [M]^{-1}[[C] + [G]]\{Y_2\} + [M]^{-1}[[K] + [H]]\{Y_1\}$$
$$= [M]^{-1}\{F\} \tag{8.55a}$$

or

$$\{\dot{Y}_2\} = -[M]^{-1}[[C] + [G]]\{Y_2\} - [M]^{-1}[[K]$$
$$+ [H]]\{Y_1\} + [M]^{-1}\{F\} \tag{8.55b}$$

provided that the inverse of the inertia matrix $[M]^{-1}$ exists. Upon combining the two sets of first-order differential equations given by Eqs. (8.54) and (8.55b), we arrive at the following system of $2N$ first-order differential equations

$$\{\dot{Y}\} = [A]\{Y\} + [B]\{F\} \tag{8.56}$$

where the $(2N \times 1)$ *state vector* $\{Y\}$ and its time derivative $\{\dot{Y}\}$ are given by

$$\{Y\} = \begin{Bmatrix} Y_1 \\ Y_2 \end{Bmatrix} = \begin{Bmatrix} q_1 \\ q_2 \\ \vdots \\ q_N \\ \dot{q}_1 \\ \dot{q}_2 \\ \vdots \\ \dot{q}_N \end{Bmatrix} \quad \text{and} \quad \{\dot{Y}\} = \begin{Bmatrix} \dot{Y}_1 \\ \dot{Y}_2 \end{Bmatrix} = \begin{Bmatrix} \dot{q}_1 \\ \dot{q}_2 \\ \vdots \\ \dot{q}_N \\ \ddot{q}_1 \\ \ddot{q}_2 \\ \vdots \\ \ddot{q}_N \end{Bmatrix} \tag{8.57}$$

In Eq. (8.56), the $(2N \times 2N)$ *state matrix* $[A]$ and the $(2N \times N)$ matrix $[B]$ are, respectively, given by

$$[A] = \begin{bmatrix} [0] & [I] \\ -[M]^{-1}[[K] + [H]] & -[M]^{-1}[[C] + [G]] \end{bmatrix}$$

$$[B] = \left\{ \begin{matrix} [0] \\ [M]^{-1} \end{matrix} \right\} \tag{8.58}$$

where $[I]$ is an $(N \times N)$ identity matrix, $[0]$ is an $(N \times N)$ null matrix; that is, a matrix whose elements are all zero. Equation (8.56) is referred to as the state-space form[3] of Eq. (8.50).

The initial displacements and the initial velocities of the inertial elements are given by

$$\{Y_1(0)\} = \left\{ \begin{matrix} q_1(0) \\ q_2(0) \\ \vdots \\ q_N(0) \end{matrix} \right\} \quad \text{and} \quad \{Y_2(0)\} = \left\{ \begin{matrix} \dot{q}_1(0) \\ \dot{q}_2(0) \\ \vdots \\ \dot{q}_N(0) \end{matrix} \right\} \tag{8.59}$$

The form of Eqs. (8.56) to (8.59) are well suited to numerical evaluation by standard numerical procedures.[4] Alternatively, a solution for Eq. (8.56) subject to the initial conditions given by Eqs. (8.59) can be found by using the Laplace transform method, as discussed in Appendix A. A third choice for solving Eq. (8.56) is based on the eigenvalues and eigenvectors of the state matrix and its transpose. This approach parallels the normal-mode approach.[5] It is noted that since the matrix $[A]$ is not symmetric, its eigenvalues are complex valued.

In the vibrations literature, the words "mode of a system" are reserved for the eigenvectors determined from the eigenvalue problem associated with the second-order form of the governing equations; that is, eigensystems such as Eq. (7.79) associated with Eq. (7.73). However, in the broader literature, it is common to use the words "modes of a system" for the eigenvectors determined from the eigenvalue problem associated with the state-space form of the governing equations. This is further explored in Example 8.8.

EXAMPLE 8.6 State-space form of equations for a gyro-sensor

We illustrate how the governing equations obtained for the gyro-sensor in Example 7.4 can be written in state-space form. For convenience, the governing equations determined in Example 7.4 are repeated below.

$$[M] \left\{ \begin{matrix} \ddot{x} \\ \ddot{y} \end{matrix} \right\} + [C] \left\{ \begin{matrix} \dot{x} \\ \dot{y} \end{matrix} \right\} + [G] \left\{ \begin{matrix} \dot{x} \\ \dot{y} \end{matrix} \right\} + [K] \left\{ \begin{matrix} x \\ y \end{matrix} \right\} = \left\{ \begin{matrix} f_x \\ 0 \end{matrix} \right\} \tag{a}$$

---

[3]It is noted that the state-space form of a system is not unique; one can obtain different state-space forms by considering equations different from Eq. (8.54).

[4]For example, the MATLAB function ode45 can be used to numerically solve these equations.

[5]L. Meirovitch, *Fundamentals of Vibrations,* McGraw Hill, NY, Chapter 7 (2001) or L. Meirovitch (1980), *ibid.*

where

$$[M] = \begin{bmatrix} m & 0 \\ 0 & m \end{bmatrix}, \quad [C] = \begin{bmatrix} c_x & 0 \\ 0 & c_y \end{bmatrix}, \quad [G] = \begin{bmatrix} 0 & -2m\omega_z \\ 2m\omega_z & 0 \end{bmatrix}$$

$$[K] = \begin{bmatrix} k_x - m\omega_z^2 & 0 \\ 0 & k_y - m\omega_z^2 \end{bmatrix} \tag{b}$$

Since we have a two degree-of-freedom system, the state vector is a (4×1) vector. Noting that $N = 2$ in Eq. (8.56) and identifying q_1 as x and q_2 as y, the state vector and its time derivative are given by

$$\{Y\} = \begin{Bmatrix} Y_1 \\ Y_2 \end{Bmatrix} = \begin{Bmatrix} x \\ y \\ \dot{x} \\ \dot{y} \end{Bmatrix}, \quad \{\dot{Y}\} = \begin{Bmatrix} \dot{Y}_1 \\ \dot{Y}_2 \end{Bmatrix} = \begin{Bmatrix} \dot{x} \\ \dot{y} \\ \ddot{x} \\ \ddot{y} \end{Bmatrix}, \quad \text{and} \quad \{F\} = \begin{Bmatrix} f_x \\ 0 \end{Bmatrix} \tag{c}$$

In order to construct the (4×4) state matrix $[A]$ and the (4×2) matrix $[B]$, we first determine the inverse of the inertia matrix $[M]$. Since the mass matrix given by Eqs. (b) is a diagonal matrix, the inverse is given by

$$[M]^{-1} = \begin{bmatrix} 1/m & 0 \\ 0 & 1/m \end{bmatrix} \tag{d}$$

Making use of Eqs. (d), (b), and (8.58) and recognizing that the chosen example is free of circulatory forces—that is, $[H] = [0]$—we arrive at the state matrix.

$$[A] = \begin{bmatrix} \begin{bmatrix} 0 & 0 \\ 0 & 0 \end{bmatrix} & \begin{bmatrix} 1 & 0 \\ 0 & 1 \end{bmatrix} \\ -\begin{bmatrix} 1/m & 0 \\ 0 & 1/m \end{bmatrix}\begin{bmatrix} k_x - m\omega_z^2 & 0 \\ 0 & k_y - m\omega_z^2 \end{bmatrix} & -\begin{bmatrix} 1/m & 0 \\ 0 & 1/m \end{bmatrix}\left(\begin{bmatrix} c_x & 0 \\ 0 & c_y \end{bmatrix} + \begin{bmatrix} 0 & -2m\omega_z \\ 2m\omega_z & 0 \end{bmatrix}\right) \end{bmatrix}$$

$$= \begin{bmatrix} 0 & 0 & 1 & 0 \\ 0 & 0 & 0 & 1 \\ -(k_x - m\omega_z^2)/m & 0 & -c_x/m & 2\omega_z \\ 0 & -(k_y - m\omega_z^2)/m & -2\omega_z & -c_y/m \end{bmatrix} \tag{e}$$

From Eqs. (8.58) and (d), the matrix $[B]$ is found to be

$$[B] = \begin{bmatrix} 0 & 0 \\ 0 & 0 \\ -1/m & 0 \\ 0 & -1/m \end{bmatrix} \tag{f}$$

From Eqs. (8.56) and (8.57), and Eqs. (c), (e), and (f), we arrive at the following state-space form of Eq. (a):

$$\begin{Bmatrix} \dot{x} \\ \dot{y} \\ \ddot{x} \\ \ddot{y} \end{Bmatrix} = \begin{bmatrix} 0 & 0 & 1 & 0 \\ 0 & 0 & 0 & 1 \\ -(k_x - m\omega_z^2)/m & 0 & -c_x/m & 2\omega_z \\ 0 & -(k_y - m\omega_z^2)/m & -2\omega_z & -c_y/m \end{bmatrix} \begin{Bmatrix} x \\ y \\ \dot{x} \\ \dot{y} \end{Bmatrix}$$

$$+ \begin{bmatrix} 0 & 0 \\ 0 & 0 \\ -1/m & 0 \\ 0 & -1/m \end{bmatrix} \begin{Bmatrix} f_x \\ 0 \end{Bmatrix} \tag{g}$$

EXAMPLE 8.7 State-space form of equations for a model of a milling system

We revisit the three degree-of-system treated in Example 7.1 for a milling system and determine the state-space form of the governing equations. The approach follows along the lines of Example 8.6. First, from Eqs. (8.57) and Eqs. (b) of Example 7.1, we identify the state vector as a (6×1) vector and construct this vector and its time derivative as follows.

$$\{Y\} = \begin{Bmatrix} Y_1 \\ Y_2 \end{Bmatrix} = \begin{Bmatrix} x_1 \\ x_2 \\ x_3 \\ \dot{x}_1 \\ \dot{x}_2 \\ \dot{x}_3 \end{Bmatrix} \quad \text{and} \quad \{\dot{Y}\} = \begin{Bmatrix} \dot{Y}_1 \\ \dot{Y}_2 \end{Bmatrix} = \begin{Bmatrix} \dot{x}_1 \\ \dot{x}_2 \\ \dot{x}_3 \\ \ddot{x}_1 \\ \ddot{x}_2 \\ \ddot{x}_3 \end{Bmatrix} \tag{a}$$

Then, the inverse of the inertia matrix $[M]$ is obtained. Since the mass matrix given in Eqs. (b) of Example 7.1 is a diagonal matrix, the inverse of this matrix is

$$[M]^{-1} = \begin{bmatrix} 1/m_1 & 0 & 0 \\ 0 & 1/m_2 & 0 \\ 0 & 0 & 1/m_3 \end{bmatrix} \tag{b}$$

Making use of Eqs. (8.58) and Eqs. (b) of Example 7.1 and noting that this system is free of gyroscopic and circulatory forces—that is, $[G] = [0]$ and $[H] = [0]$—we find the following state matrix:

$$[A] = \begin{bmatrix} \begin{matrix} 0 & 0 & 0 \\ 0 & 0 & 0 \\ 0 & 0 & 0 \end{matrix} \\[2em] -\begin{bmatrix} 1/m_1 & 0 & 0 \\ 0 & 1/m_2 & 0 \\ 0 & 0 & 1/m_3 \end{bmatrix}\begin{bmatrix} k_1 & -k_1 & 0 \\ -k_1 & k_1 + k_2 & -k_2 \\ 0 & -k_2 & k_2 + k_3 \end{bmatrix} \end{bmatrix}$$

$$\begin{bmatrix} 1 & 0 & 0 \\ 0 & 1 & 0 \\ 0 & 0 & 1 \end{bmatrix}$$

$$-\begin{bmatrix} 1/m_1 & 0 & 0 \\ 0 & 1/m_2 & 0 \\ 0 & 0 & 1/m_3 \end{bmatrix}\begin{bmatrix} c_1 & -c_1 & 0 \\ -c_1 & c_1 + c_2 & -c_2 \\ 0 & -c_2 & c_2 + c_3 \end{bmatrix} \tag{c}$$

Carrying out the matrix multiplication operations, the matrix in Eq. (c) is reduced to the form

$$[A] = \begin{bmatrix} 0 & 0 & 0 & 1 & 0 & 0 \\ 0 & 0 & 0 & 0 & 1 & 0 \\ 0 & 0 & 0 & 0 & 0 & 1 \\ -k_1/m_1 & k_1/m_1 & 0 & -c_1/m_1 & c_1/m_1 & 0 \\ k_1/m_2 & -(k_1+k_2)/m_2 & k_2/m_2 & c_1/m_2 & -(c_1+c_2)/m_2 & c_2/m_2 \\ 0 & k_2/m_3 & -(k_2+k_3)/m_3 & 0 & c_2/m_3 & -(c_2+c_3)/m_3 \end{bmatrix} \quad \text{(d)}$$

In this case, from Eqs. (8.58) and Eqs. (b), we find that the (6×3) matrix $[B]$ is given by

$$[B] = \begin{bmatrix} 0 & 0 & 0 \\ 0 & 0 & 0 \\ 0 & 0 & 0 \\ -1/m_1 & 0 & 0 \\ 0 & -1/m_2 & 0 \\ 0 & 0 & -1/m_3 \end{bmatrix} \quad \text{(e)}$$

Hence, the state-space form of Eqs. (b) of Example 7.1 is given by

$$\begin{Bmatrix} \dot{x}_1 \\ \dot{x}_2 \\ \dot{x}_3 \\ \ddot{x}_1 \\ \ddot{x}_2 \\ \ddot{x}_3 \end{Bmatrix} = [A] \begin{Bmatrix} x_1 \\ x_2 \\ x_3 \\ \dot{x}_1 \\ \dot{x}_2 \\ \dot{x}_3 \end{Bmatrix} + [B] \begin{Bmatrix} -f(t) \\ 0 \\ 0 \end{Bmatrix} \quad \text{(f)}$$

where the matrices $[A]$ and $[B]$ are given by Eqs. (d) and (e), respectively.

EXAMPLE 8.8   Eigenvalues and eigenvectors of a proportionally damped system from the state matrix

In this example, we revisit Examples 7.17 and 7.18 and illustrate by using the state-space form of equations for the undamped system and the proportionally damped system that the eigenvectors have a similar structure in both cases. This was pointed out in Section 7.2.3. Let us suppose that the matrices given in Example 7.17 correspond to the system shown in Figure 7.1. The governing equations of motion of this two degree-of-freedom system are repeated below after setting the forces to zero.

$$\begin{bmatrix} m_1 & 0 \\ 0 & m_2 \end{bmatrix} \begin{Bmatrix} \ddot{x}_1 \\ \ddot{x}_2 \end{Bmatrix} + \begin{bmatrix} c_1+c_2 & -c_2 \\ -c_2 & c_2+c_3 \end{bmatrix} \begin{Bmatrix} \dot{x}_1 \\ \dot{x}_2 \end{Bmatrix}$$

$$+ \begin{bmatrix} k_1+k_2 & -k_2 \\ -k_2 & k_2+k_3 \end{bmatrix} \begin{Bmatrix} x_1 \\ x_2 \end{Bmatrix} = \begin{Bmatrix} 0 \\ 0 \end{Bmatrix} \quad \text{(a)}$$

To find the eigenvalues and eigenvectors associated with the two degree-of-freedom system, we write Eq. (a) in the following state-space form:

$$\begin{Bmatrix} \dot{x}_1 \\ \dot{x}_2 \\ \ddot{x}_1 \\ \ddot{x}_2 \end{Bmatrix} = [A] \begin{Bmatrix} x_1 \\ x_2 \\ \dot{x}_1 \\ \dot{x}_2 \end{Bmatrix} = [A]\{x\} \tag{b}$$

Following the procedure outlined in Example 8.5, we find that the state matrix is given by

$$[A] = \begin{bmatrix} 0 & 0 & 1 & 0 \\ 0 & 0 & 0 & 1 \\ -(k_1 + k_2)/m_1 & k_2/m_1 & -(c_1 + c_2)/m_1 & c_2/m_1 \\ k_2/m_2 & -(k_2 + k_3)/m_2 & c_2/m_2 & -(c_2 + c_3)/m_2 \end{bmatrix} \tag{c}$$

Since Eq. (b) is a set of linear ordinary differential equations with constant coefficients, one can assume a solution of the form

$$\{x\} = \{X\}e^{\lambda t} \tag{d}$$

On substituting Eq. (d) into Eq. (b) and canceling the common factor of $e^{\lambda t}$ on both sides, we arrive at the following system

$$[A]\{X\} = \lambda\{X\} \tag{e}$$

The algebraic system of equations given by Eq. (e) constitutes an eigenvalue problem, since we are seeking those special values of λ for which the vector $\{X\}$ will be nontrivial. For the chosen physical system, we have a system of four algebraic equations, which will mean that we will have four eigenvalues and a set of four corresponding eigenvectors. These eigenvalues are determined from

$$\det[[A] - \lambda[I]] = 0 \tag{f}$$

Since matrix $[A]$ is not a symmetric matrix, the eigenvalues of this matrix can no longer be expected to be real. Furthermore, from Eqs. (b) and (d), it is seen that the first two entries in each eigenvector are associated with the displacement states, which are x_1 and x_2, and that the next two entries in each eigenvector are associated with the velocity states, which are $\dot{x}_1$ and $\dot{x}_2$. From Eq. (d), we see that $\dot{x}_1 = \lambda x_1$ and $\dot{x}_2 = \lambda x_2$; hence, the third entry of the eigenvector associated with the eigenvalue λ is λ times the first entry of this eigenvector and the fourth entry of this eigenvector is λ times the second entry of this eigenvector.

We numerically determine the eigenvalues and eigenvectors associated with the state matrix $[A]$ for the parameter values given in Example 7.17 and compare the results obtained in the undamped and damped cases presented in Example 7.18.

Eigenvalues and Eigenvectors in the Undamped Case

In the undamped case, we set the damping coefficients $c_i = 0$ in Eq. (c). From Eq. (a) of Example 7.17, we then make the substitutions $k_1 + k_2 = 2$,

$k_2 = 1$, $k_2 + k_3 = 1$, $m_1 = 2$, and $m_2 = 1$ and find that the state-space matrix is given by

$$[A] = \begin{bmatrix} 0 & 0 & 1 & 0 \\ 0 & 0 & 0 & 1 \\ -1 & 1/2 & 0 & 0 \\ 1 & -1 & 0 & 0 \end{bmatrix} \tag{g}$$

Upon substituting Eq. (g) into Eq. (f), we find that the eigenvalues[6] are

$$\lambda_{11,12} = \mp j0.541$$
$$\lambda_{21,22} = \mp j1.307 \tag{h}$$

and the corresponding eigenvectors are

$$\{X\}_{11} = \begin{Bmatrix} 0.508 \\ 0.718 \\ -j0.275 \\ -j0.389 \end{Bmatrix}, \quad \{X\}_{12} = \begin{Bmatrix} 0.508 \\ 0.718 \\ j0.275 \\ j0.389 \end{Bmatrix},$$

$$\{X\}_{21} = \begin{Bmatrix} -0.351 \\ 0.496 \\ j0.459 \\ -j0.648 \end{Bmatrix}, \quad \{X\}_{22} = \begin{Bmatrix} -0.351 \\ 0.496 \\ -j0.459 \\ j0.648 \end{Bmatrix} \tag{i}$$

The imaginary eigenvalues are identical to those given in Eqs. (c) and (d) of Example 7.18. Upon comparing the results of Example 7.18 and this example, it is clear that in the undamped case

$$\lambda_{11,12} = \mp j\omega_1 \quad \text{and} \quad \lambda_{21,22} = \mp j\omega_2 \tag{j}$$

or

$$(\lambda_{11,12})^2 = -\omega_1^2 \quad \text{and} \quad (\lambda_{21,22})^2 = -\omega_2^2$$

where the ω_i are the natural frequencies of the system. Equations (j) are consistent with Eqs. (7.35). The four eigenvalues occur in the form of two complex conjugate pairs. The eigenvectors associated with a complex conjugate pair of eigenvalues are complex conjugates of each other; for example, the eigenvector $\{X\}_{11}$ associated with the eigenvalue λ_{11} is the complex conjugate of the eigenvector $\{X\}_{12}$ associated with the eigenvalue λ_{12}. In addition, it can be verified that the third entry of each eigenvector is the corresponding eigenvalue times the first entry of this eigenvector and that the fourth entry of each eigenvector is the corresponding eigenvalue times the second entry of this eigenvector.

Eigenvalues and Eigenvectors in the Damped Case

In the damped case, we find from Eq. (a) of Example 7.17 that $c_1 + c_2 = 0.8$, $c_2 = 0.2$, and $c_2 + c_3 = 0.4$. Upon substituting these values and the values for k_i and m_i used to obtain Eq. (g), we find that the state-space matrix is given by

---

[6]The MATLAB function eig was used.

$$[A] = \begin{bmatrix} 0 & 0 & 1 & 0 \\ 0 & 0 & 0 & 1 \\ -1 & 1/2 & -0.4 & 0.1 \\ 1 & -1 & 0.2 & -0.4 \end{bmatrix} \quad \text{(k)}$$

After substituting Eq. (k) into Eq. (f), we find that the eigenvalues are:[7]

$$\lambda_{d1_{1,2}} = -0.129 \mp j0.526$$
$$\lambda_{d2_{1,2}} = -0.271 \mp j1.278 \quad \text{(l)}$$

and the associated eigenvectors are

$$\{X\}_{d1_1} = \begin{Bmatrix} -0.141 - j0.488 \\ -0.200 - j0.690 \\ -0.238 + j0.137 \\ -0.337 + j0.194 \end{Bmatrix}, \quad \{X\}_{d1_2} = \begin{Bmatrix} -0.141 + j0.488 \\ -0.200 + j0.690 \\ -0.238 - j0.137 \\ -0.337 - j0.194 \end{Bmatrix},$$

$$\{X\}_{d2_1} = \begin{Bmatrix} 0.233 - j0.263 \\ -0.329 + j0.371 \\ -0.399 - j0.227 \\ -0.564 + j0.321 \end{Bmatrix}, \quad \{X\}_{d2_2} = \begin{Bmatrix} 0.233 + j0.263 \\ -0.329 - j0.371 \\ -0.399 + j0.227 \\ -0.564 - j0.321 \end{Bmatrix} \quad \text{(m)}$$

On comparing the eigenvalues determined for the damped system with those determined in Example 7.18, we see that the eigenvalues determined in both cases are equal and that they are in the form

$$\lambda_{d1_{1,2}} = -\zeta_1\omega_1 \mp j\omega_{d1} \quad \text{and} \quad \lambda_{d2_{1,2}} = -\zeta_2\omega_2 \mp j\omega_{d2} \quad \text{(n)}$$

where ζ_i are the damping factors and ω_{di} are the damped natural frequencies. At first glance, the eigenvectors determined for the undamped case and presented in Eqs. (i) appear to have a different form from the eigenvectors determined for the damped case and presented in Eqs. (m). To compare them, we normalize each eigenvector by dividing throughout with the first entry of that eigenvector; this means that the first entry of each eigenvector will be 1 in both the undamped and damped cases. The normalized eigenvectors for the undamped and damped cases are shown below.

Normalized eigenvectors in the undamped case

$$\{X\}_{11} = \begin{Bmatrix} 1.000 \\ 1.413 \\ -j0.541 \\ -j0.766 \end{Bmatrix}, \quad \{X\}_{12} = \begin{Bmatrix} 1.000 \\ 1.413 \\ j0.541 \\ j0.766 \end{Bmatrix},$$

$$\{X\}_{21} = \begin{Bmatrix} 1.000 \\ -1.413 \\ j1.307 \\ -j1.846 \end{Bmatrix}, \quad \{X\}_{22} = \begin{Bmatrix} 1.000 \\ -1.413 \\ j1.307 \\ -j1.846 \end{Bmatrix} \quad \text{(o)}$$

---

[7]The MATLAB function `eig` was used. However, the eigenvalues and associated eigenvectors presented here are not in the same order as that determined by MATLAB.

Normalized eigenvectors in the damped case

$$\{X\}_{d11} = \begin{Bmatrix} 1.000 \\ 1.414 \\ -0.129 - j0.525 \\ -0.183 - j0.743 \end{Bmatrix}, \quad \{X\}_{d12} = \begin{Bmatrix} 1.000 \\ 1.414 \\ -0.129 + j0.525 \\ -0.183 + j0.743 \end{Bmatrix},$$

$$\{X\}_{d21} = \begin{Bmatrix} 1.000 \\ -1.411 \\ -0.270 - j1.278 \\ -1.748 - j0.598 \end{Bmatrix}, \quad \{X\}_{d22} = \begin{Bmatrix} 1.000 \\ -1.411 \\ -0.270 + j1.278 \\ -1.748 + j0.598 \end{Bmatrix} \quad \text{(p)}$$

Comparing the eigenvectors given in Eqs. (o) and (p), we see that the first two entries of an eigenvector determined for the undamped case are identical to the first two entries of the corresponding eigenvector in the proportionally damped case; this means that the ratio of the displacement states for modes of the undamped system and the ratio of the displacement states in the corresponding modes of the proportionally damped system are the same. For this reason, it is often loosely stated in the literature that the modes of the undamped system are "identical" to the corresponding modes of the proportionally damped system. From Eqs. (o) and (p), it is clear that since the first entry of each eigenvector is one, the third entry of each eigenvector is the corresponding eigenvalue. Recall the discussion following Eq. (f).

EXAMPLE 8.9 Free oscillation comparison for a system with an arbitrary damping model and a system with a constant modal damping model

We compare the free oscillations of two systems with identical mass and stiffness matrices, but with different damping matrices. In one case, the damping matrix product term

$$[C_t] = [M_D]^{-1}[\Phi]^T [C][\Phi] \quad \text{(a)}$$

is a diagonal matrix while in the other case, this matrix product is not diagonal. For the diagonal matrix case, the damping factors are the same for each of the two modes and, hence, this case is referred to as a constant modal damping case. In this case, as illustrated in Example 8.2, one can determine the solution for free oscillations by using the normal-mode approach. However, this is not possible to do when the damping matrix $[C]$ is arbitrary or the matrix product $[C_t]$ is not a diagonal matrix. Here, we consider the constant modal damping case of Example 8.2 as a baseline case and make changes to the damping matrix $[C]$ determined for this baseline case so that the matrix product $[C_t]$ is not a diagonal matrix. In both cases, the solutions for free oscillations of the displacement states are determined by making use of the state-space formulation; that is, Eq. (8.56). The initial conditions for both cases are given by Eqs. (a) of Example 8.2.

Constant Modal Damping Case

In Example 8.2, we considered damped oscillations of a two degree-of-freedom system. In this example, which is a continuation of Examples 7.12 and 7.16, it is assumed that the damping matrix product term can be approximated by the following equation, which is obtained by assuming constant modal damping and dropping the off-diagonal terms because they are "small."

$$[M_D]^{-1} [\Phi]^T [C][\Phi] \approx [(2\zeta\omega)] = \begin{bmatrix} 2\zeta\omega_1 & 0 \\ 0 & 2\zeta\omega_2 \end{bmatrix} \tag{b}$$

To determine the damping matrix $[C]$, which will lead to the diagonal matrix shown in Eq. (b), we carry out a series of matrix multiplications and find from Eq. (b) that the damping matrix $[C]$ needs to take the form

$$[C] = \begin{bmatrix} c_1 + c_2 & -c_2 \\ -c_2 & c_2 + c_3 \end{bmatrix} = [[\Phi]^T]^{-1}[M_D][(2\zeta\omega)][\Phi]^{-1} \tag{c}$$

Equation (c) can be used to determine the damping coefficients c_1, c_2, and c_3 that will provide constant modal damping and a matrix $[C_t]$ that will be a diagonal matrix.

Previously, in Example 7.16, the modal matrix and the modal masses were determined along with the system natural frequencies. On using Eq. (7.68a), Eqs. (a), (b), and (g) of Example 7.16, and the value of $\zeta = 0.05$ used in Example 8.2, Eq. (c) results in

$$\begin{bmatrix} c_1 + c_2 & -c_2 \\ -c_2 & c_2 + c_3 \end{bmatrix} = \begin{bmatrix} 0.893 & 1 \\ -2.518 & 1 \end{bmatrix}^{-1} \begin{bmatrix} 3.658 & 0 \\ 0 & 10.31 \end{bmatrix}$$

$$\times \begin{bmatrix} 0.1 \times 2.519 & 0 \\ 0 & 0.1 \times 5.623 \end{bmatrix} \begin{bmatrix} 0.893 & -2.518 \\ 1 & 1 \end{bmatrix}^{-1}$$

$$= \begin{bmatrix} 0.577 & -0.246 \\ -0.246 & 0.900 \end{bmatrix} \text{Ns/m} \tag{d}$$

from which it is found that

$$c_1 = 0.332 \text{ N·s/m}$$
$$c_2 = 0.246 \text{ N·s/m}$$
$$c_3 = 0.654 \text{ N·s/m} \tag{e}$$

For these values, it can be verified that

$$[M_D]^{-1} [\Phi]^T [C][\Phi] = \begin{bmatrix} 0.252 & 0 \\ 0 & 0.562 \end{bmatrix} \tag{f}$$

which, as expected, is a diagonal matrix.

In order to determine the associated free oscillations, we first used Eq. (d) and the matrices $[M]$ and $[K]$ that are given by Eqs. (a) of Example 7.16 to form the matrices $[A]$ and $[B]$ given by Eqs. (8.58). These matrices are used, in turn, in Eq. (8.56) to numerically obtain the state vector $\{Y\}$.[8] The

---

[8]The MATLAB function ode45 was used.

FIGURE 8.7
Displacements of a two degree-of-freedom system with two different damping models when both masses are subjected to equal, but opposite, initial displacements: (a) displacement of m_1 and (b) displacement of m_2. Solid lines represent constant damping model of Example 8.2 with $\zeta = 0.05$ and dashed lines represent an arbitrary damping model whose damping matrix is given by Eq. (g).

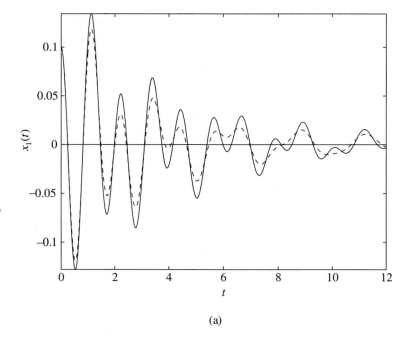

(a)

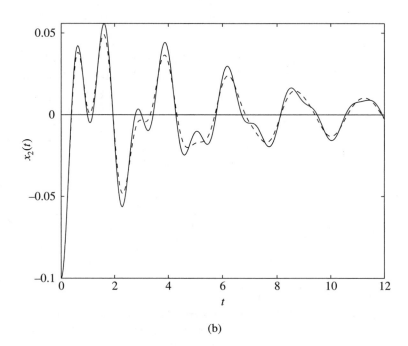

(b)

results obtained for free oscillations of the displacement states, which are in agreement with those obtained analytically in Example 8.2 and shown in Figure 8.1, are presented in Figure 8.7.

Arbitrary Damping Case

If we increase the values of c_1, c_2, and c_3, from those given by Eqs. (e), then the resulting damping matrix $[C]$ will not necessarily result in a diagonal matrix on carrying out the operations given in Eq. (a). This can help assess how the free oscillation characteristics change when there are non-zero off-diagonal terms in the matrix $[C_t]$.

It is found that doubling the values of either c_1 or c_3 has "little" effect on the responses $x_j(t)$ compared to those obtained with the damping matrix product given in Eq. (f). However, doubling the value of c_2 from 0.246 N·s/m to 0.49 N·s/m does result in noticeable differences in the responses of the inertia elements. In this case, the damping matrix takes the form

$$[C] = \begin{bmatrix} 0.821 & -0.246 \\ -0.246 & 1.144 \end{bmatrix} \text{N·s/m} \tag{g}$$

and evaluating the damping matrix product we obtain

$$[M_D]^{-1} [\Phi]^T [C][\Phi] = \begin{bmatrix} 0.253 & 0.025 \\ 0.009 & 0.855 \end{bmatrix} \tag{h}$$

which is not a diagonal matrix. In this case, the matrix $[C]$ given by Eq. (g) is used along with the matrices $[M]$ and $[K]$ from Eqs. (a) of Example 7.16 to form the matrices $[A]$ and $[B]$ in Eqs. (8.58). These matrices are used, in turn, in Eq. (8.56) to numerically obtain the state vector $\{Y\}$. The results for the displacement states are shown in Figure 8.7.

On comparing the results of Figure 8.7, it is seen that there are discernible differences between the free oscillations of the two cases in which the only difference between the models of the two systems is due to the damping matrix.

8.4 LAPLACE TRANSFORM APPROACH

In this section, the general solution for the response of a system such as that shown in Figure 8.8 is determined by using Laplace transforms. This method is applicable to linear systems of differential equations with constant coefficients, either in the second-order form of Eq. (7.3) or in the first-order form of Eq. (8.56). The procedure to determine the solution follows along the lines of what was illustrated in Chapter 4 for single degree-of-freedom systems. First, the governing system of differential equations is transformed into an algebraic system of equations by using Laplace transforms. Next, we solve this algebraic system to determine the responses of the different inertial elements in the Laplace domain. In the final step, these responses are transformed back to the time domain by using the inverse Laplace transform. As had been done

FIGURE 8.8
System with two degrees of freedom.

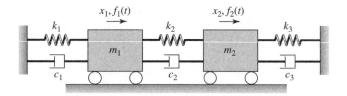

in Chapter 4, this final step is not explicitly carried out here. Instead, we either take recourse to tables such as those presented in Appendix A or numerically evaluate the appropriate MATLAB functions.

For purposes of illustration and algebraic ease, the discussion is restricted to two degree-of-freedom systems. The governing equations of motion of the system shown in Figure 8.8 are given by Eq. (7.1b), which are repeated below.

$$m_1 \frac{d^2x_1}{dt^2} + (c_1 + c_2)\frac{dx_1}{dt} + (k_1 + k_2)x_1 - c_2\frac{dx_2}{dt} - k_2x_2 = f_1(t)$$

$$\frac{m_2\,d^2x_2}{dt^2} + (c_2 + c_3)\frac{dx_2}{dt} + (k_2 + k_3)x_2 - c_2\frac{dx_1}{dt} - k_2x_1 = f_2(t) \quad (8.60)$$

Introducing the nondimensional quantities from Eqs. (7.41) and the following additional quantities for the nondimensional time, damping factor, and damping coefficient ratio, respectively,

$$\tau = \omega_{n1}t, \quad 2\zeta_j = \frac{c_j}{m_j\omega_{nj}}, \quad \text{and} \quad c_{32} = \frac{c_3}{c_2} \quad (8.61)$$

Eqs. (8.60) are rewritten as

$$\frac{d^2x_1}{d\tau^2} + (2\zeta_1 + 2\zeta_2 m_r\omega_r)\frac{dx_1}{d\tau}$$
$$+ (1 + m_r\omega_r^2)x_1 - 2\zeta_2 m_r\omega_r\frac{dx_2}{d\tau} - m_r\omega_r^2 x_2 = \frac{f_1(\tau)}{k_1}$$

$$\frac{d^2x_2}{d\tau^2} + 2\zeta_2\omega_r(1 + c_{32})\frac{dx_2}{d\tau}$$
$$+ \omega_r^2(1 + k_{32})x_2 - 2\zeta_2\omega_r\frac{dx_1}{d\tau} - \omega_r^2 x_1 = \frac{f_2(\tau)}{k_1 m_r} \quad (8.62)$$

Carrying out the Laplace transforms of the different terms on each side of Eqs. (8.62) and making use of Laplace transform pair 2 in Table A of Appendix A, we arrive at

$$A(s)X_1(s) - B(s)X_2(s) = K_1(s)$$
$$-C(s)X_1(s) + E(s)X_2(s) = K_2(s) \quad (8.63)$$

where the coefficients $A(s)$, $B(s)$, $C(s)$, and $E(s)$ are given by

$$A(s) = s^2 + 2(\zeta_1 + \zeta_2 m_r\omega_r)s + 1 + m_r\omega_r^2$$
$$B(s) = 2\zeta_2 m_r\omega_r s + m_r\omega_r^2$$
$$C(s) = 2\zeta_2\omega_r s + \omega_r^2$$
$$E(s) = s^2 + 2\zeta_2\omega_r(1 + c_{32})s + \omega_r^2(1 + k_{32}) \quad (8.64)$$

and

$$K_1(s) = \frac{F_1(s)}{k_1} + \dot{x}_1(0) + [s + 2\zeta_1 + 2\zeta_2 m_r \omega_r] x_1(0) - 2\zeta_2 m_r \omega_r x_2(0)$$

$$K_2(s) = \frac{F_2(s)}{k_1 m_r} + \dot{x}_2(0) + [s + 2\zeta_2 \omega_r(1 + c_{32})] x_2(0) - 2\zeta_2 \omega_r x_1(0) \quad (8.65)$$

In Eqs. (8.65), the transforms $K_1(s)$ and $K_2(s)$ are determined by the forcing and the initial conditions, the overdot indicates the time derivative with respect to the nondimensional time τ, $X_1(s)$ and $X_2(s)$ are the Laplace transforms of $x_1(\tau)$ and $x_2(\tau)$, respectively, and $F_1(s)$ and $F_2(s)$ are the Laplace transforms of the force inputs, $f_1(\tau)$ and $f_2(\tau)$, respectively. Furthermore, $x_1(0)$ and $\dot{x}_1(0)$ are the initial displacement and the initial velocity of mass m_1, respectively, and $x_2(0)$ and $\dot{x}_2(0)$ are the initial displacement and the initial velocity of mass m_2, respectively.

Solving for $X_1(s)$ and $X_2(s)$ from Eqs. (8.63) yields

$$X_1(s) = \frac{K_1(s)E(s)}{D_1(s)} + \frac{K_2(s)B(s)}{D_1(s)}$$

$$X_2(s) = \frac{K_1(s)C(s)}{D_1(s)} + \frac{K_2(s)A(s)}{D_1(s)} \quad (8.66)$$

where the denominator $D_1(s)$ is given by

$$\begin{aligned} D_1(s) = s^4 &+ [2\zeta_1 + 2\zeta_2 \omega_r m_r + 2\zeta_2 \omega_r (1 + c_{32})]s^3 \\ &+ [1 + m_r \omega_r^2 + \omega_r^2 + 4\zeta_1 \zeta_2 \omega_r + \omega_r^2 k_{32} + 4\zeta_2 \omega_r c_{32}(\zeta_1 + \zeta_2 \omega_r m_r)]s^2 \\ &+ [2\zeta_2 \omega_r + 2\zeta_1 \omega_r^2 + 2k_{32}\omega_r^2(\zeta_1 + \zeta_2 \omega_r m_r) + 2c_{32}\zeta_2 \omega_r(1 + m_r \omega_r^2)]s \\ &+ \omega_r^2[1 + k_{32}(1 + m_r \omega_r^2)] \end{aligned} \quad (8.67)$$

Here, the polynomial $D_1(s)$ is the same as the polynomial in the characteristic equation associated with free oscillations of a two degree-of-freedom system. In fact, setting the damping factors $\zeta_1 = \zeta_2 = 0$ and $s = j\Omega$, we obtain the characteristic equation, Eq. (7.45), which was obtained in the context of free oscillations of the undamped system.

The desired displacement responses $x_1(\tau)$ and $x_2(\tau)$ are determined by executing the inverse Laplace transforms of $X_1(s)$ and $X_2(s)$ given by Eqs. (8.66). The solution

$$x_j(\tau) = L^{-1}[X_j(s)] \quad \text{for} \quad j = 1, 2 \quad (8.68)$$

is referred to as the *general solution* for the response of the two degree-of-freedom system given by Eqs. (8.62). The symbol L^{-1} denotes the inverse Laplace transform. To determine the inverse Laplace transforms in Eqs. (8.68), the method of partial fractions and the table provided in Appendix A can be used. Alternatively, readily available algorithms such as the ones in the MATLAB Controls Toolbox and Symbolic Math Toolbox can be used to determine the responses based on Eqs. (8.66). In the next two subsections, we illustrate how the responses can be determined for arbitrary forcing and arbitrary initial conditions.

8.4.1   Response to Arbitrary Forcing

If we assume that the initial conditions are zero and that we have arbitrary forcing, the transforms $K_1(s)$ and $K_2(s)$ in Eqs. (8.65) reduce to

$$K_1(s) = \frac{F_1(s)}{k_1}$$

$$K_2(s) = \frac{F_2(s)}{k_1 m_r} \tag{8.69}$$

where $F_i(s)$ is the Laplace transform of the force input $f_i(t)$. We now consider two cases of forcing.

Impulse Excitation

As a first case, we determine the response of the vibratory system shown in Figure 8.8, when the second mass is subjected to an impulse; that is,

$$f_1(t) = 0 \quad \text{and} \quad f_2(t) = F_o \delta(t) \tag{8.70}$$

Upon using the Laplace transform pair 5 in Table A of Appendix A, we determine that

$$F_2(s) = F_o \tag{8.71}$$

Then, from Eqs. (8.69), the transforms $K_1(s)$ and $K_2(s)$ are

$$K_1(s) = \frac{F_1(s)}{k_1} = 0$$

$$K_2(s) = \frac{F_2(s)}{k_1 m_r} = \frac{F_o}{k_1 m_r} \tag{8.72}$$

Based on Eqs. (8.66) and (8.72), the displacement responses in the Laplace domain are given by

$$X_1(s) = \frac{F_o B(s)}{k_1 m_r D_1(s)}$$

$$X_2(s) = \frac{F_o A(s)}{k_1 m_r D_1(s)} \tag{8.73}$$

For the special case where $k_3 = c_3 = 0$ in Figure 8.8—that is $k_{32} = c_{32} = 0$—the polynomial $D_1(s)$ reduces to

$$D_2(s) = s^4 + [2\zeta_1 + 2\zeta_2\omega_r m_r + 2\zeta_2\omega_r]s^3 + [1 + m_r\omega_r^2 + \omega_r^2 + 4\zeta_1\zeta_2\omega_r]s^2 \\ + [2\zeta_2\omega_r + 2\zeta_1\omega_r^2]s + \omega_r^2 \tag{8.74}$$

and Eqs. (8.73) reduce to

$$X_1(s) = \frac{F_o B(s)}{k_1 m_r D_2(s)}$$

$$X_2(s) = \frac{F_o A(s)}{k_1 m_r D_2(s)} \tag{8.75}$$

Step Input

As a second case, we consider the determination of the response of the vibratory system shown in Figure 8.8 when the second mass is subjected to a step input; that is,

$$f_1(t) = 0 \quad \text{and} \quad f_2(t) = F_o u(t) \tag{8.76}$$

Upon using the Laplace transform pair 6 in Table A of Appendix A we determine that

$$F_2(s) = \frac{F_o}{s} \tag{8.77}$$

Then, from Eqs. (8.69), the transforms $K_1(s)$ and $K_2(s)$ are given by

$$K_1(s) = 0$$

$$K_2(s) = \frac{F_o}{sk_1 m_r} \tag{8.78}$$

Based on Eqs. (8.66) and (8.78), the displacement responses in the Laplace domain are

$$X_1(s) = \frac{F_o B(s)}{sk_1 m_r D_1(s)}$$

$$X_2(s) = \frac{F_o A(s)}{sk_1 m_r D_1(s)} \tag{8.79}$$

For the special case where $k_3 = c_3 = 0$ in Figure 8.8—that is $k_{32} = c_{32} = 0$—the polynomial $D_1(s)$ reduces to $D_2(s)$ given by Eq. (8.74) and the responses given by Eqs. (8.79) reduce to

$$X_1(s) = \frac{F_o B(s)}{k_1 m_r s D_2(s)}$$

$$X_2(s) = \frac{F_o A(s)}{k_1 m_r s D_2(s)} \tag{8.80}$$

To determine the time-domain responses, the inverse Laplace transforms of the impulse response given by Eqs. (8.75) and the step response given by Eqs. (8.80) were evaluated numerically,[9] and the results obtained are shown in Figures 8.9 and 8.10. In each of Figures 8.9 and 8.10, the response of the mass m_1 is graphed by using a dashed line and the response of mass m_2 is graphed by using a solid line. The response of the second mass is more pronounced compared to the response of the first mass, since the forcing is directly applied to the second mass. However, due to the coupling in the stiffness and damping matrices, the mass m_1 also responds to the forcing. In the case of the impulse excitation applied to the second mass, at the higher value of the mass ratio m_r, the responses of the first and second masses are characterized by damped oscillations with the same period for $\omega_r \geq 1$. In the case

---

[9]The MATLAB functions `tf`, `step`, and `impulse` from the Controls Toolbox were used.

FIGURE 8.9
Normalized displacement responses of a two degree-of-freedom system when m_2 is subjected to an impulse force, $\zeta_1 = \zeta_2 = 0.2$, $k_3 = c_3 = 0$, and $\tau = \omega_{n1}t$: (a) $m_r = 0.1$ and (b) $m_r = 0.5$. [Dashed line: $x_1(\tau)$; solid line $x_2(\tau)$.]

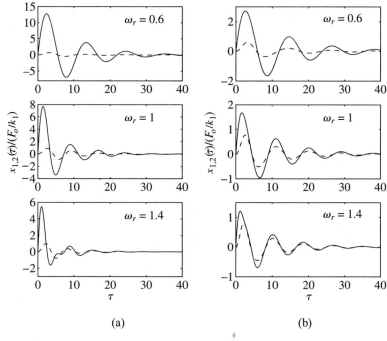

FIGURE 8.10
Normalized displacement responses of two degree-of-freedom system when m_2 is subjected to a step force, $\zeta_1 = \zeta_2 = 0.2$, $k_3 = c_3 = 0$, and $\tau = \omega_{n1}t$: (a) $m_r = 0.1$ and (b) $m_r = 0.5$. [Dashed line: $x_1(\tau)$; solid line $x_2(\tau)$.]

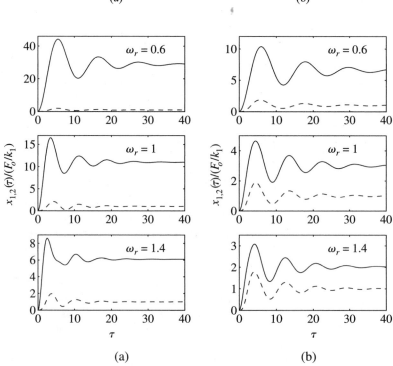

of the step input, this observation is true of the transient oscillations. On examining Figure 8.10, it is seen that the settling positions of the two masses are different for the given step input.

8.4.2 Response to Initial Conditions

We use the general solution given by Eqs. (8.66) to examine free oscillations of the system shown in Figure 8.8 for different initial conditions. In order to isolate the response to initial conditions, we set the forcing $f_1(t)$ and $f_2(t)$ in Eqs. (8.62) to zero. Therefore, the corresponding Laplace transforms are $F_1(s) = 0$ and $F_2(s) = 0$, and the transforms $K_1(s)$ and $K_2(s)$ in Eqs. (8.65) reduce to

$$K_1(s) = \dot{x}_1(0) + [s + 2\zeta_1 + 2\zeta_2 m_r \omega_r]x_1(0) - 2\zeta_2 m_r \omega_r x_2(0)$$

$$K_2(s) = \dot{x}_2(0) + [s + 2\zeta_2 \omega_r(1 + c_{32})]x_2(0) - 2\zeta_2 \omega_r x_1(0) \qquad (8.81)$$

For the special case where the spring k_3 and the damper c_3 are not present in Figure 8.8—that is, $k_{32} = c_{32} = 0$—the coefficient $E(s)$ in Eqs. (8.64) reduces to

$$E_2(s) = s^2 + 2\zeta_2 \omega_r s + \omega_r^2 \qquad (8.82)$$

and the function $K_2(s)$ in Eqs. (8.81) reduces to

$$K_{22}(s) = \dot{x}_2(0) + [s + 2\zeta_2 \omega_r]x_2(0) - 2\zeta_2 \omega_r x_1(0) \qquad (8.83)$$

The polynomial $D_1(s)$ in Eq. (8.67) reduces to the polynomial $D_2(s)$ given by Eq. (8.74) and, therefore, the responses of the masses m_1 and m_2 given by Eqs. (8.66) in the Laplace domain reduce to

$$X_1(s) = \frac{K_1(s)E_2(s)}{D_2(s)} + \frac{K_{22}(s)B(s)}{D_2(s)}$$

$$X_2(s) = \frac{K_1(s)C(s)}{D_2(s)} + \frac{K_{22}(s)A(s)}{D_2(s)} \qquad (8.84)$$

where $K_1(s)$ is given by Eqs. (8.81), $K_{22}(s)$ is given by Eq. (8.83), $E_2(s)$ is given by Eq. (8.82), $D_2(s)$ is given by Eq. (8.74), and $A(s)$ and $B(s)$ are given by Eqs. (8.64).

We now consider the special case where the masses m_1 and m_2 in Figure 8.8 are both subjected to the same initial velocity V_o; that is, the initial conditions are

$$x_1(0) = 0, \quad \frac{dx_1(0)}{dt} = V_o, \quad x_2(0) = 0, \quad \text{and} \quad \frac{dx_2(0)}{dt} = V_o \qquad (8.85)$$

Noting from Eqs. (8.61) that the nondimensional time $\tau = \omega_{n1}t$, the transforms $K_1(s)$ and $K_{22}(s)$ given by Eqs. (8.81) and (8.83), respectively, reduce to

$$K_1(s) = \dot{x}_1(0) = V_o/\omega_{n1}$$

$$K_{22}(s) = \dot{x}_2(0) = V_o/\omega_{n1} \qquad (8.86)$$

and, therefore, Eqs. (8.84) reduce to

$$X_1(s) = \frac{V_o}{\omega_{n1}D_2(s)}\,[E_2(s) + B(s)]$$

$$X_2(s) = \frac{V_o}{\omega_{n1}D_2(s)}\,[C(s) + A(s)] \tag{8.87}$$

The time-domain responses $x_1(\tau)$ and $x_2(\tau)$ are the inverse transforms of Eqs. (8.87). These have been obtained numerically[10] and they are shown in Figure 8.11. In Figure 8.11, solid lines are used to depict the response of the mass m_1 and broken lines are used to depict the response of the mass m_2. As expected, the free oscillations of the two masses show characteristics of damped oscillations, and the long-time responses of these two masses settle down to the equilibrium position; that is,

$$\lim_{\tau \to \infty} x_1(\tau) = 0 \quad \text{and} \quad \lim_{\tau \to \infty} x_2(\tau) = 0 \tag{8.88}$$

For $m_r = 0.1$, it is seen that the periods of the damped oscillations of the two masses are different. As the mass ratio m_r increases—that is, the mass m_2 increases in comparison to the mass m_1—the periods of damped oscillations of both masses approach each other. As seen for single degree-of-freedom systems, the responses to initial velocity seen in Figure 8.11 and the responses to impulses seen in Figure 8.9 have similar characteristics.

FIGURE 8.11
Responses of m_1 and m_2 when the masses are each subjected to the same initial velocity for different mass ratios m_r, $\zeta_1 = 0.1$, $\zeta_2 = 0.2$, and different values of ω_r: (a) $\omega_r = 0.3$ and (b) $\omega_r = 0.865$. [Solid line: $x_1(\tau)$; dashed line $x_2(\tau)$.]

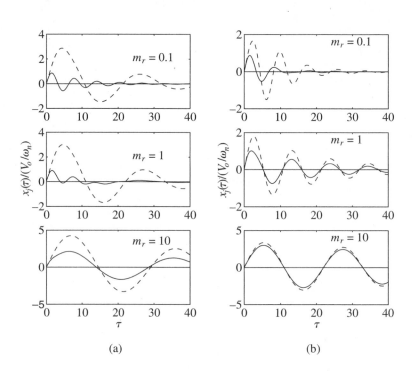

(a)   (b)

[10]The MATLAB function `ilaplace` from the Symbolic Toolbox was used.

EXAMPLE 8.10 Damped free oscillations of a spring-mass system revisited

We return to Example 8.2 and solve for the free-oscillation response of a two degree-of-freedom system by using Laplace transforms. The mass, stiffness, and damping matrices used in Examples 7.16 and 8.9 are used to generate the numerical results for the following initial conditions:

$$x_1(0) = d \text{ m}, \quad \dot{x}_1(0) = 0 \text{ m/s}, \quad x_2(0) = -d \text{ m}, \quad \text{and} \quad \dot{x}_2(0) = 0 \text{ m/s} \quad \text{(a)}$$

In Eqs. (a) of Example 8.2, $d = 0.1$.

For the initial conditions given in Eqs. (a), Eqs. (8.81) reduce to

$$\begin{aligned}
K_1(s) &= [s + 2\zeta_1 + 2\zeta_2 m_r \omega_r]d + 2\zeta_2 m_r \omega_r d \\
&= d\{s + 2\zeta_1 + 4\zeta_2 m_r \omega_r\} \\
K_2(s) &= -[s + 2\zeta_2 \omega_r (1 + c_{32})]d - 2\zeta_2 \omega_r d \\
&= -d\{s + 2\zeta_2 \omega_r (2 + c_{32})\} \quad \text{(b)}
\end{aligned}$$

From Eqs. (a) of Example 7.16 and Eqs. (7.41), we determine the following quantities:

$$\omega_{n1} = 2.887 \text{ rad/s}, \quad \omega_{n2} = 2.722 \text{ rad/s}$$

$$\omega_r = 0.943, \quad m_r = 2.25, \quad \text{and} \quad k_{32} = \frac{15}{20} = 0.75 \quad \text{(c)}$$

Furthermore, for the constant modal damping case of Example 8.9, we found that

$$c_1 = 0.332 \text{ N·s/m}$$
$$c_2 = 0.246 \text{ N·s/m}$$
$$c_3 = 0.654 \text{ N·s/m} \quad \text{(d)}$$

By using Eqs. (8.61), we determine that

$$\zeta_1 = \frac{0.332}{2 \times 1.2 \times 2.887} = 0.048$$

$$\zeta_2 = \frac{0.246}{2 \times 2.7 \times 2.722} = 0.017$$

$$c_{32} = \frac{0.654}{0.246} = 2.662 \quad \text{(e)}$$

Upon substituting the values given in Eqs. (c), (d), and (e) into Eqs. (8.64), (8.67) and (b), we obtain

$$A(s) = s^2 + 0.167s + 3$$
$$B(s) = 0.071s + 2$$
$$C(s) = 0.032s + 0.889$$
$$E(s) = s^2 + 0.115s + 1.556$$
$$D_1(s) = s^4 + 0.282s^3 + 4.573s^2 + 0.479s + 2.889$$
$$K_1(s) = 0.1s + 0.0238$$
$$K_2(s) = -0.1s - 0.0147 \quad \text{(f)}$$

By using Eqs. (f) in Eqs. (8.66), which give the displacements of the individual inertial elements in the Laplace transform domain, we arrive at

$$X_1(s) = \frac{0.1s^3 + 0.0282s^2 - 0.0427s + 0.0076}{s^4 + 0.282s^3 + 4.573s^2 + 0.4792s + 2.889}$$

$$X_2(s) = \frac{-0.1s^3 - 0.0282s^2 - 0.213s - 0.0229}{s^4 + 0.282s^3 + 4.573s^2 + 0.4792s + 2.889} \tag{g}$$

Upon taking the inverse Laplace transform of Eqs. (g) numerically,[11] we obtain the results shown in Figure 8.1.

In the absence of damping, $c_1 = c_2 = c_3 = 0$, and Eqs. (f) and (g) become, respectively,

$$A(s) = s^2 + 3$$
$$B(s) = 2$$
$$C(s) = 0.889$$
$$E(s) = s^2 + 1.556$$
$$D_1(s) = s^4 + 4.556s^2 + 2.889$$
$$K_1(s) = 0.1s$$
$$K_2(s) = -0.1s \tag{h}$$

and

$$X_1(s) = \frac{(0.1s^2 - 0.0444)s}{s^4 + 4.556s^2 + 2.889}$$

$$X_2(s) = \frac{-(0.1s^3 + 0.211)s}{s^4 + 4.556s^2 + 2.889} \tag{i}$$

Again taking the inverse Laplace transform of Eqs. (i) numerically, we obtain the results shown in Figure 8.2.

We now consider the two degree-of-freedom system with the same mass and stiffness matrices as before, but with a different damping matrix. In particular, we use the damping values of the arbitrary damping case of Example 8.9; that is, the values of the values of c_1 and c_3 remain the same as in Eqs. (d) while $c_2 = 0.49$ N·s/m. In this case, we find that $c_{32} = 1.33$ and $\zeta_2 = 0.0334$. Making the appropriate substitutions and taking the inverse Laplace transform numerically produces the same results as those shown in Figure 8.7b. We note that the Laplace transform method does not require any restrictions to be placed on the form of the viscous damping matrix.

8.4.3   Force Transmitted to a Boundary

Once the responses to certain forcing conditions and/or certain initial conditions are determined, one may also determine the force transmitted to a fixed boundary such as the left end in Figure 8.8. In this case, from a free-body diagram of mass m_1, the force transmitted is

---

[11]The function `ilaplace` from MATLAB's Symbolic Toolbox was used.

$$f_{base}(t) = c_1 \frac{dx_1}{dt} + k_1 x_1 \tag{8.89a}$$

which, in terms of the nondimensional time τ takes the form

$$f_{base}(\tau) = k_1 \left[2\zeta_1 \frac{dx_1}{d\tau} + x_1 \right] \tag{8.89b}$$

Taking the Laplace transforms of both sides of Eq. (8.89b), we obtain

$$F_{base}(s) = k_1[(2\zeta_1 s + 1)X_1(s) - 2\zeta_1 x_1(0)] \tag{8.90}$$

where $X_1(s)$ is given by Eqs. (8.66). Once $F_{base}(s)$ is determined, the corresponding time information is determined from

$$f_{base}(\tau) = L^{-1}[F_{base}(s)] \tag{8.91}$$

The transmitted force given by Eqs. (8.89) is used in Section 8.7, where we address the notion of transmissibility ratio.

8.5 TRANSFER FUNCTIONS AND FREQUENCY-RESPONSE FUNCTIONS

In Sections 5.3 and 6.2, we considered transfer functions associated with single degree-of-freedom systems and explained the relationship between a frequency-response function and a transfer function. In Section 8.2.3, we illustrated how frequency-response functions associated with a two degree-of-freedom system were constructed from the response to a harmonic excitation. Here, we discuss further frequency-response functions for two degree-of-freedom systems. The nature of the relationship between a frequency-response function and a transfer function is also addressed. The fundamental nature of this relationship remains the same as that discussed for a single degree-of-freedom system. However, in the case of a multi-degree-of-freedom system, due to the presence of more than one inertial element, one deals with more than one transfer function and one frequency-response function.

As in the case of vibratory systems described by linear single degree-of-freedom systems, the responses of vibratory systems described by linear multi-degree-of-freedom systems can be determined if the transfer functions for the systems are known. For this reason, the material presented here provides a basis for the discussions[12] on mechanical filters later in this section, vibration absorbers in Section 8.6, transmissibility ratio in Section 8.7, and the moving base model in Section 8.8.

We now obtain expressions for the transfer functions associated with the inertial elements m_1 and m_2 shown in Figure 8.8. We first set the initial

---

[12]If one goes back to Section 8.4 after completing this section, it will be seen that the notion of a transfer function was used in determining the numerical results for responses in the presence of different initial conditions and different forcing conditions.

conditions to zero, assume that $k_{32} = c_{32} = 0$ for convenience, and use Eqs. (8.65), (8.66), (8.74), and (8.82) to arrive at

$$X_1(s) = \frac{1}{k_1 D_2(s)} [F_1(s)E_2(s) + F_2(s)B(s)/m_r]$$

$$X_2(s) = \frac{1}{k_1 D_2(s)} [F_1(s)C(s) + F_2(s)A(s)/m_r] \tag{8.92}$$

Equations (8.92) will be used to determine four transfer functions, one pair associated with the forcing applied to one inertial element (say, m_1 in Figure 8.8) and the other pair associated with the forcing applied to the other inertial element. As discussed in Section 6.2 for a single degree-of-freedom system, an impulse force can be used to determine a transfer function. We determine the transfer functions $G_{ij}(s)$ where the subscript i refers to the response (or output) location and the subscript j refers to the force (or input) location. Therefore, to determine the first pair of transfer functions, an impulse forcing is applied to mass m_1; that is,

$$f_1(\tau) = F_o\delta(\tau)$$

$$f_2(\tau) = 0 \tag{8.93}$$

Then, we have[13]

$$G_{11}(s) = \frac{X_1(s)}{F_1(s)}$$

$$G_{21}(s) = \frac{X_2(s)}{F_1(s)} \tag{8.94}$$

where

$$F_1(s) = F_o \tag{8.95}$$

Making use of Eqs. (8.92), (8.94), and (8.95), we find that

$$k_1 G_{11}(s) = \frac{E_2(s)}{D_2(s)}$$

$$k_1 G_{21}(s) = \frac{C(s)}{D_2(s)} \tag{8.96}$$

Similarly, we determine the other pair of transfer functions by applying an impulse forcing to mass m_2; that is,

$$f_1(\tau) = 0$$

$$f_2(\tau) = F_o\delta(\tau) \tag{8.97}$$

Then, we have

$$G_{12}(s) = \frac{X_1(s)}{F_2(s)}$$

$$G_{22}(s) = \frac{X_2(s)}{F_2(s)} \tag{8.98}$$

---

[13]It should be clear from the form of Eqs. (8.94) that excitations other than impulse excitations can also be used to determine the transfer functions $G_{jk}(s)$.

where

$$F_2(s) = F_o \tag{8.99}$$

Making use of Eqs. (8.92), (8.98), and (8.99), we find that

$$k_1 G_{12}(s) = \frac{B(s)}{m_r D_2(s)}$$

$$k_1 G_{22}(s) = \frac{A(s)}{m_r D_2(s)} \tag{8.100}$$

From the form of Eqs. (8.96) and (8.100), it is evident that the polynomial $D_2(s)$ appears in the denominator of each transfer function; this is the same polynomial that is associated with the characteristic equation of this two degree-of-freedom system.

Note that Eqs. (8.62) are written in terms of the nondimensional time τ instead of t before the Laplace transforms were executed. Therefore, the frequency-response functions are given by $G_{il}(j\omega/\omega_{n1})$ or $G_{il}(j\Omega)$, where $\Omega = \omega/\omega_{n1}$ is the nondimensional frequency ratio. (See Laplace transform pair 1 in Table A of Appendix A.) Hence, the frequency-response functions are determined from Eqs. (8.96) and (8.100) to be

$$k_1 G_{11}(j\Omega) = \frac{E_2(j\Omega)}{D_2(j\Omega)}$$

$$k_1 G_{21}(j\Omega) = \frac{C(j\Omega)}{D_2(j\Omega)}$$

$$k_1 G_{12}(j\Omega) = \frac{B(j\Omega)}{m_r D_2(j\Omega)} = k_1 G_{21}(j\Omega)$$

$$k_1 G_{22}(j\Omega) = \frac{A(j\Omega)}{m_r D_2(j\Omega)} \tag{8.101}$$

where the terms in the numerators and the denominators are given by

$$A(j\Omega) = -\Omega^2 + 2(\zeta_1 + \zeta_2 m_r \omega_r)j\Omega + 1 + m_r \omega_r^2$$
$$B(j\Omega) = 2\zeta_2 m_r \omega_r j\Omega + m_r \omega_r^2$$
$$C(j\Omega) = 2\zeta_2 \omega_r j\Omega + \omega_r^2$$
$$E_2(j\Omega) = -\Omega^2 + 2\zeta_2 \omega_r j\Omega + \omega_r^2$$
$$D_2(j\Omega) = \Omega^4 - j[2\zeta_1 + 2\zeta_2 \omega_r m_r + 2\zeta_2 \omega_r]\Omega^3$$
$$\quad -[1 + m_r \omega_r^2 + \omega_r^2 + 4\zeta_1 \zeta_2 \omega_r]\Omega^2 + j[2\zeta_2 \omega_r$$
$$\quad + 2\zeta_1 \omega_r^2]\Omega + \omega_r^2 \tag{8.102}$$

When the damping is absent, $D_2(j\Omega)$ given by the last of Eqs. (8.102) reduces to the characteristic equation, Eq. (7.45), when the spring k_3 is absent.

The magnitudes of the frequency-response functions are given by

$$k_1 H_{il}(\Omega) = |G_{il}(j\Omega)| \quad i,l = 1, 2 \tag{8.103}$$

and the associated phase responses are given by

$$\varphi_{il}(\Omega) = \tan^{-1} \frac{\text{Im}[G_{il}(j\Omega)]}{\text{Re}[G_{il}(j\Omega)]} \tag{8.104}$$

As discussed in Section 5.3 for a single degree-of-freedom system and in Section 8.2.3 for a two degree-of-freedom system, frequency-response functions can also be constructed from responses to harmonic excitations. This can be used to interpret Eqs. (8.103) and (8.104) in the following manner. Let us suppose that a harmonic excitation of the following form acts on the system shown in Figure 8.8:

$$f_1(t) = F_o \cos(\omega t) \quad \text{or} \quad f_1(\tau) = F_o \cos(\Omega \tau)$$
$$f_2(t) = 0 \tag{8.105}$$

Then, $H_{11}(\Omega)$ and $H_{21}(\Omega)$ represent the amplitude-response functions of the inertia elements m_1 and m_2, respectively. These amplitude-response functions are functions of the excitation frequency ω or, in the nondimensional form, the frequency ratio Ω. The associated phase-response functions of the inertia elements m_1 and m_2 are given by $\varphi_{11}(\Omega)$ and $\varphi_{21}(\Omega)$, respectively. Similarly, the amplitude-response functions $H_{12}(\Omega)$ and $H_{22}(\Omega)$ provide the amplitudes of the responses of the inertia elements m_1 and m_2, respectively, when a harmonic excitation of the following form is imposed on the system shown in Figure 8.8.

$$f_1(t) = 0$$
$$f_2(t) = F_o \cos(\omega t) \quad \text{or} \quad f_2(\tau) = F_o \cos(\Omega \tau) \tag{8.106}$$

The associated phase-response functions of the inertia elements m_1 and m_2 are given by $\varphi_{12}(\Omega)$ and $\varphi_{22}(\Omega)$, respectively.

In Figure 8.12, for the case where $k_3 = c_3 = 0$, the nondimensional amplitude-response functions $k_1 H_{ij}(\Omega)$ are plotted as a function of the excitation frequency ratio Ω and the system frequency ratio ω_r. These plots are graphs of the functions given by Eqs. (8.103). Since the system is damped, the amplitude responses of the inertia elements m_1 and m_2 have finite values for all values of the excitation frequency. For "small" values of ω_r, it is seen that the response of one of the inertia elements is pronounced at and close to the lower resonance, while the response of the other inertia element is pronounced at and close to the higher resonance. As the excitation-frequency ratio Ω is increased past the higher resonance value, the responses of both inertia elements are relatively uniform, as was the case for responses of damped single degree-of-freedom systems forced by harmonic excitations.

In Figure 8.13, the amplitude responses are shown along with the associated phase responses. These responses have been generated by using Eqs. (8.103) and (8.104). Phase shift characteristics as seen in Chapter 5 for a harmonically forced single degree-of-freedom system at resonance can be seen at the two resonance frequency locations for the two degree-of-freedom system. The phase at each resonance is either $-90°$ or $90°$.

System Identification

For constructing Figures 8.12 and 8.13, the system parameters were assumed and substituted into Eqs. (8.103) and (8.104). In practice, one often measures frequency-response functions and then determines the system parameters, as discussed in Section 5.3 for single degree-of-freedom systems. This approach

FIGURE 8.12
Magnitudes of frequency-response functions for the two masses shown in Figure 8.8 when $k_3 = c_3 = 0$, $\zeta_1 = \zeta_2 = 0.05$, and $m_r = 0.1$ or 0.5: (a) $k_1 H_{n1}(\Omega)$, $n = 1, 2$ and (b) $k_1 H_{n2}(\Omega)$, $n = 1, 2$.

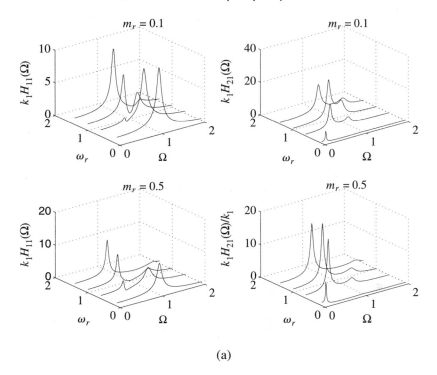

(a)

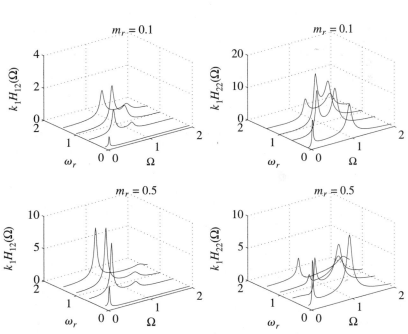

(b)

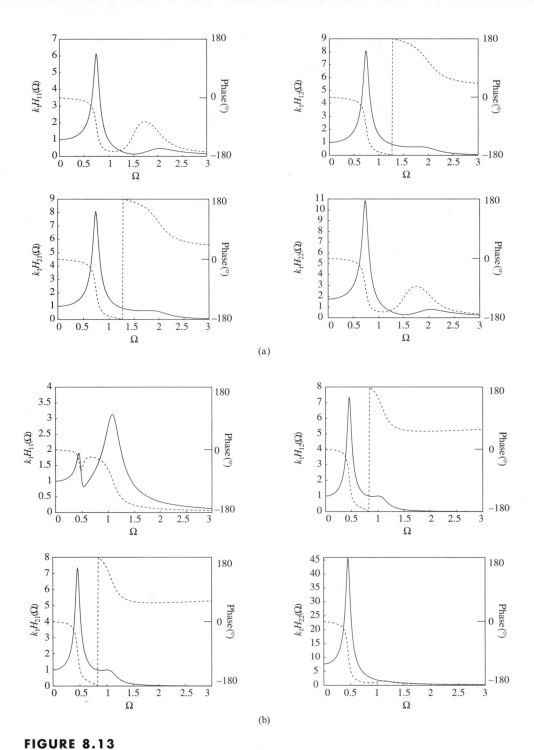

FIGURE 8.13

Frequency-responses functions for the two masses shown in Figure 8.8 as a function of $\Omega = \omega/\omega_{n1}$ when $k_3 = c_3 = 0$, $\zeta_1 = \zeta_2 = 0.1$, and $m_r = 0.6$: (a) $\omega_r = 1.5$ and (b) $\omega_r = 0.5$. [The solid lines represent amplitude responses and the dashed lines represent phase responses.]

to the problem is the inverse of the approach that we have taken so far; that is, going from measurements to the determination of system parameters rather than going from system parameters to response predictions. Thus, if a linear vibratory model is assumed to represent the physical system, then one fits the measurement data with functions such as those given by Eqs. (8.103) and (8.104) to obtain an estimate for the system parameters. This topic falls under the purview of modal analysis[14] and system identification.

To illustrate the determination of the parameters of a two degree-of-freedom system from measured data, consider the experimental data shown by the squares in Figure 8.14. They were obtained by applying a harmonically varying force of measured magnitude $F(\omega)$ at the frequency ω to mass m_1 and measuring the displacement response of mass m_2 at this frequency. The ratio of these two measured quantities at each value of ω is represented by the squares in Figure 8.14. In order to determine the system's properties, we use the model represented by $k_1 H_{21}(\omega)$ and fit the data to this model. From Eqs. (8.103), the quantity $H_{21}(\omega)$ is

$$H_{21}(\omega/\omega_{n1}) = \frac{1}{k_1} \left| \frac{C(j\omega/\omega_{n1})}{D_2(j\omega/\omega_{n1})} \right|$$

From this equation and Eqs. (8.102), we see that there are six parameters to be determined from the curve-fitting procedure: k_1, ω_{n1}, ζ_1, ζ_2, m_r, and ω_r. The known values are $H_{21}(\omega)$ and ω. By using the results of standard curve fitting procedures[15] we find that $k_1 = 9797.3$ N/m, $\omega_n = 250.5$ rad/s, $\zeta_1 =$

FIGURE 8.14

Experimentally obtained data values and the result after fitting a model to them.

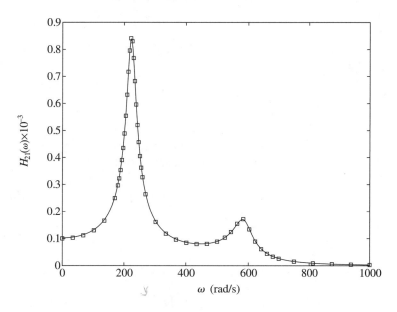

[14]D. Ewins, *Modal Testing: Theory and Practice,* John Wiley & Sons, NY (1984).

[15]The function `lsqnonlin` from MATLAB's Optimization Toolbox was used. For this model, the procedure is very sensitive to the initial guesses and must be employed interactively in order to obtain a satisfactory residual.

0.082, $\zeta_2 = 0.041$, $m_r = 0.204$, and $\omega_r = 2.095$. We now use the relations given by Eqs. (7.41) and (8.61) to determine the remaining parameters. Thus, from k_1 and ω_{n1} we obtain $m_1 = 0.156$ kg; from m_1 and m_r we obtain $m_2 = 0.032$ kg; from ω_r and ω_{n1} we obtain $\omega_{n2} = 524.8$ rad/s; from m_2 and ω_{n2} we obtain $k_2 = 8765.3$ N/m; from ζ_1, ω_{n1}, and m_1 we obtain $c_1 = 6.45$ N·s/m; and from ζ_2, ω_{n2}, and m_2 we obtain $c_2 = 1.37$ N·s/m.

EXAMPLE 8.11 Frequency-response functions for system with bounce and pitch motions

In this example, we revisit the physical system shown in Figure 7.5 when a force $f(t)$ and a moment $m(t)$ are imposed at the location G. Although the assumed forcing and moment inputs are not realistic, they have been considered to illustrate how frequency-response functions can be constructed. To determine these frequency-response functions, we first obtain the governing equations of the system for "small" oscillations. We then take the Laplace transforms of these equations to obtain the associated transfer functions. The frequency-response functions are determined from these transfer functions. We also illustrate how a sensor measurement can be represented in terms of the system frequency-response functions.

The governing equations for the unforced case have been obtained in Eq. (k) of Example 7.3. We modify Eq. (k) to include the force $f(t)$ and the moment $m(t)$ to obtain the following governing equations:

$$\begin{bmatrix} m & 0 \\ 0 & J_G \end{bmatrix} \begin{Bmatrix} \ddot{\hat{y}} \\ \ddot{\hat{\theta}} \end{Bmatrix} + \begin{bmatrix} c_1 + c_2 & -(c_1 L_1 - c_2 L_2) \\ -(c_1 L_1 - c_2 L_2) & (c_1 L_1^2 + c_2 L_2^2) \end{bmatrix} \begin{Bmatrix} \dot{\hat{y}} \\ \dot{\hat{\theta}} \end{Bmatrix}$$
$$+ \begin{bmatrix} k_1 + k_2 & -(k_1 L_1 - k_2 L_2) \\ -(k_1 L_1 - k_2 L_2) & (k_1 L_1^2 + k_2 L_2^2) \end{bmatrix} \begin{Bmatrix} \hat{y} \\ \hat{\theta} \end{Bmatrix} = \begin{Bmatrix} f(t) \\ m(t) \end{Bmatrix} \tag{a}$$

If we denote the Laplace transforms of, $\hat{y}(t)$, $\hat{\theta}(t)$, $f(t)$, and $m(t)$ by $\hat{Y}(s)$, $\hat{\Theta}(s)$, $F(s)$, and $M(s)$, respectively, then upon taking the Laplace transform of Eq. (a) and assuming that the initial conditions are zero, we arrive at

$$\begin{bmatrix} ms^2 + (c_1 + c_2)s + k_1 + k_2 & -(c_1 L_1 - c_2 L_2)s - (k_1 L_1 - k_2 L_2) \\ -(c_1 L_1 - c_2 L_2)s - (k_1 L_1 - k_2 L_2) & J_G s^2 + (c_1 L_1^2 + c_2 L_2^2)s + (k_1 L_1^2 + k_2 L_2^2) \end{bmatrix}$$
$$\cdot \begin{Bmatrix} \hat{Y}(s) \\ \hat{\Theta}(s) \end{Bmatrix} = \begin{Bmatrix} F(s) \\ M(s) \end{Bmatrix} \tag{b}$$

From Eq. (b), we find that

$$\begin{Bmatrix} \hat{Y}(s) \\ \hat{\Theta}(s) \end{Bmatrix} = \begin{bmatrix} G_{11}(s) & G_{12}(s) \\ G_{21}(s) & G_{22}(s) \end{bmatrix} \begin{Bmatrix} F(s) \\ M(s) \end{Bmatrix} \tag{c}$$

where the transfer functions $G_{ij}(s)$ are given by the following expressions:

$$G_{11}(s) = \frac{J_G s^2 + (c_1 L_1^2 + c_2 L_2^2)s + (k_1 L_1^2 + k_2 L_2^2)}{D(s)}$$

$$G_{12}(s) = G_{21}(s) = \frac{(c_1 L_1 + c_2 L_2)s + (k_1 L_1 - k_2 L_2)}{D(s)}$$

$$G_{22}(s) = \frac{ms^2 + (c_1 + c_2)s + k_1 + k_2}{D(s)} \tag{d}$$

and the polynomial $D(s)$ is given by

$$D(s) = [J_G s^2 + (c_1 L_1^2 + c_2 L_2^2)s + (k_1 L_1^2 + k_2 L_2^2)][ms^2 + (c_1 + c_2)s + k_1 + k_2]$$
$$- [(c_1 L_1 - c_2 L_2)s + (k_1 L_1 - k_2 L_2)]^2 \tag{e}$$

The frequency-response functions $G_{ik}(j\omega)$ are determined by setting $s = j\omega$ in Eqs. (d) and (e).

Let us suppose that an acceleration sensor is located at a distance L_{sensor} to the right of the point G in Figure 7.5. Then the acceleration measured by the sensor is

$$a_s(t) = \ddot{y}(t) + L_{sensor}\ddot{\theta}(t) \tag{f}$$

We shall determine how the frequency information in the sensor measurement is related to the frequency information in the forcing by making use of the frequency-response functions given by Eqs. (d). Upon taking the Laplace transform of Eq. (f) and assuming zero initial conditions, we arrive at

$$A_s(s) = s^2(\hat{Y}(s) + L_{sensor}\hat{\Theta}(s)) \tag{g}$$

For illustrative purposes, we assume that the excitation moment $m(t) = 0$; hence, $M(s) = 0$. Making use of Eq. (c) to express the responses $\hat{Y}(s)$ and $\hat{\Theta}(s)$ in terms of the applied forcing, we arrive at

$$\hat{Y}(s) = G_{11}(s)F(s)$$
$$\hat{\Theta}(s) = G_{21}(s)F(s) \tag{h}$$

where the transfer functions $G_{ik}(s)$ in Eq. (h) are given by Eqs. (d). Upon substituting Eqs. (h) into Eq. (g), we arrive at

$$A_s(s) = s^2[G_{11}(s) + L_{sensor}G_{21}(s)]F(s) \tag{i}$$

To obtain the frequency response function, we set $s = j\omega$ in Eq. (i), which leads to

$$A_s(j\omega) = -\omega^2[G_{11}(j\omega) + L_{sensor}G_{21}(j\omega)]F(j\omega) \tag{j}$$

Equation (j) relates the frequency information in the accelerometer measurement to the applied forcing and the system frequency-response functions. If the excitation moment $m(t) \neq 0$, then it will be found that the sensor measurement also depends on the frequency-response functions $G_{12}(j\omega)$ and $G_{22}(j\omega)$.

EXAMPLE 8.12 Amplitude response of a micromechanical filter

There is a class of devices called micromechanical filters that have been used in signal processing for the last 60 years.[16] These are resonant electro-mechanical systems that exhibit such characteristics as narrow bandwidth, low loss, and good stability. These devices are getting considerable attention again with the advent of small-scale systems being developed in the field of MEMS.[17] One form of these devices consists of two masses and three springs as shown in Figure 8.15a. The equivalent vibratory model is shown in Figure 8.15b. The device shown in Figure 8.15 consists of an inertial element m_1 being driven by an electrostatic comb transducer. An electrostatic comb transducer senses the motion of the other inertial element m_2. Between the two masses is a coupling spring k_2. The masses and the geometric parameters are chosen so that $m_1 = m_2$, $k_1 = k_3$, and $c_1 = c_3$. The goal is to select the appropriate values for m_j, k_j, and c_j to create a filter whose amplitude response of m_2 is relatively uniform within a specified bandwidth. The degree of uniformity is called *pass band ripple,* and the corresponding magnitudes are usually expressed in dB.

The filter is described by the transfer function involving the displacement response of m_2 and the force applied to mass m_1; that is, we are interested in the transfer function $G_{21}(s) = X_2(s)/F_1(s)$. To determine this function, we note that the governing equations of the system shown in Figure 8.13b are a special case of the system given by Eqs. (8.60) with $c_2 = 0$. Hence, we start from the responses in the Laplace domain given by Eqs. (8.66). Recognizing that the initial conditions are zero and $F_2(s) = 0$, we arrive at

$$X_2(s) = \frac{F_1(s)C(s)}{k_1 D_4(s)} \tag{a}$$

where, from Eqs. (8.64), we have that for $c_2 = 0$

$$C(s) = \omega_r^2 \tag{b}$$

and from Eqs. (8.67), we have

$$\begin{aligned}
D_4(s) = s^4 &+ [2\zeta_1 + 2\zeta_{32}\omega_r]s^3 \; [1 + m_r\omega_r^2 + \omega_r^2 + k_{31}/m_r + 4\zeta_1\zeta_{32}\omega_r]s^2 \\
&+ [2\zeta_1\omega_r^2 + 2k_{31}\zeta_1/m_r + 2\zeta_{32}\omega_r(1 + m_r\omega_r^2)]s \\
&+ \omega_r^2 + k_{31}(1 + m_r\omega_r^2)/m_r
\end{aligned} \tag{c}$$

In Eq. (c), we have introduced the quantities

$$\zeta_{32} = \frac{c_3}{m_2\omega_{n2}} \quad \text{and} \quad k_{31} = \frac{k_3}{k_1} \tag{d}$$

[16]R. A. Johnson, *Mechanical Filters in Electronics,* John Wiley & Sons, NY (1983).

[17]L. Lin et al., "Microelectromechanical Filters for Signal Processing," *J. Microelectromechanical Systems,* Vol. 7, No. 3, pp. 286–294 (September 1998); and K. Wang and C. T.-C. Nguyen, "High-Order Microelectromechanical Electronic Filters," Proceedings, 10th Annual International Workshop on Micro Electro Mechanical Systems, IEEE Robotics and Automation Society, pp. 25–30 (January 1997).

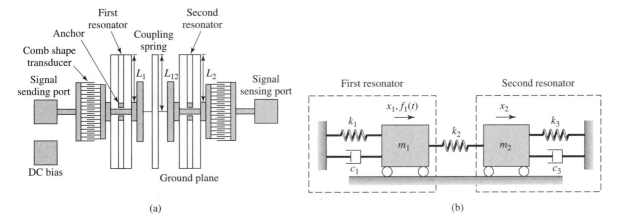

FIGURE 8.15

(a) Layout of a series two resonator micromechanical filter and (b) vibratory model.
Source: From L. Lin, et al., "Microelectromechanical Filters for Signal Processing," Journal of Microelectromechanical Systems, Vol. 7, No. 3, pp. 286–294 (September 1998). Copyright © 1998 IEEE. Reprinted with permission.

The nondimensional transfer function $k_1 G_{21}(s)$ is given by

$$k_1 G_{21}(s) = \frac{X_2(s)}{F_1(s)} = \frac{C(s)}{D_4(s)} = \frac{\omega_r^2}{D_4(s)} \tag{e}$$

The amplitude-response function is obtained by setting $s = j\Omega$ and considering the magnitude of $k_1 G_{21}(j\Omega)$. Thus,

$$k_1 H_{21}(\Omega) = |k_1 G_{21}(j\Omega)| = \left| \frac{\omega_r^2}{D_4(j\Omega)} \right| \tag{f}$$

where

$$\begin{aligned} D_4(j\Omega) = \ &\Omega^4 + j[2\zeta_1 + 2\zeta_{32}\omega_r]\Omega^3 \\ &- [1 + m_r\omega_r^2 + \omega_r^2 + k_{31}/m_r + 4\zeta_1\zeta_{32}\omega_r]\Omega^2 \\ &+ j[2\zeta_1\omega_r^2 + 2\zeta_1 k_{31}/m_r + 2\zeta_{32}\omega_r(1 + m_r\omega_r^2)]\Omega \\ &+ \omega_r^2 + k_{31}(1 + m_r\omega_r^2)/m_r \end{aligned} \tag{g}$$

The damping coefficients for micromechanical filters vary, but typically they are very low and on the order of 0.001, or less. To give an idea of the amplitude response of m_2, we select $m_r = k_{31} = 1$, $\omega_r = 0.25$, and $\zeta_1 = \zeta_{32} = 0.001$. The numerically obtained results are shown in Figure 8.16.

In Section 5.3.3, we discussed how a single degree-of-freedom system could be viewed as a mechanical filter. It was shown that the filter parameters are dependent on the system resonance and the system's damping factor. In this example, we have used a two degree-of-freedom system to construct a mechanical filter. Examining Figure 8.16, it is seen that the system parameters have been chosen to place the two system natural frequencies and, hence, their resonance locations, with a certain frequency separation. This separation determines the center frequency of the filter and the bandwidth BW of the filter. Recalling the definitions introduced in Section 5.3.3 for a filter, the bandwidth BW is determined by the -3 dB values of the amplitude response, which occur at Ω_{cl} and Ω_{cu}, the nondimensional lower and upper cutoff frequencies, respectively. In

FIGURE 8.16
Amplitude response of a micromechanical filter for $\omega_r = 0.25$, $m_r = k_{31} = 1$, and $\zeta_1 = \zeta_{32} = 0.001$. The data have been normalized with respect to the maximum value of $k_1 H_{21}(\Omega)$.

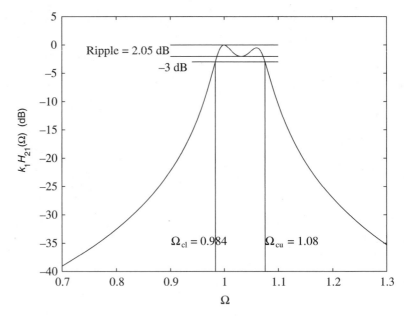

Section 5.3.3, we discussed that the lower the damping factor ζ, the higher the Q factor associated with a filter. Due to the low damping levels associated with microelectromechanical systems, they are attractive candidates for mechanical filters. However, unlike in Section 5.3.3 where certain explicit relationships between the filter parameters and the system parameters of a single degree-of-freedom system could be obtained, when a multi-degree-of-freedom system is designed to act as a filter, one has to resort to numerical means to determine how a change in a certain system parameter affects the filter design.

8.6 VIBRATION ABSORBERS

Vibration absorbers are used in many applications, which include power transmission lines, automobiles, aircraft, optical platforms, and rotating machinery. In Section 8.2.3, we introduced the undamped vibration absorber and showed how the absorber system could be designed based on the zeros of the forced response or the frequency-response function. In this section, we revisit this problem and broaden the discussion to include damping and different types of dampers. In Sections 8.6.1 and 8.6.2, we discuss the design of two very different types of vibration absorbers that are based on linear-system principles. In Section 8.6.3, we introduce nonlinear vibration absorbers and give an indication of the types of absorber designs that are possible.

8.6.1 Linear Vibration Absorber

As an illustration of a linear vibration absorber, we consider the system shown in Figure 8.17. In this system, it is assumed that the primary system with damping

FIGURE 8.17
System with vibration
absorber.

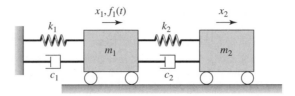

has attached to it a secondary system, the vibration absorber. A disturbance $f_1(t)$ acts on the primary system and it is assumed that there is no force acting on the absorber mass m_2. The governing equations are determined from Eqs. (8.62) by setting $f_2(\tau) = 0$, $k_{32} = 0$ ($k_3 = 0$), and $c_{32} = 0$ ($c_3 = 0$). This leads to

$$\frac{d^2 x_1}{d\tau^2} + (2\zeta_1 + 2\zeta_2 m_r \omega_r) \frac{dx_1}{d\tau} + (1 + m_r \omega_r^2) x_1$$

$$- 2\zeta_2 m_r \omega_r \frac{dx_2}{d\tau} - m_r \omega_r^2 x_2 = \frac{f_1(\tau)}{k_1}$$

$$\frac{d^2 x_2}{d\tau^2} + 2\zeta_2 \omega_r \frac{dx_2}{d\tau} + \omega_r^2 x_2 - 2\zeta_2 \omega_r \frac{dx_1}{d\tau} - \omega_r^2 x_1 = 0 \qquad (8.107)$$

where the different parameters in Eqs. (8.107) are given by Eqs. (7.41).

The design question that is posed is the following: How can a combination of parameters k_2, c_2, and m_2 of the secondary system be chosen so that the response amplitude of the primary system is at a minimum (or zero) in the specified frequency range of the excitation? To answer this question, we start from the responses in the Laplace domain given by Eqs. (8.66). The frequency-response functions are derived from these expressions by setting $s = j\Omega$, where Ω is a nondimensional frequency parameter. Thus, we arrive at

$$\frac{X_1(j\Omega)}{F_1(j\Omega)} = G_{11}(j\Omega) = \frac{E_2(j\Omega)}{k_1 D_2(j\Omega)} \qquad (8.108)$$

where $D_2(j\Omega)$ and $E_2(j\Omega)$ are given by Eqs. (8.102). The relation for $E_2(j\Omega)$ is repeated here for convenience as

$$E_2(j\Omega) = -\Omega^2 + j(2\zeta_2 \omega_r \Omega) + \omega_r^2 \qquad (8.109)$$

In Example 8.3, we considered an undamped primary system and an undamped vibration absorber, and it was shown that when this primary system is subjected to a harmonic excitation at the natural frequency of the primary system, the vibration absorber can be designed to provide an equal and opposite force to the excitation force on the primary system. Therefore, the effective excitation experienced by the primary system is cancelled at this excitation frequency. In the present case, the primary and secondary systems are damped. The question is whether the same canceling effect can be accomplished.

Special Case of Absorber System: $\zeta_2 = 0$

From Eqs. (8.108), we see that when $E_2(j\Omega) = 0$ the response of the primary system given by $X_1(j\Omega)$ is also zero. When ζ_2 is zero, it is seen from Eq. (8.109) that $E_2(j\Omega) = 0$ if

$$\Omega = \omega_r \qquad (8.110a)$$

or equivalently from Eqs. (7.41) that

$$\omega = \omega_{n2} \tag{8.110b}$$

Equation (8.110a) is the same as that previously determined from Eq. (h) of Example 8.3. Thus, if we choose k_2 and m_2 so that Eq. (8.110b) is satisfied, then we can have a zero of the forced response of the primary system at the chosen excitation frequency. The implication of this observation is as follows. Suppose that a harmonic excitation is imposed on the primary mass m_1 at a frequency $\omega = \omega_{n1}$, where ω_{n1} is the natural frequency of the undamped primary system; that is, the system of Figure 8.17 without the secondary system. In the absence of the secondary system, since we are exciting the primary system at its undamped natural frequency, we expect the response of this linear system to be "large." With the inclusion of an undamped secondary system, and for the choice of k_2 and m_2 satisfying Eqs. (8.110), we find that the response of the primary system is zero at $\omega = \omega_{n1}$. The absorber parameters k_2 and m_2 need to be chosen so that

$$\omega_{n2} = \omega = \omega_{n1} \tag{8.111}$$

where ω_{n2} is the natural frequency of the undamped, uncoupled secondary system. As discussed in Example 8.4, another equation or condition apart from Eq. (8.111) will be needed to determine the absorber parameters.

In Example 8.3, we found that the response of the primary mass is zero despite having an excitation acting directly on it, since the force produced by the absorber on the primary mass is equal and opposite to it. To see if this holds true for this case, we consider Eqs. (8.63) to (8.65), set $k_3 = c_2 = c_3 = f_2(\tau) = 0$, assume that the initial conditions are zero, and set $s = j\Omega$ to obtain

$$(-\Omega^2 + 1 + m_r\omega_r^2 + 2j\zeta_1\Omega)X_1(j\Omega) - m_r\omega_r^2 X_2(j\Omega) = \frac{F_1(j\Omega)}{k_1}$$

$$-\omega_r^2 X_1(j\Omega) + (-\Omega^2 + \omega_r^2)X_2(j\Omega) = 0 \tag{8.112}$$

In Eqs. (8.112), $X_1(j\Omega)$ and $X_2(j\Omega)$ are the complex amplitudes of the frequency responses of the inertia elements m_1 and m_2, respectively. From the second of Eqs. (8.112), if the vibration absorber is chosen to satisfy Eq. (8.110a), then

$$X_1(j\Omega) = 0 \tag{8.113a}$$

and, hence, it follows from the first of Eqs. (8.112) and Eqs. (7.41) that

$$X_2(j\Omega) = -\frac{F_1(j\Omega)}{k_1 m_r \omega_r^2} = -\frac{F_1(j\Omega)}{k_2} \tag{8.113b}$$

which is identical to Eq. (j) of Example 8.3. Thus, the force generated by the spring k_2 on the primary mass m_1 opposes the disturbing force and is equal to it in magnitude. As a result, the mass m_1 does not move and instead all of the energy provided to the system through the forcing $f_1(t)$ is absorbed by the secondary system.

Although we would like to take advantage of the zero of the forced response at a selected excitation frequency, it is not possible to realize this in practice. This can occur because of poor frequency tuning of the absorber after installation or frequency detuning over time caused by changes to the primary system stiffness and inertia characteristics or by changes to the absorber system stiffness and inertia characteristics. The possible variations in the system parameters poses the question: How can one design an absorber to be effective over a broad frequency range? Before answering this question, we return to the free oscillation problem. With the introduction of the secondary system, the absorber, the two degree-of-freedom system has two natural frequencies. The two natural frequencies of the undamped two degree-of-freedom system are given by the roots of the characteristic equation, which is

$$D_2(j\Omega) = \Omega^4 - [1 + m_r\omega_r^2 + \omega_r^2]\Omega^2 + \omega_r^2 = 0 \tag{8.114}$$

The solution of Eq. (8.114), which is a quadratic equation in Ω^2, is

$$\Omega_{1,2} = \sqrt{\frac{1}{2}\left[1 + \omega_r^2(1 + m_r) \mp \sqrt{(1 + \omega_r^2(1 + m_r))^2 - 4\omega_r^2}\right]} \tag{8.115}$$

The variations of the two nondimensional natural frequencies Ω_1 and Ω_2 are plotted in Figure 8.18 with respect to the mass ratio $m_r = m_2/m_1$ and the frequency ratio ω_r. As the mass ratio is increased, the separation between the two natural frequencies increases. In the presence of a harmonic disturbance acting on the mass of the primary system, we would like the response of mass m_1 not to be large when we are operating at frequencies close to either of the two resonance frequencies of the undamped two degree-of-freedom system. To address this, we return to the forced oscillation problem.

FIGURE 8.18

Natural frequencies of a two degree-of-freedom system with a vibration absorber as a function of the mass ratio m_r.

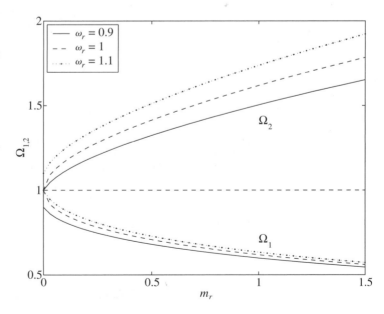

General Case of Absorber System

In Figures 8.19 and 8.20, the nondimensional amplitude response

$$k_1 H_{11}(\Omega) = |k_1 G_{11}(j\Omega)| \tag{8.116}$$

determined from Eq. (8.108) is plotted as a function of the nondimensional excitation frequency ratio Ω. The mass ratio m_r, the ratio frequency ω_r, and the damping of the primary system expressed by ζ_1 are held fixed in each case, and the damping factor ζ_2 is varied. The ratio ω_r is decreased from 1 to 0.97 (i.e., by 3%) in going from Figure 8.19 to Figure 8.20. In both figures, we have one scenario with an undamped vibration absorber and two scenarios with two different damped vibration absorbers. The responses seen in Figures 8.19 and 8.20 are characteristic of two degree-of-freedom systems excited by a harmonic excitation, where the excitation frequency range includes the two resonance frequencies. Based on Eq. (8.110a), it is clear that when $\zeta_2 = 0$, the response of the primary system is zero. However, when $\zeta_2 \neq 0$ and/or if the nondimensional excitation frequency is different from $\Omega = 1$, the response of the primary system may not have a "small" magnitude. Therefore, we would like to choose the absorber system parameters so that the response of the primary system is as small as possible over a wide frequency range that includes $\Omega = 1$.

For the general case, when the damping factor ζ_1 of the primary system is not zero, the choice of the secondary system parameters cannot be put in explicit form such as Eq. (8.111) and one has to resort to numerical means. However, when $\zeta_1 = 0$, one can obtain[18] optimal values for the secondary

FIGURE 8.19

Amplitude response of primary mass m_1 with a damped vibration absorber: $m_r = 0.1$, $\zeta_1 = 0.1$, and $\omega_r = 1$.

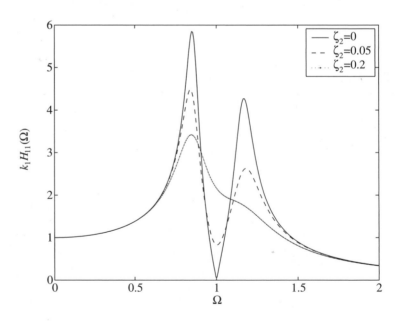

[18]J. P. Den Hartog, *Mechanical Vibrations*, Dover, NY, pp. 87–113 (1985).

FIGURE 8.20
Amplitude response of primary mass m_1 with a damped vibration absorber and when $\omega_r \neq 1$: $m_r = 0.1$, $\zeta_1 = 0.1$, and $\omega_r = 0.97$.

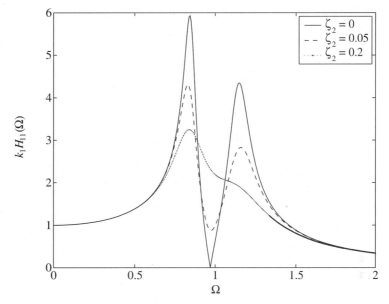

system natural frequency and the secondary system damping ratio to tailor the amplitude response $H_{11}(\Omega)$ of the primary system. The optimal values for the absorber system parameters are obtained when the values of the peaks shown in Figure 8.21 are equal; that is, when

$$H_{11}(\Omega_A) = H_{11}(\Omega_B) \tag{8.117}$$

Upon carrying out the derivation, the following optimal values are obtained:

$$\omega_{r,opt} = \frac{1}{1 + m_r} \quad \text{and} \quad \zeta_{2,opt} = \sqrt{\frac{3m_r}{8(1 + m_r)^3}} \tag{8.118}$$

The ratio ω_r is optimal or there is optimal tuning when the first of Eqs. (8.118) is satisfied. The second of Eqs. (8.118) provides the optimal damping value for the secondary system. It is important to note that the mass ratio is the sole determining factor for the optimal values. For a general case, when $\zeta_1 \neq 0$, although the optimal condition expressed by Eq. (8.117) is valid, explicit forms such as Eqs. (8.118) cannot be obtained.

The objective is to determine optimal secondary system parameters for a damped absorber (i.e., $\zeta_2 \neq 0$) so that there is an operating region including $\Omega = 1$ wherein the variation of the amplitude of the primary system m_1 as a function of frequency can remain relatively constant. That is, we would like to reduce the absorber's sensitivity to small variations in frequency. This can be realized through optimization solution techniques.[19] Referring to Figure 8.21, the goal is to find the secondary system parameters for which the peak amplitudes A and B are equal and are as "small" as possible, while the

[19]E. Pennestri, "An Application of Chebyshev's Min-Max Criterion to the Optimal Design of a Damped Dynamic Vibration Absorber," *J. Sound Vibration,* Vol. 217, No. 4, pp. 757–765, 1998.

FIGURE 8.21
Identification of amplitude response functions to be minimized.

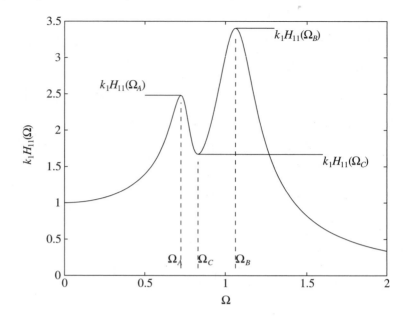

minimum between these peaks C is as close to A and B as possible in terms of magnitude. In other words, we would like to find the system parameters that minimize each of the following three maximum values simultaneously: $H_{11}(\Omega_A)$, $H_{11}(\Omega_B)$, and $1/H_{11}(\Omega_C)$. This can be stated as follows.

$$\min_{\omega_r, \zeta_2} \{H_{11}(\Omega_A)\}$$
$$\min_{\omega_r, \zeta_2} \{H_{11}(\Omega_B)\}$$
$$\min_{\omega_r, \zeta_2} \{1/H_{11}(\Omega_C)\}$$
$$\text{subject to: } \omega_r > 0$$
$$\zeta_2 \geq 0 \tag{8.119}$$

For illustration, we assume that $\zeta_1 = 0.1$ and $m_r = 0.1, 0.2, 0.3$ and 0.4 and use readily available numerical procedures[20] to determine the optimum values of $\zeta_{2,opt}$ and $\omega_{r,opt}$. The results are shown in Figure 8.22 for four different values of the mass ratio m_r. It is seen that these results satisfy the condition given by Eq. (8.117). Comparing the results shown in the four cases in Figure 8.22, it is clear that the last case is preferable, since the primary system's response is attenuated to the smallest magnitude over the chosen frequency range. For a given primary system mass, when $m_r = 0.1$, the secondary system is the smallest and the displacement amplitude of the secondary system is likely to be the largest. Since there are restrictions on the amplitude of motion of the secondary system, if the displacement amplitude of the secondary system is required to be less than a certain value, then this

---

[20]The MATLAB function `fminimax` from the Optimization Toolbox was used.

FIGURE 8.22

Optimum values for the parameters of a vibration absorber and the resulting amplitude responses of m_1 for $\zeta_1 = 0.1$. The dashed lines are the responses obtained for the optimum values given by Eqs. (8.118).

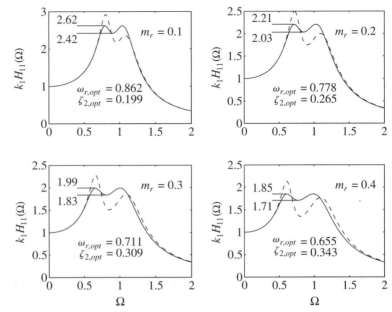

requirement can be included as a constraint for Eqs. (8.119). In terms of design, there are still other choices one can come up with if Eq. (8.119) is replaced with another set of objectives. However, here, we have pointed out that for a practical design of an absorber, one will have to resort to numerical means and take advantage of algorithms such as that used in the present context. Furthermore, it is noted that when one carries out design, usually there is more than one design solution to be reckoned with. In addition, the number of degrees of freedom is likely to be higher than two in a practical situation.

Upon comparing the results obtained from Eqs. (8.118), we see that there are differences. For example, when $m_r = 0.1$, Eqs. (8.118) give that $\omega_{r,opt} = 0.909$ and $\zeta_{2,opt} = 0.168$, whereas, from Figure 8.22 we see that the numerical optimization procedure gives $\omega_{r,opt} = 0.862$ and $\zeta_{2,opt} = 0.199$. Similarly, for $m_r = 0.4$, Eqs. (8.118) give that $\omega_{r,opt} = 0.714$ and $\zeta_{2,opt} = 0.234$, whereas, from Figure 8.22 we see that the numerical optimization procedure gives $\omega_{r,opt} = 0.655$ and $\zeta_{2,opt} = 0.343$. However, it is seen from Figure 8.22 that the results given by Eqs. (8.118) provide good estimations of the design parameters.

In designing a vibration absorber, there are other aspects such as the static deflection of the two degree-of-freedom system, stresses in the absorber springs, displacement of the absorber mass, etc. that one has to deal with. Many practical hints for designing absorbers are available in the literature.[21]

[21]H. Bachmann et al., *Vibration Problems in Structures: Practical Guidelines,* Birkhäuser Verlag, Basel, Germany, Appendix D (1995).

> **Design Guideline:**   In many practical situations, a vibration absorber system should have a ratio of the absorber mass to the primary system mass of about one-tenth, a damping factor for the absorber system of about 0.2 when the primary system is lightly damped, and an uncoupled natural frequency of the absorber system that is about 15% less than that of the primary system.

EXAMPLE 8.13   Absorber design for a rotating system with mass unbalance

A rotating system with an unbalanced mass is modeled as a single degree-of-freedom system with a mass of 10 kg, a natural frequency of 40 Hz, and a damping factor of 0.2. The system requires an undamped absorber so that the natural frequencies of the resulting two degree-of-freedom system are outside the frequency range of 30 Hz to 50 Hz. In addition, the absorber is to be designed such that for force amplitudes of up to 6000 N at 40 Hz, the steady-state amplitude of the absorber's response will not exceed 20 mm.

The parameters provided for the primary system are

$$m_1 = 10 \text{ kg}$$
$$\omega_{n1} = 40 \times 2\pi = 251.33 \text{ rad/s}$$
$$\zeta_1 = 0.2 \tag{a}$$

For this system, the absorber must be such that the resulting two degree-of-freedom system's lowest natural frequency ω_1 and the highest natural frequency ω_2 have, respectively, the following limits.

$$\omega_1 < 30 \times 2\pi = 188.50 \text{ rad/s}$$
$$\omega_2 > 50 \times 2\pi = 314.16 \text{ rad/s} \tag{b}$$

or

$$\Omega_1 = \frac{\omega_1}{\omega_{n1}} < \frac{188.50 \text{ rad/s}}{251.33 \text{ rad/s}} = 0.75$$
$$\Omega_2 = \frac{\omega_2}{\omega_{n1}} > \frac{314.16 \text{ rad/s}}{251.33 \text{ rad/s}} = 1.25 \tag{c}$$

where Ω_1 and Ω_2 are the nondimensional natural frequency limits of the rotating system with the absorber. The requirement on the steady-state amplitude of the absorber at 40 Hz is expressed as

$$|X_2(j\omega)|_{\omega=251.33} \leq 20 \times 10^{-3} \tag{d}$$

For the absorber to meet the requirements given by Eqs. (b), (c) and (d), we choose an undamped spring-mass system of stiffness k_2 and mass m_2. To meet the requirement given by Eqs. (c), we note from Figure 8.18 that the larger the mass ratio, the larger the separation between the nondimensional natural frequencies Ω_1 and Ω_2. The frequency ratio ω_r is chosen to satisfy Eq. (8.111); that is,

$$\omega_r = \frac{\omega_{n2}}{\omega_{n1}} = 1 \tag{e}$$

Then, Eq. (8.115) is used to determine the nondimensional natural frequencies Ω_1 and Ω_2 for a chosen mass ratio m_r. Let

$$m_r = \frac{m_2}{m_1} = 0.6 \tag{f}$$

From Eqs. (8.115), (e), and (f), we arrive at

$$\begin{aligned}
\Omega_{1,2} &= \sqrt{\frac{1}{2}[1 + \omega_r^2(1 + m_r) \mp \sqrt{(1 + \omega_r^2(1 + m_r))^2 - 4\omega_r^2}]} \\
&= \sqrt{\frac{1}{2}[1 + 1^2 \times (1 + 0.6) \mp \sqrt{(1 + 1^2 \times (1 + 0.6))^2 - 4 \times 1^2}]} \\
&= 0.685, \ 1.460 \tag{g}
\end{aligned}$$

Thus, for the chosen mass ratio m_r and frequency ratio ω_r, the requirement given by Eq. (b) is satisfied; that is, the two degree-of-freedom system's natural frequencies are outside the range of 30 Hz to 50 Hz.

It remains to be determined if the requirement (d) is satisfied. Making use of Eqs. (a), (e), and (f), we find that the absorber stiffness k_2 is given by

$$\begin{aligned}
k_2 &= m_2\omega_{n2}^2 = (m_r m_1)(\omega_r \omega_{n1})^2 \\
&= (0.6 \times 10 \text{ kg}) \times (1 \times 251.33 \text{ rad/s})^2 \\
&= 379 \times 10^3 \text{ N/m} \tag{h}
\end{aligned}$$

Since we have chosen an undamped vibration absorber (i.e., $\zeta_2 = 0$) and $\omega_r = 1$, we can use Eq. (8.113b) to determine if Eq. (d) is satisfied. Making use of the given value for the forcing amplitude of 6000 N and the stiffness k_2 calculated in Eq. (h), we find that

$$|X_2|_{\omega=251.33} = \left| -\frac{6000 \text{ N}}{379 \times 10^3 \text{ N/m}} \right| = 15.83 \times 10^{-3} \text{ m} \tag{i}$$

Hence, comparing Eqs. (d) and (h), we find that the second requirement has also been satisfied. Had we chosen a frequency ratio ω_r different from 1, we could not have used Eq. (8.113b). Instead, we would have to use Eqs. (8.112) to determine the steady-state amplitude of the absorber response.

EXAMPLE 8.14 Absorber design for a machine system

A machine has a mass of 200 kg and a natural frequency of 130 rad/s. An absorber mass of 20 kg and a spring-damper combination is to be attached to this machine so that the machine can be operated in as wide a frequency range as possible around the machine's natural frequency. We shall determine the values for the absorber spring constant k_2 and damping coefficient c_2.

For the given parameters, the mass ratio m_r is

$$m_r = \frac{m_2}{m_1} = \frac{20 \text{ kg}}{200 \text{ kg}} = 0.1 \tag{a}$$

The optimal frequency ratio $\omega_{r,opt}$ and the optimal damping ratio $\zeta_{2,opt}$ are determined from Eqs. (8.118) as

$$\omega_{r,opt} = \frac{1}{1 + m_r} = \frac{1}{1 + 0.1} = 0.909$$

$$\zeta_{2,opt} = \sqrt{\frac{3m_r}{8(1 + m_r)^3}} = \sqrt{\frac{3 \times 0.1}{8(1 + 0.1)^3}} = 0.168 \tag{b}$$

The absorber's natural frequency is

$$\omega_{n2} = \omega_{r,opt}\omega_{n1} = 0.909 \times (130 \text{ rad/s}) = 118.17 \text{ rad/s} \tag{c}$$

Hence, the absorber stiffness is

$$k_2 = m_2\omega_{n2}^2$$
$$= (20 \text{ kg}) \times (118.17 \text{ rad/s})^2 = 279.28 \times 10^3 \text{ N/m} \tag{d}$$

and the absorber damping coefficient is

$$c_2 = 2\zeta_2 m_2\omega_{n2}$$
$$= 2 \times 0.168 \times (20 \text{ kg}) \times (118.17 \text{ rad/s})$$
$$= 794.10 \text{ N/(m/s)} \tag{e}$$

8.6.2 Centrifugal Pendulum Vibration Absorber[22]

Centrifugal pendulum absorbers, which are also known as rotating pendulum vibration absorbers, are widely used in many rotary applications. For example, in rotating shafts, the centrifugal pendulum absorber is often employed to reduce undesirable vibrations at frequencies that are proportional to the rotational speed of the shaft. Since the rotational speed can vary over a wide range, in order to have an effective absorber, the natural frequency of the absorber should be proportional to the rotational speed. This can be realized by using the centrifugal pendulum vibration absorber.

A conceptual representation of rotary system with this absorber is illustrated in Figure 8.23, where the pendulum absorber moves in the plane containing the unit vectors e_1 and e_2. These unit vectors are fixed to the pendulum, which has a mass m and length r. The unit vectors e_1' and e_2' are fixed to the rotary system. For this two degree-of-freedom system, we will derive the

[22]See, for example: R. G. Mitchiner and R. G. Leonard, "Centrifugal Pendulum Vibration Absorbers—Theory and Practice," *J. Vibration Acoustics*, Vol. 113, pp. 503–507 (1991); M. Sharif-Bakhtiar and S. W. Shaw, "Effects of Nonlinearities and Damping on the Dynamic Response of a Centrifugal Pendulum Vibration Absorber," *J. Vibration Acoustics*, Vol. 114, pp. 305–311 (1992); and M. Hosek, H. Elmali, and N. Olgac, "A tunable vibration absorber: the centrifugal delayed resonator," *J. Sound Vibration*, Vol. 205, No. (2), pp. 151–165 (1997).

governing nonlinear equations of the system by using the generalized coordinates θ and φ, linearize these equations, and use these equations to show how this absorber can be designed. In Figure 8.23, the angle φ is measured relative to the rotary system, which has a rotary inertia J_O about the point O, which is fixed for all time. An external moment $M_T(t)$ acts on the rotary system.

Equations of Motion

The Lagrange equations given by Eqs. (7.7) are used to get the governing equations of motion. From Figure 8.23, we see that the velocity of the pendulum with respect to the point O is given by

$$V_m = R\dot\theta\, e_1' + r(\dot\varphi + \dot\theta)e_2 = (R\dot\theta\,\sin\varphi)e_1 + (R\dot\theta\,\cos\varphi + r(\dot\varphi + \dot\theta))e_2 \quad (8.120)$$

Hence, the kinetic energy of the system is

$$\begin{aligned}
T &= \frac{1}{2}J_O\dot\theta^2 + \frac{1}{2}m(V_m \cdot V_m)\\
&= \frac{1}{2}J_O\dot\theta^2 + \frac{1}{2}m[(R\dot\theta\,\sin\varphi)^2 + (R\dot\theta\,\cos\varphi + r(\dot\varphi + \dot\theta))^2]\\
&= \frac{1}{2}J_O\dot\theta^2 + \frac{1}{2}m[R^2\dot\theta^2 + r^2(\dot\varphi + \dot\theta)^2 + 2rR(\dot\theta^2 + \dot\varphi\dot\theta)\cos\varphi]\\
&= \frac{1}{2}[J_O + m(R^2 + r^2 + 2rR\cos\varphi)]\dot\theta^2 + \frac{1}{2}mr^2\dot\varphi^2\\
&\quad + mr(r + R\cos\varphi)\dot\varphi\dot\theta
\end{aligned} \quad (8.121)$$

If we ignore the effects of gravity and the stiffness of the rotary system,[23] then there is no contribution to the system potential energy. In addition, since we have not taken any form of damping into account, there is no dissipation. For the generalized coordinates we have $q_1 = \theta$ and $q_2 = \varphi$. Then, setting $D = V = 0$ and recognizing that $Q_1 = M_T(t)$ and $Q_2 = 0$, the Lagrange equations given by Eqs. (7.7) reduce to the following:

$$\begin{aligned}
\frac{d}{dt}\left(\frac{\partial T}{\partial \dot\theta}\right) - \frac{\partial T}{\partial \theta} &= M_T(t)\\
\frac{d}{dt}\left(\frac{\partial T}{\partial \dot\varphi}\right) - \frac{\partial T}{\partial \varphi} &= 0
\end{aligned} \quad (8.122)$$

After substituting Eqs. (8.121) into Eqs. (8.122) and carrying out the indicated differentiations, we obtain the following coupled, nonlinear equations:

$$\begin{aligned}
[J_O + m(R^2 + r^2 + 2rR\cos\varphi)]\ddot\theta &+ mr(r + R\cos\varphi)\ddot\varphi\\
- mrR\dot\theta\dot\varphi\sin\varphi - mrR\dot\varphi^2\sin\varphi &= M_T(t)\\
mr(r + R\cos\varphi)\ddot\theta + mr^2\ddot\varphi + mrR\dot\theta^2\sin\varphi &= 0
\end{aligned} \quad (8.123)$$

Dividing each of the terms in Eqs. (8.123) by mR^2 and introducing the nondimensional pendulum length ratio $\gamma = r/R$, we arrive at

$$\begin{aligned}
J_1(\varphi)\ddot\theta + J_2(\varphi)\ddot\varphi - \gamma(2\dot\varphi\dot\theta + \dot\varphi^2)\sin\varphi &= M_T(t)/(mR^2)\\
J_2(\varphi)\ddot\theta + \gamma^2\ddot\varphi + \gamma\sin\varphi\dot\theta^2 &= 0
\end{aligned} \quad (8.124)$$

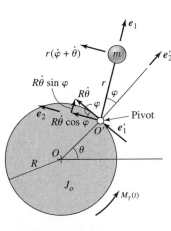

FIGURE 8.23
Centrifugal pendulum absorber.

[23]For consideration of rotary system stiffness, see C. Genta, Chapter 5, *ibid.*

where

$$J_1(\varphi) = J_O/(mR^2) + 1 + \gamma^2 + 2\gamma \cos \varphi$$
$$J_2(\varphi) = \gamma^2 + \gamma \cos \varphi \tag{8.125}$$

The rotation θ of the rotary system is assumed to consist of two parts, one, a steady-state rotation ωt, due to rotation at a constant angular speed ω, and another part $\psi(t)$ due to a disturbance about this rotation. Thus, we arrive at

$$\theta(t) = \omega t + \psi(t) \tag{8.126}$$

After substituting Eq. (8.126) into Eqs. (8.124), we obtain

$$J_1(\varphi)\ddot{\psi} + J_2(\varphi)\ddot{\varphi} - \gamma(2\dot{\varphi}\dot{\psi} + \dot{\varphi}^2)\sin \varphi - 2\omega\gamma\dot{\varphi} \sin \varphi = M_T(t)/(mR^2)$$
$$J_2(\varphi)\ddot{\psi} + \gamma^2\ddot{\varphi} + \gamma \sin \varphi(\omega + \dot{\psi})^2 = 0 \tag{8.127}$$

Linearized System

We consider "small" oscillations of the pendulum about the position $\varphi = 0$ by linearizing the trigonometric terms using the approximations $\sin \varphi \approx \varphi$ and $\cos \varphi \approx 1$ and neglecting the quadratic nonlinearities in Eqs. (8.127). After performing this linearization and making use of Eqs. (8.125), we arrive at the following linear system of equations:

$$[J_O/mR^2 + (1 + \gamma^2)]\ddot{\psi} + \gamma(1 + \gamma)\ddot{\varphi} = M_T(t)/(mR^2)$$
$$(1 + \gamma)\ddot{\psi} + \gamma\ddot{\varphi} + \omega^2\varphi = 0 \tag{8.128}$$

The second of Eqs. (8.128) has the form of a pendulum equation, and from this equation, the natural frequency ω_p of the pendulum is identified as

$$\omega_p = \sqrt{\frac{\omega^2}{\gamma}} = \frac{\omega}{\sqrt{\gamma}} = \omega\sqrt{R/r} \tag{8.129}$$

where we have used the fact that $\gamma = r/R$. Therefore, the pendulum frequency is linearly proportional to the rotation speed ω.

Let us suppose that the external moment applied to the rotating system is in the form of a harmonic excitation; that is,

$$M_T(t) = M_o \sin \omega_o t \tag{8.130}$$

Since there is no damping in this case, we use the procedure of Section 8.2.3 and assume a harmonic solution for Eqs. (8.128) of the form

$$\psi(t) = \Psi \sin \omega_o t$$
$$\varphi(t) = \Phi \sin \omega_o t \tag{8.131}$$

Upon substituting Eqs. (8.131) and (8.130) into Eqs. (8.128) and canceling the $\sin \omega_o t$ term, we obtain the following system of equations:

$$\omega_o^2 (J_O/mR^2 + (1 + \gamma^2))\Psi + \omega_o^2\gamma(1 + \gamma)\Phi = -M_o/mR^2$$
$$-\omega_o^2\gamma(1 + \gamma)\Psi + (\gamma\omega^2 - \omega_o^2\gamma^2)\Phi = 0 \tag{8.132}$$

From the second of Eqs. (8.132), it is seen that if the nondimensional pendulum length is chosen so that

$$\gamma = \frac{\omega^2}{\omega_o^2} \tag{8.133}$$

then the response amplitude of the rotary system Ψ is zero regardless of the excitation imposed on the system. When γ is given by Eq. (8.133), the response amplitude of the pendulum Φ is determined from the first of Eqs. (8.132) as

$$\Phi = -\frac{M_o}{mR^2\omega_o^2\gamma(1 + \gamma)} = -\frac{M_o}{mR^2\omega^2(1 + \omega^2/\omega_o^2)} \tag{8.134}$$

When the tuning of the pendulum absorber is chosen in accordance with Eq. (8.133), then the zero response of the rotary system in the presence of an external moment can be explained as follows. The pendulum exerts an inertial moment equal and opposite to the applied external moment on the rotary system so that the effective external moment felt by the rotary system has zero magnitude at the disturbance frequency ω_o.

Comparing Eqs. (8.133) and (8.129), it is clear that the pendulum frequency ω_p is equal to the disturbance frequency ω_o; that is, we have a tuned pendulum absorber.

EXAMPLE 8.15 Design of a centrifugal pendulum vibration absorber for an internal combustion engine[24]

In a four-stroke-cycle internal combustion engine operating at the speed ω, it is necessary to suppress oscillations at the frequency ω_o, which is three times the operating speed; that is,

$$\omega_o = 3\omega \tag{a}$$

In this case, based on Eqs. (8.133), one can choose a pendulum of the nondimensional length

$$\gamma = \frac{\omega^2}{\omega_o^2} = \frac{1}{9} \tag{b}$$

which means that the pendulum length r needs to be 1/9 the radius R.

8.6.3 Nonlinear Vibration Absorber

We provide a brief introduction to vibration absorbers designed based on nonlinear-system behavior and to the different types of nonlinear vibration absorbers that can be designed. We also point out that the oscillations of nonlinear

---

[24]C. Genta, *ibid.* This reference also provides details of many possible designs of pendulum absorbers.

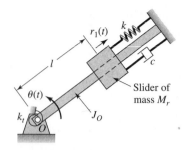

FIGURE 8.24
Bar-slider model.
Source: From A. Khajepour,
M. F. Golnaraghi, and
K. A. Morris, "Applications of
Center Manifold Theory to
Regulation of a Flexible
Beam." *ASME Journal of
Vibrations Acoustics,* Vol.
119, pp. 158–165 (1997).
Copyright © 1997 ASME.
Reprinted with permission.

systems can display vibration characteristics remarkably different from those seen in the oscillations of linear systems.[25] For instance, consider the long-time response of a damped linear system when the system is subjected to a harmonic forcing. The response of this system can never display aperiodic oscillations. On the other hand, in a similar situation, a nonlinear system can display aperiodic characteristics such as chaos.[26] These aperiodic characteristics are illustrated in a pendulum absorber system discussed in this section. The other system treated here is bar-slider system, which will be discussed first.

Bar-Slider System

Consider the bar-slider system[27] shown in Figure 8.24 where M_r is a sliding point mass on the bar that has a mass moment of inertia J_O about the fixed point O. The bar-slider system is assumed to be in a plane perpendicular to gravity. The slider is restrained by a linear spring-damper combination and there is a torsion spring attached to the bar at the point O, which restrains the motions of the bar. To describe the oscillations of this two degree-of-freedom system, the generalized coordinates r_1 and θ are chosen.

We shall derive the governing nonlinear equations of motion and show that there are quadratic and cubic coupling nonlinearities in the governing equations. By generating numerical results, it is shown that the slider mass can absorb the energy input to this two degree-of-freedom system and this helps attenuate the angular oscillations of the bar. This attenuation of angular oscillations of the bar is enhanced in the presence of a two-to-one frequency relationship between the natural frequencies ω_r and ω_θ of the linear system. This type of frequency relationship involving just the natural frequencies of a multi-degree-of-freedom system is called an *internal resonance;* in the particular case considered here, the internal resonance is a *two-to-one internal resonance.*[28]

Lagrange's equations given by Eqs. (7.7) are used to establish the governing equations in terms of the generalized coordinates r_1 and θ. Based on Figure 8.24, the system kinetic energy, the system potential energy, and the dissipation function D are constructed as follows:

$$T = \frac{1}{2} J_O \dot{\theta}^2 + \frac{1}{2} M_r \dot{r}_1^2 + \frac{1}{2} M_r (l + r_1)^2 \dot{\theta}^2$$

$$V = \frac{1}{2} k_t \theta^2 + \frac{1}{2} k r_1^2$$

$$D = \frac{1}{2} c \dot{r}_1^2 \tag{8.135}$$

[25] A. H. Nayfeh and D. T. Mook, 1979, *ibid.*

[26] A. H. Nayfeh and B. Balachandran, 1995, *ibid.*

[27] A. Khajepour, M. F. Golnaraghi, and K. A. Morris, "Application of Center Manifold Theory to Regulation of a Flexible Beam." *ASME J. Vibrations Acoustics,* Vol. 119, pp. 158–165 (1997).

[28] Vibration absorbers based on internal resonances have been studied since the 1960s; see, for example, the following articles: E. Sevin, "On the Parametric Excitation of a Pendulum-type Vibration Absorber," *ASME J. Applied Mechanics,* Vol. 28, pp. 330–334 (1961); and R. S. Haxton and A. D. S. Barr, "The Autoparametric Vibration Absorber," *ASME J. Eng. Industry.,* Vol. 94, pp. 119–125 (1972).

where the position given by length l corresponds to the unstretched position of the spring to which the slider is attached. Making use of Eqs. (7.7), and noting that the generalized forces $Q_1 = 0$ and $Q_2 = 0$, we arrive at

$$\frac{d}{dt}\left(\frac{\partial T}{\partial \dot{r}_1}\right) - \frac{\partial T}{\partial r_1} + \frac{\partial D}{\partial \dot{r}_1} + \frac{\partial V}{\partial r_1} = 0_1$$

$$\frac{d}{dt}\left(\frac{\partial T}{\partial \dot{\theta}}\right) - \frac{\partial T}{\partial \theta} + \frac{\partial D}{\partial \dot{\theta}} + \frac{\partial V}{\partial \theta} = 0 \tag{8.136}$$

On substituting Eqs. (8.135) into Eqs. (8.136) and performing the indicated operations, the result is the following coupled, nonlinear equations of motion:

$$M_r\ddot{r}_1 + c\dot{r}_1 + kr_1 - \underbrace{M_r l\dot{\theta}^2}_{\substack{\text{quadratic} \\ \text{nonlinearity}}} - \underbrace{M_r r_1 \dot{\theta}^2}_{\substack{\text{cubic} \\ \text{nonlinearity}}} = 0$$

$$[J_O + M_r l^2]\ddot{\theta} + k_t\theta + \underbrace{2M_r l(\dot{r}_1\dot{\theta} + r_1\ddot{\theta})}_{\text{quadratic nonlinearities}} + \underbrace{2M_r r_1(\dot{r}_1\dot{\theta} + r_1\ddot{\theta})}_{\text{cubic nonlinearities}} = 0 \tag{8.137}$$

In Eqs. (8.137), the source of the coupling is the quadratic and cubic nonlinearities. For small oscillations about the system equilibrium position $r_1 = 0$ and $\theta = 0$, the nonlinear equations are linearized to obtain

$$M_r\ddot{r}_1 + c\dot{r}_1 + kr_1 = 0$$

$$[J_O + M_r l^2]\ddot{\theta} + k_t\theta = 0 \tag{8.138}$$

From the linearized system of Eqs. (8.138), it is seen that the radial motions of the slider and the angular oscillations of the bar are uncoupled. Hence, the undamped system natural frequencies are readily identified as

$$\omega_r = \sqrt{\frac{k}{M_r}}$$

$$\omega_\theta = \sqrt{\frac{k_t}{(J_O + M_r)l^2}} \tag{8.139}$$

The ratio of the two natural frequencies given in Eqs. (8.139) is critical for the design of nonlinear vibration absorber.

We rewrite the nonlinear equations given by Eqs. (8.137) in terms of the following nondimensional quantities:

$$\tau = \omega_\theta t, \quad \omega_c = \frac{\omega_r}{\omega_\theta}, \quad \zeta = \frac{c}{2M_r\omega_r},$$

$$r(\tau) = \frac{r_1(\tau)}{l}, \quad m = \frac{M_r}{J_O/l^2 + M_r} \tag{8.140}$$

Then Eqs. (8.137) become

$$\ddot{r} + 2\zeta\omega_c\dot{r} + \omega_c^2 r - (1 + r)\dot{\theta}^2 = 0$$

$$[1 + m(2 + r)r]\ddot{\theta} + \theta + 2m(1 + r)\dot{r}\dot{\theta} = 0 \tag{8.141}$$

where the overdot now indicates the derivative with respect to the nondimensional time τ. From these equations, it is seen that the motion of the system can be studied in terms of the following nondimensional parameters: a) the frequency ratio ω_c, b) the mass ratio m, and c) the damping ratio ζ.

For arbitrary magnitudes of oscillations, the nonlinear equations (8.141) can only be solved numerically.[29] To illustrate the absorber action, we assume that the system is subjected to an initial rotation $\theta(0) = 0.3$ rad (i.e., 17.2°), an initial radial displacement $r(0) = 0$, and that the initial radial velocity and the initial angular velocity are zero. For $m = 0.3$, $\zeta = 0.15$, and $\omega_c = 0.5, 1.5, 2$, and 2.5, the numerically generated results are presented in Figure 8.25.

It is seen from Figure 8.25 that when $\omega_c = 2.0$, the angular oscillations decay more rapidly than in the other cases considered. In this special case, the coupling between the radial oscillations of the slider mass and the angular oscillations of the bar are enhanced due to the nonlinear coupling terms. These types of frequency relationships have been used to construct vibration controllers for free oscillations and forced oscillations.[30]

Pendulum Absorber[31]

As a companion system to the slider-mass system, we consider the pendulum absorber system shown in Figure 8.26. This system consists of a mass M mounted on a combination of a linear spring and a linear viscous damper. A pendulum of mass m and length l is attached to the mass M. A linear torsion damper with a damping coefficient c_t is also included. To describe the motions of this two degree-of-freedom system, the generalized coordinates x and φ are used. The coordinate x is measured from the system static-equilibrium position.

When the system is forced harmonically along the vertical direction, the up and down motions of the system can be undesirable, especially when the excitation frequency is close to the system natural frequency associated with the vertical motions. To attenuate these up and down oscillations, the pendulum is designed as an absorber, so that the input energy along the vertical direction is absorbed as angular oscillations of the pendulum.

The harmonic forcing acting on the mass M is assumed to have an amplitude F_o and a frequency ω. Starting from the Lagrange equations given by Eqs. (7.7) and recognizing that the generalized forces are $Q_1 = F_o \cos \omega t$ and $Q_2 = 0$, we find that the governing equations are obtained from

$$\frac{d}{dt}\left(\frac{\partial T}{\partial \dot{\varphi}}\right) - \frac{\partial T}{\partial \varphi} + \frac{\partial D}{\partial \dot{\varphi}} + \frac{\partial V}{\partial \varphi} = F_o \cos \omega t$$

$$\frac{d}{dt}\left(\frac{\partial T}{\partial \dot{x}}\right) - \frac{\partial T}{\partial x} + \frac{\partial D}{\partial \dot{x}} + \frac{\partial V}{\partial x} = 0 \tag{8.142}$$

The system kinetic energy T is given by

---

[29]The MATLAB function `ode45` was used.

[30]A. H. Nayfeh, *Nonlinear Interactions,* John Wiley & Sons, NY, Chapter 2 (2000).

[31]A. Tondl, T. Ruijgork, F. Verhulst, and R. Nabergoj, *Autoparametric Resonance in Mechanical Systems,* Cambridge University Press, Cambridge, England, Chapter 4 (2000).

FIGURE 8.25
Response of the slider-mass system for $\zeta = 0.15$, $m = 0.3$, and $\omega_c = 0.5$, 1.5, 2.0, and 2.5.

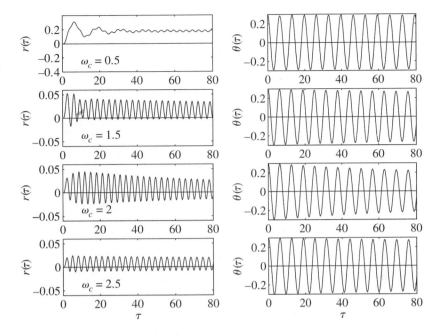

$$T = \frac{1}{2} M\dot{x}^2 + \frac{1}{2} m(\mathbf{V}_m \cdot \mathbf{V}_m)$$

$$= \frac{1}{2} M\dot{x}^2 + \frac{1}{2} m[(l\dot{\varphi}\cos\varphi)^2 + (\dot{x} + l\dot{\varphi}\sin\varphi)^2] \qquad (8.143)$$

where, in arriving at Eq. (8.143), we have made use of the pendulum velocity

$$\mathbf{V}_m = (-l\dot{\varphi}\cos\varphi)\mathbf{i} + (\dot{x} + l\dot{\varphi}\sin\varphi)\mathbf{j} \qquad (8.144)$$

The system potential energy V and the dissipation function D take the form

$$V = \frac{1}{2} kx^2 + mgl(1 - \cos\varphi)$$

$$D = \frac{1}{2} c\dot{x}^2 + \frac{1}{2} c_t\dot{\varphi}^2 \qquad (8.145)$$

FIGURE 8.26
System with pendulum absorber.

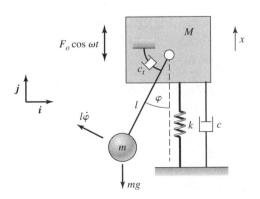

After substituting Eqs. (8.143) and (8.145) in Eqs. (8.142) and carrying out the indicated operations, the result is the following system of coupled nonlinear equations:

$$(M + m)\ddot{x} + c\dot{x} + kx + \underbrace{ml[\ddot{\varphi} \sin \varphi + \dot{\varphi}^2 \cos \varphi]}_{\text{nonlinear terms}} = F_o \cos \omega t$$

$$ml^2\ddot{\varphi} + c_t\dot{\varphi} + \underbrace{ml\ddot{x} \sin \varphi + mgl \sin \varphi}_{\text{nonlinear terms}} = 0 \qquad (8.146)$$

For small oscillations about the system equilibrium position $x = 0$ and $\varphi = 0$, Eqs. (8.146) are linearized by using the approximations $\sin \varphi \approx \varphi$ and $\cos \varphi \approx 1$ and dropping the nonlinear terms. The result is

$$(M + m)\ddot{x} + c\dot{x} + kx = F_o \cos \omega t$$

$$ml^2\ddot{\varphi} + c_t\dot{\varphi} + mgl\varphi = 0 \qquad (8.147)$$

From Eqs. (8.147) it is seen that the vertical motions of the system are uncoupled from the angular oscillations of the pendulum in the linear system. Therefore, the undamped system natural frequencies is readily identified as

$$\omega_x = \sqrt{\frac{k}{m + M}}$$

$$\omega_\varphi = \sqrt{\frac{g}{l}} \qquad (8.148)$$

The ratio of the two natural frequencies given in Eqs. (8.148) is critical for the considered nonlinear vibration absorber design.

We introduce the following nondimensional quantities

$$z = \frac{x}{l}, \quad \tau = \omega_x t, \quad \omega_r = \frac{\omega_\varphi}{\omega_x}, \quad \Omega = \frac{\omega}{\omega_x}, \quad f_o = \frac{F_o}{(M + m)l\omega_x^2}$$

$$2\zeta_x = \frac{c}{(M + m)\omega_x}, \quad 2\zeta_t = \frac{c_t}{ml^2\omega_x}, \quad m_r = \frac{m}{(M + m)} \qquad (8.149)$$

into Eqs. (8.146) and arrive at

$$\ddot{z} + 2\zeta_x\dot{z} + z + m_r[\ddot{\varphi} \sin \varphi + \dot{\varphi}^2 \cos \varphi] = f_o \cos \Omega\tau$$

$$\ddot{\varphi} + 2\zeta_t\dot{\varphi} + (\omega_r^2 + \ddot{z})\sin \varphi = 0 \qquad (8.150)$$

where the overdot now indicates the derivative with respect to τ. It is seen from the form of Eqs. (8.150) that the motions of this nonlinear system can be studied in terms of the following nondimensional quantities: a) the mass ratio m_r, b) the frequency ratio ω_r, c) the damping ratios ζ_x and ζ_t, d) the forcing amplitude f_o, and e) the forcing frequency Ω.

For arbitrary magnitudes of oscillations, the nonlinear system given by Eqs. (8.150) can only be solved numerically. Furthermore, from the second of Eqs. (8.150), it is seen that $\varphi = 0$ is always a solution of this system. Therefore, if nontrivial solutions for φ are sought, one has to be careful in picking an initial condition for which $\varphi(0) \neq 0$. As pointed out in Section 4.6.2, initial conditions are critical in determining the response of a nonlinear system. Since we would like the absorber to be effective, when the forcing frequency is in resonance with the system natural frequency for vertical motions, we choose $\Omega = 1$. Furthermore, to enhance the nonlinear coupling between the angular oscillations and the up and down translations, we set $\omega_r = 0.5$; that is, the natural frequency of the pendulum oscillations is one half of the natural frequency of vertical translations of the system. Light damping is also considered. The numerical results[32] generated are presented in Figure 8.27 for two different values of non-dimensional forcing magnitude f_o. The

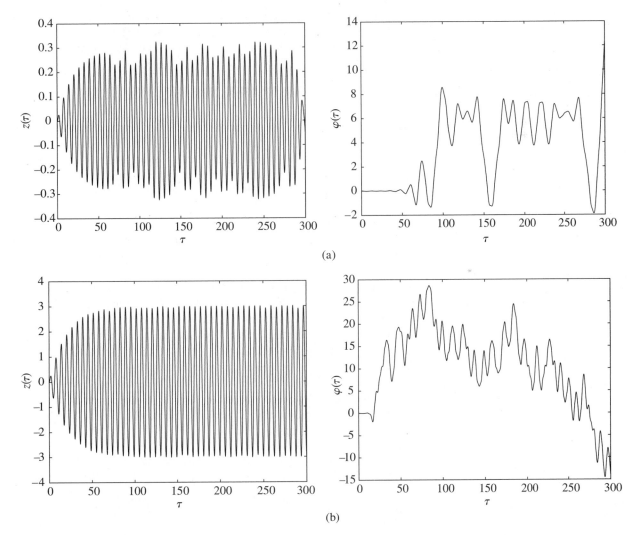

(a)

(b)

FIGURE 8.27
Response of the system with the pendulum absorber for $\Omega = 1$, $\omega_r = 0.5$, $\zeta_x = 0.05$, $\zeta_t = 0.005$, and $m_r = 0.05$: (a) $f_o = 0.03$ and (b) $f_o = 0.3$.

[32]The MATLAB function `ode45` was used.

initial condition for the angular state is picked as $\varphi(0) = 0.02$ rad (1.15°) to avoid converging to the trivial solution $\varphi = 0$. The initial translation state z, the initial angular velocity $\dot{\varphi}$, and the initial translation velocity $\dot{z}$ are all zero.

It is seen that although the system is being driven at the resonance of the system's translation motions, the pendulum absorber is effective in limiting them. However, the translation motions and the angular motions have an aperiodic character, despite a harmonic input into this system. The pendulum motion is an irregular mixture of oscillations and rotation. [Recall that one complete revolution of the pendulum is 2π radians.] The aperiodic motions seen in Figure 8.27 are examples of chaotic motions, which have some special characteristics. For more details on how these characteristics can be identified, the reader is referred to the nonlinear dynamics literature.[33]

8.7 VIBRATION ISOLATION: TRANSMISSIBILITY RATIO

In many machinery-mounting situations, the single degree-of-freedom model given in Section 5.7 is inadequate because the formulation does not take into account the stiffness of the flooring. To have a more realistic model, let us suppose that m_1 is the mass of the flooring and m_2 is the mass of the machinery as shown in Figure 8.28. In some instances, the machinery is connected to a seismic mass, which in turn is connected via springs to the structure, usually the ground.

The objective is to determine the various parameters of the system so that the force transmitted to the ground or the structure that is supporting the flooring[34] is as small as practical. To do this, we need to examine the ratio of the magnitude of the force transmitted to the ground $f_{\text{base}}(t)$ to the magnitude of the force applied to the machinery $f_2(t)$. If motions about the static-equilibrium position are considered, then the equations governing the system shown in Figure 8.28 are represented by Eqs. (8.60) when we set the spring constant k_3, the damping coefficient c_3, and the force $f_1(t)$ to zero. Then, the force transmitted to the ground is given by Eq. (8.89b).

The required transfer function is $F_{\text{base}}(s)/F_2(s)$. To obtain this transfer function, we set all initial conditions to zero and assume that the force applied to m_2 is an impulse; that is, $f_2(\tau) = F_o\delta(\tau)$. Then, using Eqs. (8.90), (8.71), and (8.75), we obtain

$$T_R(s) = \frac{F_{base}(s)}{F_2(s)} = \frac{F_{base}(s)}{F_o} = \frac{k_1(2\zeta_1 s + 1)X_1(s)}{F_o} = \frac{(2\zeta_1 s + 1)B(s)}{m_r D_2(s)} \quad (8.151)$$

where $D_2(s)$ is given by Eq. (8.74). The transmissibility ratio is defined as

$$TR = |T_R(j\Omega)| \quad (8.152)$$

[33]A. H. Nayfeh and B. Balachandran, 1995, *ibid;* F. C. Moon, *Chaotic and Fractal Dynamics: An Introduction for Applied Scientists and Engineers,* John Wiley & Sons, NY (1992).

[34]J. A. Macinante, *Seismic Mountings for Vibration Isolation,* John Wiley & Sons, NY, Chapter 8 (1984).

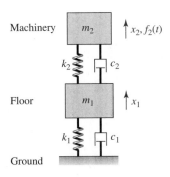

Machinery m_2 $\uparrow x_2, f_2(t)$

k_2 c_2

Floor m_1 $\uparrow x_1$

k_1 c_1

Ground

FIGURE 8.28

Representative two degree-of-freedom system for vibration transmission model.

To obtain this ratio TR, we set $s = j\Omega$ in Eq. (8.151) and substitute the result into Eq. (8.152) to obtain

$$TR = \left| \frac{(2\zeta_1 j\Omega + 1)B(j\Omega)}{m_r D_2(j\Omega)} \right| \qquad (8.153)$$

where $D_2(j\Omega)$ and $B(j\Omega)$ are given by Eqs. (8.102).

The results obtained from Eq. (8.153) are plotted in Figure 8.29 for the case where $\zeta_1 = \zeta_2 = 0.08$ and the mass ratio $m_r = 0.1$. The transmissibility ratio TR is shown as a function of the frequency ratio ω_r and the nondimensional excitation frequency ratio Ω. A point on this graph is interpreted as the magnitude of the transmissibility ratio for flooring represented by the m_1-k_1-c_1 system in Figure 8.28 and harmonic excitation acting on the mass m_2 at the nondimensional frequency Ω. The transmissibility ratio then provides the magnitude of the harmonic force transmitted to the ground. In addition, solid lines have been drawn to delineate the regions where TR is less than or equal to a certain fraction of the force applied to m_2 that reaches the ground or flooring supports. Consider the case where $TR = 0.1$. Comparing these results with the TR value of a single degree-of-freedom system given by Eq. (5.105) we see that to get a TR of 10% we need to have $\Omega > 3.32$ in a single degree-of-freedom system. On the other hand, in the case of a two degree-of-freedom system, we see that for $\Omega > 1.5$ and $\omega_r < 0.5$ we will attain the same levels. In other words, the natural frequency of m_2 by itself has to be less than half of the support/flooring. We also note from Figure 8.29 that for a given TR, the operating frequency of the machinery is almost linearly proportional to ω_r.

Design Guideline: For the transmissibility of a dynamic force through a flexible flooring system, which is modeled as a two degree-of-freedom system with $m_r = 0.1$, to be less than 10%, the nondimensional excitation frequency of the disturbing force should be greater than $1.2 + \omega_r$, where ω_r is the ratio of the system uncoupled natural frequencies.

We now examine how the system parameters can be selected to attenuate the peak magnitude of the force transmitted to the ground when an impulsive force is applied to the mass m_2. This situation arises, for example, in factories where it is necessary to isolate forges and punch presses. To determine this, we take the inverse Laplace transform of Eq. (8.151). Before doing so, however, we rewrite Eq. (8.151) using Eq. (8.64) as follows:

$$T_R(s) = \frac{\omega_r(2\zeta_1 s + 1)(2\zeta_2 s + \omega_r)}{D_2(s)} \qquad (8.154)$$

We see from Eq. (8.154) and an examination of $D_2(s)$ that when $m_r < 1$, m_r has a very small effect on T_R. Again using available numerical

FIGURE 8.29
Transmissibility ratio for two degree-of-freedom system with $m_r = 0.1$ and $\zeta_1 = \zeta_2 = 0.08$. [Solid lines represent constant values of TR.]

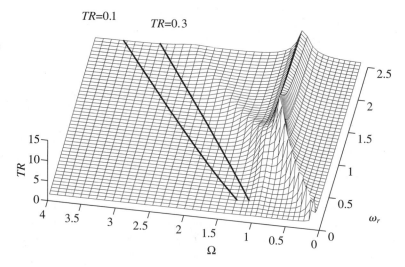

procedures,[35] representative numerically obtained results for the inverse of Eq. (8.154) are given in Figure 8.30. For small values of the frequency ratio ω_r, it is seen that the transient oscillations have a high-frequency component, which appears to "ride" on top of a low-frequency oscillation. However, as ω_r increases, this characteristic disappears. In addition, we see that the nondimensional peak amplitude is nearly equal to ω_r. Furthermore, we see that the introduction of the seismic mass m_1 can provide a much larger attenuation of

FIGURE 8.30
Magnitude of an impulse force transmitted to the base of a two degree-of-freedom system: $\zeta_1 = \zeta_2 = 0.1$.

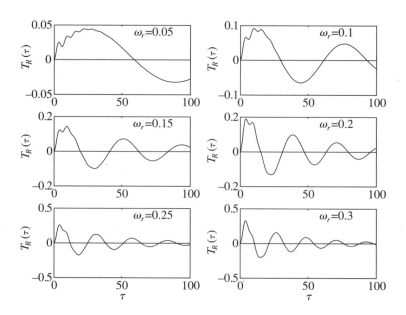

[35]The MATLAB function `ilaplace` from the Symbolic Math Toolbox was used.

the peak magnitude of the force applied to m_2, when compared to that of a single degree-of-freedom system. Recall from Section 6.2 that for a single degree-of-freedom system, the most attenuation that we could attain was about 18% for $\zeta = 0.25$.

Design Guideline: The peak transmissibility of a shock loading through a lightly damped flooring system to the ground is approximately equal to the ratio ω_r. Therefore, the ratio should be as small as possible, which is equivalent to having the uncoupled natural frequency of the system attached to the flooring be much greater than the natural frequency of the flooring by itself.

EXAMPLE 8.16 Design of machinery mounting to meet transmissibility ratio requirement

A 150 kg machine is to be operated at 500 rpm on a platform that is modeled as a single degree-of-freedom system with the following mass, stiffness, and damper values: $m_1 = 3000$ kg, $k_1 = 2 \times 10^6$ N/m, and $c_1 = 7500$ N/(m/s). The machinery mounting is to be determined so that the transmissibility ratio does not exceed 0.1.

From the given values, we have that

$$m_r = \frac{m_2}{m_1} = \frac{150 \text{ kg}}{3000 \text{ kg}} = 0.05$$

$$\omega_{n1} = \sqrt{\frac{k_1}{m_1}} = \sqrt{\frac{2 \times 10^6 \text{ N/m}}{3000 \text{ kg}}} = 25.82 \text{ rad/s}$$

$$\zeta_1 = \frac{c_1}{2m_1\omega_{n1}} = \frac{7500 \text{ N/(m/s)}}{2 \times 3000 \text{ kg} \times 25.82 \text{ rad/s}} = 0.048$$

$$\Omega = \frac{\omega}{\omega_{n1}} = \frac{(500 \text{ rev/min})((2\pi \text{ rad/rev})/(60 \text{ s/min}))}{25.82 \text{ rad/s}} = 2.03 \qquad \text{(a)}$$

Although the parameters used to construct Figure 8.29 are clearly different from the corresponding quantities in Eqs. (a), we will use the results shown in Figure 8.29 to guide our choice of mounting parameters. Thus, we choose

$$\omega_r = \frac{\omega_{n2}}{\omega_{n1}} = 0.5 \qquad \text{(b)}$$

since, for $\Omega = 2.03$, we are to the left of the solid line drawn for $TR = 0.1$. Hence, based on Eq. (b) and the machinery mass, we find that the mounting stiffness is

$$k_2 = m_2\omega_{n2}^2$$
$$= (150 \text{ kg}) \times (0.5 \times 25.82 \text{ rad/s})^2$$
$$= 25 \times 10^3 \text{ N/m} \qquad \text{(c)}$$

We can pick a mounting with negligible damping factor; that is, one for which

$$\zeta_2 \approx 0 \tag{d}$$

We now use Eq. (8.153) to determine if the transmissibility ratio is less than 0.1 for the chosen mounting stiffness at $\Omega = 2.03$. For the choice of mounting with negligible damping factor, we find from Eqs. (8.102) that

$$
\begin{aligned}
B(j\Omega) &= 2\zeta_2 m_r \omega_r j\Omega + m_r \omega_r^2 \\
&\approx m_r \omega_r^2 \\
D_2(j\Omega) &= \Omega^4 - j[2\zeta_1 + 2\zeta_2 \omega_r m_r + 2\zeta_2 \omega_r]\Omega^3 \\
&\quad - [1 + m_r \omega_r^2 + \omega_r^2 + 4\zeta_1\zeta_2\omega_r]\Omega^2 + j[2\zeta_2\omega_r + 2\zeta_1\omega_r]\Omega + \omega_r^2 \\
&\approx \Omega^4 - j2\zeta_1\Omega^3 - [1 + m_r\omega_r^2 + \omega_r^2]\Omega^2 + j2\zeta_1\omega_r^2\Omega + \omega_r^2
\end{aligned} \tag{e}
$$

On substituting Eqs. (e) into Eq. (8.153), we arrive at

$$TR = \left| \frac{(2\zeta_1 j\Omega + 1)m_r\omega_r^2}{m_r\Omega^4 - j2\zeta_1\Omega^3 - [1 + m_r\omega_r^2 + \omega_r^2]\Omega^2 + j2\zeta_1\omega_r^2\Omega + \omega_r^2} \right| \tag{f}$$

Making use of Eqs. (a) and (b) by substituting the values for Ω, ζ_1, m_r, and ω_r into Eq. (f), we arrive at

$$TR = \left| \frac{j0.002 + 0.0125}{-j0.754 - 4.1010} \right| = 0.003 \tag{g}$$

Therefore, the requirement that the transmissibility ratio should be less than 0.1 at $\Omega = 2.03$ is met for the mounting stiffness given by Eq. (c).

8.8   SYSTEMS WITH MOVING BASE

Another way to determine the efficacy of a two degree-of-freedom isolation system is to compare the magnitude of the peak (maximum) displacement of m_2 to the magnitude of the peak displacement of the ground. To analyze this type of situation, we consider the base supporting the two degree-of-freedom system to be a moving base as shown in Figure 8.31. In analyses of such systems, one usually assumes that the masses are initially at rest and that there are no applied forces directly on the inertial elements, and $x_3(t)$ is given. We return to Eqs. (8.60) and modify them to account for the moving base. Thus, Eqs. (8.60) become

$$m_1 \frac{d^2x_1}{dt^2} + (c_1 + c_2)\frac{dx_1}{dt} + (k_1 + k_2)x_1 - c_2\frac{dx_2}{dt^2} - k_2x_2 - c_1\frac{dx_3}{dt} - k_1x_3 = 0$$

$$m_2 \frac{d^2x_2}{dt^2} + c_2\frac{dx_2}{dt} + k_2x_2 - c_2\frac{dx_1}{dt} - k_2x_1 = 0 \tag{8.155}$$

FIGURE 8.31
Two degree-of-freedom system
with moving base.

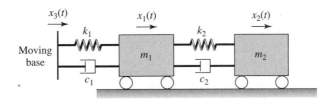

After using the nondimensional quantities of Eqs. (7.41) and (8.61), Eqs. (8.155) are rewritten as

$$\frac{d^2x_1}{d\tau^2} + (2\zeta_1 + 2\zeta_2 m_r \omega_r)\frac{dx_1}{d\tau} + (1 + m_r\omega_r^2)x_1 - 2\zeta_2 m_r\omega_r\frac{dx_2}{d\tau} - m_r\omega_r^2 x_2$$

$$- 2\zeta_1\frac{dx_3}{d\tau} - x_3 = 0$$

$$\frac{d^2x_2}{d\tau^2} + 2\zeta_2\omega_r\frac{dx_2}{d\tau} + \omega_r^2 x_2 - 2\zeta_2\omega_r\frac{dx_1}{d\tau} - \omega_r^2 x_1 = 0 \qquad (8.156)$$

Then, taking the Laplace transform of Eqs. (8.156) and solving for $X_1(s)$ and $X_2(s)$, which are the transforms of $x_1(\tau)$ and $x_2(\tau)$, respectively, we find that

$$X_1(s) = \frac{K_3(s)E_2(s)}{D_2(s)}$$

$$X_2(s) = \frac{K_3(s)C(s)}{D_2(s)} \qquad (8.157)$$

where $D_2(s)$ is given by Eq. (8.74), $C(s)$ is given by Eqs. (8.64), $E_2(s)$ is given by Eq. (8.82), and

$$K_3(s) = (2\zeta_1 s + 1)X_3(s) - 2\zeta_1 x_3(0) \qquad (8.158)$$

Comparing Eqs. (8.158) and (8.90), we see that $k_1 K_3(s)$ is the Laplace transform of the total force generated by the moving base.

To compare the responses of the single degree-of-freedom system with a moving base to that of a two degree-of-freedom system with a moving base, we assume that the displacement of the base is a half-sine wave as shown in Figure 6.24 and given by Eq. (b) of Example 6.7; that is,

$$x_3(\tau) = X_o \sin(\Omega_o\tau)[u(\tau) - u(\tau - \tau_o)] \qquad (8.159)$$

where $\Omega_o = \omega_o/\omega_{n1}$, $\tau_o = \omega_{n1}t_o$, and $t_o = \pi/\omega_o$. We shall determine the response of m_2. Assuming that $x_3(0) = 0$ for convenience, and using transform pair 10 in Table A of Appendix A to obtain the Laplace transform of Eq. (8.159), it is found that

$$X_3(s) = \frac{X_o\Omega_o(1 + e^{-\pi s/\Omega_o})}{s^2 + \Omega_o^2} \qquad (8.160)$$

Then $K_3(s)$, which is given by Eq. (8.158), becomes

$$K_3(s) = \frac{X_o\Omega_o(2\zeta_1 s + 1)(1 + e^{-\pi s/\Omega_o})}{s^2 + \Omega_o^2} \qquad (8.161)$$

FIGURE 8.32
Displacement response of m_2 when a half-sine wave displacement is applied to the system base for $m_r = 0.1$ and $\zeta_1 = \zeta_2 = 0.1$: (a) $\omega_r = 0.05$ and (b) $\omega_r = 0.2$.

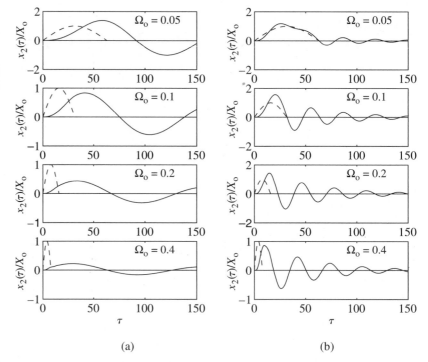

(a)   (b)

After substituting Eq. (8.161) in Eq. (8.157), the response of the mass m_2 is

$$X_2(s) = \frac{\Omega_o X_o C(s)(2\zeta_1 s + 1)(1 + e^{-\pi s/\Omega_o})}{(s^2 + \Omega_o^2)D_2(s)} \tag{8.162}$$

The numerically obtained inverse Laplace transform[36] of Eq. (8.162) is shown in Figure 8.32. We see that as the duration of the half-sine wave pulse decreases, the amplitude of m_2 decreases. This behavior is opposite to what takes place during the base excitation of a single degree-of-freedom system, where as the pulse duration decreased the peak displacement of the mass increased. (Recall Figure 6.22.) We see, then, that the interposition of m_1 and its spring and damper act as a mechanical filter, decreasing the amount of relatively high frequency energy generated by the half-sine wave pulse from being transferred to m_2. Thus, a two degree-of-freedom system with appropriately chosen parameters can be an effective isolator of ground vibrations compared to a single degree-of-freedom system.

EXAMPLE 8.17 Isolation of an electronic assembly

We shall now consider the isolation of electronic components.[37] In this two degree-of-freedom system, m_1 is the outer casing of an electronic assembly as shown in Figure 8.33. Inside the casing, there is a flexibly supported elec-

[36]The MATLAB function `ilaplace` from the Symbolic Math Toolbox was used.

[37]A. M. Veprik and V. I. Babitsky, "Vibration Protection of Sensitive Electronic Equipment from Harsh Harmonic Vibration," *J. Sound Vibration*, Vol. 238, No. 1, pp. 19–30 (2000).

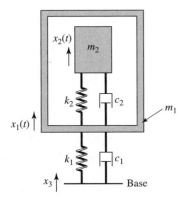

FIGURE 8.33
Model of an electronic system contained in a package that is subjected to a base disturbance.

tronic component modeled as a system with a mass m_2, stiffness k_2, and viscous damping c_2. The outer casing is elastically mounted as shown in Figure 8.33, usually to a base that is a relatively stiff structural member. The movement of the structural member subjects the entire system to a displacement. Typically, the mass ratio m_r is much less than one and the uncoupled natural frequencies of the systems are such that $\omega_r > 3$.

Let the relative displacement between masses m_1 and m_2 be defined by

$$z_2(\tau) = x_2(\tau) - x_1(\tau) \tag{a}$$

Then, for a given value d that specifies the maximum amplitude ratio,

$$d = |X_{13}(\Omega)| = \left| \frac{X_1(j\Omega)}{X_3(j\Omega)} \right| \tag{b}$$

we shall find the value of the damping factor ζ_1 so that the ratio

$$|Z_{23}(\Omega)| = \left| \frac{Z_2(j\Omega)}{X_3(j\Omega)} \right| \tag{c}$$

is a minimum. We assume that ω_r, m_r, and ζ_2 are given. In practical cases, $\zeta_2 \ll 1$. Then, from Eqs. (8.157) and (8.158) we have that

$$X_{13}(s) = \frac{X_1(s)}{X_3(s)} = \frac{(2\zeta_1 s + 1)E_2(s)}{D_2(s)}$$

$$Z_{23}(s) = \frac{X_2(s)}{X_3(s)} - \frac{X_1(s)}{X_3(s)} = \frac{(2\zeta_1 s + 1)}{D_2(s)}(C(s) - E_2(s)) \tag{d}$$

FIGURE 8.34
Optimum value of ζ_1 for several combinations of m_r and ζ_2 when $\omega_r = 3.5$ and $|X_{13}(\Omega_1)| = d = 2.0$.

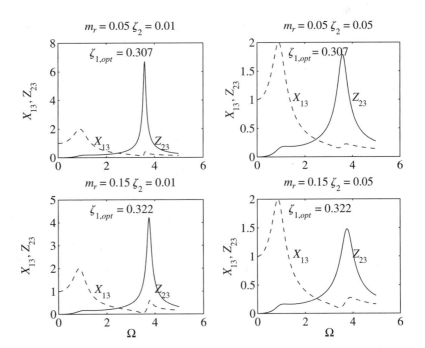

The respective amplitude-response functions are obtained by setting $s = j\Omega$ in Eq. (d) and then taking the absolute values. Thus, we obtain

$$|X_{13}(\Omega)| = d = \left| \frac{(2\zeta_1 \Omega j + 1)E_2(j\Omega)}{D_2(j\Omega)} \right|$$

$$|Z_{23}(\Omega)| = \left| \frac{(2\zeta_1 \Omega j + 1)}{D_2(j\Omega)} (C(j\Omega) - E_2(j\Omega)) \right| \qquad \text{(e)}$$

We shall again use the optimization methods mentioned in Section 8.6. If the maximum value of the X_{13} occurs at Ω_1 and that of Z_{23} at Ω_2, then the objective is to find the value of ζ_1 that makes the maximum value $Z_{23}(\Omega_2)$ a minimum value subject to the requirement that $X_{13}(\Omega_1) = d$, when the values of d, ζ_2, ω_r, and m_r are given. This is stated as follows:

$$\min_{\zeta_1} \max_{Z_{23}(\Omega)} \{Z_{23}(\Omega)\}$$

subject to: $\zeta_1 \geq 0$

$$X_{13}(\Omega_1) = d \qquad \text{(f)}$$

The results for several combinations of d, ζ_2, ω_r, and m_r are shown in Figures 8.34 and 8.35. We see that we get large attenuation of the maximum relative amplitude when we compare these results to those obtained for a single degree-of-system given by Eq. (5.82) for the same damping ratio ζ_2. For the small values of ζ_2 used in these numerical examples, 0.05 and 0.01, we see that the maximum values are approximately proportional to $1/(2\zeta_2)$. Therefore, the maximum amplitudes would be 10 and 50, respectively. Also, the values of ξ_1 that are required to obtain these reductions are attainable in practice.

FIGURE 8.35
Optimum value of ζ_1 for several combinations of m_r and ζ_2 when $\omega_r = 3.5$ and $|X_{13}(\Omega_1)| = d = 2.5$.

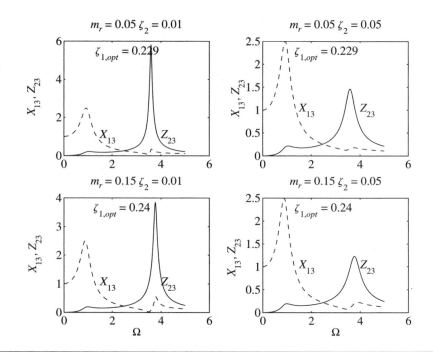

8.9 SUMMARY

In this chapter, we illustrated how the responses of linear multi-degree-of-freedom systems could be determined by using the normal-mode approach, Laplace transforms, and the state-space formulation. For the nonlinear multi-degree-of-freedom systems treated in this chapter, the solutions were numerically determined. The responses of multi-degree-of-freedom systems to harmonic, step, and impulse excitations were also examined. The notions of resonance, frequency-response functions, and transfer functions introduced in the earlier chapters for single-degree-of-freedom systems were revisited and discussed for multi-degree-of-freedom systems. For linear vibratory systems, it was shown how the frequency-response functions can be used to design vibration absorbers, mechanical filters, and address issues such as base isolation and transmission ratio. A brief introduction to the design of vibration absorbers based on nonlinear-system behavior was also presented.

EXERCISES

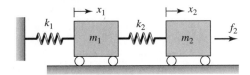

FIGURE E8.1

8.1 Consider the two degree-of-freedom system shown in Figure E8.1, where a forcing f_2 is imposed on the mass m_2. For $m_1 = m_2 = 1$ kg, $k_1 = 2$ N/m, $k_2 = 1$ N/m, and $f_2 = F_2\sin\omega t$ N, derive the governing equations of motion of the system and carry out the following:

a) Determine a solution for the response of this system by using the direct approach discussed in Section 8.2.3.

b) Construct the frequency-response functions $H_{12}(\Omega)$ and $H_{22}(\Omega)$ and plot these functions as a function of the nondimensional excitation frequency Ω.

8.2 Consider the two degree-of-freedom system shown in Figure E8.1, and apply a harmonic forcing $f_1 = F_1\sin\omega t$ to mass m_1. For $m_1 = m_2 = 2$ kg, $k_1 = 2$ N/m, $k_2 = 1$ N/m, derive the governing equations of motion of the system and carry out the following:

a) Determine a solution for the response of this system by using the direct approach discussed in Section 8.2.3.

b) Construct the frequency-response functions $H_{11}(\Omega)$ and $H_{21}(\Omega)$ and plot these functions as a function of the nondimensional excitation frequency Ω.

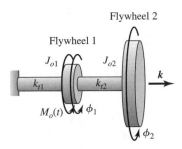

FIGURE E8.3

8.3 Consider the system of flywheels shown in Figure E8.3. For an applied moment $M_o(t) = M_1\cos\omega t$, determine a solution for the response of the system by using the direct approach presented in Section 8.2.3.

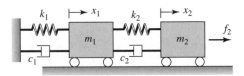

FIGURE E8.4

8.4 Consider the two degree-of-freedom system shown in Figure E8.4, where $m_1 = m_2 = 1$ kg, $k_1 = 2$ N/m, $k_2 = 1$ N/m, $c_1 = 0.4$ N/m/s, and $c_2 = 0.2$ N/m/s. To determine the response of this damped system to a harmonic excitation, one can assume the excitation to be of the form

$$f_2(t) = F_2 e^{j\omega t}$$

which is referred to as a *complex-valued excitation* because of the $e^{j\omega t}$ term. To determine the steady-state solution for the response of this system, we assume a solution of the form

$$\begin{Bmatrix} x_1(t) \\ x_2(t) \end{Bmatrix} = \begin{Bmatrix} X_1(j\omega) \\ X_2(j\omega) \end{Bmatrix} e^{j\omega t}$$

and then solve for the response amplitudes $X_1(j\omega)$ and $X_2(j\omega)$ from the governing equations of motion.[38] Carry out these operations for the system shown in Figure E8.4.

8.5 Based on the response amplitudes determined in Exercise 8.4, construct the frequency-response functions

$$H_{12}(j\Omega) = X_1(j\Omega)/F_2$$
$$H_{22}(j\Omega) = X_2(j\Omega)/F_2$$

where Ω is the nondimensional excitation frequency. Plot graphs of the amplitude and phase of the each of these functions as a function of Ω, compare them to the plots of frequency-response functions determined in Exercise 8.1, and discuss their differences and similarities.

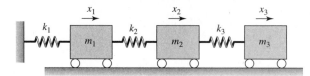

FIGURE E8.6

8.6 Consider the three degree-of-freedom system shown in Figure E8.6. Assume that a harmonic forcing $f_2 = F_2 \cos \omega t$ is imposed on the mass m_2. Determine if mass m_1 can have a zero response at any of the excitation frequencies.

---

[38]This approach used to determine the solution of a damped system to a harmonic excitation complements the direct approach discussed in Section 8.2.3 for undamped systems; also see Appendix D.

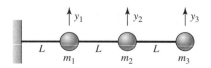

FIGURE E8.7

8.7 Consider the system shown in Figure E8.7 in which the three masses m_1, m_2, and m_3 are located on a uniform cantilever beam with flexural rigidity EI. The inverse of the stiffness matrix for this system $[K]$, which is called the flexibility matrix, is given by

$$[K]^{-1} = \frac{L^3}{3EI} \begin{bmatrix} 27 & 14 & 4 \\ 14 & 8 & 2.5 \\ 4 & 2.5 & 1 \end{bmatrix}$$

If the masses of the system are all identical—that is, $m_1 = m_2 = m_3 = m$—then determine the response of this system when it is forced sinusoidally at the location of mass m_2 with a forcing amplitude of F_2 and an excitation frequency of ω.

8.8 Use the normal-mode approach to determine the solution of the system given in Exercise 8.1. Assume that all of the initial conditions are zero.

8.9 For the two degree-of-freedom system shown in Figure E8.4 and for the system parameter values given in Exercise 8.4, derive the governing equations of motion of the system, uncouple these equations by using the modal matrix, and present the uncoupled system of equations.

8.10 Consider the damped system of Exercise 8.9 and set the forcing amplitude to zero. Examine free oscillations of this system by using the normal-mode for the following sets of initial conditions and discuss the participation of different modes in each case:
   a) $x_1(0) = x_2(0) = 0.2$ m, $\dot{x}_1(0) = \dot{x}_2(0) = 0$ m/s
   b) $x_1(0) = -0.2$ m, $x_2(0) = 0.2$ m, $\dot{x}_1(0) = \dot{x}_2(0) = 0$ m/s

8.11 Use the normal-mode approach to determine a solution for the response of the system discussed in Exercise 8.9, when the harmonic forcing

$$f_2 = 10\sin(20t)$$

N is imposed on mass m_2. Assume that all of the initial conditions are zero.

8.12 Use the normal-mode approach to determine a solution for the response of system discussed in Exercise 8.6, when the impulse $f_2 = F_2\delta(t)$ is imposed on mass m_2. Assume that all of the initial conditions are zero, $m_1 = m_2 = m_3 = m$, and $k_1 = k_2 = k_3 = k$.

8.13 Repeat Exercise 8.12 with an impulse $f_1 = F_1\delta(t)$ imposed on mass m_1 instead of mass m_2.

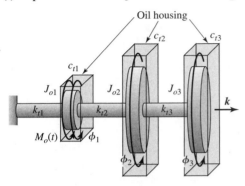

FIGURE E8.14

8.14 Consider the three degree-of-freedom system shown in Figure E8.14. The different system parameter values are as follows: $J_{o1} = 1$ kg·m², $J_{o2} = 4$ kg·m², $J_{o3} = 1$ kg·m², $k_{t1} = k_{t2} = 10$ N·m/rad, and $k_{t3} = 5$ N·m/rad. Assume that the damping matrix is proportional to the inertia matrix; that is,

$$\begin{bmatrix} c_{t1} & 0 & 0 \\ 0 & c_{t2} & 0 \\ 0 & 0 & c_{t3} \end{bmatrix} = \alpha \begin{bmatrix} J_{O1} & 0 & 0 \\ 0 & J_{O2} & 0 \\ 0 & 0 & J_{O3} \end{bmatrix}$$

where $\alpha = 0.5$. When the external moment $M_o(t)$ is absent, determine the response of this system by using the normal-mode approach for each of the following sets of initial conditions and discuss the participation of different modes in each case:

a) $\phi_1(0) = \phi_2(0) = \phi_3(0) = 1$ rad and $\dot{\phi}_1(0) = \dot{\phi}_2(0) = \dot{\phi}_3(0) = 0$ rad/s
b) $\phi_1(0) = \phi_3(0) = 1$ rad, $\phi_2(0) = -0.5$ rad, and
$\dot{\phi}_1(0) = \dot{\phi}_2(0) = \dot{\phi}_3(0) = 0$ rad/s

8.15 An optical platform is found to experience undesirable vibrations at a frequency of 100 Hz. The associated magnitude of the disturbance acting on the platform is estimated to be 100 N. Design a spring-mass system as an absorber for this system, with the constraint that the absorber response amplitude cannot exceed 5 mm.

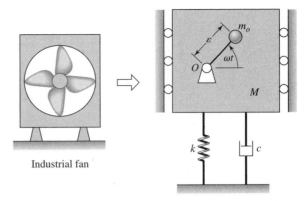

Industrial fan

FIGURE E8.16

8.16 An industrial-fan system, a model of which is shown in Figure E8.16, is found to experience undesirable vibrations when operated at 600 rpm. It is assessed that these undesirable vibrations are due to a mass unbalance in the fan, and it is estimated the mass unbalance $m_o = 2$ kg is located at a distance $\epsilon = 0.5$ m from the point O shown in Figure E8.16. Design a vibration absorber for this industrial-fan system, with the restriction that the mass of the absorber cannot exceed 75 kg.

8.17 For the system of Exercise 8.1, verify that the frequency-response functions determined by using the unit impulse excitation $f_2(\tau) = \delta(\tau)$ agree with those determined in that problem.

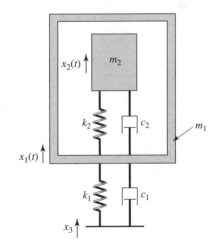

FIGURE E8.18

8.18 A model of an electronic system m_2 contained in a package m_1 is shown in Figure E8.18. Determine the transfer function $X_2(s)/X_3(s)$ and show that it has the form

$$\frac{X_2(s)}{X_3(s)} = \frac{(2\zeta_1 s + 1)C(s)}{D_2(s)}$$

where the polynomials $C(s)$ and $D_2(s)$ are given by Eqs. (8.64) and (8.74), respectively.

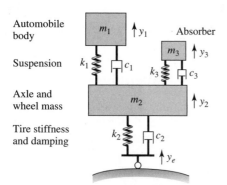

Automobile body

Absorber

Suspension

Axle and wheel mass

Tire stiffness and damping

FIGURE E8.19

8.19 Obtain the transfer function between the absorber response and the base excitation in the vehicle model shown in Figure E8.19.

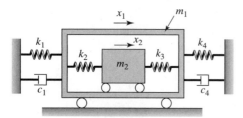

FIGURE E8.20

8.20 Derive the governing equations of motion for the system shown in Figure E8.20 and present them in state-space form.

8.21 Put the linearized system governing "small" amplitude motions of the pendulum-absorber system of Exercise 7.3 in state-space form.

8.22 Obtain a state-space form of the equations of motion governing "small" amplitude motions of the pendulum absorber system shown in Figure E7.5. Assume that oscillations about the static-equilibrium

position corresponding to the bottom position of the pendulum are of interest.

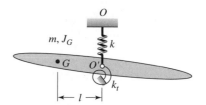

FIGURE E8.23

8.23 Present the governing equations of motion for "small" oscillations of the airfoil system shown in Figure E8.23 in state-space form. In this system, the stiffness of the translation spring is k, the stiffness of the torsion spring is k_t, G is the center of the mass of the airfoil located a distance l from the attachment point, m is the mass of the airfoil, and J_G is the mass moment of inertia of the airfoil about the center of mass. Use the generalized coordinates y and θ to obtain the equations of motion, where y is the vertical translation of the point O' and θ is the angular oscillation about an axis normal to the airfoil plane.

8.24 Consider the two degree-of-freedom system with two different damping models treated in Example 8.9 and obtain the state matrix for this system and study the eigenvalues and eigenvectors of this state matrix for the systems with constant modal damping and arbitrary damping models. Also, compare these eigenvalues and eigenvectors with those for the undamped system and draw comparisons and discuss them.

8.25 If the damping matrix given in Exercise 8.14 is altered so that the damping matrix now has the following structure

$$\begin{bmatrix} c_{t1} & 0 & 0 \\ 0 & c_{t2} & 0 \\ 0 & 0 & c_{t3} \end{bmatrix} = \begin{bmatrix} 0.2J_{O1} & 0 & 0 \\ 0 & 0.6J_{O2} & 0 \\ 0 & 0 & 0.2J_{O3} \end{bmatrix}$$

determine the response of the flywheels for the inertia and stiffness parameters given in Exercise 8.14 and the initial conditions in case (a) of this exercise.

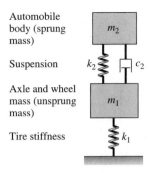

Automobile body (sprung mass) — m_2

Suspension — k_2, c_2

Axle and wheel mass (unsprung mass) — m_1

Tire stiffness — k_1

FIGURE E8.26

8.26 The vertical motions of an automobile are described by using the quarter-car model shown in Figure E8.26. The different system parameters in this two degree-of-freedom model are as follows: $m_1 = 80$ kg, $m_2 = 1100$ kg, $k_2 = 30$ kN/m, $k_1 = 300$ kN/m, and $c_2 = 5000$ N/(m/s). When this vehicle is traveling with a constant speed of 40 m/s on a flat road, it hits a bump, which produces an initial vertical displacement of 0.2 m and an initial vertical velocity of 0.2 m/s at the base of the system. Determine the ensuing response of this system.

8.27 Study the micromechanical filter of Example 8.12 and graph the amplitude response of the micromechanical filter similar to that shown in Figure 8.14 for the following system parameter values: $\omega_r = 0.3$, $m_r = 0.5$, $k_{31} = 1$, and $\zeta_1 = \zeta_{32} = 0.001$.

8.28 For the industrial-fan system of Exercise 8.16, if the original restriction on the absorber mass is removed, design the absorber with the following new restrictions: a) the natural frequencies of the system with the absorber should lie outside the range of 575 rpm to 625 rpm and b) the response amplitude of the absorber mass should not exceed 25 mm.

8.29 A machine has a mass of 150 kg and a natural frequency of 150 rad/s. An absorber mass of 30 kg and a spring-damper combination is to be attached to this machine, so that the machine can be operated in as wide a frequency range as possible around the machine's natural frequency. Determine the optimal parameters for the absorber.

8.30 In order to attenuate oscillations of telecommunication towers, tuned vibration absorbers are typ-

ically used. The second natural frequency of a representative telecommunication tower is 0.6 Hz and the third natural frequency of this tower is 1.5 Hz. Design two optimal absorbers, one to be effective in a frequency range that includes 0.6 Hz and another to be effective in a frequency range that includes 1.5 Hz. The mass ratio for each absorber can be picked to lie in the range from 0.02 to 0.05.

8.31 A rotary engine experiences disturbances at the sixth harmonic of the rotating speed of the system. Determine the nondimensional pendulum length of a centrifugal absorber for this system.

8.32 Determine the free responses of the bar-slider system described by Eqs. (8.141) for the following parameter values: $m = 0.5$, $\zeta = 0.20$, and $\omega_c = 0.5$, 1.5, 2, and 2.5. Assume that in each case, the motions are initiated from $\theta(0) = 0.5$ rad with all of the other initial conditions being zero. Graph the time histories for the radial displacements of the slider and the angular motions of the bar and discuss the effectiveness of the nonlinear vibration absorber in suppressing the angular motions of the bar.

8.33 Determine the free responses of the pendulum-absorber system described by Eqs. (8.151) for the following parameter values: $\Omega = 1$, $\omega_r = 0.5$, $\zeta = 0.10$, $\zeta_t = 0.05$, $m_r = 0.10$ and $f_o = 0.4$. Assume that in each case, the motions are initiated from $\varphi(0) = 0.2$ rad with all of the other initial conditions being zero. Graph the time histories for the vertical motions of the system and the angular motions of the pendulum and discuss the effectiveness of the nonlinear vibration absorber in suppressing the vertical translations.

8.34 A 200 kg machine is to be operated at 500 rpm on a flexible platform that can be modeled as a single degree-of-freedom system with the following mass, stiffness, and damper values: $m_1 = 3000$ kg, $k_1 = 2 \times 10^6$ N/m, and $c_1 = 7500$ N/(m/s). Determine the properties of the machinery mounting so that the transmissibility ratio does not exceed 0.1.

8.35 Construct the transmissibility ratio plot of Figure 8.16 for the following parameters of a two degree-of-freedom system: $m_r = 0.2$ and $\zeta_1 = \zeta_2 = 0$.

f elastic structures can be modeled as thin elastic beams. Towers, drills, and baseball bats are
such systems.

9

Vibrations of Beams

9.1 INTRODUCTION

In Chapters 3 through 8, vibrations of systems with finite degrees of freedom were treated. As mentioned in Section 2.5, elements with distributed inertia and stiffness properties, such as beams, are used to model many physical systems such as the ski of Section 2.5.4, the work-piece-tool system of Section 2.5.5, and the MEMS accelerometer of Section 2.5.2. As noted previously, *distributed-parameter systems,* which are also referred to as *spatially continuous systems,* have an infinite number of degrees of freedom. Apart from beams, distributed systems that one could use in vibratory models include strings, cables, bars undergoing axial vibrations, shafts undergoing torsional vibrations, plates, and shells. Except for the last two systems mentioned, the descriptions of all of the other systems require one spatial coordinate. The equations of motion, which govern vibratory systems with finite degrees of freedom, are ordinary differential equations, and these equations are in the form of an initial-value problem. By contrast, the equations of motion governing distributed-parameter systems are in the form of partial differential equations, and the determination of the solution for the vibratory response of a distributed-parameter system requires the use of additional mathematical techniques. However, notions such as natural frequencies, mode shapes, orthogonality of modes, and normal-mode solution procedures used in the context of finite degree-of-freedom systems apply equally well to infinite degree-of-freedom systems. An infinite degree-of-freedom system has an infinite number of natural frequencies and a mode shape associated with the free oscillations at each one of these frequencies.

In this chapter, the free and forced vibrations of beams are considered at length. As illustrated by the diverse examples of Section 2.5, vibratory models of many physical systems require the use of beam elements. In addition to these examples, other examples where beam elements are used to model physical systems include models of rotating machinery, ship hulls, aircraft wings, and vehicular and railroad bridges. Propeller blades in a turbine and the rotor blades of a helicopter are modeled by using beam elements. Since the vibratory behavior of beams is of practical importance for these different systems, the focus of this chapter will be on beam vibrations.

In each of the applications cited above, and in many others, the beams are acted upon by dynamically varying forces. Depending on the frequency content of these forces, the forces have the potential to excite the beam at one or more of its natural frequencies. One of the frequent requirements of a design engineer is to create an elastic structure that responds minimally to the imposed dynamic loading, so that large displacement amplitudes, high stresses and structural fatigue are minimized, and wear and radiated noise are decreased.

The governing equations of motion for beams are obtained by using the mechanics of elastic beams and Hamilton's principle. Free oscillations of unforced and undamped beams are treated and various factors that influence the natural frequencies and modes are examined. This examination includes the treatment of inertial elements and springs attached at an intermediate location and beam geometry variation. The limitations of the models used in the previous chapters are also pointed out in the context of systems where a flexible structure supports systems with one or two degrees of freedom. The use of the normal-mode approach to determine the forced response of a beam is also presented.

9.2 GOVERNING EQUATIONS OF MOTION

In this section, we illustrate how the governing equations of an elastic beam undergoing small transverse vibrations are obtained for arbitrary loading conditions and boundary conditions. A beam element in a deformed configuration is shown in Figure 9.1. The x-axis runs along the span of the beam, and the y-axis and z-axis run along transverse directions to the x-axis. End moments of magnitude M are shown acting along the j direction, and it is assumed that the beam displacement is confined to the x-z plane. The displacement $w(x,t)$ denotes the transverse displacement at a location along the beam.

The derivation of the governing equations of motion is based on the *extended Hamilton's principle*. To use this principle, one first needs to deter-

FIGURE 9.1
Deformation of a beam
subjected to end moments.

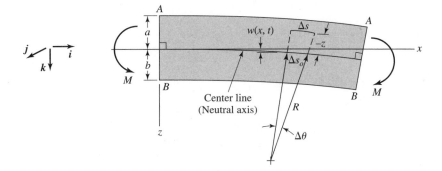

mine the system potential energy, the system kinetic energy, and the work
done on the system. For determining the system kinetic energy, each element
of length Δx is treated like a rigid body and for determining the system potential energy, stress-strain relationships in the beam material are used. To
this end, preliminaries from solid mechanics are presented in Section 9.2.1,
and then the expressions for the kinetic energy, potential energy, and work
are obtained in Section 9.2.2.

9.2.1 Preliminaries from Solid Mechanics

In Figure 9.1, it is seen that the face of the beam located toward the center of
curvature will be contracted while that on the opposite face will be extended;
that is, face AA will be extended, while the face BB will be contracted. The
line passing through the centroids of the cross-section of the beam is called
the *central line*. Here, a fiber along the central line is assumed to experience
zero axial strain. Hence, this central line is the *neutral axis*. The deformation
of the beam is assumed to be described by Bernoulli-Euler beam theory,[1]
which is applicable to thin elastic beams whose length to depth ratio is
greater than 10. In keeping with this theory, it is assumed that the neutral axis
remains unaltered, that the plane sections of the beam normal to the neutral
axis remain plane and normal to the deformed central line, and that the transverse normals such as BA experience zero strain along the normal direction.
For a fiber located at a distance z from the neutral axis, as shown in Figure
9.1, the strain experienced along the length of the beam is given by

$$\epsilon = \frac{\Delta s - \Delta s_o}{\Delta s_o} = -\frac{z}{R} \tag{9.1}$$

where R is the *radius of curvature*, Δs_o is the length of a fiber along the neutral axis, Δs is the length of a fiber that is located at a distance z from the neutral axis, and we have used geometry to write

$$\Delta s = (R - z)\Delta\theta \quad \text{and} \quad \Delta s_o = R\Delta\theta \tag{9.2}$$

[1]E. P. Popov, *Engineering Mechanics of Solids,* Prentice Hall, Upper Saddle River, NJ, Chapter
6 (1990).

From Hooke's law, the corresponding axial stress σ acting on the fiber is

$$\sigma = E\epsilon = -\frac{Ez}{R} \tag{9.3}$$

where E is the Young's modulus of the material. According to the convention shown in Figure 9.1, a positive displacement w is in the direction of the unit vector $\mathbf{k}$. Therefore, the fibers above the neutral axis experience a positive σ, which denotes tension, and the fibers below the neutral axis experience a negative σ, which denotes compression.

At an internal section of the beam, a moment balance about the y-axis leads to

$$M = -\int_{y_1}^{y_2} \int_{-a}^{b} \sigma z \, dz \, dy = \frac{EI}{R} \tag{9.4}$$

where y_1 and y_2 are the spatial limits corresponding to integration along the y direction, we have used Eq. (9.3), and

$$I = \int_{y_1}^{y_2} \int_{-a}^{b} z^2 \, dz \, dy \tag{9.5}$$

The quantity I represents the area moment of inertia of the beam's cross-section about the y-axis, which is through the centroid. In general, the limits of the double integral in Eq. (9.5) do not have to be constants; that is, $a = a(x)$, $b = b(x)$, $y_1 = y_1(x)$, and $y_2 = y_2(x)$. In this case, the area moment of inertia varies along the length, and therefore, in general, $I = I(x)$. The *curvature* $\kappa = 1/R$, which is assumed to be positive for concave curvature downwards, is

$$\kappa = \frac{1}{R} = \frac{\partial^2 w}{\partial x^2}\left[1 + \left(\frac{\partial w}{\partial x}\right)^2\right]^{-3/2} \tag{9.6}$$

If it is assumed that the displacement is small—that is, $|\partial w/\partial x| \ll 1$, where $\partial w/\partial x$ is the slope of the neutral axis at location x—then Eq. (9.6) simplifies to

$$\kappa = \frac{1}{R} \cong \frac{\partial^2 w}{\partial x^2} \tag{9.7}$$

Upon substituting Eq. (9.7) into Eqs. (9.1) and (9.4), we obtain

$$\epsilon = -z\frac{\partial^2 w}{\partial x^2}$$

$$M = EI\frac{\partial^2 w}{\partial x^2} \tag{9.8}$$

Thus, the magnitude of the strain and bending moment are proportional to the second spatial derivative of the beam displacement. The statement that the bending moment is linearly proportional to the second spatial derivative of the beam displacement is the *Bernoulli-Euler law*, which is the underlying basis for the theory of linear elastic thin beams.

Equation (9.8) was obtained by considering only the effects of moments on the ends of the beam. If, in addition, there is a transverse load $f(x,t)$, then there are vertical shear forces within the beam that resist this force. In Figure 9.2, if the sum of the moments about point o is taken along the j direction, and if the rotary inertia of the beam element is neglected, the result is

$$M + (V + \Delta V)\Delta x = M + \Delta M$$

which leads to

$$\frac{\Delta M}{\Delta x} = V + \Delta V$$

In the limit $\Delta x \to 0$, the shear force increment $\Delta V \to 0$, and we have

$$\lim_{\Delta x \to 0} \frac{\Delta M}{\Delta x} = \frac{\partial M}{\partial x} = V \tag{9.9a}$$

which, after making use of Eq. (9.8), results in

$$V = \frac{\partial M}{\partial x} = \frac{\partial}{\partial x}\left(EI \frac{\partial^2 w}{\partial x^2}\right) \tag{9.9b}$$

Thus, the shear force is equal to the change of the bending moment along the x-axis. Consequently, if $M(x)$ is constant along x, then $V = 0$.

9.2.2 Potential Energy, Kinetic Energy, and Work

We construct the system potential energy, the system kinetic energy, and determine the work done by external forces for further use in Section 9.2.4.

Potential Energy

The potential energy of a deformed beam has contributions from different sources, including the strain energy. For a beam undergoing axial strains due

FIGURE 9.2
Deformation of an element of a beam subjected to a transverse load.

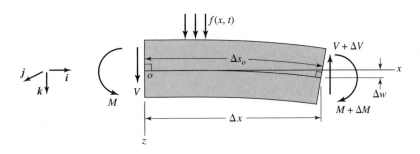

to bending, if the strain energy is the only contribution to the system potential energy, the beam's potential energy is written as[2]

$$U(t) = \frac{1}{2} \int_0^L \int_{-a}^b \int_{y_1}^{y_2} \sigma \epsilon \, dy \, dz \, dx = \frac{1}{2} \int_0^L \int_{-a}^b \int_{y_1}^{y_2} \frac{Ez^2}{R^2} \, dy \, dz \, dx$$

$$= \frac{1}{2} \int_0^L EI \left(\frac{\partial^2 w}{\partial x^2} \right)^2 dx \tag{9.10}$$

where we have used Eqs. (9.1), (9.3), and (9.7).[3]

Kinetic Energy

Assuming that the translation kinetic energy of the beam is the only contribution to the system kinetic energy, one can write it as

$$T(t) = \frac{1}{2} \int_0^L \int_{-a}^b \int_{y_1}^{y_2} \rho \left(\frac{\partial w}{\partial t} \right)^2 dy \, dz \, dx = \frac{1}{2} \int_0^L \rho A \left(\frac{\partial w}{\partial t} \right)^2 dx \tag{9.11}$$

where $A = A(x)$ is the cross-sectional area of the beam and $\rho = \rho(x)$ is the mass density of the beam's material. If the rotary inertia of the beam element is also taken into account, an additional term corresponding to the rotational kinetic energy will have to be included in Eq. (9.11), as described by Eq. (1.23).

Work

The work done by the applied transverse conservative load per unit length $f_c(x,t)$ is given by

$$W_f(t) = \int_0^L f_c(x,t) w(x,t) dx \tag{9.12a}$$

If gravity is the only distributed conservative load acting on the beam, then

$$f_c(x,t) = \rho(x)A(x)g \tag{9.12b}$$

If the beam is also under the action of an axial tensile force[4] $p(x,t)$, as shown in Figure 9.3, then the length of the central line no longer remains constant, but extends to a new length. If we assume that the deformation is small in

---

[2]See, for example, I. S. Sokolonikoff, *Mathematical Theory of Elasticity,* McGraw Hill, NY, Chapters 1–3 (1956).

[3]Since the symbol V is used for the shear force, the symbol U is used for the potential energy, which differs from the notation used in the earlier chapters.

[4]Axial loads are common in rotating blades, pipes with flow, and structural columns.

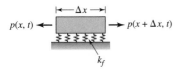

FIGURE 9.3
Beam element under axial tensile load and on an elastic foundation.

magnitude and does not affect the loading $p(x,t)$, then the change in length of an element of the beam is $(\Delta s - \Delta x)$ where[5]

$$\Delta s \approx \left[1 + \frac{1}{2} \left(\frac{\Delta w}{\Delta x} \right)^2 \right] \Delta x \qquad (9.13)$$

Therefore, the external work of the axial force is given by[6]

$$W_p = -\frac{1}{2} \int_0^L p(x,t) \left(\frac{\partial w}{\partial x} \right)^2 dx \qquad (9.14)$$

where we have used Eq. (9.13). Since the tensile force acts to oppose the beam transverse displacement w, the work done has a minus sign. If the axial force is compressive, then $p(x,t)$ is replaced by $-p(x,t)$.

Finally, we consider the linear elastic foundation[7] on which the beam is resting, as shown in Figure 9.3. The transverse displacement of the beam creates a force in the foundation with the magnitude $f_f(x,t) = k_f w(x,t)$, where k_f is the spring constant per unit length of the foundation. This spring force opposes the motion of the beam. The external work done by the elastic foundation is

$$W_k(t) = -\frac{1}{2} \int_0^L k_f w^2(x,t) dx \qquad (9.15)$$

where, again, we have introduced a minus sign to account for the fact that the foundation force acting on the beam opposes the beam displacement.

---

[5]Note that from geometry, we have

$$\Delta s^2 = \Delta w^2 + \Delta x^2$$

which leads to

$$\frac{\Delta s}{\Delta x} = \sqrt{ 1 + \left(\frac{\Delta w}{\Delta x} \right)^2 }$$

For small slope—that is, $|\Delta w/\Delta x| \ll 1$—this leads to

$$\frac{\Delta s}{\Delta x} \approx 1 + \frac{1}{2} \left(\frac{\Delta w}{\Delta x} \right)^2$$

or equivalently,

$$\Delta s - \Delta x \approx \frac{1}{2} \left(\frac{\Delta w}{\Delta x} \right)^2 \Delta x$$

[6]To construct this integral, it is noted that the work done on a segment is $-p(x,t)(\Delta s - \Delta x)$, the limit $\Delta x \to 0$ is considered, and in this limit, Δx is replaced by dx.

[7]This type of elastic foundation is frequently used to model structures on soil and it is often referred to as a *Pasternak foundation*.

System Lagrangian

With the objective of constructing the system Lagrangian L_T, we construct the function $G_B(x,t,w,\dot{w},w',\dot{w}',w'')$ from the expression

$$\int_0^L G_B(x,t,w,\dot{w},w',\dot{w}',w'')dx = T(t) - U(t) + W_c(t) \qquad (9.16)$$

where

$$W_c(t) = W_f(t) + W_p(t) + W_k(t) \qquad (9.17)$$

and we have introduced the compact notation

$$\dot{w} = \frac{\partial w}{\partial t}, \quad w' = \frac{\partial w}{\partial x},$$

$$\dot{w}' = \frac{\partial^2 w}{\partial x \partial t}, \quad w'' = \frac{\partial^2 w}{\partial x^2} \qquad (9.18)$$

In Eqs. (9.18), $\dot{w}$, w', $\dot{w}'$, and w'' represent the beam velocity, beam slope, beam angular velocity, and beam curvature, respectively. In Eq. (9.17), $W_c(t)$ is the work done by conservative forces and it has been constructed assuming that the work $W_f(t)$ done by the external loading $f_c(x,t)$ is conservative and that the work $W_p(t)$ done by the axial loading $p(x,t)$ is conservative.[8]

After collecting the spatial integrals given by Eqs. (9.10), (9.11), (9.12a), (9.14), and (9.15) for $U(t)$, $T(t)$, $W_f(t)$, $W_p(t)$, and $W_k(t)$, respectively, we find from Eq. (9.16) that

$$G_B(x,t,w,\dot{w},w',\dot{w}',w'') = \frac{1}{2}[\rho A \dot{w}^2 - EI w''^2 - pw'^2 - k_f w^2] + f_c w \qquad (9.19)$$

In Eq. (9.19), there is no $\dot{w}'$ term, since we have neglected the rotary inertia of the beam cross-section in the development.

On the boundaries of the beam at $x = 0$ and $x = L$, one can have discrete external elements that contribute to the total kinetic energy and the total potential energy of the system. Consider, for example, the beam shown in Figure 9.4. At the left boundary ($x = 0$), there is a linear translation spring with stiffness k_1 and a linear torsion spring with stiffness k_{t1}. Similarly, at the right boundary ($x = L$), there is a linear translation spring with stiffness k_2 and a linear torsion spring with stiffness k_{t2}. There is also an inertia element with mass M_1 and rotary inertia J_1 at the left boundary and an inertia element with mass M_2 and rotary inertia J_2 at the right boundary. There are also linear viscous dampers with damping coefficients c_1 at $x = 0$ and c_2 at $x = L$. Taking the difference between the kinetic energy of the inertia element at the left

---

[8]Conservative forces were first mentioned in Section 2.3, where it was noted that forces expressed in terms of a potential function, such as a spring force or a gravity loading, are conservative. On the other hand, dissipative loads such as those due to dampers and time-dependent loads such as harmonic excitations are nonconservative.

FIGURE 9.4
Beam with discrete elements at the boundaries.

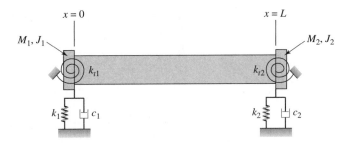

boundary and the potential energy of the stiffness elements at the left boundary in Figure 9.4, we obtain the discrete Lagrangian function

$$G_0(t,w_0,\dot{w}_0,w'_0,\dot{w}'_0) = \underbrace{\frac{1}{2}M_1\dot{w}_0^2}_{\substack{\text{Translational}\\\text{kinetic energy}\\\text{of rigid body}}} + \underbrace{\frac{1}{2}J_1\dot{w}_0'^2}_{\substack{\text{Rotational}\\\text{kinetic energy}\\\text{of rigid body}}} - \underbrace{\frac{1}{2}k_1w_0^2}_{\substack{\text{Potential}\\\text{energy of}\\\text{translation}\\\text{spring}}} - \underbrace{\frac{1}{2}k_{t1}w_0'^2}_{\substack{\text{Potential}\\\text{energy of}\\\text{torsion spring}}} \quad (9.20)$$

where the subscript 0 has been used to denote that the quantity is evaluated at $x = 0$; that is, $w_0 = w(0,t)$ indicates the displacement at the boundary $x = 0$, $w'_0 = \partial w(0,t)/\partial x$ indicates the slope at the boundary $x = 0$, and so on. Also, the translational velocity of the center of mass of the rigid body is denoted as $\dot{w}_0$, the angular displacement of the torsion spring is denoted as w'_0, and the angular velocity of the rigid body is denoted as $\dot{w}'_0$. Similarly, the discrete Lagrangian function corresponding to the right boundary $x = L$ is given by

$$G_L(t,w_L,\dot{w}_L,w'_L,\dot{w}'_L) = \underbrace{\frac{1}{2}M_2\dot{w}_L^2}_{\substack{\text{Translational}\\\text{kinetic energy}\\\text{of rigid body}}} + \underbrace{\frac{1}{2}J_2\dot{w}_L'^2}_{\substack{\text{Rotational}\\\text{kinetic energy}\\\text{of rigid body}}} - \underbrace{\frac{1}{2}k_2w_L^2}_{\substack{\text{Potential}\\\text{energy of}\\\text{translation}\\\text{spring}}} - \underbrace{\frac{1}{2}k_{t2}w_L'^2}_{\substack{\text{Potential}\\\text{energy of}\\\text{torsion spring}}} \quad (9.21)$$

where the subscript L has been used to denote that the quantity is evaluated at $x = L$; that is, $w_L = w(L,t)$ indicates the displacement at the boundary $x = L$, $w_L' = \partial w(L,t)/\partial x$ indicates the slope at the boundary $x = L$, and so on. Also, the translational velocity of the center of mass of the rigid body is denoted as $\dot{w}_L$, the angular rotation of the torsion spring is denoted as w'_L, and the angular velocity of the rigid body is denoted as $\dot{w}'_L$.

Recalling that the Lagrangian L_T is the difference between the system kinetic energy and system potential energy, for the beam system it is given by[9]

$$L_T = \int_0^L G_B(x,t,w,\dot{w},w',\dot{w}',w'')dx + G_0 + G_L \quad (9.22a)$$

where G_B is given by Eq. (9.19), G_0 is given by Eq. (9.20), and G_L is given by Eq. (9.21). These last two terms on the right-hand side of Eq. (9.22a) are

[9]Since the system is conservative, the work done by conservative forces in Eq. (9.16) is replaced by the equivalent potential energy.

included in the spatial integral by using the delta function introduced in Chapter 6. This leads to

$$L_T = \int_0^L [G_B(x,t,w,\dot{w},w',\dot{w}',w'') + G_0\delta(x) + G_L\delta(x - L)]dx$$

$$= \int_0^L G(x,t,w,\dot{w},w',\dot{w}',w'')dx \qquad (9.22b)$$

where

$$G(x,t,w,\dot{w},w',\dot{w}',w'') = G_B + G_0\delta(x) + G_L\delta(x - L) \qquad (9.23)$$

In Eqs. (9.22b) and (9.23), the delta function is used to represent contributions from the discrete attachments at the spatial locations $x = 0$ and $x = L$. Specifically, $\delta(x)$ assumes a zero value when x is different from zero, and $\delta(x - L)$ assumes a zero value when x is different from L. The use of these functions enables us to include the discrete contributions in an expression involving the whole domain, as in the spatial integral of Eq. (9.22b).

The damper elements at the boundaries introduce the following discrete nonconservative forces at the boundaries:

$$f_{nc0}(t) = -c_1\dot{w}_0 \quad \text{and} \quad f_{ncL}(t) = -c_2\dot{w}_L \qquad (9.24)$$

9.2.3 Extended Hamilton's Principle and Derivation of Equations of Motion

To derive the governing equations of motion of the beam, we employ the *extended Hamilton's principle*. A general formulation applicable to spatially one-dimensional continua is first presented, and the beam equation is obtained as one application of this formulation. First, conservative systems—that is, systems without losses due to damping and other dissipation sources—are considered and then nonconservative systems are considered.

Conservative Systems

The statement of the extended Hamilton's principle[10] for holonomic and conservative systems is as follows. Of all possible paths of motion to be taken

---

[10]The extended Hamilton's principle is expressed as

$$\int_{t_1}^{t_2} (\delta T + \delta W)dt = 0$$

where δ is the variation operator, the times t_1 and t_2 are the initial time and the final time, respectively, T is the total system kinetic energy, and δW includes work done by both conservative and nonconservative forces. In the absence of nonconservative forces, $\delta W = -\delta V$, where V is the total system potential energy. Hence, for conservative systems, the extended Hamilton's principle is written as

$$\int_{t_1}^{t_2} (\delta T - \delta V)dt = 0$$

or, equivalently as

$$\int_{t_1}^{t_2} \delta L_T dt = 0$$

where $L_T = T - V$ is the Lagrangian of the system.

between two instants of time t_1 and t_2, the actual path to be taken by the system corresponds to a stationary value of the integral I_H; that is,

$$\delta I_H = 0 \qquad (9.25)$$

where δ is the variation operator[11] and

$$
\begin{aligned}
I_H &= \int_{t_1}^{t_2} L_T\, dt \\
&= \int_{t_1}^{t_2} \left[\int_0^L G_B(x,t,w,\dot{w},w',\dot{w}',w'')\, dx + G_0 + G_L \right] dt \qquad (9.26)
\end{aligned}
$$

In arriving at Eq. (9.26), we have made use of Eqs. (9.22) for the Lagrangian of the system.

The first term inside the integral on the right-hand side of Eq. (9.26) represents the contributions to the system kinetic energy, the system potential energy, and the work done in the continuum (beam) interior $0 < x < L$. The second and third terms on the right-hand side of Eq. (9.26) represent contributions from the discrete elements attached at the boundaries $x = 0$ and $x = L$, respectively. The damper elements at the boundaries, which are shown in Figure 9.4, are not considered in the conservative case.

By using calculus of variations, it can be shown[12] that Eq. (9.26) has a stationary value when the following conditions are satisfied in the interior $0 < x < L$ and at the boundaries $x = 0$ and $x = L$.

Continuum interior ($0 < x < L$)

$$\frac{\partial G_B}{\partial w} - \frac{\partial}{\partial x}\left(\frac{\partial G_B}{\partial w'}\right) + \frac{\partial^2}{\partial x^2}\left(\frac{\partial G_B}{\partial w''}\right) - \frac{\partial}{\partial t}\left(\frac{\partial G_B}{\partial \dot{w}}\right) + \frac{\partial^2}{\partial x \partial t}\left(\frac{\partial G_B}{\partial \dot{w}'}\right) = 0 \quad (9.27)$$

Boundary conditions at $x = 0$

$$w(0,t) = 0 \qquad (9.28a)$$

or

$$\frac{\partial G_0}{\partial w_0} - \frac{\partial}{\partial t}\left(\frac{\partial G_0}{\partial \dot{w}_0}\right) - \left[\frac{\partial G_B}{\partial w'} - \frac{\partial}{\partial x}\left(\frac{\partial G_B}{\partial w''}\right) - \frac{\partial}{\partial t}\left(\frac{\partial G_B}{\partial \dot{w}'}\right)\right]_{x=0} = 0 \quad (9.28b)$$

and

$$w'(0,t) = 0 \qquad (9.29a)$$

or

$$\frac{\partial G_0}{\partial w_0'} - \frac{\partial}{\partial t}\left(\frac{\partial G_0}{\partial \dot{w}_0'}\right) - \left(\frac{\partial G_B}{\partial w''}\right)_{x=0} = 0 \qquad (9.29b)$$

---

[11]The symbol δ is used everywhere else in the book to mean the delta function; this is the only exception.

[12]E. B. Magrab, *Vibrations of Elastic Structural Members,* Sijthoff & Noordhoff International Publishing Co., The Netherlands, pp. 5–12 (1979).

Boundary conditions at x = L

$$w(L,t) = 0 \tag{9.30a}$$

or

$$\frac{\partial G_L}{\partial w_L} - \frac{\partial}{\partial t}\left(\frac{\partial G_L}{\partial \dot{w}_L}\right) + \left[\frac{\partial G_B}{\partial w'} - \frac{\partial}{\partial x}\left(\frac{\partial G_B}{\partial w''}\right) - \frac{\partial}{\partial t}\left(\frac{\partial G_B}{\partial \dot{w}'}\right)\right]_{x=L} = 0 \tag{9.30b}$$

and

$$w'(L,t) = 0 \tag{9.31a}$$

or

$$\frac{\partial G_L}{\partial w'_L} - \frac{\partial}{\partial t}\left(\frac{\partial G_L}{\partial \dot{w}'_L}\right) + \left(\frac{\partial G_B}{\partial w''}\right)_{x=L} = 0 \tag{9.31b}$$

Equations (9.27) to (9.31) represent the general form of the equations of motion for beams subjected to conservative forces.

Nonconservative Systems

In the nonconservative case, the work done by the nonconservative force per unit length $f_{nc}(x,t)$ acting on the system in the continuum interior and the nonconservative forces $f_{nc0}(t) = f_{nc}(0,t)$ and $f_{ncL}(t) = f_{nc}(L,t)$ acting at the boundaries $x = 0$ and $x = L$, respectively, also need to be taken into account. In this case, Eqs. (9.27) to (9.31) are modified to the following.

Continuum interior (0 < x < L)

$$\frac{\partial G_B}{\partial w} - \frac{\partial}{\partial x}\left(\frac{\partial G_B}{\partial w'}\right) + \frac{\partial^2}{\partial x^2}\left(\frac{\partial G_B}{\partial w''}\right) - \frac{\partial}{\partial t}\left(\frac{\partial G_B}{\partial \dot{w}}\right)$$

$$+ \frac{\partial^2}{\partial x \partial t}\left(\frac{\partial G_B}{\partial \dot{w}'}\right) + f_{nc} = 0 \tag{9.32}$$

Equation (9.32) is referred to as the *Lagrange differential equation* of motion for spatially one-dimensional continuous systems.

Boundary conditions at x = 0

$$w(0,t) = 0 \tag{9.33a}$$

or

$$\frac{\partial G_0}{\partial w_0} - \frac{\partial}{\partial t}\left(\frac{\partial G_0}{\partial \dot{w}_0}\right) - \left[\frac{\partial G_B}{\partial w'} - \frac{\partial}{\partial x}\left(\frac{\partial G_B}{\partial w''}\right)\right.$$

$$\left. - \frac{\partial}{\partial t}\left(\frac{\partial G_B}{\partial \dot{w}'}\right)\right]_{x=0} + f_{nc0} = 0 \tag{9.33b}$$

and

$$w'(0,t) = 0 \tag{9.34a}$$

or

$$\frac{\partial G_0}{\partial w_0'} - \frac{\partial}{\partial t}\left(\frac{\partial G_0}{\partial \dot{w}_0'}\right) - \left(\frac{\partial G_B}{\partial w''}\right)_{x=0} = 0 \tag{9.34b}$$

Boundary conditions at $x = L$

$$w(L,t) = 0 \tag{9.35a}$$

or

$$\frac{\partial G_L}{\partial w_L} - \frac{\partial}{\partial t}\left(\frac{\partial G_L}{\partial \dot{w}_L}\right) + \left[\frac{\partial G_B}{\partial w'} - \frac{\partial}{\partial x}\left(\frac{\partial G_B}{\partial w''}\right)\right.$$
$$\left. - \frac{\partial}{\partial t}\left(\frac{\partial G_B}{\partial \dot{w}'}\right)\right]_{x=L} + f_{ncL} = 0 \tag{9.35b}$$

and

$$w'(L,t) = 0 \tag{9.36a}$$

or

$$\frac{\partial G_L}{\partial w_L'} - \frac{\partial}{\partial t}\left(\frac{\partial G_L}{\partial \dot{w}_L'}\right) + \left(\frac{\partial G_B}{\partial w''}\right)_{x=L} = 0 \tag{9.36b}$$

9.2.4 Beam Equation for a General Case

We now derive the governing equation for the beam considered in Section 9.2.1 and shown in Figure 9.5. This system consists of an external transverse load $f(x,t)$, which for purposes of generality is assumed to be composed of a conservative load and a nonconservative load. It is expressed as

$$f(x,t) = f_c(x,t) + f_{nc}(x,t) \tag{9.37}$$

where the conservative part is due to gravitational loading, as in Eq. (9.12b). The load due to the spring foundation with stiffness k_f per unit length is a conservative load, and, for convenience, the load due to the axial load $p(x,t)$ is assumed to be a conservative load. At the left boundary, there is a linear translation spring with stiffness k_1 and a linear torsion spring with stiffness k_{t1}. Similarly, at the right boundary, there is a linear translation spring with stiffness k_2 and a linear torsion spring with stiffness k_{t2}. There is also an inertia element with mass M_1

FIGURE 9.5

Beam on an elastic foundation, under axial and transverse loads, and with discrete elements at the boundaries.

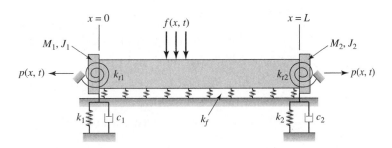

and rotary inertia J_1 at the left boundary and an inertia element with mass M_2 and rotary inertia J_2 at the right boundary. There are also linear viscous dampers with damping coefficients c_1 at $x = 0$ and c_2 at $x = L$, respectively. Since the system is nonconservative, we will make use of Eqs. (9.32) through (9.36) to derive the equations of motion and the boundary conditions.

The functions G_B, G_0, and G_L are given by Eqs. (9.19), (9.20), and (9.21), respectively. Since we have viscous dampers at the boundaries, there are nonconservative forces that are given by Eqs. (9.24); they are reproduced below.

$$f_{nc0}(t) = -c_1\dot{w}_0 \quad \text{and} \quad f_{ncL}(t) = -c_2\dot{w}_L \tag{9.38}$$

On using Eq. (9.19) to evaluate the individual terms in Eq. (9.32), we arrive at

$$\frac{\partial G_B}{\partial w} = -k_f w(x,t) + f_c$$

$$\frac{\partial}{\partial x}\left(\frac{\partial G_B}{\partial w'}\right) = -\frac{\partial}{\partial x}\left(p(x,t)\frac{\partial w}{\partial x}\right)$$

$$\frac{\partial^2}{\partial x^2}\left(\frac{\partial G_B}{\partial w''}\right) = -\frac{\partial^2}{\partial x^2}\left(EI(x)\frac{\partial^2 w}{\partial x^2}\right)$$

$$\frac{\partial}{\partial t}\left(\frac{\partial G_B}{\partial \dot{w}}\right) = \rho A(x)\frac{\partial^2 w}{\partial t^2}$$

$$\frac{\partial^2}{\partial x \partial t}\left(\frac{\partial G_B}{\partial \dot{w}'}\right) = 0 \tag{9.39}$$

After substituting Eqs. (9.39) into Eq. (9.32) and taking into account Eq. (9.37), we obtain the governing equation of motion for the beam in $0 < x < L$:

$$\underbrace{\frac{\partial^2}{\partial x^2}\left(EI(x)\frac{\partial^2 w}{\partial x^2}\right)}_{\substack{\text{Force per unit length due to}\\\text{beam flexural stiffness } EI(x)}} - \underbrace{\frac{\partial}{\partial x}\left(p(x,t)\frac{\partial w}{\partial x}\right)}_{\substack{\text{Force per unit length due}\\\text{to tensile axial force}\\p(x,t)}} + \underbrace{k_f w(x,t)}_{\substack{\text{Force per unit}\\\text{length due to}\\\text{elastic foundation}\\\text{with stiffness}\\\text{per unit area } k_f}} + \underbrace{\rho A(x)\frac{\partial^2 w}{\partial t^2}}_{\substack{\text{Force per unit}\\\text{length due to}\\\text{beam mass per}\\\text{unit length } \rho A(x)}} = \underbrace{f(x,t)}_{\substack{\text{External}\\\text{transverse}\\\text{loading per}\\\text{unit length}}} \tag{9.40}$$

where

$$\underbrace{f(x,t)}_{\substack{\text{External}\\\text{transverse}\\\text{loading per}\\\text{unit length}}} = \underbrace{f_c(x,t)}_{\substack{\text{Conservative part}\\\text{of external}\\\text{transverse}\\\text{loading per}\\\text{unit length}}} + \underbrace{f_{nc}(x,t)}_{\substack{\text{Nonconservative}\\\text{part of external}\\\text{transverse loading}\\\text{per unit length}}} \tag{9.41a}$$

If the gravitational loading is the only conservative loading acting on the system, then making use of Eqs. (9.12b) and (9.37), the transverse loading $f(x,t)$ is written as

$$f(x,t) = \rho(x)A(x)g + f_{nc}(x,t) \tag{9.41b}$$

Furthermore, the conservative load due to gravity is a static load; that is, a time-independent load. If distributed damping is present, where the associated damping coefficient is c, then one possible form of this nonconservative force is

$$f_{nc}(x,t) = -c\,\frac{\partial w}{\partial t} \tag{9.41c}$$

where the units of this term are force/length. The minus sign in Eq. (9.41c) indicates that this loading opposes the motion. In Example 9.6, consideration of such a term is illustrated.

The four boundary conditions are obtained as follows. The two boundary conditions at $x = 0$ are obtained by using Eqs. (9.19) and (9.20) in Eqs. (9.33) and (9.34), and the two boundary conditions at $x = L$ are obtained by using Eqs. (9.19) and (9.21) in Eqs. (9.35) and (9.36). Performing the indicated operations, we obtain:

$x = 0$

$$w(0,t) = 0 \tag{9.42a}$$

or

$$\left[k_1 w + c_1 \dot{w} + M_1 \frac{\partial^2 w}{\partial t^2} + \frac{\partial}{\partial x}\left(EI(x)\frac{\partial^2 w}{\partial x^2} \right) - p\,\frac{\partial w}{\partial x} \right]_{x=0} = 0 \tag{9.42b}$$

and

$$w'(0,t) = 0 \tag{9.43a}$$

or

$$\left[-k_{t1}\frac{\partial w}{\partial x} - J_1 \frac{\partial^3 w}{\partial x \partial t^2} + EI(x)\frac{\partial^2 w}{\partial x^2} \right]_{x=0} = 0 \tag{9.43b}$$

$x = L$

$$w(L,t) = 0 \tag{9.44a}$$

or

$$\left[k_2 w + c_2 \dot{w} + M_2 \frac{\partial^2 w}{\partial t^2} - \frac{\partial}{\partial x}\left(EI(x)\frac{\partial^2 w}{\partial x^2} \right) + p\,\frac{\partial w}{\partial x} \right]_{x=L} = 0 \tag{9.44b}$$

and

$$w'(L,t) = 0 \tag{9.45a}$$

or

$$\left[k_{t2}\frac{\partial w}{\partial x} + J_2 \frac{\partial^3 w}{\partial x \partial t^2} + EI(x)\frac{\partial^2 w}{\partial x^2} \right]_{x=L} = 0 \tag{9.45b}$$

Referring to Figure 9.2, it is recalled that the quantity w' is the slope of the beam with respect to the x-axis or the rotation of the neutral axis of the beam about the y axis. The shear force V and the bending moment M are present in the boundary conditions as noted from Eqs. (9.8) and (9.9); that is,

$$M = EI \frac{\partial^2 w}{\partial x^2}$$

$$V = \frac{\partial M}{\partial x} = \frac{\partial}{\partial x} \left(EI \frac{\partial^2 w}{\partial x^2} \right)$$

(9.46)

Thus, each of Eqs. (9.42b) and (9.44b) represents a force balance at an end of the beam and each of Eqs. (9.43b) and (9.45b) represents a moment balance at an end of the beam. From the boundary conditions given by Eqs. (9.42) to (9.45), we see that if the magnitudes of the stiffness and inertia elements are different from zero, then the displacement and the slope cannot be zero at either end of the beam. Hence, in the general case, the four boundary conditions for the system shown in Figure 9.5 are given by

x = 0

$$\left[k_1 w + c_1 \dot{w} + M_1 \frac{\partial^2 w}{\partial t^2} + \frac{\partial}{\partial x} (EI(x) \frac{\partial^2 w}{\partial x^2}) - p \frac{\partial w}{\partial x} \right]_{x=0} = 0$$

$$\left[-k_{t1} \frac{\partial w}{\partial x} - J_1 \frac{\partial^3 w}{\partial x \partial t^2} + EI(x) \frac{\partial^2 w}{\partial x^2} \right]_{x=0} = 0$$

(9.47a)

x = L

$$\left[k_2 w + c_2 \dot{w} + M_2 \frac{\partial^2 w}{\partial t^2} - \frac{\partial}{\partial x} (EI(x) \frac{\partial^2 w}{\partial x^2}) + p \frac{\partial w}{\partial x} \right]_{x=L} = 0$$

$$\left[k_{t2} \frac{\partial w}{\partial x} + J_2 \frac{\partial^3 w}{\partial x \partial t^2} + EI(x) \frac{\partial^2 w}{\partial x^2} \right]_{x=L} = 0$$

(9.47b)

From Eqs. (9.42) to (9.45), it is clear that the boundary conditions are, in general, specified in terms of either displacement and/or force and either slope and/or moment. Boundary conditions expressed only in terms of displacement or slope are referred to as *geometric boundary conditions,* and boundary conditions expressed in terms of shear force or bending moment are referred to as *dynamic boundary conditions.* Therefore, boundary conditions obtained through force balance and moment balance are dynamic boundary conditions. The different boundary conditions, which are obtained from Eqs. (9.47), are summarized in Table 9.1. In presenting these boundary conditions, we have omitted the subscript convention employed so far to denote a specific boundary. This convention should be included as appropriate, depending on the boundary being considered.

To illustrate how the different boundary conditions in Table 9.1 are obtained, consider the first entry, which specifies the boundary conditions at a clamped end. One can write these boundary conditions directly from geometry; that is, at a clamped end, the displacement and the slope are zero. Alternatively, we consider Eqs. (9.47a) and divide the first of Eqs. (9.47a) by

TABLE 9.1
Boundary Conditions
for Beams

Case	Description	Boundary Conditions	Remarks
1	Clamped	$w = 0$ $\dfrac{\partial w}{\partial x} = 0$	
2	Pinned (hinged, simply supported)	$w = 0$ $\dfrac{\partial^2 w}{\partial x^2} = 0$	
3	Free, with axial force	$\dfrac{\partial^2 w}{\partial x^2} = 0$ $\dfrac{\partial}{\partial x}\left(EI \dfrac{\partial^2 w}{\partial x^2}\right) - p\dfrac{\partial w}{\partial x} = 0$	p is tensile. Replace p with $-p$ for a compressive force. For no axial force, $p = 0$. Valid at either end of the beam.
4	Free, with massless rigid constraint and axial force	$\dfrac{\partial w}{\partial x} = 0$ $\dfrac{\partial}{\partial x}\left(EI \dfrac{\partial^2 w}{\partial x^2}\right) - p\dfrac{\partial w}{\partial x} = 0$	Rigid constraint does not permit rotation, but can move unimpeded vertically. p is tensile. Replace p with $-p$ for a compressive force. For no axial force, $p = 0$. Valid at either end of the beam.
5	Free, with a translation spring	$\dfrac{\partial^2 w}{\partial x} = 0$ $\dfrac{\partial}{\partial x}\left(EI \dfrac{\partial^2 w}{\partial x^2}\right) = s_o k w$	k: spring constant (no resistance to torsion) $s_o = -1$ at $x = 0$ $s_o = +1$ at $x = L$ $k \rightarrow \infty$; that is, $w = 0$, Case 2 $k = 0$; that is, $(EIw'')' = 0$, Case 3 if $p = 0$
6	Free, with a torsion spring	$EI \dfrac{\partial^2 w}{\partial x^2} = s_o k_t \dfrac{\partial w}{\partial x}$ $\dfrac{\partial}{\partial x}\left(EI \dfrac{\partial^2 w}{\partial x^2}\right) = 0$	k_t: spring constant (no resistance to vertical motion) $s_o = +1$ at $x = 0$ $s_o = -1$ at $x = L$ $k_t \rightarrow \infty$; that is, $w' = 0$, Case 2 if $p = 0$ $k_t = 0$; that is, $w'' = 0$, Case 3 if $p = 0$
7	Pinned, with a torsion spring	$w = 0$ $EI \dfrac{\partial^2 w}{\partial x^2} = s_o k_t \dfrac{\partial w}{\partial x}$	k_t: spring constant (no resistance to vertical motion) $s_o = +1$ at $x = 0$; $s_o = -1$ at $x = L$ $k_t \rightarrow \infty$; that is, $w' = 0$, Case 1 $k_t = 0$; that is, $w'' = 0$, Case 2
8	Free, with torsion and translation springs	$EI \dfrac{\partial^2 w}{\partial x^2} = s_o' k_t \dfrac{\partial w}{\partial x}$ $\dfrac{\partial}{\partial x}\left(EI \dfrac{\partial^2 w}{\partial x^2}\right) = s_o k w$	k: spring constant (no resistance to torsion) k_t: spring constant (no resistance to vertical motion) $s_o = -1$ at $x = 0$; $s_o = +1$ at $x = L$ $s_o' = +1$ at $x = 0$; $s_o' = -1$ at $x = L$

(continue

TABLE 9.1
(continued)

Case	Description	Boundary Conditions	Remarks
			$k \to \infty$; Case 7; $k = 0$, Case 6 $k_t = 0$; Case 5; $k \to \infty$ and $k_t \to \infty$, Case 1 $k_t \to \infty$ and $k = 0$, Case 4 if $p = 0$
9	Free, with mass attached	$EI\dfrac{\partial^2 w}{\partial x^2} = s_o J_o \dfrac{\partial^3 w}{\partial x \partial t^2}$ $\dfrac{\partial}{\partial x}\left(EI\dfrac{\partial^2 w}{\partial x^2}\right) = s_o' M_o \dfrac{\partial^2 w}{\partial t^2}$	M_o: attached mass J_o: mass moment of inertia of M_o $s_o = +1$ at $x = 0$ $s_o = -1$ at $x = l$ $s_o' = -1$ at $x = 0$ $s_o' = +1$ at $x = l$ $M_o = 0$ and $J_o = 0$, Case 3 if $p = 0$

the translation stiffness k_1 and the second of Eqs. (9.47a) by the torsion stiffness k_{t1}. The result is

$$\left[w + \frac{c_1}{k_1}\dot{w} + \frac{M_1}{k_1}\frac{\partial^2 w}{\partial t^2} + \frac{1}{k_1}\frac{\partial}{\partial x}\left(EI(x)\frac{\partial^2 w}{\partial x^2}\right) - \frac{p}{k_1}\frac{\partial w}{\partial x}\right]_{x=0} = 0$$

$$\left[-\frac{\partial w}{\partial x} - \frac{J_1}{k_{t1}}\frac{\partial^3 w}{\partial x \partial t^2} + \frac{EI(x)}{k_{t1}}\frac{\partial^2 w}{\partial x^2}\right]_{x=0} = 0$$

Upon taking the limits $k_1 \to \infty$ and $k_{t1} \to \infty$, these equations lead to the boundary conditions given by Eqs. (9.42a) and (9.43a), respectively; that is, the displacement is zero and the slope is zero. Similarly, if one were to use Eqs. (9.47b) and considers the limits the translation stiffness $k_2 \to \infty$ and the torsion stiffness $k_{t2} \to \infty$, then we arrive at the boundary conditions given by Eqs. (9.44a) and (9.45a). Thus, we can think of a clamped end as a boundary with infinite translation stiffness and infinite rotation stiffness.

If we consider the boundary conditions for a pinned end, which is the second entry of Table 9.1, this boundary is thought of as having infinite translation stiffness and zero torsion stiffness and zero rotary inertia. In the case of the free end without an axial force—that is, a special case of the third entry of Table 9.1—we see that this boundary condition is thought of as having zero translation stiffness, zero torsion stiffness, zero translation inertia, and zero rotary inertia. In practice, most boundary conditions lie between the two limiting cases; that is, between a free end and a fixed end.

In Section 9.3, we examine the free oscillations of beams. Before doing so, we make several simplifying assumptions to reduce the algebra in the development. First, we assume that the beam is homogeneous and that it has a uniform cross-section along its length; that is,

$$EI(x) = EI, \quad \rho(x) = \rho, \quad \text{and} \quad A(x) = A \tag{9.48}$$

In addition, in Sections 9.3.1 and 9.3.2, we assume that the axial load is absent and that the elastic foundation is not present; that is,

$$p(x,t) = 0 \quad \text{and} \quad k_f = 0 \tag{9.49}$$

The boundary conditions simplify accordingly.

The effect of axial load and elastic foundation on the free oscillations of beams is considered in Section 9.3.3, and beams with varying cross-sections are considered in Section 9.3.5. In light of the assumptions given by Eqs. (9.48) and (9.49), Eq. (9.40) reduces to

$$EI \frac{\partial^4 w}{\partial x^4} + \rho A \frac{\partial^2 w}{\partial t^2} = f(x,t) \tag{9.50}$$

Equation (9.50), along with the appropriate boundary conditions given by Eqs. (9.47) or chosen from Table 9.1, represent the governing equations of a damped beam[13] subjected to transverse loading. We study the free response of the undamped system first, and then use this as a basis to determine the response of the damped system subjected to dynamic forcing in Section 9.4.

Before proceeding to the next section, a few comments about Eq. (9.50) are in order. This equation is a partial differential equation with a fourth-order spatial derivative and a second-order time derivative. Since the highest spatial derivative is fourth order, four boundary conditions are needed. Similarly, since the highest time derivative is second order, two initial conditions are needed. In the rest of this chapter, it is assumed that appropriate information is available to completely define the response determined as a solution of Eq. (9.50).

EXAMPLE 9.1 Boundary conditions for a cantilever beam with an extended mass[14]

Consider a uniform cantilever beam that has an extended rigid mass M_2 attached to its free end as shown in Figure 9.6. Comparing this system to the system shown in Case 9 in Table 9.1, we note the mass center is located away from $x = L$ and the discrete springs are attached away from $x = L$. The mass has a mass moment of inertia J_2 about its center of mass. In addition, a linear spring with stiffness k_2 and a torsion spring with stiffness k_{t2} are attached to the free end of the mass. In Figure 9.6b, the transverse displacements are shown at several locations on the extended mass. With this information, we shall obtain an expression for the discrete Lagrangian G_L at the right end of the beam. Using Eq. (9.21) as a guide, we obtain the following expression for G_L at $x = L$

$$G_L(t, w_L, \dot{w}_L, w'_L, \dot{w}'_L) = \frac{1}{2} M_2 (\dot{w}_L + d_0 \dot{w}'_L)^2 + \frac{1}{2} J_2 \dot{w}'^2_L$$

$$- \frac{1}{2} k_2 (w_L + [d_0 + d_1] w'_L)^2 - \frac{1}{2} k_{t2} w'^2_L \tag{a}$$

---

[13]Although the beam interior is undamped, due to the presence of damping elements at the boundaries, the beam system is considered a damped system.

[14]D. Zhou, "The vibrations of a cantilever beam carrying a heavy tip mass with elastic supports," *J. Sound Vibration,* Vol. 206, No. 2, pp. 275–279 (1997); H. Seidel and L. Csepregi, "Design optimization for cantilever-type accelerometers," *Sensors and Actuators,* Vol. 6, pp. 81–92 (1984

FIGURE 9.6
(a) Cantilever beam with an extended mass attached to its free end and (b) displacement at several locations on the extended mass.

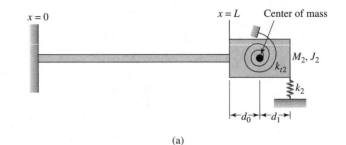

(a)

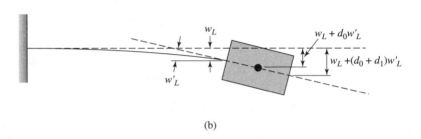

(b)

where the prime denotes the derivative with respect to x and the overdot denotes the derivative with respect to time. For the system of Figure 9.6, the function G_B is obtained from Eq. (9.19) by setting $k_f = p = f_c = 0$; that is,

$$G_B(x,t,w,\dot{w},w',\dot{w}',w'') = \frac{1}{2}[\rho A \dot{w}^2 - EI w''^2] \tag{b}$$

Upon substituting Eqs. (a) and (b) into Eqs. (9.30b) and (9.31b) and performing the indicated operations, we obtain, respectively,

$$EI \frac{\partial^3 w}{\partial x^3}\bigg|_{x=L} = \left[k_2 w + M_2 \frac{\partial^2 w}{\partial t^2} + (d_0 + d_1)k_2 \frac{\partial w}{\partial x} + M_2 d_0 \frac{\partial^3 w}{\partial x \partial t^2}\right]_{x=L}$$

$$EI \frac{\partial^2 w}{\partial x^2}\bigg|_{x=L} = -\bigg[k_2(d_0 + d_1)w + M_2 d_0 \frac{\partial^2 w}{\partial t^2} + \left(k_{t2} + k_2(d_0 + d_1)^2\right)\frac{\partial w}{\partial x}$$

$$+ (J_2 + M_2 d_0^2)\frac{\partial^3 w}{\partial x \partial t^2}\bigg]_{x=L} \tag{c}$$

When $d_1 = d_0 = 0$, Eqs. (c) reduce to those given by Eqs. (9.47b) with $p = c_2 = 0$. The other two boundary conditions are given by the first entry of Table 9.1.

9.3 FREE OSCILLATIONS

9.3.1 Introduction

In this section, free oscillations of undamped, homogeneous and uniform beams are examined in detail. To this end, we use Eqs. (9.50) and (9.41b) and

FIGURE 9.7
Beam with spring elements at left and right boundaries and an inertia element at the right boundary.

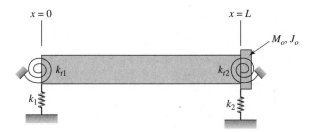

set the external nonconservative loading $f_{nc}(x,t)$ to zero and obtain the following governing equation of the beam for $0 < x < L$:

$$EI \frac{\partial^4 w}{\partial x^4} + \rho A \frac{\partial^2 w}{\partial t^2} = \rho A g \tag{9.51}$$

In order to keep the discussion general in terms of boundary conditions, we consider the system shown in Figure 9.7. We place a translation spring with constant k_1 and a torsion spring with constant k_{t1} at $x = 0$ and place a translation spring with constant k_2 and a torsion spring with constant k_{t2} at $x = L$. In addition, at $x = L$ there is a rigid body with a mass M_o that has a mass moment of inertia J_o. Then, from Eqs. (9.47a) with $p = c_1 = M_1 = J_1 = 0$, the boundary conditions at $x = 0$ are

$$EIw''(0,t) = k_{t1}w'(0,t)$$
$$EIw'''(0,t) = -k_1 w(0,t) \tag{9.52a}$$

From Eqs. (9.47a) with $p = c_2 = 0$, $M_2 = M_o$, $J_2 = J_o$, the boundary conditions at $x = L$ are

$$EIw''(L,t) = -k_{t2}w'(L,t) - J_o \ddot{w}'(L,t)$$
$$EIw'''(L,t) = k_2 w(L,t) + M_o \ddot{w}(L,t) \tag{9.52b}$$

In writing Eqs. (9.52), we have employed the compact notation that

$$w|_{x=0} = w(0,t), \quad \frac{\partial w}{\partial x}\bigg|_{x=0} = w'(0,t), \quad \frac{\partial^2 w}{\partial x^2}\bigg|_{x=0} = w''(0,t),$$
$$\frac{\partial^3 w}{\partial x \partial t^2}\bigg|_{x=0} = \ddot{w}'(0,t)$$

and so on. In addition, we have used the fact that EI is constant.

The response of the beam governed by Eqs. (9.51) and (9.52) and the initial conditions is expressed as

$$w(x,t) = w_{st}(x) + w_{dyn}(x,t) \tag{9.53}$$

where $w_{st}(x)$ is the static response or the static-equilibrium position and $w_{dyn}(x,t)$ is the displacement measured from the static-equilibrium position. The static-equilibrium position satisfies the *static-equilibrium equation*

$$EI \frac{d^4 w_{st}}{dx^4} = \rho A g \tag{9.54}$$

and the following boundary conditions at $x = 0$

$$EIw_{st}''(0) = k_{t1}w_{st}'(0)$$
$$EIw_{st}'''(0) = -k_1w_{st}(0) \qquad (9.55a)$$

and the following boundary conditions at $x = L$

$$EIw_{st}''(L) = -k_{t2}w_{st}'(L)$$
$$EIw_{st}'''(L) = k_2w_{st}(L) \qquad (9.55b)$$

Although the solution of Eq. (9.54) subject to the boundary conditions Eqs. (9.55) can be found, this is not explicitly carried out here. Nevertheless, the droop of a beam due to its own weight is given by $w_{st}(x)$. Any constant loading acting on a distributed-parameter system will influence the static-equilibrium position of that system. For most commonly used structural materials and geometries, the static deformation $w_{st}(x)$ due to gravity and other static loading is assumed to be small in magnitude and this deformation is often ignored in the analysis. However, there are cases such as an aircraft wing, where the static deformation is pronounced and cannot be ignored. In MEMS, residual stresses due to the fabrication process can lead to a static deformation.[15]

In Sections 3.2 and 7.2, it was shown the static-equilibrium position of a single degree-of-freedom system is determined by solving an algebraic equation and the static-equilibrium position of a multi-degree-of-freedom system is determined by solving a system of algebraic equations, respectively. Here, by contrast, we find that the static-equilibrium position of a beam is determined by solving a differential equation.

On substituting Eq. (9.53) into Eqs. (9.51) and (9.52) and making use of Eqs. (9.54) and (9.55), we find that the oscillations about the static equilibrium position are described by

$$EI\,\frac{\partial^4 w_{dyn}}{\partial x^4} + \rho A\,\frac{\partial^2 w_{dyn}}{\partial t^2} = 0 \qquad (9.56)$$

subject to the following boundary conditions at $x = 0$

$$EIw_{dyn}''(0,t) = k_{t1}w_{dyn}'(0,t)$$
$$EIw_{dyn}'''(0,t) = -k_1w_{dyn}(0,t) \qquad (9.57a)$$

and the following boundary conditions at $x = L$

$$EIw_{dyn}''(L,t) = -k_{t2}w_{dyn}'(L,t) - J_o\ddot{w}_{dyn}'(L,t)$$
$$EIw_{dyn}'''(L,t) = k_2w_{dyn}(L,t) + M_o\ddot{w}_{dyn}(L,t) \qquad (9.57b)$$

Equation (9.56) is a linear, homogeneous partial-differential equation that is used to describe the free oscillations of the beam about the static-equilibrium position. As discussed in the context of finite degree-of-freedom

[15]Y. Yee, M. Park, and K. Chun, "A sticking model of suspended polysilicon microstructure including residual stress gradient and post release temperature," *J. Microelectromechanical Systems,* Vol. 7, No. 3, pp. 339–344 (1998).

systems, gravity loading does not appear in the equation governing the dynamic response $w_{dyn}(x,t)$ of the beam. Based on the form of Eq. (9.56), we assume that there is a separable solution; that is,

$$w_{dyn}(x,t) = W(x)G(t) \tag{9.58}$$

where $W(x)$ is a function that depends only on the spatial variable x and $G(t)$ is a function that depends only on the temporal variable t. By assuming this form of solution, we are assuming that every point on the beam has the same time dependence. In other words, the ratio of displacements at two different points on a beam is independent of time.

On substituting Eq. (9.58) into (9.56) and rearranging the terms, we obtain

$$\frac{1}{G}\frac{d^2G}{dt^2} = -\frac{EI}{\rho AW}\frac{d^4W}{dx^4} \tag{9.59}$$

Since the left-hand side consists only of terms that are functions of time and the right-hand side consists only of terms that are functions of the spatial variable x, the ratio on each side must be a constant; that is,

$$\frac{1}{G}\frac{d^2G}{dt^2} = -\frac{EI}{\rho AW}\frac{d^4W}{dx^4} = \lambda \tag{9.60}$$

where λ is a constant. Based on the experience gained with free oscillations of undamped finite degree-of-freedom systems in Sections 4.1 and 7.4, it is known that the free oscillations of vibratory systems are described by sine or cosine harmonic functions.[16] Noting that the beam is undamped, and assuming that there is no damping in the boundary conditions, we set the constant λ as

$$\lambda = -\omega^2 \tag{9.61}$$

Upon making use of Eqs. (9.60) and (9.61), we obtain the *temporal eigenvalue equation*

$$\frac{d^2G}{dt^2} + \omega^2 G = 0 \tag{9.62}$$

which has a solution of the form

$$G(t) = G_{os} \sin \omega t + G_{oc} \cos \omega t \tag{9.63}$$

In the harmonic solution given by Eq. (9.63), G_{os} and G_{oc} are arbitrary constants that will be determined by the initial conditions and ω is an unknown quantity to be determined. The quantity ω, which describes the frequency of oscillation of the beam in its free state, is called the *natural frequency*. As we will see later in this section, there is an infinite number of natural frequencies

---

[16]As discussed in Chapter 4, this will not be true of an unstable vibratory system. In general, for an undamped linear vibratory system in the absence of rotation, it can be said that unbounded oscillations do not occur and that the bounded oscillations that do occur are harmonic in form.

for the beam. On substituting Eq. (9.58) into Eq. (9.56) and making use of Eq. (9.62), we arrive at

$$\left(EI\frac{d^4W}{dx^4} - \rho A\omega^2 W\right)G(t) = 0 \tag{9.64a}$$

Since Eq. (9.64a) has to be satisfied for all time and for arbitrary $G(t)$ not necessarily zero, the only way this is possible is if we have

$$EI\frac{d^4W}{dx^4} - \rho A\omega^2 W = 0 \tag{9.64b}$$

Similarly, substituting Eq. (9.58) into Eq. (9.57) and making use of Eq. (9.62), we arrive at the following boundary conditions at $x = 0$

$$EIW''(0) = k_{t1}W'(0)$$
$$EIW'''(0) = -k_1W(0) \tag{9.65a}$$

and the following boundary conditions at $x = L$

$$EIW''(L) = -k_{t2}W'(L) + \omega^2 J_o W'(L) = (-k_{t2} + \omega^2 J_o)W'(L)$$
$$EIW'''(L) = k_2 W(L) - \omega^2 M_o W(L) = (k_2 - \omega^2 M_o)W(L) \tag{9.65b}$$

Equations (9.64b) and (9.65) represent the *spatial eigenvalue problem.* As in the eigenvalue problems encountered in Chapters 4 and 7, the trivial solution $W(x) = 0$ is a solution of Eqs. (9.64b) and (9.65). However, we are not interested in this trivial solution, but in determining the special values of ω, called the *eigenvalues,* for which $W(x)$ will be nontrivial. In the terminology of spatial eigenvalue problems, these nontrivial functions are called the *eigenfunctions.* In the case of the beam, the eigenvalues provide us the natural frequencies and the eigenfunctions provide us the mode shapes. The natural frequencies and mode shapes depend on the boundary conditions and the geometry of the beam. This dependence will be explored in the following sections.

9.3.2 Natural Frequencies, Mode Shapes, and Orthogonality of Modes

We shall now determine the natural frequencies and mode shapes for a general set of boundary conditions and discuss an important property of mode shapes, the *orthogonality property.* To elaborate on this property and to determine the conditions that go with this property, it is not necessary to determine explicitly the natural frequencies and mode shapes. Therefore, we first present a discussion on this property, and then we illustrate how Laplace transforms can be used to determine the natural frequencies and mode shapes for uniform beams subjected to arbitrary boundary conditions.

Orthogonal Functions

An orthogonal function is defined as follows. If a sequence of real functions $\{\varphi_n(x)\}$, $n = 1, 2, \ldots$, has the property that over some interval

$$\int_a^b \left[p_1(x)\varphi_n(x)\varphi_m(x) + p_2(x)\frac{d\varphi_n(x)}{dx}\frac{d\varphi_m(x)}{dx}\right]dx = \delta_{nm}N_n \tag{9.66}$$

where δ_{nm} is the Kronecker delta function,[17] then the functions are said to form an *orthogonal set* with respect to the weighting functions $p_1(x)$ and $p_2(x)$ on that interval. The quantity $\sqrt{N_n}$ is called the *norm* of the function $\varphi_n(x)$. Hence, if m and n are different from each other—that is, we have two different orthogonal functions—the integral given by Eq. (9.66) evaluates to zero. This property is analogous to the orthogonality property associated with vectors; that is, the scalar dot product of two identical vectors is the square of the vector's magnitude, while the scalar dot product of two orthogonal vectors is zero.

In the present context, the mode shapes $W_n(x)$ will take the place of $\varphi_n(x)$ in Eq. (9.66) and the weighting function will be determined by the stiffness and inertia properties of the system. The material presented here parallels that presented in Section 7.3.2 for systems with multiple degrees of freedom.

Spatial Eigenvalue Problem in Terms of Nondimensional Quantities

Before proceeding further, we introduce the following notation:

$$\eta = x/L, \quad \Omega^4 = \omega^2 \frac{\rho A L^4}{EI} = \frac{\omega^2 L^4}{c_b^2 r^2} = \omega^2 t_o^2$$

$$c_b^2 = E/\rho, \quad r^2 = I/A, \quad t_o = \frac{L^2}{c_b r}$$

$$K_j = \frac{k_j L^3}{EI}, \quad B_j = \frac{k_{tj} L}{EI} \quad j = 1, 2$$

$$m_o = \rho A L, \quad j_o = m_o L^2 \tag{9.67}$$

In Eq. (9.67), η is a nondimensional spatial variable, Ω is a nondimensional frequency coefficient, c_b is the longitudinal (or bar) speed along the x-axis of the beam, r is the radius of gyration of the beam's cross-section about the neutral axis, m_o is the mass of the beam, and K_j and B_j are nondimensional translation and torsion spring constants, respectively. On substituting the appropriate expressions from Eqs. (9.67) into Eqs. (9.64b) and (9.65), the result is

$$\frac{d^4 W}{d\eta^4} - \Omega^4 W = 0 \tag{9.68}$$

with the following boundary conditions at $\eta = 0$

$$W''(0) = B_1 W'(0)$$
$$W'''(0) = -K_1 W(0) \tag{9.69a}$$

and the following boundary conditions at $\eta = 1$

$$W''(1) = \left(-B_2 + \frac{J_o \Omega^4}{j_o}\right) W'(1)$$

$$W'''(1) = \left(K_2 - \frac{M_o \Omega^4}{m_o}\right) W(1) \tag{9.69b}$$

---

[17]$\delta_{nm} = 1$, $n = m$, $\delta_{nm} = 0$, otherwise.

where the prime ($'$) will be used from this point on to denote the derivative with respect to η. It is noted that in Eqs. (9.68) and (9.69), $W(\eta)$ has the units of displacement.

Orthogonality Property of Mode Shapes

We assume that Ω_n is an eigenvalue and $W_n(\eta)$ is the associated eigensolution or mode shape of the system given by Eqs. (9.68) and (9.69); that is,

$$\frac{d^4 W_n}{d\eta^4} - \Omega_n^4 W_n = 0 \tag{9.70}$$

and at $\eta = 0$

$$W_n''(0) = B_1 W_n'(0)$$
$$W_n'''(0) = -K_1 W_n(0) \tag{9.71a}$$

and at $\eta = 1$

$$W_n''(1) = \left(-B_2 + \frac{J_o \Omega_n^4}{j_o}\right) W_n'(1)$$
$$W_n'''(1) = \left(K_2 - \frac{M_o \Omega_n^4}{m_o}\right) W_n(1) \tag{9.71b}$$

Let $W_m(\eta)$ be another solution of Eqs. (9.68) and (9.69), which corresponds to the mth natural frequency coefficient Ω_m, where $\Omega_m \neq \Omega_n$. We now multiply Eq. (9.70) by $W_m(\eta)$, integrate over the interval $0 \leq \eta \leq 1$, and use integration by parts to obtain

$$\int_0^1 W_m W_n^{IV} d\eta - \Omega_n^4 \int_0^1 W_m W_n d\eta = W_m W_n''' \Big|_0^1 - W_m' W_n'' \Big|_0^1$$
$$+ \int_0^1 W_m'' W_n'' d\eta - \Omega_n^4 \int_0^1 W_m W_n d\eta = 0 \tag{9.72}$$

Performing the same set of operations, but reversing the order of m and n, we get

$$\int_0^1 W_n W_m^{IV} d\eta - \Omega_m^4 \int_0^1 W_n W_m d\eta = W_n W_m''' \Big|_0^1 - W_n' W_m'' \Big|_0^1$$
$$+ \int_0^1 W_n'' W_m'' d\eta - \Omega_m^4 \int_0^1 W_n W_m d\eta = 0 \tag{9.73}$$

After subtracting Eq. (9.73) from Eq. (9.72), the resulting expression is

$$(\Omega_m^4 - \Omega_n^4) \int_0^1 W_m W_n d\eta + W_m W_n''' \Big|_0^1 - W_n W_m''' \Big|_0^1 + W_n' W_m'' \Big|_0^1 - W_m' W_n'' \Big|_0^1 = 0$$

which is written in expanded form as

$$(\Omega_m^4 - \Omega_n^4) \int_0^1 W_m W_n d\eta + W_m(1)W_n'''(1) - W_m(0)W_n'''(0) - W_n(1)W_m'''(1)$$

$$+ W_n(0)W_m'''(0) + W_n'(1)W_m''(1) - W_n'(0)W_m''(0) - W_m'(1)W_n''(1)$$

$$+ W_m'(0)W_n''(0) = 0 \tag{9.74}$$

Upon substituting the boundary conditions given by Eqs. (9.71) into Eq. (9.74), we arrive at

$$(\Omega_m^4 - \Omega_n^4)\left[\int_0^1 W_m W_n d\eta + \frac{M_o}{m_o} W_m(1)W_n(1) + \frac{J_o}{j_o} W_m'(1)W_n'(1)\right] = 0 \quad (9.75)$$

Since $\Omega_n \neq \Omega_m$, the expression inside the brackets must equal zero. This leads to

$$\int_0^1 W_m W_n d\eta + \frac{M_o}{m_o} W_m(1)W_n(1) + \frac{J_o}{j_o} W_m'(1)W_n'(1) = \delta_{nm}N_n \tag{9.76a}$$

where δ_{nm} is the Kronecker delta and

$$N_n = \int_0^1 W_n^2(\eta)d\eta + \frac{M_o}{m_o} W_n^2(1) + \frac{J_o}{j_o} W_n'^2(1) \tag{9.76b}$$

Equation (9.76a) is written as

$$\int_0^1 \left[1 + \frac{M_o}{m_o}\delta(\eta - 1)\right] W_m(\eta)W_n(\eta)d\eta$$

$$+ \frac{J_o}{j_o} \int_0^1 W_m'(\eta)W_n'(\eta)\delta(\eta - 1)d\eta = \delta_{nm}N_n \tag{9.77}$$

where $\delta(\eta - 1)$ is the delta function. Equation (9.77) is a special case of Eq. (9.66) if we identify x with η, $p_1(x)$ with $1 + (M_o/m_o)\delta(\eta - 1)$, $p_2(x)$ with $(J_o/j_o)\delta(\eta - 1)$, $W_n(\eta)$ with $\varphi_n(x)$, and $W_n'(\eta)$ with $d\varphi_n(x)/dx$.

Equation (9.76a) represents the *orthogonality condition* for the beam and the boundary conditions, and the functions $W_m(\eta)$ that satisfy this condition are called *orthogonal functions.* They can be shown to form a complete set of orthogonal functions.[18] It is also pointed out that the form of the expression on the left-hand side of Eq. (9.77) is symmetric in the spatial functions $W_n(\eta)$ and $W_m(\eta)$; that is, the form of the equation does not change if the functions $W_n(\eta)$ and $W_m(\eta)$ are interchanged. This type of symmetry is characteristic of self-adjoint systems whose eigenfunctions are known to be orthogonal functions.[19]

[18]E. B. Magrab, *ibid,* Chapter 1.

[19]L. Meirovitch, *Principles and Techniques of Vibrations,* Prentice Hall, Upper Saddle River, NJ, Chapter 7 (1997).

Based on Eqs. (9.76) and (9.77), it should be clear that for all of the boundary conditions shown in Table 9.1, the mode shapes are orthogonal functions. The orthogonality property of the modes was established for beams with uniform and homogeneous properties; that is, beams with constant flexural rigidity EI, constant mass density ρ, and constant area of cross-section A. It can be shown that an orthogonality condition similar in form to Eqs. (9.76) also applies to beams with nonuniform and inhomogeneous properties.

Solutions for Natural Frequencies and Mode Shapes

We shall now determine the specific form of $W_n(\eta)$ by solving Eqs. (9.68) and (9.69). In order to find this specific form, we first need to solve for the eigenvalue Ω_n. To determine this eigenvalue, we need to determine the characteristic equation whose roots will provide us the eigenvalues of the system. The eigenvalue Ω_n and the associated eigenfunction $W_n(\eta)$ are determined by solving the boundary-value problem using Laplace transforms. Instead of solving for the different boundary conditions shown in Table 9.1 on a case-by-case basis, a solution is obtained for a general set of boundary conditions. Then, in Section 9.3.3, this general solution is appropriately modified for various special cases of the general boundary conditions.

In order to determine the solution by using the Laplace transform it is recognized that Eq. (9.68) is a special case of Eq. (A.7) of Appendix A, with the function W taking the place of y, the nondimensional spatial variable η taking the place of the independent variable x and $\beta = k = f(x) = 0$ in Eq. (A.7). Denoting the Laplace transform of $W(\eta)$ by $\tilde{W}(s)$—that is,

$$\tilde{W}(s) = \int\limits_0^\infty W(\eta)e^{-s\eta}d\eta$$

—and making use of Eqs. (A.8) and (A.9), we find that the Laplace transform of Eq. (9.68) results in

$$\tilde{W}(s) = \frac{1}{(s^4 - \Omega^4)}[W(0)s^3 + W'(0)s^2 + W''(0)s + W'''(0)] \qquad (9.78)$$

where $W(0)$, $W'(0)$, $W''(0)$, and $W'''(0)$ are the displacement, slope, second derivative of $W(\eta)$, and third derivative of $W(\eta)$, respectively, at $\eta = 0$. These four quantities and the nondimensional frequency coefficient Ω represent the five unknown quantities that need to be determined. Noting that we have the four boundary conditions given by Eqs. (9.69), one can at best solve for four of these quantities, one of which will be the nondimensional frequency coefficient. This means that the resulting solution for the mode shape will have an arbitrary constant. This situation is similar to what we encountered in Chapter 7, where we found that an eigenvector is arbitrary with respect to a scaling constant.

We see from the boundary conditions given by Eqs. (9.69a) that $W''(0)$ and $W'''(0)$ are expressed in terms of $W'(0)$ and $W(0)$, respectively. Thus, one of the advantages of the Laplace transform solution method is that two of the four unknowns are directly specified by the boundary conditions at $\eta = 0$, thereby leaving only three unknown quantities to be determined; in this case, $W'(0)$, $W(0)$, and Ω. Making use of Laplace transform pairs 23 through 26 in Table A of Appendix A, or alternatively, directly making use of Eqs. (A.16) and (A.17) of Appendix A with $\delta = \epsilon = \Omega$, leads to the following inverse Laplace transform of Eq. (9.78)

$$W(\eta) = W(0)Q(\Omega\eta) + W'(0)R(\Omega\eta)/\Omega$$
$$+ W''(0)S(\Omega\eta)/\Omega^2 + W'''(0)T(\Omega\eta)/\Omega^3 \quad (9.79)$$

where the nondimensional spatial functions are given by

$$Q(\xi) = 0.5[\cosh(\xi) + \cos(\xi)]$$
$$R(\xi) = 0.5[\sinh(\xi) + \sin(\xi)]$$
$$S(\xi) = 0.5[\cosh(\xi) - \cos(\xi)]$$
$$T(\xi) = 0.5[\sinh(\xi) - \sin(\xi)] \quad (9.80)$$

If one were to consider, for example, a cantilever beam without a rigid mass at the free end, then Eq. (9.79) will reduce to

$$W(\eta) = W''(0)S(\Omega\eta)/\Omega^2 + W'''(0)T(\Omega\eta)/\Omega^3$$

because $W(0) = W'(0) = 0$ at the fixed end. This simplification always applies to the case of a beam clamped at the end $x = 0$. Requiring that the mode shape satisfy the boundary conditions $W''(1) = W'''(1) = 0$ at the free end, two equations are obtained in terms of $W''(0)$ and $W'''(0)$. Setting the determinant of the coefficients of $W''(0)$ and $W'''(0)$ to zero, the characteristic equation for the cantilever beam is determined. Instead of solving for each set of boundary conditions on a case-by-case basis, we shall find a general form of the characteristic equation based on the boundary conditions given by Eqs. (9.69).

Upon substituting the boundary conditions at $\eta = 0$ given by Eq. (9.69a) into Eq. (9.79), we obtain

$$W(\eta) = [Q(\Omega\eta) - K_1 T(\Omega\eta)/\Omega^3]W(0)$$
$$+ [R(\Omega\eta)/\Omega + B_1 S(\Omega\eta)/\Omega^2]W'(0) \quad (9.81)$$

On substituting Eq. (9.81) into the boundary conditions at $\eta = 1$, which are given by Eq. (9.69b), the following two equations are obtained in terms of the three unknown quantities[20] $W(0)$, $W'(0)$, and Ω:

$$[(1 + a_1 b_2)S(\Omega) - a_1 R(\Omega) - b_2 T(\Omega)]\Omega W(0)$$
$$+ [T(\Omega) + (b_1 - b_2)Q(\Omega) - b_1 b_2 R(\Omega)]W'(0) = 0$$
$$[R(\Omega) - (a_1 + a_2)Q(\Omega) + a_1 a_2 T(\Omega)]\Omega W(0)$$
$$+ [(1 - a_2 b_1)S(\Omega) + b_1 T(\Omega) - a_2 R(\Omega)]W'(0) = 0 \quad (9.82)$$

---

[20]Equations (9.82) represent a set of two algebraic equations in three unknowns, namely, $W(0)$, $W'(0)$, and Ω. This is an algebraic eigenvalue problem whose solution will provide us Ω and a ratio of the other two unknowns.

In Eq. (9.82), the nondimensional quantities a_j and $b_j, j = 1, 2$ are given by

$$a_1 = \frac{K_1}{\Omega^3}, \qquad a_2 = \frac{1}{\Omega^3}\left(K_2 - \frac{M_o\Omega^4}{m_o}\right),$$

$$b_1 = \frac{B_1}{\Omega}, \qquad b_2 = \frac{1}{\Omega}\left(-B_2 + \frac{J_o\Omega^4}{j_o}\right) \qquad (9.83)$$

In obtaining Eq. (9.82), we made use of the relations

$$Q'(\Omega\eta) = \Omega T(\Omega\eta)$$
$$T'(\Omega\eta) = \Omega S(\Omega\eta)$$
$$R'(\Omega\eta) = \Omega Q(\Omega\eta)$$
$$S'(\Omega\eta) = \Omega R(\Omega\eta) \qquad (9.84)$$

where the prime denotes the derivative with respect to η. To obtain higher derivatives, Eqs. (9.84) are used as follows. For example,

$$\frac{d^2Q(\Omega\eta)}{d\eta^2} = \frac{d}{d\eta}\left(\frac{dQ(\Omega\eta)}{d\eta}\right) = \Omega\frac{dT(\Omega\eta)}{d\eta} = \Omega^2 S(\Omega\eta)$$

The natural frequency coefficient Ω_n is obtained by setting the determinant of coefficients of Eqs. (9.82) to zero, which results in the following *characteristic equation:*

$$
\begin{aligned}
z_1[&\cos\Omega_n\sinh\Omega_n + \sin\Omega_n\cosh\Omega_n] \\
&+ z_2[\cos\Omega_n\sinh\Omega_n - \sin\Omega_n\cosh\Omega_n] \\
&- 2z_3\sin\Omega_n\sinh\Omega_n + z_4(\cos\Omega_n\cosh\Omega_n - 1) \\
&+ z_5(\cos\Omega_n\cosh\Omega_n + 1) + 2z_6\cos\Omega_n\cosh\Omega_n = 0 \qquad (9.85)
\end{aligned}
$$

In Eqs. (9.85), the coefficients z_i are given by

$$z_1 = [b_{1n}b_{2n}(a_{1n} + a_{2n}) + (b_{1n} - b_{2n})]$$
$$z_2 = [a_{1n}a_{2n}(b_{1n} - b_{2n}) - (a_{1n} + a_{2n})]$$
$$z_3 = (a_{1n}a_{2n} + b_{1n}b_{2n})$$
$$z_4 = (1 - a_{1n}a_{2n}b_{1n}b_{2n})$$
$$z_5 = (a_{2n}b_{2n} - a_{1n}b_{1n})$$
$$z_6 = (a_{1n}b_{2n} - a_{2n}b_{1n}) \qquad (9.86)$$

and the nondimensional coefficients $a_{1n}, a_{2n}, b_{1n},$ and b_{2n} are given by

$$a_{1n} = \frac{K_1}{\Omega_n^3}, \qquad a_{2n} = \frac{1}{\Omega_n^3}\left(K_2 - \frac{M_o\Omega_n^4}{m_o}\right),$$

$$b_{1n} = \frac{B_1}{\Omega_n}, \qquad b_{2n} = \frac{1}{\Omega_n}\left(-B_2 + \frac{J_o\Omega_n^4}{j_o}\right) \qquad (9.87)$$

The transcendental equation given by Eq. (9.85) has an infinite number of roots. Thus, a continuous system has an infinite number of natural frequencies and associated mode shapes.

The mode shapes are obtained as follows. In Eq. (9.81), we let $\Omega = \Omega_n$ and rewrite it as

$$W_n(\eta) = \frac{W'(0)}{\Omega_n} \left\{ [Q(\Omega_n\eta) - a_{1n}T(\Omega_n\eta)] \frac{\Omega_n W(0)}{W'(0)} + R(\Omega_n\eta) + b_{1n}S(\Omega_n\eta) \right\}$$

$$(9.88)$$

We note that the natural frequency coefficients determined from Eq. (9.85) are a function of a_{jn} and b_{jn}, $j = 1, 2$ and $n = 1, 2 \ldots$, and that these quantities can vary between the limits of 0 and ∞. For instance, for a beam with a free end at $\eta = 0$, a_{1n} is 0 while for a beam with a fixed end at $\eta = 0$, a_{1n} is ∞. Similarly, in the presence of a free end at $\eta = 1$, a_{2n} is 0, and so forth. Due to the wide range of the parameters' values, the form of the mode shape must be specialized for the limiting cases. Therefore, we determine the expression for $\Omega W(0)/W'(0)$ from the second of Eqs. (9.82) after setting $\Omega = \Omega_n$. Furthermore, since the magnitude of $W'(0)$ is indeterminate, for convenience we set $W'(0)/\Omega_n = 1$ in Eq. (9.88). Equivalently, if the left-hand side and the right-hand side are scaled by $W'(0)/\Omega_n$, the resulting equation is nondimensional; that is, $W(\eta)/(W'(0)/\Omega_n)$, is nondimensional. Therefore, we shall consider that $W(\eta)$ is divided by $W'(0)/\Omega_n$ and that $W'(0)/\Omega_n$ is set equal to 1. The mode shape is given by one of the following expressions, depending on the upper limit of b_{1n}. The different cases to be studied in Section 9.3.3 are grouped under one of the two cases shown below.

Case 1: $0 \leq a_{jn} \leq \infty$, $0 \leq b_{2n} \leq \infty$, $0 \leq b_{1n} < \infty$

The mode shape given by Eq. (9.88) is rewritten as

$$W_n(\eta) = C_n[Q(\Omega_n\eta) - a_{1n}T(\Omega_n\eta)] + R(\Omega_n\eta) + b_{1n}S(\Omega_n\eta) \qquad (9.89)$$

where the nondimensional coefficient C_n is given by

$$C_n = \frac{a_{2n}R(\Omega_n) + (a_{2n}b_{1n} - 1)S(\Omega_n) - b_{1n}T(\Omega_n)}{R(\Omega_n) - (a_{1n} + a_{2n})Q(\Omega_n) + a_{1n}a_{2n}T(\Omega_n)} \qquad (9.90)$$

Case 2: $0 \leq a_{jn} \leq \infty$, $0 \leq b_{2n} \leq \infty$, and $b_{1n} \to \infty$ (infinite torsion stiffness at $\eta = 0$)

The mode shape given by Eq. (9.88) is rewritten as

$$W_n(\eta) = C_n[Q(\Omega_n\eta) - a_{1n}T(\Omega_n\eta)] + S(\Omega_n\eta) \qquad (9.91)$$

where the nondimensional coefficient C_n is given by

$$C_n = \frac{a_{2n}S(\Omega_n) - T(\Omega_n)}{R(\Omega_n) - (a_{1n} + a_{2n})Q(\Omega_n) + a_{1n}a_{2n}T(\Omega_n)} \qquad (9.92)$$

Note that $W_n(\eta)$ is a nondimensional quantity.

It is seen that Eqs. (9.90) and (9.92) are independent of b_{2n}; however, b_{2n} does appear in the characteristic equation, Eq. (9.85), and it does affect the numerical value of the natural frequency coefficient Ω_n. As seen subsequently, these results provide us with a single solution from which we will be able to examine a very large combination of different boundary conditions by simply letting the a_{jn} and b_{jn}, $j = 1, 2$, take on any value from 0 and infinity.

The boundary conditions given by Eqs. (9.71) are restated by using the notation introduced in Eqs. (9.87). In this notation, $W_n(\eta)$ satisfies the following boundary conditions at $\eta = 0$

$$W_n''(0) = b_{1n}\Omega_n W_n'(0)$$
$$W_n'''(0) = -a_{1n}\Omega_n^3 W_n(0) \tag{9.93a}$$

and the following boundary conditions at $\eta = 1$

$$W_n''(1) = b_{2n}\Omega_n W_n'(1)$$
$$W_n'''(1) = a_{2n}\Omega_n^3 W_n(1) \tag{9.93b}$$

Several special cases of Eqs. (9.85) and (9.89) or (9.91) are given in the next section. However, since these results will be presented in terms of the frequency coefficient Ω_n, some clarifying remarks are in order. The natural frequency coefficient is related to the physical frequency f_n, in Hertz as given by Eqs. (9.67), which upon rearrangement yield

$$f_n = \frac{\omega_n}{2\pi} = \frac{c_b r \Omega_n^2}{2\pi L^2} \text{ Hz} \tag{9.94}$$

where ω_n is the natural frequency with units of rad/s. Since the physical frequency is related to the square of the frequency coefficient, seemingly small differences in the natural frequency coefficient can result in significant differences in f_n. As we shall see, the boundary conditions can greatly affect Ω_n. Based on Eq. (9.94), one can formulate the following design guideline.

Design Guideline: The different parameters that affect the natural frequencies of a beam system are the longitudinal speed c_b, the radius of gyration r, and the length of the beam L. The geometric parameter that influences f_n the most is the beam's length, L. On the other hand, changing from one common structural material to another has a much smaller influence on the natural frequency coefficient. This change in material is captured by the longitudinal speed c_b, which is approximately 5000 m/s for both steel and aluminum. The natural frequency is directly proportional to the radius of gyration, which for a beam of rectangular cross section is $r = h/\sqrt{12}$, and h is the depth of the beam. For a beam with a solid circular cross-section of radius r_o, the radius of gyration is $r = r = r_o/\sqrt{2}$.

The results presented in this chapter are valid only for thin beams; that is, beams whose ratio of its radius of gyration to its length is such that $r/L < 0.1$. When this ratio is exceeded, one should consider using the Timoshenko beam theory.[21]

[21]See, for example, E. B. Magrab, *ibid,* Chapter 5; and S. M. Han, H. Benaroya, and T. Wei, "Dynamics of Transversely Vibrating Beams Using Four Engineering Theories," *J. Sound Vibration,* Vol. 225, No. 5, pp. 935–988 (1999).

9.3.3 Effects of Boundary Conditions

In this section, we consider nine different boundary conditions based on the different conditions provided in Table 9.1 and obtain expressions for the natural frequencies and mode shapes associated with each of those cases. Based on these expressions, the dependence of free oscillation characteristics on the boundary conditions is discussed.

1) Beam clamped at each end. For this case, we provide the details for reducing the general results given by Eqs. (9.85) through (9.92) for this special case. The subsequent cases are treated in a similar manner. The procedure involves first examining the boundary conditions at each end of the beam and then determining which values of a_{jn} and b_{jn} go to zero and which ones go to ∞. This is carried out with the help of Table 9.1. In the present case, the boundary conditions at $\eta = 0$ are

$$W_n(0) = 0$$
$$W'_n(0) = 0 \tag{9.95a}$$

and the boundary conditions at $\eta = 1$ are

$$W_n(1) = 0$$
$$W'_n(1) = 0 \tag{9.95b}$$

Since a clamped end corresponds to a boundary with infinite translation stiffness and infinite rotation stiffness, from Eqs. (9.87), one can obtain the clamped-end boundary conditions by considering the following limits:

$$\eta = 0: a_{1n} \to \infty \quad \text{and} \quad b_{1n} \to \infty$$
$$\eta = 1: a_{2n} \to \infty \quad \text{and} \quad b_{2n} \to \infty \tag{9.96}$$

The next step is to make use of Eqs. (9.96) in Eqs. (9.85) and (9.86) to obtain the characteristic equation. In order to realize this, we divide Eq. (9.85) by the product of all those a_{jn} and b_{jn} that tend to ∞, which is the same as dividing each z_j, $j = 1, 2, \ldots, 6$ given in Eq. (9.86) by each of these quantities. After dividing by the appropriate quantities, the limit is taken. In the present case, all four of the parameters go to ∞. Therefore, we divide all the z_j by $a_{1n}a_{2n}b_{1n}b_{2n}$ and take the limit. We notice that in the limit all the $z_j/(a_{1n}a_{2n}b_{1n}b_{2n}) \to 0$ except $z_4/(a_{1n}a_{2n}b_{1n}b_{2n})$, which equals -1. Thus, Eq. (9.85), the characteristic equation, simplifies to

$$\cos \Omega_n \cosh \Omega_n - 1 = 0 \tag{9.97}$$

The roots[22] of Eq. (9.97) provide the natural frequency coefficients Ω_n, from which the natural frequencies f_n are obtained by using Eq. (9.94).

The mode shape is given by Eqs. (9.91) and (9.92), since $b_{1n} \to \infty$. The procedure to determine the mode shape is similar to that used to obtain the characteristic equation. We divide the numerator and denominator of C_n by

---

[22]The roots of this and all subsequent transcendental equations are obtained using the MATLAB function `fzero`.

any a_{jn} that $\rightarrow \infty$. Notice that a_{1n} appears in the term in the brackets that multiplies C_n. Therefore, for this case, we divide the numerator and denominator by $a_{1n}a_{2n}$. Thus, combining Eqs. (9.91) and (9.92), we obtain

$$W_n(\eta) = \frac{S(\Omega_n) - T(\Omega_n)/a_{2n}}{R(\Omega_n)/a_{1n}a_{2n} - (1/a_{1n} + 1/a_{2n})Q(\Omega_n) + T(\Omega_n)} [Q(\Omega_n\eta)/a_{1n}$$
$$- T(\Omega_n\eta)] + S(\Omega_n\eta) \tag{9.98}$$

Taking the limits $a_{1n} \rightarrow \infty$ and $a_{2n} \rightarrow \infty$, we obtain

$$W_n(\eta) = \frac{S(\Omega_n)}{T(\Omega_n)} T(\Omega_n\eta) + S(\Omega_n\eta) \tag{9.99}$$

The first four natural frequency coefficients obtained by solving Eq. (9.97) are tabulated in Table 9.2 and the mode shapes associated with the first four natural frequencies are shown in Figure 9.8. In each case, the mode shape is normalized by the maximum absolute value of $W_n(\eta)$. Although Eq. (9.97) admits the solution $\Omega_n = 0$, it is not considered here because the associated mode shape given by Eq. (9.99) is trivial.

A *node point* is a point on the beam where the mode shape has a zero value. For a beam clamped at each end, the mode shape corresponding to the nondimensional frequency coefficient Ω_n has $(n - 1)$ node points, excluding the boundary points. The locations of the node points for each mode shape are listed in Figure 9.8. We see from the locations of these node points that for $n = 1, 3, \ldots$ the mode shapes are *symmetric*; that is, for the *odd-numbered modes*

$$W(\eta) = W(1 - \eta) \quad 0 \leq \eta \leq 0.5$$

In other words, when the beam is "folded" about $\eta = 0.5$ (the beam's midpoint), the mode shape from both halves overlap. For $n = 2, 4, \ldots$ the node shapes are *asymmetric*; that is, for the *even-numbered modes*

$$W(\eta) = -W(1 - \eta) \quad 0 \leq \eta \leq 0.5$$

The symmetry and asymmetry of the mode shapes are due to the symmetry in the boundary conditions, which are the same at each end of the beam. In general, mode shapes do not exhibit these symmetric and asymmetric properties.

		Ω_n/π		
TABLE 9.2 Natural Frequency Coefficients for Beams with Different Boundary Conditions	Clamped/clamped or Free/free [Eq. (9.97) or (9.111)]	Clamped/pinned [Eq. (9.105)]	Pinned/pinned [Eq. (9.102)]	Clamped/free (cantilever) [Eq. (9.108)]
n				
1	1.5056	1.2499	1	0.5969
2	2.4998	2.2500	2	1.4942
3	3.5000	3.2500	3	2.5002
≥ 4	$n + 0.5$	$n + 0.25$	n	$n - 0.5$

FIGURE 9.8
Mode shapes and the location of node points for a beam clamped at each end.

Node points:
$\eta = 0, 1$

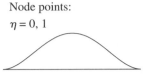

First mode: $\Omega_1 = 4.73$

Node points:
$\eta = 0, 0.5, 1$

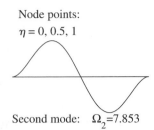

Second mode: $\Omega_2 = 7.853$

Node points:
$\eta = 0, 0.358, 0.642, 1$

Third mode: $\Omega_3 = 10.996$

Node points:
$\eta = 0, 0.279, 0.5, 0.721, 1$

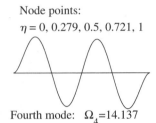

Fourth mode: $\Omega_4 = 14.137$

Nodes of the different beam modes are important for determining the locations of sensors and actuators on the considered system. If one wishes to actuate a certain mode, then one should locate the actuation source away from the nodes of a mode. Alternatively, if one wishes to sense a certain mode by using a sensor, then one should avoid placing the sensor at the nodes of the mode of interest. Hence, if one would like to use a displacement sensor to sense the first three modes of a beam clamped at each end, it is clear from Figure 9.8 that the sensor should be located away from $\eta = 0, 0.358, 0.5, 0.642$, and 1.0.

Thus far, we have discussed displacement mode shapes. One can also plot a *strain mode shape,* which shows how the axial strain of a mode due to bending vibrations changes along the length of the beam. Based on Eq. (9.8), we note that the axial strain is proportional to the second derivative of the displacement. Hence, the strain mode shape associated with the nth nondimensional frequency coefficient Ω_n is determined from Eq. (9.99) as

$$\frac{d^2 W_n(\eta)}{d\eta^2} = -\frac{S(\Omega_n)}{T(\Omega_n)} \frac{d^2 T_n(\Omega_n \eta)}{d\eta^2} + \frac{d^2 S_n(\Omega_n \eta)}{d\eta^2} \qquad (9.100a)$$

where the different derivatives on the right-hand side are evaluated using Eqs. (9.84), to obtain

$$\frac{d^2 W_n(\eta)}{d\eta^2} = -\Omega_n^2 \frac{S(\Omega_n)}{T(\Omega_n)} R_n(\Omega_n \eta) + \Omega_n^2 Q_n(\Omega_n \eta) \qquad (9.100b)$$

To determine the appropriate locations of strain sensors, one first uses Eq. (9.100) to determine the node points and then chooses locations away from these node points. From a numerical evaluation of Eq. (9.100b) for the first mode, the strain node points are at $\eta = 0.224$ and $\eta = 0.776$; for the second mode, they are at $\eta = 0.132$, $\eta = 0.500$, and $\eta = 0.868$. On comparing these node points of the strain mode shapes with those of the displacement mode shapes provided in Figure 9.8, it is seen that they occur at different locations.

EXAMPLE 9.2 Determination of the properties of a beam supporting rotating machinery

An engineer has to mount a piece of rotating machinery that is spinning at 1800 rpm to the center of a steel beam that is 3 m long and clamped at both ends. (Refer to the system shown in Figure 9.9.) In order to ensure that the beam is not excited at its fundamental natural frequency, the engineer would like the beam's first natural frequency to be three times that of the excitation frequency. We shall determine the minimum radius of gyration of the beam's cross-section so that these requirements are met. The static deflection requirements for the beam will be ignored and the weight of the rotating machinery is neglected; it is only considered as a source of disturbance to the beam. The beam's Young's modulus is $E = 1.9625 \times 10^{11}$ N/m^2 and the mass density is $\rho = 7850$ kg/m^3. This means that the longitudinal speed $c_b = \sqrt{E/\rho}$ in the bar is approximately 5000 m/s.

The excitation frequency f_e due to the rotating machinery is

$$f_e = \frac{1800 \text{ rev/min}}{60 \text{ s/min}} = 30 \text{ Hz} \tag{a}$$

Therefore, we need to determine the beam geometry so that the first natural frequency

$$f_1 \geq 3f_e = 90 \text{ Hz} \tag{b}$$

From Eq. (9.94) and Eq. (b), we have that

$$f_1 = \frac{c_b r \Omega_1^2}{2\pi L^2} \geq 3f_e \tag{c}$$

which, upon using Eq. (a), leads to

$$r \geq \frac{6\pi L^2 f_e}{c_b \Omega_1^2} = \frac{180\pi L^2}{c_b \Omega_1^2} \tag{d}$$

From Table 9.2, we see that for a beam clamped at each end the first natural frequency coefficient $\Omega_1 = 1.5056\pi$. Therefore, since $L = 3$ m and $c_b = 5000$ m/s, we obtain from Eq. (d) that

$$r \geq \frac{180 \times \pi \times 3^2}{5000 \times (1.5056\pi)^2} = 0.0455 \text{ m}$$

$$\geq 4.55 \text{ cm} \tag{e}$$

If the cross-section of the beam were rectangular, with width b and depth h, then

$$r^2 = \frac{I}{A} = \frac{bh^3}{12bh} = \frac{h^2}{12} \tag{f}$$

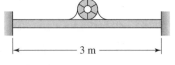

FIGURE 9.9
Beam clamped at each boundary and excited by a rotating machine mounted at its midpoint.

3 m

and the depth of the beam is determined from Eqs. (e) and (f) to be

$$h \geq 2r\sqrt{3} = 2 \times 4.55 \times \sqrt{3} = 15.76 \text{ cm} \tag{g}$$

Notice that for a beam of rectangular cross-section, the width of the beam b does not affect the natural frequency. However, it does affect the static displacement of the beam. Refer to Case 6 in Table 2.3, where the equivalent stiffness of a beam clamped at each boundary is provided. From this expression, it is seen that the displacement is proportional to I, which depends on the width b.

If the mass of the machinery were taken into account, then the value of Ω_1 that is obtained from Table 9.2 would be replaced by the appropriate value of Ω_1 obtained from Table 9.5 or computed from Eq. (9.169).

2) Beam pinned at each end. In this case, the boundary conditions are equivalent to having infinite translation stiffness at each end and zero rotation stiffness and zero inertia at each end. From Eqs. (9.87), we have the conditions that

$$\eta = 0: a_{1n} \rightarrow \infty \quad \text{and} \quad b_{1n} = 0$$
$$\eta = 1: a_{2n} \rightarrow \infty \quad \text{and} \quad b_{2n} = 0 \tag{9.101}$$

Making use of Eqs. (9.101), we reduce the general characteristic equation given by Eq. (9.85) to

$$\sin(\Omega_n) = 0 \tag{9.102}$$

whose roots provide the nondimensional frequency coefficients Ω_n. The associated mode shape is determined from Eqs. (9.89) and (9.90) to be

$$W_n(\eta) = \sin(\Omega_n \eta) \tag{9.103}$$

The first four natural frequency coefficients are tabulated in Table 9.2 and the corresponding mode shapes are shown in Figure 9.10. The mode shape corresponding to the nondimensional frequency coefficient Ω_n has $(n - 1)$ node points, excluding the boundary points. From the numerical values of the node points, we see that the beam has symmetric and asymmetric mode shapes, which is again due to the symmetry in the boundary conditions. Although Eq. (9.102) admits the solution $\Omega_n = 0$, it is not considered here because the associated mode shape given by Eq. (9.103) is trivial. The strain mode shapes are given by

$$\frac{d^2 W_n(\eta)}{d\eta^2} = -\Omega_n^2 \sin(\Omega_n \eta)$$

Thus, the node points for the strain are identical to the node points of the displacement mode shapes given in Figure 9.10.

3) Beam clamped at end $x = 0$ and pinned at end $x = L$. In this case, the boundary conditions are equivalent to having infinite translation stiffness and infinite rotation stiffness at $x = 0$ and infinite translation stiffness and zero

FIGURE 9.10

Mode shapes and the location of node points for a beam pinned at each end.

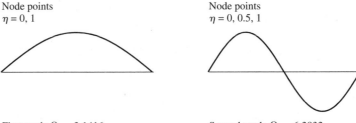

Node points
$\eta = 0, 1$

First mode $\Omega_1 = 3.1416$

Node points
$\eta = 0, 0.5, 1$

Second mode $\Omega_2 = 6.2832$

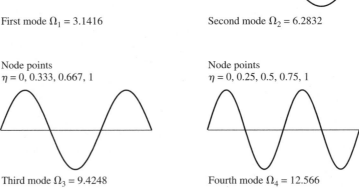

Node points
$\eta = 0, 0.333, 0.667, 1$

Third mode $\Omega_3 = 9.4248$

Node points
$\eta = 0, 0.25, 0.5, 0.75, 1$

Fourth mode $\Omega_4 = 12.566$

inertia and zero rotation stiffness at $x = L$. In terms of the expressions given in Eqs. (9.87), the boundary conditions are expressed in terms of the limits

$$\eta = 0: a_{1n} \rightarrow \infty \text{ and } b_{1n} \rightarrow \infty$$
$$\eta = 1: a_{2n} \rightarrow \infty \text{ and } b_{2n} = 0 \tag{9.104}$$

Then making use of Eqs. (9.104) in Eq. (9.85), we find that the characteristic equation has the form

$$\cos(\Omega_n)\sinh(\Omega_n) - \sin(\Omega_n)\cosh(\Omega_n) = 0 \tag{9.105}$$

whose roots provide the nondimensional frequency coefficients Ω_n. Upon substituting Eqs. (9.104) in Eqs. (9.91) and (9.92), which represent the general form for a mode shape when $b_{1n} \rightarrow \infty$, we obtain

$$W_n(\eta) = -\frac{S(\Omega_n)}{T(\Omega_n)} T(\Omega_n \eta) + S(\Omega_n \eta) \tag{9.106}$$

The first four natural frequency coefficients are tabulated in Table 9.2 and the corresponding mode shapes are plotted in Figure 9.11. On comparing the characteristic equation and the form of the mode shape with those presented in Condition 1 for a beam with clamped ends at both boundaries, we find that the characteristic equation is different, but the form of the mode shape is the same. However, since the characteristic equation has different roots in each case, the numerical values of the natural frequency coefficients are different in each case, and hence, the mode shapes are different as is seen from the results shown in Table 9.2 and the graphs shown in Figures 9.8 and 9.11. The mode shape corresponding to the nondimensional frequency coefficient Ω_n has $(n - 1)$ node points, excluding the boundary points. Since the

FIGURE 9.11
Mode shapes and the location of node points for a beam clamped at one end and pinned at the other.

Node points
$\eta = 0, 1$

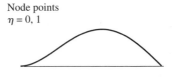

First mode $\Omega_1 = 3.927$

Node points
$\eta = 0, 0.557, 1$

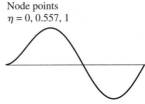

Second mode $\Omega_2 = 7.0686$

Node points
$\eta = 0, 0.386, 0.692, 1$

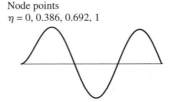

Third mode $\Omega_3 = 10.21$

Node points
$\eta = 0, 0.295, 0.529, 0.765, 1$

Fourth mode $\Omega_4 = 13.352$

boundary conditions are not the same at each end of the beam, the mode shapes do not have the symmetry properties discussed in the previous two cases. This is also borne out by the locations of the node points for the different modes.

4) Beam clamped at end $x = 0$ and free at end $x = L$ (cantilever).   In this case, the boundary conditions are equivalent to a boundary with infinite translation stiffness and infinite rotation stiffness at the end $x = 0$ and the free end is equivalent to a boundary with zero stiffness and zero inertia. In terms of the expressions given in Eqs. (9.87), the boundary conditions are expressed in terms of the limits

$$\eta = 0: a_{1n} \to \infty \quad \text{and} \quad b_{1n} \to \infty$$
$$\eta = 1: a_{2n} = 0 \quad \text{and} \quad b_{2n} = 0 \tag{9.107}$$

Then, making use of Eqs. (9.107) in Eq. (9.85), the characteristic equation is obtained from Eq. (9.85) as

$$\cos(\Omega_n)\cosh(\Omega_n) + 1 = 0 \tag{9.108}$$

whose roots provide the nondimensional frequency coefficients Ω_n. The mode shape associated with Ω_n is determined from Eqs. (9.91) and (9.92) by making use of Eqs. (9.107), which results in

$$W_n(\eta) = -\frac{T(\Omega_n)}{Q(\Omega_n)} T(\Omega_n\eta) + S(\Omega_n\eta) \tag{9.109}$$

The first four natural frequency coefficients determined by solving Eq. (9.108) are tabulated in Table 9.2 and the corresponding mode shapes are shown in Figure 9.12. The mode shape corresponding to the nth frequency

FIGURE 9.12
Mode shapes and the location of node points for a cantilever beam.

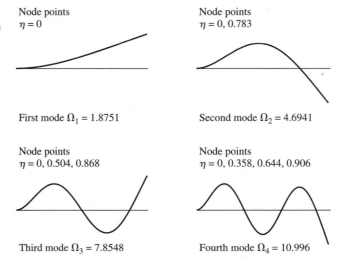

Node points
$\eta = 0$

First mode $\Omega_1 = 1.8751$

Node points
$\eta = 0, 0.783$

Second mode $\Omega_2 = 4.6941$

Node points
$\eta = 0, 0.504, 0.868$

Third mode $\Omega_3 = 7.8548$

Node points
$\eta = 0, 0.358, 0.644, 0.906$

Fourth mode $\Omega_4 = 10.996$

coefficient Ω_n has $(n - 1)$ node points, excluding the boundary points. However, since the boundary conditions are not the same at each end of the beam, the mode shapes do not have any symmetry or asymmetry. The largest relative displacement for all modes occurs at the free end of the beam. However, the largest axial strain due to bending vibrations occurs at the clamped end of the beam.

While displacement mode shapes are important for determining locations of displacement sensors, velocity sensors, and accelerometer sensors, strain mode shapes are important for determining the locations of strain sensors. For design purposes, the information about locations of sensors and actuators with reference to nodes of a mode is put forth in the form of the following guideline.

Design Guideline: For a system modeled as a beam, if a sensor needs to be located so as to sense the vibrations in a certain mode, then this sensor should be located away from the nodes of this mode. Similarly, in order to actuate or excite a certain vibration mode, the actuator should be away from the node of this mode. Alternatively, if a sensor should not sense a certain vibration mode or an actuator should not excite a certain vibration mode, then they should be located at a node of the mode of interest.

Before proceeding to the next case, we examine the percentage differences Δ in the lowest natural frequencies in beams with different boundary conditions. We assume that the beam material and geometry are the same in each case and that only the boundary conditions are different from one case to another. The natural frequencies of the beam clamped at both ends are used as reference values for the calculations. Then, from Eq. (9.94) and Table 9.2, we determine the following

$$\Delta_{c/p} = 100 \left(\frac{f_{1c/p}}{f_{1c/c}} - 1 \right) = 100 \left(\frac{\Omega^2_{1c/p}}{\Omega^2_{1c/c}} - 1 \right) = 100 \left(\frac{(1.2499\pi)^2}{(1.5056\pi)^2} - 1 \right)$$
$$= -31.1\%$$

$$\Delta_{p/p} = 100 \left(\frac{\Omega^2_{1p/p}}{\Omega^2_{1c/c}} - 1 \right) = 100 \left(\frac{(\pi)^2}{(1.5056\pi)^2} - 1 \right) = -55.9\%$$

$$\Delta_{c/f} = 100 \left(\frac{\Omega^2_{1c/f}}{\Omega^2_{1c/c}} - 1 \right) = 100 \left(\frac{(0.5969\pi)^2}{(1.5056\pi)^2} - 1 \right) = -84.3\%$$

where $\Delta_{c/p}$ is the measure of how much the first natural frequency $f_{1c/p}$ of a clamped-pinned beam is below the first natural frequency $f_{1c/c}$ of the clamped-clamped beam, $\Delta_{p/p}$ is the measure of how much the first natural frequency $f_{1p/p}$ of a pinned-pinned beam is below the first natural frequency $f_{1c/c}$ of the clamped-clamped beam, and $\Delta_{c/f}$ is the measure of how much the first natural frequency $f_{1c/f}$ of a cantilever beam is below the first natural frequency $f_{1c/c}$ of the clamped-clamped beam. Thus, the clamped-pinned beam's natural frequency is 31% lower than that of a beam clamped at each end and the cantilever beam's natural frequency is 84% lower than that of a beam clamped at each end. These observations lend themselves to the following design guideline.

Design Guideline: For uniform beams having the same material and same geometry, the beam clamped at both ends has the highest fundamental natural frequency.

---

5) Beam free at each end. For this case, the boundary conditions are equivalent to a boundary with zero inertia and zero stiffness at each end. From Eqs. (9.87), we find that

$$\eta = 0: a_{1n} = 0 \quad \text{and} \quad b_{1n} = 0$$
$$\eta = 1: a_{2n} = 0 \quad \text{and} \quad b_{2n} = 0 \tag{9.110}$$

Then, making use of Eqs. (9.85) and (9.110), we find that the characteristic equation is

$$\cos(\Omega_n)\cosh(\Omega_n) - 1 = 0 \tag{9.111}$$

which is identical to the characteristic equation for a beam clamped at both ends; that is, Eq. (9.97). The mode shape associated with the nth frequency coefficient Ω_n is determined from Eq. (9.89) and (9.110) as

$$W_n(\eta) = -\frac{S(\Omega_n)}{R(\Omega_n)} Q(\Omega_n \eta) + R(\Omega_n \eta) \tag{9.112}$$

The first four natural frequency coefficients determined from the roots of Eq. (9.111) are tabulated in Table 9.2 and the corresponding mode shapes are plotted in Figure 9.13. In this case, there are no nodes at the boundaries. The mode shape corresponding to the nth nondimensional frequency coefficient Ω_n has $(n + 1)$ node points; that is, there are two node points in the first mode, and so forth. Thus, although the natural frequency coefficients are the same as those for a beam clamped at both ends, the mode shapes are different. Examples of practical applications that use models of beams free at both ends are the study of the vibrations of launch vehicles[23] and ships.

In addition to the mode shapes shown in Figure 9.13, a beam free at both ends also has *rigid-body modes* whose form is given by

$$W_n(\eta) = C_0 \quad \text{and} \quad W_n(\eta) = D_0 + E_0\eta \tag{9.113}$$

Both modes in Eq. (9.113) are associated with $\Omega_n = 0$; that is, the natural frequency in each case is zero. Unlike Cases 1, 2, and 3, the zero eigenvalue corresponds to a nontrivial mode shape in this case. The first of the expressions in Eqs. (9.113) represents a case where every point on the beam experiences the same translation, and the second of these expressions represents a case where every point on the beam experiences a combination of translation and rotation. Unlike the flexural modes of vibration given by Eq. (9.112), for the rigid-body modes given by Eq. (9.113), $d^2W_n/d\eta^2$ is zero everywhere on the beam and hence, the strain associated with vibrations in a rigid-body mode is zero everywhere on the beam.

6) Beam clamped at one end and carrying a mass at its other end.

For this case, we assume that the boundary at $x = 0$ is clamped and the boundary with the mass is at $x = L$. From Eqs. (9.87), we arrive at

$$\eta = 0: a_{1n} \to \infty \quad \text{and} \quad b_{1n} \to \infty$$

$$\eta = 1: a_{2n} = -\frac{\Omega_n M_0}{m_0} \quad \text{and} \quad b_{2n} = 0 \tag{9.114}$$

Then Eqs. (9.85) and (9.114) are used to determine the characteristic equation

$$\frac{M_o\Omega_n}{m_o}[\cos(\Omega_n)\sinh(\Omega_n) - \sin(\Omega_n)\cosh(\Omega_n)]$$

$$+ \cos(\Omega_n)\cosh(\Omega_n) + 1 = 0 \tag{9.115}$$

To determine the mode shape associated with the nth nondimensional frequency coefficient Ω_n, we use Eqs. (9.114) in Eqs. (9.91) and (9.92) and obtain

$$W_n(\eta) = -\frac{T(\Omega_n) + (M_o/m_o)\Omega_n S(\Omega_n)}{(M_o/m_o)\Omega_n T(\Omega_n) + Q(\Omega_n)} T(\Omega_n\eta) + S(\Omega_n\eta) \tag{9.116}$$

From Eqs. (9.115) and (9.116), we see that when the ratio of the attached mass to the mass of the beam M_o/m_o becomes very large, the characteristic

[23]A. Joshi, "Prediction of Free-Free Modes from Single Point Support Ground Vibration Test of Launch Vehicles," *J. Sound Vibration*, Vol. 216, No. 4, pp. 739–747 (1998).

FIGURE 9.13

Mode shapes and the location of node points for a beam free at each end.

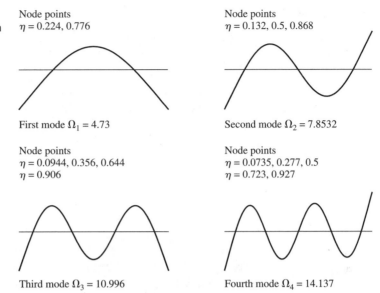

Node points
$\eta = 0.224, 0.776$

First mode $\Omega_1 = 4.73$

Node points
$\eta = 0.132, 0.5, 0.868$

Second mode $\Omega_2 = 7.8532$

Node points
$\eta = 0.0944, 0.356, 0.644$
$\eta = 0.906$

Third mode $\Omega_3 = 10.996$

Node points
$\eta = 0.0735, 0.277, 0.5$
$\eta = 0.723, 0.927$

Fourth mode $\Omega_4 = 14.137$

equation and mode shapes approach the characteristic equation and mode shapes of a beam clamped at one end and pinned at the other, which are given by Eqs. (9.105) and (9.106), respectively. When M_o/m_o becomes very small, the characteristic equation and mode shapes approach the characteristic equation and mode shapes of a beam clamped at one end and free at the other, as given by Eqs. (9.108) and (9.109), respectively. The first four natural frequency coefficients determined by numerically solving Eq. (9.115) for several values of M_o/m_o are tabulated in Table 9.3, and the corresponding mode shapes determined from Eq. (9.116) are presented in Figure 9.14. The mode shape corresponding to the nth frequency coefficient Ω_n has $(n - 1)$ node points, not including the boundary at $\eta = 0$. The variation in the lowest three natural frequency coefficients with respect to M_o/m_o is plotted in Figure 9.15. The nondimensional natural frequency coefficient Ω_1 approaches zero as M_o/m_o approaches infinity. As would be expected, an increase in the magnitude of the end mass results in decreases in the magnitudes of the natural frequencies. An example of a system that is modeled as a cantilever beam with a mass on its free end is a water tower.

In the literature, the boundary condition considered here for the mass at the end—that is, the shear-force boundary condition given by the second of Eqs. (9.71b)—has been extended to consider the case where the center of gravity of the attached mass is not concentrated at $\eta = 1$, but is located a small distance beyond this point, as shown in Example 9.1. In another extension, a single degree-of-freedom spring-mass system attached to the beam end mass has been considered.[24]

---

[24]M. Gürgöze, "On the Eigenfrequencies of a Cantilever Beam with Attached Tip Mass and a Spring-Mass System," *J. Sound Vibration*, Vol. 190, No. 2, pp. 149–162 (1996).

TABLE 9.3

Natural Frequency Coefficients of a Cantilever Beam Carrying a Mass at the Free End

		Ω_n/π			
n	$M_o/m_o = 0$	$M_o/m_o = 0.1$	$M_o/m_o = 1$	$M_o/m_o = 10$	$M_o/m_o = \infty$
1	0.5969	0.5484	0.3972	0.2342	0
2	1.4942	1.4004	1.2832	1.2537	1.2499*
3	2.5002	2.3717	2.2709	2.2522	2.2500*
4	$n - 0.5$	3.3492	3.2648	3.2516	$n - 0.75$*
	Eq. (9.108) [Cantilever]	Eq. (9.115)	Eq. (9.115)	Eq. (9.115)	Eq. (9.105) [Clamped/pinned]

*In the limit, the nth mode of a cantilever beam with a mass at the free end becomes $(n-1)$th mode of a clamped/pinned beam.

In many instances where a beam carries an end mass, the inertia of the beam is neglected and the system is represented by an equivalent single degree-of-freedom system. This situation is revisited to point out when it is reasonable to neglect the beam inertia and when it is not. For a beam with an end mass, the natural frequency of the equivalent single degree-of-freedom system can be determined by making use of the stiffness expressions listed in Table 2.3. For a cantilever beam with an end mass, we make use of Case 4 of Table 2.3 and find that the natural frequency is

$$\omega_n^2 = \frac{k}{M_o} = \frac{3}{M_o}\left(\frac{EI}{L^3}\right) \tag{9.117}$$

Equation (9.117) is the result obtained by approximating the first natural frequency of a cantilever beam with an end mass as the natural frequency of a single degree-of-freedom system where only the beam stiffness is taken into

FIGURE 9.14

Mode shapes for a cantilever beam carrying a mass at the free end for different values of M_o/m_o: ······ $M_o/m_o = 0.1$; —— $M_o/m_o = 1$; – – – $M_o/m_o = 10$. (For the first mode, the modes shapes are normalized to the maximum value of $|W_n(\eta)|$ for $M_o/m_o = 0.1$.)

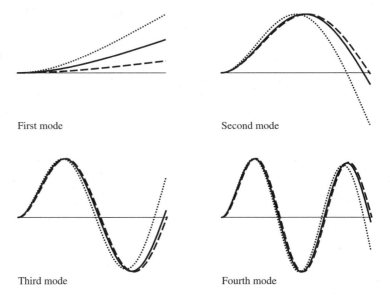

First mode   Second mode

Third mode   Fourth mode

FIGURE 9.15

First three natural frequency coefficients for a cantilever beam carrying a mass at its free end as a function of M_o/m_o.

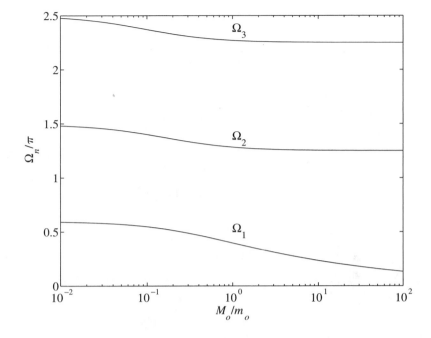

account and the beam inertia is neglected. However, when the beam inertia is not neglected, the first natural frequency ω_e is obtained from Eqs. (9.67) as

$$\omega_e^2 = \frac{EI\Omega_1^4}{\rho AL^4} = \left(\frac{EI}{L^3}\right)\frac{\Omega_1^4}{m_o} \tag{9.118}$$

where Ω_1 is the lowest natural frequency coefficient obtained by solving Eq. (9.115). On comparing Eqs. (9.117) and (9.118), the percentage error one incurs in using Eq. (9.117) is

$$\epsilon = 100\left(\frac{\omega_n}{\omega_e} - 1\right) = 100\left(\frac{1}{\Omega_1^2}\sqrt{\frac{3m_o}{M_o}} - 1\right)\% \tag{9.119}$$

From the numerical solutions of Eqs. (9.115) and (9.119) it is found that we have an error $\epsilon < 5\%$ when the mass ratio $M_o/m_o > 2.3$, and that we have an error $\epsilon < 1\%$ when the mass ratio $M_o/m_o > 11.7$. For $M_o/m_o = 1$, the error is 11.2%. This leads to the following design guideline.

Design Guideline: When a cantilever beam of mass m_o and an end mass M_o is approximated by a single degree-of-freedom system where the beam stiffness is taken into account and the beam inertia is neglected, the approximation for the first natural frequency obtained by using the single degree-of-freedom model is reasonable to use only when the ratio of the end mass to the beam mass M_o/m_o is greater than 2.3.

---

7) Beam clamped at one end and restrained by a translation spring at the other end. For this case, we assume that the clamped end is located at $x = 0$ and that a translation spring is located at the end $x = L$. Then, from Eqs. (9.87), we find that

$$\eta = 0: a_{1n} \to \infty \quad \text{and} \quad b_{1n} \to \infty$$

$$\eta = 1: a_{2n} = \frac{K_2}{\Omega_n^3} \quad \text{and} \quad b_{2n} = 0 \tag{9.120}$$

Making use of Eqs. (9.85) and (9.120), we determine the characteristic equation as

$$-(K_2/\Omega_n^3)[\cos(\Omega_n)\sinh(\Omega_n) - \sin(\Omega_n)\cosh(\Omega_n)]$$

$$+ \cos(\Omega_n)\cosh(\Omega_n) + 1 = 0 \tag{9.121}$$

From Eqs. (9.91), (9.92), and (9.120), we find that the mode shapes are given by

$$W_n(\eta) = \frac{T(\Omega_n) - K_2 S(\Omega_n)/\Omega_n^3}{K_2 T(\Omega_n)/\Omega_n^3 - Q(\Omega_n)} T(\Omega_n \eta) + S(\Omega_n \eta) \tag{9.122}$$

From Eqs. (9.121) and (9.122), we see that as K_2 becomes very large, the characteristic equation and mode shapes approach the characteristic equation and mode shapes of a beam clamped at one end and pinned at the other, as given by Eqs. (9.105) and (9.106), respectively. On the other hand, as K_2 becomes very small, the characteristic equation and mode shapes approach those of a beam clamped at one end and free at the other, as given by Eqs. (9.108) and (9.109), respectively. Comparing these limiting cases to the case of a cantilever carrying a mass at its free end, we see that although the beams have very different physical constraints, they have identical limiting cases. The first four natural frequency coefficients determined for several values of K_2 are tabulated in Table 9.4, and the corresponding mode shapes are presented in Figure 9.16. The mode shape corresponding to the nth frequency coefficient Ω_n has $(n - 1)$ node points, not including the boundary $\eta = 0$. We have plotted the variation in the lowest three natural frequency coefficients as a function of K_2 in Figure 9.17. As would be expected, as the stiff-

TABLE 9.4

Natural Frequency Coefficients of a Cantilever Beam with a Translation Spring at the Free End

				Ω_n/π			
n	$K_2 = 0$	$K_2 = 1$	$K_2 = 10$	$K_2 = 100$	$K_2 = 1000$	$K_2 = \infty$
1	0.5969	0.6398	0.8400	1.1588	1.2407	1.2499
2	1.4942	1.4973	1.5259	1.7876	2.1888	2.2500
3	2.5002	2.5009	2.5069	2.5732	3.0407	3.2500
4	$n - 0.5$	3.5002	3.5024	3.5252	3.8041	$n + 0.25$
	Eq. (9.108) [Cantilever]	Eq. (9.121)	Eq. (9.121)	Eq. (9.121)	Eq. (9.121)	Eq. (9.105) [Clamped/pinned]

FIGURE 9.16
Mode shapes for a cantilever beam with a translation spring at its free end for different values of K_2: ······ $K_2 = 1$; —— $K_2 = 100$; – – – $K_2 = 10,000$. (For the first mode, the modes shapes are normalized to the maximum value of $|W_n(\eta)|$ for $K_2 = 1$.)

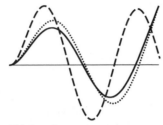

First mode

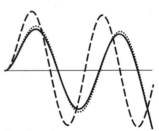

Second mode

Third mode

Fourth mode

ness of the end spring increases, the magnitudes of the natural frequencies increase. However, at the two limits, $k_2 \to 0$ and $k_2 \to \infty$, the graph of each frequency coefficient versus k_2 reaches a constant value.

8) Beam restrained by translation springs at each end. For this case, we assume that the spring with stiffness k_1 is located at $x = 0$ and that the

FIGURE 9.17
First three natural frequency coefficients for a cantilever beam with a translation spring at the free end as a function of the spring stiffness K_2.

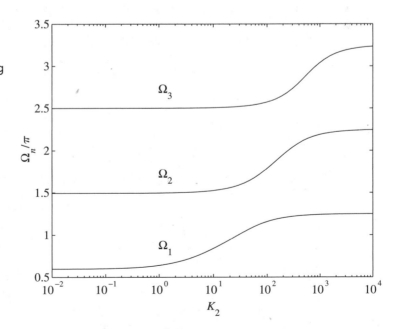

translation spring with stiffness k_2 is located at $x = L$. Then, from Eqs. (9.87), we find that

$$\eta = 0: a_{1n} = \frac{K_1}{\Omega_n^3} \quad \text{and} \quad b_{1n} = 0$$

$$\eta = 1: a_{2n} = \frac{K_2}{\Omega_n^3} \quad \text{and} \quad b_{2n} = 0 \tag{9.123}$$

Making use of Eqs. (9.85) and (9.123), we determine the characteristic equation as

$$(a_{1n} + a_{2n})[\cos(\Omega_n)\sinh(\Omega_n) - \sin(\Omega_n)\cosh(\Omega_n)]$$

$$+ 2a_{1n}a_{2n} \sin(\Omega_n)\sinh(\Omega_n) + 1 - \cos(\Omega_n)\cosh(\Omega_n) = 0 \tag{9.124}$$

From Eqs. (9.89), (9.90), and (9.123), we find that the expression for the mode shape takes the form

$$W_n(\eta) = \frac{a_{2n}R(\Omega_n) - S(\Omega_n)}{R(\Omega_n) - (a_{1n} + a_{2n})Q(\Omega_n) + a_{1n}a_{2n}T(\Omega_n)} [Q(\Omega_n\eta)$$

$$- a_{1n}T(\Omega_n\eta)] + R(\Omega_n\eta) \tag{9.125}$$

From Eqs. (9.124) and (9.125), we see that when K_1 and K_2 become very large, the characteristic equation and mode shapes approach the characteristic equation and mode shapes of a beam pinned at each end, which are given by Eqs. (9.102) and (9.103), respectively. At the other extreme, as both K_1 and K_2 become very small in magnitude, the characteristic equation and mode shapes approach the characteristic equation and mode shapes of a beam free at both ends, which are given by Eqs. (9.111) and (9.112), respectively.

9) Beam supported by torsion springs at each end.   For this case, we assume that the torsion spring with stiffness k_{t1} is at the end $x = 0$ and the torsion spring with stiffness k_{t2} is at the end $x = L$. Then, from Eqs. (9.87), we find that

$$\eta = 0: a_{1n} \rightarrow \infty \quad \text{and} \quad b_{1n} = \frac{B_1}{\Omega_n}$$

$$\eta = 1: a_{2n} \rightarrow \infty \quad \text{and} \quad b_{2n} = -\frac{B_2}{\Omega_n} \tag{9.126}$$

Making use of Eqs. (9.85) and (9.126) leads to the characteristic equation

$$(B_1 + B_2)[\cos(\Omega_n)\sinh(\Omega_n) - \sin(\Omega_n)\cosh(\Omega_n)]/\Omega_n$$

$$- 2 \sin(\Omega_n)\sinh(\Omega_n) - B_1B_2[1 - \cos(\Omega_n)\cosh(\Omega_n)]/\Omega_n^2 = 0 \tag{9.127}$$

The expression for the mode shape is determined from Eqs. (9.89), (9.90), and (9.126) as

$$W_n(\eta) = -\frac{R(\Omega_n) + B_1S(\Omega_n)/\Omega_n}{T(\Omega_n)} T(\Omega_n\eta)$$

$$+ R(\Omega_n\eta) + B_1S(\Omega_n\eta)/\Omega_n \tag{9.128}$$

From Eqs. (9.127) and (9.128), it is seen that as both B_1 and B_2 become very large, the characteristic equation and mode shapes approach those of a beam clamped at each end, as given by Eqs. (9.97) and (9.99), respectively. When both B_1 and B_2 become very small, the characteristic equation and mode shapes approach those of a beam pinned at both ends, as given by Eqs. (9.102) and (9.103), respectively. The results of this case and the preceding one, Case 8, have been extended[25] to include the effects of an elastic foundation k_f.

9.3.4 Effects of Stiffness and Inertial Elements Attached at an Interior Location[26]

In the previous sections, we considered free vibrations of beams that had springs and inertia elements attached at the boundaries. In this section, we shall extend these results and examine the free vibrations of beams that have springs and inertia elements attached to an interior point of the beam. In particular, we shall determine the natural frequency coefficients and modes shapes for the three configurations shown in Figure 9.18: i) a beam with an attached translation spring, ii) a beam with an attached mass, and iii) a beam with an attached undamped single degree-of-freedom system. In all three cases, it is assumed that the beam has a uniform cross-section. We shall first give the governing equation for each of these systems and then determine the complete solution only for the case of a beam with an attached single degree-of-freedom system. It will be shown that special cases of this system are the beam with an attached mass and the beam with an attached spring.

For each of the beams shown in Figure 9.18, at the left boundary $x = 0$, there is a linear translation spring with stiffness k_1 and a linear torsion spring with stiffness k_{t1}. At the right boundary $x = L$, there are a linear translation spring with stiffness k_2 and a linear torsion spring with stiffness k_{t2}. In addition, as shown Figure 9.18a, a linear translation spring with stiffness k_s is attached at $x = L_1$, $0 < L_1 < L$. In Figure 9.18b, a mass M_s is attached at $x = L_1$, $0 < L_1 < L$. Finally, in Figure 9.18c, an undamped single degree-of-freedom system of mass M_s and stiffness k_s is attached to the beam[27] at $x = L_1$, $0 < L_1 < L$.

---

[25]M. A. De Rosa and M. J. Maurizi, "The Influence of Concentrated Masses and Pasternak Soil on the Free Vibrations of Euler Beams—Exact Solution," *J. Sound Vibration,* Vol. 212, No. 4, pp. 573–581 (1998).

[26]This type of analysis has been extended to include two degree-of-freedom systems. See, for example, M. U. Jen and E. B. Magrab, "Natural Frequencies and Modes Shapes of Beams Carrying a Two Degree-of-Freedom Spring-Mass System," *J. Vibrations Acoustics,* Vol. 115, pp. 202–209 (April 1993) and H. Ashrafiuon, "Optimal Design of Vibration Absorber Systems Supported by Elastic Base," *J. Vibrations Acoustics,* Vol. 114, pp. 280–283 (April 1992).

[27]It is mentioned that these limits on L_1 can be removed in certain cases. Consider a cantilever beam with a mass attached to its free end. One could formulate this situation by considering the beam to be uniform and to include the mass as part of the boundary condition. On the other hand, one could assume that the moment and shear are zero at the free end of the beam and that the attached mass on the interior of the beam is moved to $L_1 = L$.

FIGURE 9.18
Uniform and homogeneous beam with discrete elements at the boundaries and the following: (a) a spring attached at $x = L_1$; (b) a mass attached at $x = L_1$; and (c) an undamped single degree-of-freedom system attached at $x = L_1$.

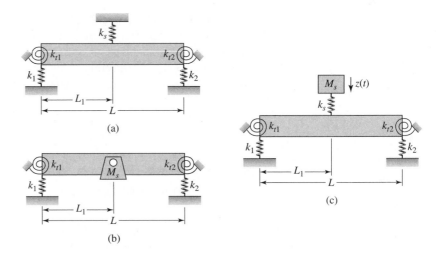

The discussion is limited to cases where the axial load $p(x,t) = 0$ and the externally applied transverse load $f(x,t) = 0$. Furthermore, the oscillations are assumed to be about the equilibrium position in each case. A primary feature of the equations in each case is the representation of the discrete element as an equivalent distributed element by using the delta function $\delta(x)$.

Linear Spring Attached to an Interior Point $x = L_1$

The governing equation of motion for the system shown in Figure 9.18a is obtained from Eq. (9.40) by representing the discrete spring k_s as a distributed elastic spring. This is done by replacing the elastic foundation term $k_f w(x,t)$ by

$$k_f w(x,t) \to \frac{k_s}{L} w(x,t)\delta(x - L_1) \tag{9.129}$$

where $\delta(x - L_1)$ is the delta function. Taking into account that the beam is uniform and homogeneous, Eq. (9.40) is written as

$$EI \frac{\partial^4 w}{\partial x^4} + \frac{k_s}{L} w(x,t)\delta(x - L_1) + \rho A \frac{\partial^2 w}{\partial t^2} = 0 \tag{9.130}$$

Mass Attached to an Interior Point $x = L_1$

The governing equation of motion for the system shown in Figure 9.18b is obtained from Eq. (9.40) by representing the discrete mass M_s as a distributed mass. This is done by replacing the mass density of the beam by

$$\rho \to \rho + \frac{M_s}{AL} \delta(x - L_1) \tag{9.131}$$

Then Eq. (9.40) is reduced to

$$EI \frac{\partial^4 w}{\partial x^4} + \left[A\rho + \frac{M_s}{L} \delta(x - L_1) \right] \frac{\partial^2 w}{\partial t^2} = 0 \tag{9.132}$$

Undamped Single Degree-of-Freedom System Attached to an Interior Point $x = L_1$

The governing equation of motion for the system shown in Figure 9.18c is obtained from Eq. (9.40). If $z(t)$ is the displacement of the mass M_s about its equilibrium position, then the term that represents the force per unit length of the elastic foundation $k_f w(x,t)$ is replaced by

$$k_f w(x,t) \rightarrow \frac{k_s}{L}[w(x,t) - z(t)]\delta(x - L_1) \tag{9.133}$$

and Eq. (9.40) leads to

$$EI\frac{\partial^4 w}{\partial x^4} + \frac{k_s}{L}[w(x,t) - z(t)]\delta(x - L_1) + \rho A\frac{\partial^2 w}{\partial t^2} = 0 \tag{9.134a}$$

The equation describing the motion of the single degree-of-freedom system is obtained from Eq. (3.27), which is the equation of motion for a damped single degree-of-freedom system subjected to a base excitation. Here, the beam vibrations act as base excitation for the spring-mass system shown in Figure 9.18c. Setting $c = 0$ in Eq. (3.27) because damping is absent and recognizing that in the present context $x(t) = z(t)$ and $y(t) = w(L_1,t)$, Eq. (3.27) becomes

$$M_s\frac{d^2 z(t)}{dt^2} + k_s z(t) = k_s w(L_1,t) = k_s w(x,t)\delta(x - L_1) \tag{9.134b}$$

Equations (9.134), together with the boundary conditions, represent the governing equations of motion for the system shown in Figure 9.18c.

We now proceed to obtain the solution of Eqs. (9.134) for the boundary conditions shown in Figure 9.18c. It will then be shown that limiting cases of this solution also describe the responses of the systems shown in Figures 9.18a and 9.18b. Since we are considering free oscillations, we follow the procedure used in Section 9.3.2 and let

$$w(x,t) = W(x)G(t)$$
$$z(t) = Z_o G(t) \tag{9.135}$$

where $G(t)$ is a harmonic function of the form given by Eq. (9.63). Upon substituting Eqs. (9.135) into Eqs. (9.134) and using the notation of Eqs. (9.67), the spatial eigenvalue problem takes the form

$$\frac{d^4 W(\eta)}{d\eta^4} + K_s[W(\eta) - Z_o]\delta(\eta - \eta_1) - \Omega^4 W(\eta) = 0 \tag{9.136}$$

$$\left(1 - \frac{M_{so}}{K_s}\Omega^4\right)Z_o = W(\eta_1) = W(\eta)\delta(\eta - \eta_1) \tag{9.137}$$

where $\eta = x/L$, $\eta_1 = L_1/L$,

$$K_s = \frac{k_s L^3}{EI}, \quad M_{so} = \frac{M_s}{m_o}, \quad \text{and} \quad \Omega^4 = \frac{\rho A\omega^2 L^4}{EI} \tag{9.138}$$

and $m_o = \rho A L$ is the mass of the beam. It is noted that K_s is the nondimensional spring constant and M_{so} is the ratio of the mass of the single degree-of-freedom system to the mass of the beam. We now determine the characteristic equation from which the natural frequency coefficient Ω_n is obtained by making use of Eqs. (9.136) and (9.137) and the associated boundary conditions.

Upon substituting for Z_o from Eq. (9.137) into Eq. (9.136), we obtain

$$\frac{d^4W(\eta)}{d\eta^4} - B(\Omega)W(\eta)\delta(\eta - \eta_1) - \Omega^4 W(\eta) = 0 \tag{9.139}$$

where

$$B(\Omega) = \frac{M_{so}\Omega^4}{1 - M_{so}\Omega^4/K_s} \tag{9.140}$$

When the mass of the single degree-of-freedom system $M_s = 0$, then $M_{so} = 0$ in Eq. (9.140) and, therefore, $B(\Omega) = 0$. In this case, Eq. (9.139) reduces to Eq. (9.68), which describes a beam without the attached single degree-of-freedom system.

Equation (9.139) has two limiting cases that are of interest. In order to determine these two cases, we first rewrite Eq. (9.140) as

$$B(\Omega) = \Omega^4 \left[\frac{1}{M_{so}} - \frac{\Omega^4}{K_s} \right]^{-1} \tag{9.141}$$

Beam carrying a mass at $\eta = \eta_1$. When the translation spring stiffness $k_s \to \infty$, we have the case of a mass attached directly to the beam at $\eta = \eta_1$, as shown in Figure 9.18b. Then, since $K_s \to \infty$, Eq. (9.141) reduces to

$$B(\Omega) = M_{so}\Omega^4 \tag{9.142}$$

Upon substituting Eq. (9.142) into Eq. (9.139), we obtain

$$\frac{d^4W(\eta)}{d\eta^4} - \Omega^4 \left[1 + \frac{M_s}{m_o}\delta(\eta - \eta_1) \right] W(\eta) = 0 \tag{9.143}$$

which is what we would have obtained if we had used Eqs. (9.67), the first of Eqs. (9.135), and Eq. (9.132).

Beam with a spring attached at $\eta = \eta_1$. When the mass $M_s \to \infty$, we have the case where one end of a spring is attached directly to the beam at $\eta = \eta_1$ and the other end of the spring is fixed, as shown in Figure 9.18a. In this case, since $M_{so} \to \infty$, Eq. (9.141) reduces to

$$B(\Omega) = -K_s \tag{9.144}$$

Upon substituting Eq. (9.144) into Eq. (9.139), we obtain

$$\frac{d^4W(\eta)}{d\eta^4} + K_s W(\eta)\delta(\eta - \eta_1) - \Omega^4 W(\eta) = 0 \tag{9.145}$$

which could have been obtained directly from Eq. (9.130) by using Eqs. (9.67) and the first of Eqs. (9.135).

Solutions for Natural Frequencies and Mode Shapes

To solve Eq. (9.139), we take the Laplace transform of each term in Eq. (9.139) and obtain[28]

$$
\tilde{W}(s) = \frac{1}{(s^4 - \Omega^4)} [W(0)s^3 + W'(0)s^2 + W''(0)
$$
$$
+ W'''(0) + B(\Omega)W(\eta_1)e^{-s\eta_1}] \tag{9.146}
$$

where $\tilde{W}(s)$ is the Laplace transform of $W(\eta)$. In arriving at Eq. (9.146), we have made use of Eqs. (A.7) and (A.8) of Appendix A in the same manner as we did in arriving at Eq. (9.78). In comparing Eq. (9.146) with Eq. (9.78), the additional term in Eq. (9.146) that is due to the attachment of the spring-mass system at $\eta = \eta_1$ was obtained by making use of transform pair 5 from Table A in Appendix A.

The terms $W(0)$, $W(\eta_1)$, $W'(0)$, $W''(0)$, and $W'''(0)$ are the displacement at $\eta = 0$, the displacement at $\eta = \eta_1$, the slope at $\eta = 0$, the second derivative of $W(\eta)$ evaluated at $\eta = 0$, and third derivative of $W(\eta)$ evaluated at $\eta = 0$, respectively. These five quantities and the nondimensional frequency coefficient Ω represent the unknown quantities that need to be determined.

The inverse transform of the first four terms of the right-hand side of Eq. (9.146) were previously determined in obtaining Eq. (9.79), and the inverse of the last term is determined from transform pairs 3 and 23 in Table A of Appendix A. Thus, we arrive at

$$
W(\eta) = W(0)Q(\Omega\eta) + W'(0)R(\Omega\eta)/\Omega + W''(0)S(\Omega\eta)/\Omega^2
$$
$$
+ W'''(0)T(\Omega\eta)/\Omega^3 + B(\Omega)W(\eta_1)T(\Omega[\eta - \eta_1])u(\eta - \eta_1)/\Omega^3 \tag{9.147}
$$

where $u(\eta)$ is the unit step function and the spatial functions $Q(\Omega\eta)$, $R(\Omega\eta)$, $S(\Omega\eta)$, and $T(\Omega\eta)$ are given by Eqs. (9.80). To determine the six unknown quantities, we make use of the boundary conditions and the fact that Eq. (9.147) is valid at $\eta = \eta_1$. Making use of these five equations, we can at best solve for Ω and four of the other five unknowns.

Characteristic Equation and Mode Shape

The boundary conditions for the beam systems shown in Figure 9.18 follow from Eqs. (9.69) if we set $J_o = M_o = 0$. Thus, the boundary conditions at $\eta = 0$ are

$$
W''(0) = B_1 W'(0)
$$
$$
W'''(0) = -K_1 W(0) \tag{9.148a}
$$

[28]For extensions of this solution and different solution methods see: M. Gürgöze, "On the eigenfrequencies of a cantilever with attached tip mass and spring-mass system," *J. Sound Vibration,* Vol. 190, No. 2, pp. 149–162 (1996); J.-S. Wu and H.-M. Chou, "Free vibration analysis of a cantilever beam carrying any number of elastically mounted point masses with the analytical-and-numerical-combined method," *J. Sound Vibration,* Vol. 213, No. 2, pp. 317–332 (1998); and M. Gürgöze, "On the alternative formulations of the frequency equation of a Bernoulli-Euler beam to which several spring-mass systems are attached in-span," *J. Sound Vibration,* Vol. 217, No. 3, pp. 585–595 (1998).

and those at $\eta = 1$ are

$$W''(1) = -B_2 W'(1)$$
$$W'''(1) = K_2 W(1) \tag{9.148b}$$

where the prime ($'$) denotes the derivative with respect to η and the nondimensional quantities B_j and K_j are given by Eqs. (9.67). Upon substituting Eqs. (9.148a) into Eq. (9.147), we obtain

$$\begin{aligned}
W(\eta) &= [Q(\Omega\eta) - K_1 T(\Omega\eta)/\Omega^3]W(0) \\
&\quad + [R(\Omega\eta)/\Omega + B_1 S(\Omega\eta)/\Omega^2]W'(0) \\
&\quad + B(\Omega)W(\eta_1)T(\Omega[\eta - \eta_1])u(\eta - \eta_1)/\Omega^3
\end{aligned} \tag{9.149}$$

We now substitute Eq. (9.149) into the boundary conditions at $\eta = 1$, which are given by Eq. (9.148b), and perform the indicated operations to obtain the following two equations in the unknowns $W'(0)$ and $W(0)$.

$$A_{11}(\Omega)\Omega W(0) + A_{12}(\Omega)W'(0) = -B(\Omega)W(\eta_1)E_1(\Omega[1 - \eta_1])/\Omega^2$$
$$A_{21}(\Omega)\Omega W(0) + A_{22}(\Omega)W'(0) = B(\Omega)W(\eta_1)E_2(\Omega[1 - \eta_1])/\Omega^2 \tag{9.150}$$

In Eqs. (9.150),

$$\begin{aligned}
A_{11}(\Omega) &= (1 - a_1 b_2)S(\Omega) - a_1 R(\Omega) + b_2 T(\Omega) \\
A_{12}(\Omega) &= T(\Omega) + (b_1 + b_2)Q(\Omega) + b_1 b_2 R(\Omega) \\
A_{21}(\Omega) &= R(\Omega) - (a_1 + a_2)Q(\Omega) + a_1 a_2 T(\Omega) \\
A_{22}(\Omega) &= (1 - a_2 b_1)S(\Omega) + b_1 T(\Omega) - a_2 R(\Omega) \\
E_1(\Omega) &= R(\Omega[1 - \eta_1]) + b_2 S(\Omega[1 - \eta_1]) \\
E_2(\Omega) &= a_2 T(\Omega[1 - \eta_1]) - Q(\Omega[1 - \eta_1])
\end{aligned} \tag{9.151a}$$

and

$$a_j = \frac{K_j}{\Omega^3} \quad \text{and} \quad b_j = \frac{B_j}{\Omega} \quad j = 1, 2 \tag{9.151b}$$

Solving Eq. (9.150) for $W(0)$ and $W'(0)$ yields

$$W(0) = \frac{-B(\Omega)W(\eta_1)}{D_o(\Omega)\Omega^3}[E_1(\Omega)A_{22}(\Omega) + E_2(\Omega)A_{12}(\Omega)]$$
$$W'(0) = \frac{B(\Omega)W(\eta_1)}{D_o(\Omega)\Omega^2}[E_2(\Omega)A_{11}(\Omega) + E_1(\Omega)A_{21}(\Omega)] \tag{9.152}$$

where

$$D_o(\Omega) = A_{11}(\Omega)A_{22}(\Omega) - A_{12}(\Omega)A_{21}(\Omega) \tag{9.153}$$

Upon substituting Eqs. (9.152) into Eq. (9.149), we arrive at

$$\begin{aligned}
W(\eta) &= \frac{B(\Omega)W(\eta_1)}{\Omega^3}\{[-E_1(\Omega)A_{22}(\Omega) - E_2(\Omega)A_{12}(\Omega)][Q(\Omega\eta) \\
&\quad - a_1 T(\Omega\eta)]/D_o(\Omega) \\
&\quad + [E_2(\Omega)A_{11}(\Omega) + E_1(\Omega)A_{21}(\Omega)][R(\Omega\eta) + b_1 S(\Omega\eta)]/D_o(\Omega) \\
&\quad + T(\Omega[\eta - \eta_1])u(\eta - \eta_1)\}
\end{aligned} \tag{9.154}$$

Equation (9.154) must be valid at $\eta = \eta_1$. Therefore, setting $\eta = \eta_1$ in Eq. (9.154), noting that $T(0) = 0$ from Eqs. (9.80), we obtain the characteristic equation for a uniform beam carrying an undamped single degree-of-freedom system for the boundary conditions given by Eqs. (9.148):

$$B(\Omega_n)\{[-E_1(\Omega_n[1 - \eta_1])A_{22}(\Omega_n) - E_2(\Omega_n[1 - \eta_1])A_{12}(\Omega_n)]$$
$$[Q(\Omega_n\eta_1) - a_{1n}T(\Omega_n\eta_1)] + [E_2(\Omega_n[1 - \eta_1])A_{11}(\Omega_n)$$
$$+ E_1(\Omega_n[1 - \eta_1])A_{21}(\Omega_n)][R(\Omega_n\eta_1) +$$
$$b_{1n}S(\Omega_n\eta_1)]\} - D_o(\Omega_n)\Omega_n^3 = 0 \qquad (9.155)$$

where

$$a_{jn} = \frac{K_j}{\Omega_n^3} \quad \text{and} \quad b_{jn} = \frac{B_j}{\Omega_n} \quad j = 1, 2 \qquad (9.156)$$

and in the definitions of $A_{ij}(\Omega_n)$, a_i and b_i are replaced by a_{in} and b_{in}, respectively. As mentioned previously, when $M_s = 0$, $B(\Omega) = 0$ and Eq. (9.155) simplifies to

$$D_o(\Omega_n) = 0 \qquad (9.157)$$

which, upon expansion, is identical to the characteristic equation given by Eq. (9.85) with $M_o = J_o = 0$.

For the system shown in Figure 9.18c, the mode shape associated with the nondimensional frequency coefficient Ω_n is given by

$$W_n(\eta) = [-E_1(\Omega_n)A_{22}(\Omega_n) - E_2(\Omega_n)A_{12}(\Omega_n)][Q(\Omega_n\eta) - a_1T(\Omega_n\eta)]/D_o(\Omega_n)$$
$$+ [E_2(\Omega_n)A_{11}(\Omega_n) + E_1(\Omega_n)A_{21}(\Omega_n)][R(\Omega_n\eta) + b_1S(\Omega_n\eta)]/D_o(\Omega_n)$$
$$+ T(\Omega_n[\eta - \eta_1])u(\eta - \eta_1) \qquad (9.158)$$

where, for convenience, we have normalized the mode shape to remove the arbitrary constant $W(\eta_1)$; that is,

$$W_n(\eta) = \frac{W(\eta)}{B(\Omega_n)W(\eta_1)/\Omega_n^3}$$

From Eq. (9.137), the mode shape of the mass M_s is

$$Z_{on} = W_n(\eta_1)\left(1 - \frac{M_{so}}{K_s}\Omega_n^4\right)^{-1} \qquad (9.159)$$

Orthogonality of the Modes

We now determine if the mode shape given by Eq. (9.158) is an orthogonal function in the sense of Eq. (9.66). We start with Eq. (9.139) and replace $W(\eta)$ with $W_n(\eta)$ and Ω with Ω_n. Then, following the procedure used to obtain Eq. (9.72) through Eq. (9.77), we arrive at

$$(\Omega_m^4 - \Omega_n^4)\int_0^1 W_m(\eta)W_n(\eta)d\eta$$
$$+ (B(\Omega_m) - B(\Omega_n))W_m(\eta_1)W_n(\eta_1) = 0 \qquad (9.160)$$

Since Eq. (9.160) is not symmetric in $W_n(\eta)$ and $W_m(\eta)$ as in Eq. (9.76), it cannot be put in the form of Eq. (9.66). Hence, $W_n(\eta)$ is not an orthogonal function and, therefore, for a beam system with a spring-mass system attached at an interior point, the modes do not form an orthogonal set.

When $M_s \to \infty$—that is, we have a system with a spring directly attached to the beam at $\eta = \eta_1$ as shown in Figure 9.18a—we find from Eq. (9.144) that $B(\Omega_n) = B(\Omega_m) = -K_s$ and Eq. (9.160) reduces to

$$(\Omega_m^4 - \Omega_n^4) \int_0^1 W_m(\eta)W_n(\eta)d\eta = 0 \tag{9.161}$$

which leads to

$$\int_0^1 W_m(\eta)W_n(\eta)d\eta = \delta_{nm}N_n \tag{9.162a}$$

where δ_{nm} is the Kronecker delta and

$$N_n = \int_0^1 W_n^2(\eta)d\eta \tag{9.162b}$$

Since $W_n(\eta)$ and η are nondimensional quantities, N_n is a nondimensional quantity. Thus, the mode shapes for a system where a translation spring is directly attached to a beam form an orthogonal set.

When $K_s \to \infty$; that is, we have a system with a mass directly attached to the beam at $\eta = \eta_1$ as shown in Figure 9.18a, we find from Eq. (9.142) that $B(\Omega_n) = M_{so}\Omega_n^4$ and Eq. (9.160) reduces to

$$(\Omega_m^4 - \Omega_n^4)\left[\int_0^1 W_m(\eta)W_n(\eta)d\eta + M_{so}W_m(\eta_1)W_n(\eta_1) \right] = 0 \tag{9.163}$$

which leads to

$$\int_0^1 W_m(\eta)W_n(\eta)d\eta + M_{so}W_m(\eta_1)W_n(\eta_1) = \delta_{nm}N_n \tag{9.164a}$$

where

$$N_n = \int_0^1 W_n^2(\eta)d\eta + M_{so}W_n^2(\eta_1) \tag{9.164b}$$

Thus, the mode shapes for a mass directly attached to a beam are orthogonal functions.

We now examine Eq. (9.155) for three different types of boundary conditions: (i) beam pinned at each boundary; (ii) beam clamped at each boundary; and (iii) beam clamped at one end and free at the other end (cantilever). Each of these sets of results is valid for the case of a beam with an attached single degree-of-freedom system at $\eta = \eta_1$, a spring $\eta = \eta_1$, or a mass at $\eta = \eta_1$. For the three cases shown in Figure 9.18c, Figure 9.18b, and Figure 9.18a, the expression for $B(\Omega)$ is given by Eq. (9.140), Eq. (9.142), and Eq. (9.144), respectively, To specialize the general results for each set of boundary conditions, we follow the procedure given in Section 9.3.3 by setting the values of a_{jn} and b_{jn} to either zero or infinity. When any of the values of a_{jn}

and b_{jn} are to be set to infinity, we first divide the numerator and denominator of Eqs. (9.155) and (9.158) by those quantities and then take the limit.

Beam pinned at each boundary.   For this case, the boundary conditions correspond to

$$\eta = 0: a_{1n} \to \infty \quad \text{and} \quad b_{1n} = 0$$
$$\eta = 1: a_{2n} \to \infty \quad \text{and} \quad b_{2n} = 0 \tag{9.165}$$

Then Eq. (9.155), the characteristic equation in the general case, reduces to

$$B(\Omega_n)\{T(\Omega_n)[T(\Omega_n\eta_1)T(\Omega_n[1-\eta_1]) + R(\Omega_n\eta_1)R(\Omega_n[1-\eta_1])]$$
$$- R(\Omega_n)[T(\Omega_n\eta_1)R(\Omega_n[1-\eta_1]) + R(\Omega_n\eta_1)T(\Omega_n[1-\eta_1])]\}$$
$$- \Omega_n^3 [R^2(\Omega_n) - T^2(\Omega_n)] = 0 \tag{9.166}$$

where Ω_n is a root of the characteristic equation. From Eq. (9.158), the corresponding mode shape of the beam is given by

$$W_n(\eta) = C_{1n}T(\Omega_n\eta) + C_{2n}R(\Omega_n\eta) + T(\Omega_n[\eta - \eta_1])u(\eta - \eta_1) \tag{9.167a}$$

where

$$C_{1n} = \frac{T(\Omega_n)T(\Omega_n[1-\eta_1]) - R(\Omega_n)R(\Omega_n[1-\eta_1])}{R^2(\Omega_n) - T^2(\Omega_n)}$$

$$C_{2n} = \frac{T(\Omega_n)R(\Omega_n[1-\eta_1]) - R(\Omega_n)T(\Omega_n[1-\eta_1])}{R^2(\Omega_n) - T^2(\Omega_n)} \tag{9.167b}$$

Beam clamped at each boundary.   For this case, the boundary conditions are obtained by setting

$$\eta = 0: a_{1n} \to \infty \quad \text{and} \quad b_{1n} \to \infty$$
$$\eta = 1: a_{2n} \to \infty \quad \text{and} \quad b_{2n} \to \infty \tag{9.168}$$

Then Eq. (9.155), the characteristic equation in the general case, reduces to

$$B(\Omega_n)\{T(\Omega_n\eta_1)[R(\Omega_n)T(\Omega_n[1-\eta_1]) - S(\Omega_n)S(\Omega_n[1-\eta_1])]$$
$$+ S(\Omega_n\eta_1)[T(\Omega_n)S(\Omega_n[1-\eta_1]) - S(\Omega_n)T(\Omega_n[1-\eta_1])]\}$$
$$- \Omega_n^3 [S^2(\Omega_n) - R(\Omega_n)T(\Omega_n)] = 0 \tag{9.169}$$

where Ω_n is a root of the characteristic equation. From Eq. (9.158), the corresponding mode shape is given by

$$W_n(\eta) = C_{3n}T(\Omega_n\eta) + C_{4n}S(\Omega_n\eta) + T(\Omega_n[\eta - \eta_1])u(\eta - \eta_1) \tag{9.170a}$$

where

$$C_{3n} = \frac{R(\Omega_n)T(\Omega_n[1-\eta_1]) - S(\Omega_n)S(\Omega_n[1-\eta_1])}{S^2(\Omega_n) - R(\Omega_n)T(\Omega_n)}$$

$$C_{4n} = \frac{T(\Omega_n)S(\Omega_n[1-\eta_1]) - S(\Omega_n)T(\Omega_n[1-\eta_1])}{S^2(\Omega_n) - R(\Omega_n)T(\Omega_n)} \tag{9.170b}$$

Beam clamped at one end and free at the other end (cantilever). For this case, the boundary conditions are obtained by setting

$$\eta = 0: a_{1n} \to \infty \quad \text{and} \quad b_{1n} \to \infty$$
$$\eta = 1: a_{2n} = 0 \quad \text{and} \quad b_{2n} = 0 \tag{9.171}$$

Then Eq. (9.155), the characteristic equation in the general case, reduces to

$$\begin{aligned} B(\Omega_n)\{T(\Omega_n\eta_1)[T(\Omega_n)R(\Omega_n[1 - \eta_1]) - Q(\Omega_n)Q(\Omega_n[1 - \eta_1])] \\ + S(\Omega_n\eta_1)[R(\Omega_n)Q(\Omega_n[1 - \eta_1]) - Q(\Omega_n)R(\Omega_n[1 - \eta_1])]\} \\ - \Omega_n^3 [Q^2(\Omega_n) - R(\Omega_n)T(\Omega_n)] = 0 \end{aligned} \tag{9.172}$$

where Ω_n is a root of the characteristic equation. From Eq. (9.158), the corresponding mode shape is

$$W_n(\eta) = C_{5n}T(\Omega_n\eta) + C_{6n}S(\Omega_n\eta) + T(\Omega_n[\eta - \eta_1])u(\eta - \eta_1) \tag{9.173a}$$

where

$$C_{5n} = \frac{T(\Omega_n)R(\Omega_n[1 - \eta_1]) - Q(\Omega_n)Q(\Omega_n[1 - \eta_1])}{Q^2(\Omega_n) - R(\Omega_n)T(\Omega_n)}$$

$$C_{6n} = \frac{R(\Omega_n)Q(\Omega_n[1 - \eta_1]) - Q(\Omega_n)R(\Omega_n[1 - \eta_1])}{Q^2(\Omega_n) - R(\Omega_n)T(\Omega_n)} \tag{9.173b}$$

Equation (9.172) is also valid at $\eta_1 = 1$; that is, when the single degree-of-freedom system is attached to the free end of the beam. In this case, since $R(0) = 0$ and $Q(0) = 1$, Eq. (9.172) reduces to

$$B(\Omega_n)[S(\Omega_n)R(\Omega_n) - T(\Omega_n)Q(\Omega_n)]$$
$$- \Omega_n^3 [Q^2(\Omega_n) - R(\Omega_n)T(\Omega_n)] = 0 \tag{9.174}$$

Equations (9.166), (9.169), and (9.172) are used to determine the first natural frequency coefficient Ω_1 as a function of the mass ratio M_{so} for $\eta_1 = 0.5$ and $K_s = 50$ and $K_s = 5000$. The results are plotted in Figure 9.19. It is found numerically that when $K_s = 5000$, Ω_1 very closely approximates the values obtained for the case where the mass is attached directly to the beam for all three types of boundary conditions considered. Therefore, the curves shown using the solid lines in Figure 9.19 are used to estimate the first natural frequency coefficient for a beam with a mass attached at $\eta = \eta_1$. Some values[29,30] of the natural frequency coefficient for the case where the mass is

[29]Additional numerical values using different solution methods are found in the following studies: M. S. Hess, "Vibration frequencies for uniform beams with central mass," *J. Appl. Mech.* Vol. 31, pp. 556–557 (1964); K. H. Low, "On the eigenfrequencies for mass loaded beams under classical boundary conditions," *J. Sound Vib.,* Vol. 215, No. 2, pp. 381–389 (1998); S. Naguleswaran, "Lateral vibration of an Euler-Bernoulli beam carrying a particle at an intermediate point," *J. Sound Vib.,* Vol. 227, No. 1, pp. 205–214 (1999); C. H. Chang, "Free vibration of a simply supported beam carrying a rigid mass at the middle," *J. Sound Vib.,* Vol. 237, No. 4, pp. 733–744 (2000); Ö. Turhan, "On the Fundamental Frequency of Beams Carrying a Point Mass: Rayleigh Approximations Versus Exact Solutions," *J. Sound Vibration,* Vol. 230, No. 2, pp. 449–459 (2000).

[30]Extension of these results to tapered beams is found in N. M. Auciello, "On the natural frequencies of tapered beams with attached inertia elements," *J. Sound Vib.,* Vol. 199, No. 3, pp. 522–530 (1997).

FIGURE 9.19
Lowest natural frequency coefficients for a beam with an undamped single degree-of-freedom system attached at $\eta_1 = 0.5$ as a function of the properties of the attached system for different boundary conditions.

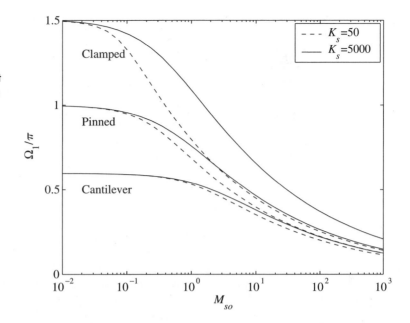

attached directly to the beam are also given in Tables 9.5 and 9.6 for two different boundary conditions. The effect of the single degree-of-freedom system on the mode shape associated with the lowest natural frequency is shown in Figure 9.20 for $K_s = 50$ and $M_{so} = 10$ as a function of location of the attachment point. These values correspond to a beam whose stiffness is about one-sixteenth that of the attached spring and whose mass is one-tenth that of the attached mass. It is seen that for this particular set of values the attachment point of the single degree-of-freedom system significantly affects the beam's mode shape, with the mode shape of the beam with the end mass being closest to the mode shape of the beam without the end mass.

As discussed in Chapter 2, in many situations where an inertia element is attached to a beam, the stiffness is taken into account to establish an equivalent single degree-of-freedom system. This situation is revisited in the context

TABLE 9.5
Lowest Natural Frequency Coefficient for Clamped/Clamped and Pinned/Pinned Beams Carrying Different Masses at Their Midpoints, $\eta_1 = 0.5$

	Ω_1/π	
M_o/m_o	Pinned/pinned [Eq. (9.166)]	Clamped/clamped [Eq. (9.169)]
0	1.0000	1.5056
0.2	0.9190	1.3582
1	0.7586	1.0943
5	0.5474	0.7783
10	0.4656	0.6603
100	0.2646	0.3743

TABLE 9.6

Lowest Natural Frequency Coefficient for a Beam Pinned at Each End and Carrying a Mass at $\eta = \eta_1$

| η_1 | Ω_1/π | | |
	$M_o/m_o = 0.1$	$M_o/m_o = 1$	$M_o/m_o = 10$
.2	0.9832	0.8691	0.5747
.3	0.9694	0.8049	0.5061
.4	0.9591	0.7697	0.4748
.5	0.9553	0.7586	0.4656

of the beam system shown in Figure 9.18c to point out when it is reasonable to neglect the beam inertia and when it is not. First, we consider the determination of the natural frequency of an equivalent single degree-of-freedom system. This is done by using the static stiffness values given by Cases 4, 5, and 6 of Table 2.3 for the cantilever, pinned/pinned and clamped/clamped beams, respectively. In each case, the approximation obtained for the first natural frequency of the system shown in Table 2.3 is compared to the natural frequency obtained when the inertia of the beam is taken into account. We note from the discussion on spring combinations in series shown in Figure 2.8a and from Eq. (a) of Example 2.3 that the equivalent spring constant k_e for a spring k_s attached to a beam of spring constant k_{beam} is

$$k_e = \left(\frac{1}{k_s} + \frac{1}{k_{beam}}\right)^{-1} = \frac{k_s}{1 + k_s/k_{beam}} \tag{9.175a}$$

From Table 2.3, we find that

$$k_{beam} = \frac{\alpha EI}{L^3}$$

where $\alpha = 3$ when a loading is applied to the end of a cantilever beam, $\alpha = 48$ when a loading is applied to the midpoint of a beam pinned at each end,

FIGURE 9.20

Mode shape associated with the lowest natural frequency of a cantilever beam as a function of the location of the attachment point of the single degree-of-freedom system for $K_s = 50$ and $M_{so} = 10$. Dashed line indicates the static equilibrium position of mass M_s.

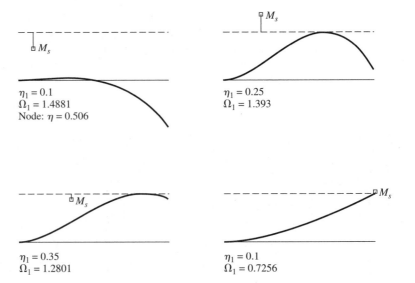

$\eta_1 = 0.1$
$\Omega_1 = 1.4881$
Node: $\eta = 0.506$

$\eta_1 = 0.25$
$\Omega_1 = 1.393$

$\eta_1 = 0.35$
$\Omega_1 = 1.2801$

$\eta_1 = 0.1$
$\Omega_1 = 0.7256$

and $\alpha = 192$ when a loading is applied to the midpoint of a beam clamped at each end. Then, from Eqs. (9.175a) and (9.138), we have

$$k_e = \frac{k_s}{1 + k_s L^3 / \alpha EI} = \frac{k_s}{1 + K_s / \alpha} \tag{9.175b}$$

Thus, the natural frequency of the equivalent single degree-of-freedom system obtained from the equivalent spring constant is given by

$$\omega_n = \sqrt{\frac{k_e}{M_s}} = \sqrt{\frac{k_s}{M_s} \left(\frac{1}{1 + K_s / \alpha} \right)} \tag{9.176}$$

To determine the first natural frequency when the beam's inertia is taken into account, we use Eq. (9.138), and find that the first natural frequency is given by

$$\omega_e = \Omega_1^2 \sqrt{\frac{EI}{m_o L^3}} \tag{9.177}$$

where Ω_1 is the first nondimensional frequency coefficient obtained by solving the appropriate characteristic equation. The percentage error ϵ between the natural frequency for a single degree-of-freedom system and the first natural frequency of the beam system in Figure 9.18c is

$$\epsilon = 100 \left(\frac{\omega_n}{\omega_e} - 1 \right) = 100 \left(\frac{1}{\Omega_1^2} \sqrt{\frac{m_o k_s L^3}{EI M_s} \left(\frac{1}{1 + K_s / \alpha} \right)} - 1 \right)$$

$$= 100 \left(\frac{1}{\Omega_1^2} \sqrt{\frac{K_s}{M_{so}} \left(\frac{1}{1 + K_s / \alpha} \right)} - 1 \right) \% \tag{9.178}$$

When K_s is very large, that is, when the mass is directly attached to the beam, Eq. (9.178) simplifies to

$$\epsilon = 100 \left(\frac{1}{\Omega_1^2} \sqrt{\frac{\alpha}{M_{so}}} - 1 \right) \% \tag{9.179}$$

The results from Eq. (9.178) are plotted in Figure 9.21, where it is seen that neglecting the beam inertia and only using the static spring constant for the beam always results in an over-estimate of the first natural frequency of the beam system shown in Figure 9.18c. In order to keep the error below, say 5%, for very large K_s one must have M_{so} greater than 4.7 for a beam pinned at each end, greater than 3.6 for a beam clamped at each end, and greater than 2.3 for a cantilever beam. As K_s decreases, it is seen from Figure 9.21 that these mass ratio requirements decrease. For the beam pinned at each end and for large K_s, it is seen that when $M_o/m_o = 1$ the error is 22%, whereas for the beam clamped at each end and for large K_s, it is seen that when $M_o/m_o = 1$ the error is 17.2%. These results lead to the following design guideline.

Design Guideline: When a beam of mass m_o is pinned at each boundary and has attached to it a single degree-of-freedom system with mass M_s and stiffness k_s, then this system is approximated by a system where the inertia of the beam is neglected when the mass ratio M_s/m_o is greater than 4.7 and K_s is very large. The error of this approximation under these conditions is less than 5% and the error decreases as K_s decreases and/or M_s/m_o increases. For a beam clamped at each boundary, these conditions hold true except that the mass ratio M_s/m_o must be greater than 3.6, and for a cantilever beam, the mass ratio M_s/m_o must be greater than 2.3.

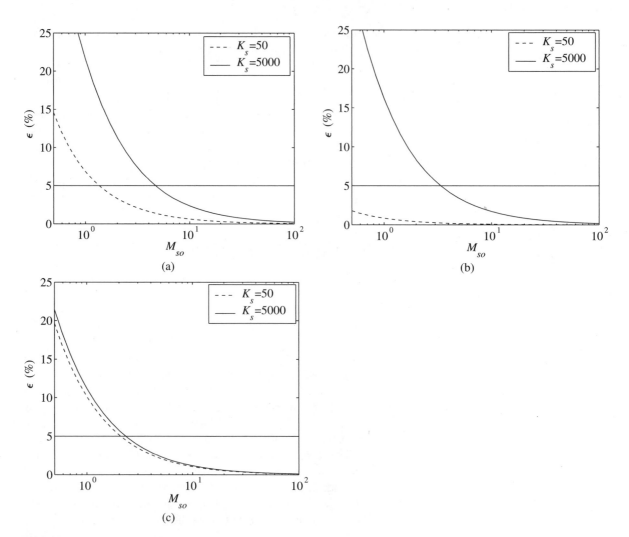

FIGURE 9.21
Percentage error in the first natural frequency when the beam inertia of the system shown in Figure 9.18c is neglected: (a) beam pinned at each end and single degree-of-freedom system attached at $\eta_1 = 0.5$; (b) beam clamped at each end and single degree-of-freedom system attached at $\eta_1 = 0.5$; and (c) cantilever beam with single degree-of-freedom system attached at the free end $\eta_1 = 1$.

We now consider the natural frequency coefficients and mode shapes of a beam restrained by a spring only, as shown in Figure 9.18a. The natural frequency coefficients Ω_1 for three sets of boundary conditions are determined from Eqs. (9.166), (9.169), and (9.172) as a function of K_s for different values of η_1. The results are plotted in Figure 9.22. Some representative examples of mode shapes for the three sets of boundary conditions are shown in Figure 9.23. In Figure 9.22a, it is seen from Table 9.2 that as K_s becomes very large and η_1 approaches 0.5, Ω_1 approaches the second natural frequency coefficient of a beam pinned at each end. In other words, the

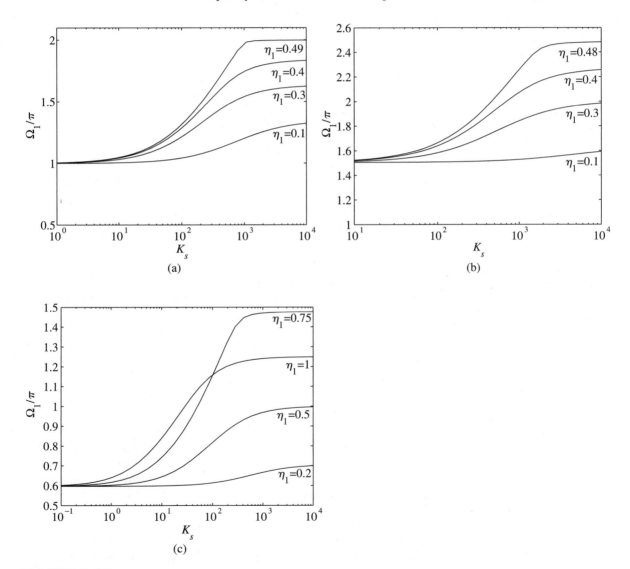

FIGURE 9.22
Lowest natural frequency coefficient of a beam restrained by a spring as a function of K_s for different attachment locations: (a) pinned/pinned beam; (b) clamped/clamped beam; and (c) cantilever beam.

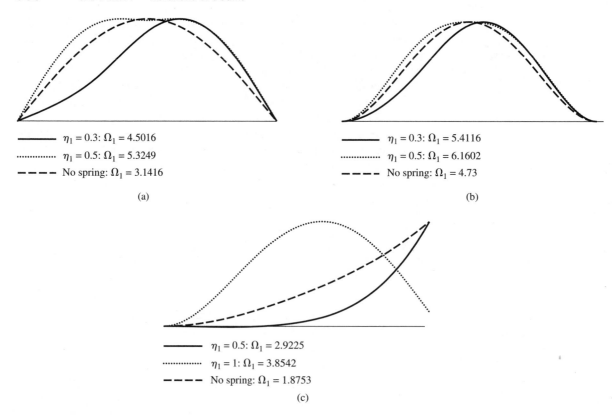

$\underline{\qquad}$ $\eta_1 = 0.3$: $\Omega_1 = 4.5016$

$\cdots\cdots$ $\eta_1 = 0.5$: $\Omega_1 = 5.3249$

$- - -$ No spring: $\Omega_1 = 3.1416$

(a)

$\underline{\qquad}$ $\eta_1 = 0.3$: $\Omega_1 = 5.4116$

$\cdots\cdots$ $\eta_1 = 0.5$: $\Omega_1 = 6.1602$

$- - -$ No spring: $\Omega_1 = 4.73$

(b)

$\underline{\qquad}$ $\eta_1 = 0.5$: $\Omega_1 = 2.9225$

$\cdots\cdots$ $\eta_1 = 1$: $\Omega_1 = 3.8542$

$- - -$ No spring: $\Omega_1 = 1.8753$

(c)

FIGURE 9.23
First mode shape of a beam restrained by a spring with $K_s = 400$ for different attachment locations:
(a) pinned/pinned beam; (b) clamped/clamped beam; and (c) cantilever beam. The dashed lines are for $K_s = 0$.

constrained beam acts like a pinned/pinned beam with half the original length. In Figure 9.22b, we have from Table 9.2 that as K_s becomes very large and η_1 approaches 0.5, Ω_1 approaches the second natural frequency coefficient of a beam clamped at each end. In Figure 9.22c, it is seen that as K_s becomes very large and η_1 approaches 1, Ω_1 approaches the first natural frequency coefficient of a beam clamped at one end and pinned at the other end. However, when the spring is located at $\eta_1 = 0.75$ and K_s is very large, Ω_1 approaches the second natural frequency coefficient of a beam clamped at one end and free at the other end. This special case is explained by noting that the spring is located very close to $\eta_1 = 0.783$, which is at the node point of the second mode of cantilever beam as shown in Figure 9.12.

9.3.5 Effects of an Axial Force and an Elastic Foundation on the Natural Frequency[31]

We shall determine the effects that an axial force $p(x,t)$ and an elastic foundation k_f have on the natural frequency coefficient. Axial forces arise in beam models of many vibratory systems including rotating machinery, where the centrifugal forces are the source of the axial forces or in vertical structures such as water towers. If we assume that $p(x,t) = p_o$ is a constant and that the beam has a uniform cross section and uniform material along its length, then Eq. (9.40) is written as

$$EI\frac{\partial^4 w}{\partial x^4} - p_o\frac{\partial^2 w}{\partial x^2} + k_f w(x,t) + \rho A\frac{\partial^2 w}{\partial t^2} = f(x,t) \qquad (9.180)$$

where p_o is a tensile force. Following the procedure used in Section 9.3.3, we assume that the externally applied transverse load is zero—that is, $f(x,t) = 0$—and that the displacement is of the form given by Eq. (9.58). Then, Eq. (9.180) leads to the spatial equation

$$\frac{d^4 W}{d\eta^4} - \hat{P}\frac{d^2 W}{d\eta^2} + (K_f - \Omega^4)W = 0 \qquad (9.181)$$

where we have employed the notation of Eq. (9.67) and introduced the quantities

$$\hat{P} = \frac{p_o L^2}{EI} \quad \text{and} \quad K_f = \frac{k_f L^4}{EI} \qquad (9.182)$$

Pinned/Pinned Beam

Rather than find a general solution to Eq. (9.181), we shall only obtain the solution to a beam that is pinned at each of its ends. This will be sufficient for us to illustrate the effects that p_o and k_f have on the natural frequency coefficient. The boundary conditions for a beam pinned at each end are written from Case 2 of Table 9.1 as

$$W(0) = W''(0) = 0$$
$$W(1) = W''(1) = 0 \qquad (9.183)$$

We determined in Eq. (9.103) that these boundary conditions are satisfied by the spatial functions

$$W(\eta) = \sin(n\pi\eta) \quad n = 1, 2, ... \qquad (9.184)$$

The substitution of Eq. (9.184) into Eq. (9.181) leads to the characteristic equation

$$(n\pi)^4 + \hat{P}(n\pi)^2 + K_f - \Omega^4 = 0 \qquad (9.185a)$$

---

[31]For a more compete treatment of this topic, see the following: F. J. Shaker, "Effect of Axial Load on Mode Shapes and Frequencies of Beams," Lewis Research Center Report NASA TN D-8109 (December 1975); G. C. Nihous, "On the continuity of the boundary value problem for vibrating free-free straight beams under axial load," *J. Sound Vibration,* Vol. 200, No. 1, pp. 110–119 (1997); and M. A. De Rosa and M. J. Maurizi, "The influence of concentrated masses and Pasternak soil on the free vibrations of Euler beams—exact solution," *J. Sound Vibration,* Vol. 212, No. 4, pp. 573–581 (1998).

which yields

$$\Omega_n = \sqrt[4]{(n\pi)^4 + \hat{P}(n\pi)^2 + K_f} \quad n = 1,2, \dots \tag{9.185b}$$

where Ω_n is the nth natural frequency coefficient. Since $K_f > 0$, we see that the presence of the elastic foundation always increases the natural frequencies of the beam, and that a tensile axial force ($\hat{P} > 0$) always increases the natural frequencies while a compressive axial force ($\hat{P} < 0$) always decreases them. For compressive axial forces one must make sure that the buckling limits of the beam are not exceeded. The buckling limits are also a function of the boundary conditions.

9.3.6 Tapered Beams[32]

We shall now remove the assumption that the beam has a constant cross-section and consider beams whose cross-section varies with the position along the length of the beam, as shown in Figure 9.24. This permits us to model such systems as fly fishing rods,[33] baseball bats, and chimneys.[34] We assume that $p = k_f = f(x,t) = 0$ in Eq. (9.40) and that the solution is of the form given by Eq. (9.58). In addition, we assume that

$$A = A_o a(\eta)$$
$$I = I_o i(\eta) \tag{9.186}$$

where A_o and I_o are constants, and $a(\eta)$ and $i(\eta)$ are nondimensional functions of η. Then, Eq. (9.40) and Eqs. (9.186) lead to the spatial equation

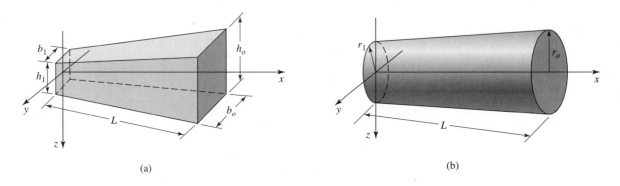

(a) (b)

FIGURE 9.24
Geometry of a tapered beam: (a) rectangular cross-section and (b) circular cross-section.

[32]For a general treatment of tapered beams see E. B. Magrab, *ibid*, pp. 153–168.

[33]J. A. Hoffmann and M. R. Hooper, "Fly Rod Response," *J. Sound Vibration*, Vol. 209, No. 3, pp. 537–541 (1998).

[34]K. Güler, "Free Vibrations and Modes of Chimneys on an Elastic Foundation," *J. Sound Vibration*, Vol. 218, No. 3, pp. 541–547 (1998).

$$\frac{d^2}{d\eta^2}\left[i(\eta)\frac{d^2W}{d\eta^2}\right] - \Omega_o^4 a(\eta)W = 0 \qquad (9.187)$$

where we have used the notation of Eqs. (9.67) and introduced the quantity

$$\Omega_o^4 = \frac{\rho A_o \omega^2 L^4}{EI_o} \qquad (9.188)$$

Consider the double-tapered beam with rectangular cross-section shown in Figure 9.24a; that is, a beam that tapers in both the xz-plane and the xy-plane. Let us assume that the taper ratios $\alpha = h_1/h_o < 1$ in the xz-plane and $\beta = b_1/b_o < 1$ in the xy-planes are equal, that is, $\alpha = \beta$, where h_1, h_o, b_1, and b_o, are as defined in Figure 9.24a. For this beam, the functions $a(\eta)$ and $i(\eta)$ are, respectively,

$$a(\eta) = [\alpha + (1 - \alpha)\eta][\beta + (1 - \beta)\eta] = [\alpha + (1 - \alpha)\eta]^2$$
$$i(\eta) = [\alpha + (1 - \alpha)\eta]^3[\beta + (1 - \beta)\eta] = [\alpha + (1 - \alpha)\eta]^4 \qquad (9.189)$$

and

$$I_o = \frac{b_o h_o^3}{12}$$

$$A_o = b_o h_o \qquad (9.190)$$

On substituting Eqs. (9.189) into (9.187), we obtain

$$\frac{d^2}{d\eta^2}\left[[\alpha + (1 - \alpha)\eta]^4 \frac{d^2W}{d\eta^2}\right] - \Omega_o^4[\alpha + (1 - \alpha)\eta]^2 W = 0 \qquad (9.191)$$

To solve Eq. (9.191), we introduce the transformation from the spatial variable η to another spatial variable φ

$$\varphi = \alpha + (1 - \alpha)\eta \qquad (9.192)$$

and note that

$$\frac{d}{d\eta} = (1 - \alpha)\frac{d}{d\varphi}$$

$$\frac{d^2}{d\eta^2} = (1 - \alpha)^2 \frac{d^2}{d\varphi^2} \qquad (9.193)$$

After substituting Eq. (9.192) into Eq. (9.191) and making use of Eqs. (9.193), we arrive at

$$\frac{d^2}{d\varphi^2}\left[\varphi^4 \frac{d^2W}{d\varphi^2}\right] - \lambda^4 \varphi^2 W = 0 \qquad (9.194)$$

where

$$\lambda = \frac{\Omega_o}{(1 - \alpha)}$$

The general solution to Eq. (9.191) is[35]

$$W(\varphi) = \varphi^{-1} [A_1 J_2(2\lambda\sqrt{\varphi}) + A_2 Y_2(2\lambda\sqrt{\varphi})$$
$$+ A_3 I_2(2\lambda\sqrt{\varphi}) + A_4 K_2(2\lambda\sqrt{\varphi})] \tag{9.195}$$

where $J_2(z)$ and $Y_2(z)$ are the Bessel functions of the first and second kind, respectively, and $I_2(z)$ and $K_2(z)$ are the modified Bessel functions of the first and second kind, respectively.

The boundary conditions for a tapered beam is chosen from those given in Table 9.1, except that we rewrite them in terms of the independent variable φ using Eqs. (9.186), (9.189), and (9.192). We now illustrate these results with two examples.

EXAMPLE 9.3 Natural frequencies of a tapered cantilever beam

We shall determine the characteristic equation for a tapered cantilever beam clamped $\eta = 1$. In terms of the spatial variable φ, the boundary conditions at $\eta = 1$ correspond to $\varphi = 1$. Thus, the boundary conditions at $\eta = 1$ are

$$W(1) = 0$$
$$\frac{dW}{d\eta}\bigg|_{\eta=1} = 0 \rightarrow (1-\alpha)\frac{dW}{d\varphi}\bigg|_{\varphi=1} = 0 \rightarrow \frac{dW(1)}{d\varphi} = 0 \tag{a}$$

At the free end, $\eta = 0$, which corresponds to $\varphi = \alpha$, we have

$$\frac{d^2W}{d\eta^2} = 0 \rightarrow (1-\alpha)^2 \frac{d^2W(\alpha)}{d\varphi^2} = 0 \rightarrow \frac{d^2W(\alpha)}{d\varphi} = 0$$

$$\frac{d}{d\eta}\left(EI\frac{d^2W}{d\eta^2}\right) = 0 \rightarrow (1-\alpha)^3 EI_o \frac{d}{d\varphi}\left(\varphi^4 \frac{d^2W}{d\varphi^2}\right) = 0$$

$$\rightarrow 4\varphi^3 \frac{d^2W(\alpha)}{d\varphi^2} + \varphi^4 \frac{d^3W(\alpha)}{d\varphi^3} = 0 \rightarrow \frac{d^3W(\alpha)}{d\varphi^3} = 0 \tag{b}$$

where we have used the first boundary condition of Eqs. (b) to simplify the second boundary condition.

On substituting Eq. (9.195) into the boundary conditions given by Eqs. (a) and (b) and setting the determinant of the coefficients A_j to zero, we obtain the following equation:

$$\begin{vmatrix} J_5(2\lambda\sqrt{\alpha}) & Y_5(2\lambda\sqrt{\alpha}) & -I_5(2\lambda\sqrt{\alpha}) & K_5(2\lambda\sqrt{\alpha}) \\ J_4(2\lambda\sqrt{\alpha}) & Y_4(2\lambda\sqrt{\alpha}) & I_4(2\lambda\sqrt{\alpha}) & K_4(2\lambda\sqrt{\alpha}) \\ J_2(2\lambda) & Y_2(2\lambda) & I_2(2\lambda) & K_2(2\lambda) \\ J_3(2\lambda) & Y_3(2\lambda) & -I_3(2\lambda) & K_3(2\lambda) \end{vmatrix} = 0 \tag{c}$$

[35]E. B. Magrab, *ibid.*, p. 26.

In arriving at Eq. (c), we used the following relations:

$$\frac{d}{d\varphi}\left[\varphi^{-n/2} J_n(2\lambda\sqrt{\varphi})\right] = -\lambda\varphi^{-(n+1)/2} J_{n+1}(2\lambda\sqrt{\varphi})$$

$$\frac{d}{d\varphi}\left[\varphi^{-n/2} Y_n(2\lambda\sqrt{\varphi})\right] = -\lambda\varphi^{-(n+1)/2} Y_{n+1}(2\lambda\sqrt{\varphi})$$

$$\frac{d}{d\varphi}\left[\varphi^{-n/2} I_n(2\lambda\sqrt{\varphi})\right] = \lambda\varphi^{-(n+1)/2} I_{n+1}(2\lambda\sqrt{\varphi})$$

$$\frac{d}{d\varphi}\left[\varphi^{-n/2} K_n(2\lambda\sqrt{\varphi})\right] = -\lambda\varphi^{-(n+1)/2} K_{n+1}(2\lambda\sqrt{\varphi})$$

The first natural frequency coefficient determined from Eq. (c) is shown in Figure 9.25 as a function of α, where

$$\Omega_{o1} = \lambda_1(1 - \alpha)$$

and λ_1 is the lowest root of Eq. (c). Upon using Eq. (9.188) and Figure 9.25, we see that a cantilever beam with a taper of $\alpha = 0.667$ ($1/\alpha = 1.5$) increases the first natural frequency coefficient by 8.8% compared to that for a uniform cantilever beam, whose first natural frequency coefficient is 1.875. From Eq. (9.94), we see that this increases the natural frequency by 18.4 %.

FIGURE 9.25
First natural frequency coefficient for a double tapered cantilever beam with $\alpha = \beta$.

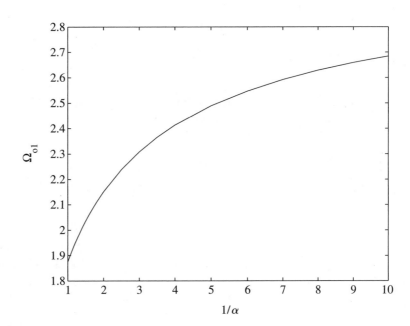

EXAMPLE 9.4 Mode shape of a baseball bat

Let us consider a baseball bat-like beam, which shall be modeled as a tapered beam that is free at each end. We shall determine the mode shape of this beam, compare it to the mode shape of a uniform beam that is free at each end, and then based on the results for the tapered beam make some judgments as to where one should strike the ball. For the conical beam shown in Figure 9.24b, we have that $\alpha = \beta = r_1/r_o < 1$, where r_o is the radius of the beam cross-section at $\eta = 1$ and r_1 is the radius of the beam cross-section at $\eta = 0$. The taper functions $a(\eta)$ and $i(\eta)$ given by Eqs. (9.189) are still applicable. In this case, Eqs. (9.190) become

$$I_o = \frac{\pi r_o^4}{2}$$
$$A_o = \pi r_o^2 \tag{a}$$

At the end $\eta = 1$, which corresponds to $\varphi = 1$ from Eq. (9.192), we have the boundary conditions

$$\frac{d^2W(1)}{d\varphi^2} = 0$$
$$\frac{d^3W(1)}{d\varphi^3} = 0 \tag{b}$$

At the end $\eta = 0$, which corresponds to $\varphi = \alpha$, we have the boundary condition

$$\frac{d^2W(\alpha)}{d\varphi^2} = 0$$
$$\frac{d^3W(\alpha)}{d\varphi^3} = 0 \tag{c}$$

On substituting Eq. (9.195) into the boundary conditions given by Eqs. (b) and (c) and setting the determinant of the coefficients A_j to zero, we obtain the following equation

$$\begin{vmatrix} J_5(2\lambda\sqrt{\alpha}) & Y_5(2\lambda\sqrt{\alpha}) & -I_5(2\lambda\sqrt{\alpha}) & K_5(2\lambda\sqrt{\alpha}) \\ J_4(2\lambda\sqrt{\alpha}) & Y_4(2\lambda\sqrt{\alpha}) & I_4(2\lambda\sqrt{\alpha}) & K_4(2\lambda\sqrt{\alpha}) \\ J_5(2\lambda) & Y_5(2\lambda) & -I_5(2\lambda) & K_5(2\lambda) \\ J_4(2\lambda) & Y_4(2\lambda) & -I_4(2\lambda) & K_4(2\lambda) \end{vmatrix} = 0 \tag{d}$$

The evaluation of Eq. (d) for a taper ratio of $\alpha = 1/2.2 = 0.4545$ leads to the frequency ratio $\Omega_{o1} = 4.081$. This is lower than the value of $\Omega_1 = 4.730$ for a uniform beam free at each end obtained from Table 9.2. The corresponding mode shape is shown in Figure 9.26 along with the mode shape for the uniform beam. In the figure, the right-hand end of the beam is the fatter end. It is seen that at the fatter end of the tapered beam, as the taper ratio is increased the node point "moves" toward the end of the beam whereas, at

FIGURE 9.26
Comparison of the first mode shape of a tapered beam with circular cross-section with $\alpha = 1/2.2$ and a uniform beam. The fatter end of the tapered beam is at the right end.

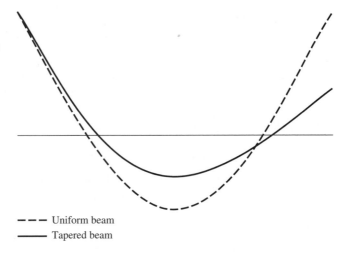

––– Uniform beam
——— Tapered beam

the thinner end, it moves toward the interior of the beam. In order for the beam to act as a rigid member, at least as far as exciting the first mode shape, one should strike the ball at the beam's (bat's) node point. That is, if the bat were to strike the ball close to a node point of the first mode, then the first mode of the bat will not be excited and there will be no energy loss due to the vibrations of the first mode of the bat. Usually, the so-called sweet spot of a baseball bat is at the node of the first mode located near the fat end of the bat. This tends to impart the maximum force to the ball, and frequently, the result is that we hear it as the "crack" of the bat.[36]

9.4 FORCED OSCILLATIONS

In Section 9.3, we studied the responses of different beam systems during free oscillations. In this section, we shall study the response of beams to externally applied dynamic transverse loading as shown in Figure 9.27. To solve for the response, we shall use the normal mode approach and the technique of separation of variables. To simplify matters, we shall assume that the boundary conditions are independent of time; that is, boundary condition 9 in Table 9.1 will not be considered. We assume that the damping force is proportional to the transverse velocity of the beam; that is, from Eq. (9.41c) we have

$$f_{nc}(t) = -c \frac{\partial w}{\partial t} \tag{9.196a}$$

[36]R. K. Adair, "The crack-of-the-bat: the acoustics of a bat hitting the ball," Paper No. 5pAA1, 141st Acoustical Society of America Meeting, Chicago, IL, June 2001.

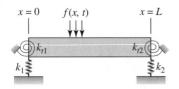

FIGURE 9.27
Forced oscillations of a beam.

Expressing the external loading $f(x,t)$ as

$$f(x,t) = f_{nc}(t) + f_d(x,t) \tag{9.196b}$$

the equation of motion for the damped vibrations of the beam is obtained from Eqs. (9.40) and (9.41c). Thus,

$$EI\frac{\partial^4 w}{\partial x^4} + c\frac{\partial w}{\partial t} + \rho A\frac{\partial^2 w}{\partial t^2} = f_d(x,t) \tag{9.197}$$

Governing Equations in Terms of Nondimensional Quantities

Employing the notation of Eqs. (9.67), Eq. (9.197) is rewritten as

$$\frac{\partial^4 w}{\partial \eta^4} + 2\zeta\frac{\partial w}{\partial \tau} + \frac{\partial^2 w}{\partial \tau^2} = g(\eta,\tau) \tag{9.198}$$

where the nondimensional spatial variable $\eta = x/L$, the nondimensional time variable $\tau = t/t_o$,

$$2\zeta = \frac{ct_o L}{m_o}, \quad g(\eta,\tau) = \frac{L^4}{EI}f_d(\eta,\tau) \tag{9.199}$$

and ζ is the nondimensional damping factor, t_o has the units of time, and $g(\eta,\tau)$ has the same units as $w(\eta,\tau)$.

For the system shown in Figure 9.27, and making use of Eqs. (9.47), the boundary conditions at $\eta = 0$ are

$$\frac{\partial^2 w(0,\tau)}{\partial \eta^2} = B_1\frac{\partial w(0,\tau)}{\partial \eta}$$

$$\frac{\partial^3 w(0,\tau)}{\partial \eta^3} = -K_1 w(0,\tau) \tag{9.200}$$

and the boundary conditions at $\eta = 1$ are

$$\frac{\partial^2 w(1,\tau)}{\partial \eta^2} = -B_2\frac{\partial w(1,\tau)}{\partial \eta}$$

$$\frac{\partial^3 w(1,\tau)}{\partial \eta^3} = K_2 w(1,\tau) \tag{9.201}$$

where B_j and K_j, $j = 1,2$, are defined in Eq. (9.67). We recall that the form of the boundary conditions given by Eqs. (9.200) and (9.201) have been chosen so that we can examine a wide range of boundary conditions by independently varying the values of B_j and K_j, $j = 1, 2$, from zero to infinity.

Solution for Response

To determine the response of the forced system described by Eqs. (9.198), (9.200), and (9.201) and the specified initial conditions $w = w(\eta,0)$ and $\dot{w}(\eta,0)$, we make use of the mode shapes obtained from the free oscillation problem. For the boundary conditions given by Eqs. (9.200) and (9.201), it is known from the material of Section 9.3 that the mode shapes $W_n(\eta)$ are orthogonal functions, the general form of which is given by Eqs. (9.89) or (9.91). In the present case, $W_n(\eta)$ is a solution of the undamped system given by Eq. (9.70); that is,

$$\frac{d^4 W_n}{d\eta^4} - \Omega_n^4 W_n = 0 \tag{9.202}$$

with the boundary conditions at $\eta = 0$ being

$$\begin{aligned} W_n''(0) &= B_1 W_n'(0) \\ W_n'''(0) &= -K_1 W_n(0) \end{aligned} \tag{9.203}$$

and those at $\eta = 1$ being

$$\begin{aligned} W_n''(1) &= -B_2 W_n'(1) \\ W_n'''(1) &= K_2 W_n(1) \end{aligned} \tag{9.204}$$

The solution for the forced response of the system is assumed to be of the form

$$w(\eta,\tau) = \sum_{n=1}^{\infty} \Psi_n(\tau) W_n(\eta) \tag{9.205}$$

where $W_n(\eta)$ are determined from the spatial eigenvalue problem associated with the undamped free oscillations of the system; that is, Eqs. (9.202) to (9.204) and the summation are taken over the infinite number of modes. Since $W_n(\eta)$ is a nondimensional quantity, $\Psi_n(\tau)$ has the same units as $w(\eta,\tau)$. The time-dependent functions $\Psi_n(\tau)$, also referred to as *modal amplitudes,* are unknown quantities that remain to be determined.

We substitute Eq. (9.205) into Eq. (9.198) to obtain

$$\sum_{n=1}^{\infty} \left[\frac{d^4 W_n}{d\eta^4} \Psi_n + 2\zeta W_n \frac{d\Psi_n}{d\tau} + W_n \frac{d^2\Psi_n}{d\tau^2} \right] = g(\eta,\tau) \tag{9.206}$$

where it is noted that due to the choice of the nondimensional variables each term in Eq. (9.206) has displacement units. Upon substituting Eq. (9.202) into Eq. (9.206) for the first term on the left-hand side, we obtain

$$\sum_{n} \left[\Omega_n^4 \Psi_n + 2\zeta \frac{d\Psi_n}{d\tau} + \frac{d^2\Psi_n}{d\tau^2} \right] W_n = g(\eta,\tau) \tag{9.207}$$

We now multiply Eq. (9.207) by $W_m(\eta)$, integrate over the interval $0 \le \eta \le 1$, and use the orthogonality property expressed by Eq. (9.77) with $M_o = J_o = 0$ to obtain

$$\frac{d^2\Psi_n}{d\tau^2} + 2\zeta \frac{d\Psi_n}{d\tau} + \Omega_n^4 \Psi_n = G_n(\tau) \quad n = 1, 2, \ldots \tag{9.208}$$

where

$$G_n(\tau) = \frac{1}{N_n} \int_0^1 W_n(\eta)g(\eta,\tau)d\eta$$

$$N_n = \int_0^1 W_n^2(\eta)d\eta \tag{9.209}$$

and $G_n(\tau)$ has displacement units and N_n is a nondimensional quantity.

Equation (9.208), which governs the variation of the modal amplitude $\Psi_n(\tau)$, is similar to the equation governing a single degree-of-freedom system discussed in Chapters 3 to 6. Therefore, the problem of determining the forced response of the system has been reduced to the problem of solving for the modal amplitudes $\Psi_n(\tau)$, each of whose governing equation is similar to the equation governing a single degree-of-freedom system. The orthogonality of the mode shapes is crucial to realizing this step.

Cases When a Mode Is Not Excited

On examining Eq. (9.208), it is clear that if $G_n(\tau) = 0$, then that particular mode corresponding to Ω_n does not get excited by the forcing. This is possible in the following two cases. In the first case, we assume that the forcing is acting at a point such that

$$g(\eta,\tau) = F(\tau)\delta(\eta - \eta_1) \tag{9.210a}$$

Then, we obtain from Eqs. (9.209) that

$$G_n(\tau) = \frac{1}{N_n} \int_0^1 W_n(\eta)F(\tau)\delta(\eta - \eta_1)d\eta = \frac{F(\tau)}{N_n} W_n(\eta_1) \tag{9.210b}$$

If η_1 is a node point, then $W_n(\eta_1) = 0$ and $G_n(\tau) = 0$. To show this, we consider a beam simply supported at each end. The mode shape is given by Eq. (9.103), and from Eq. (9.210b), $G_n(\tau)$ becomes

$$G_n(\tau) = \frac{F(\tau)}{N_n} \sin(n\pi\eta_1)$$

which is zero at $\eta_1 = 1/n$, $n = 2, 3, \dots$. Conversely, $G_n(\tau)$ is also zero, for example, when $\eta_1 = 0.5$ and $n = 2, 4, \dots$.

In the second case, we assume that the spatial distribution of the forcing is such that

$$\int_0^1 W_n(\eta)g(\eta,\tau)d\eta = 0 \tag{9.211a}$$

Equation (9.211a) is satisfied when the forcing function is orthogonal to the mode shape; that is, let

$$g(\eta,\tau) = F(\tau)W_m(\eta) \tag{9.211b}$$

where $W_m(\eta)$ is one the orthogonal functions. In this case, we obtain from Eqs. (9.209) that

$$G_n(\tau) = \frac{1}{N_n} \int_0^1 W_n(\eta)F(\tau)W_m(\eta)d\eta = F(\tau)\delta_{nm} \tag{9.211c}$$

and, therefore, all the modal amplitudes are zero except for $n = m$.

Solution for Response (*continued*)

The solution to Eq. (9.208) is obtained by using the Laplace transform, and we follow the procedure used to obtain Eq. (4.8a). In order to make use of Eq. (4.8a), we rewrite Eq. (9.208) as

$$\frac{d^2\Psi_n}{d\tau^2} + 2\zeta_n\Omega_n^2 \frac{d\Psi_n}{d\tau} + \Omega_n^4\Psi_n = G_n(\tau) \quad n = 1,2,... \tag{9.212}$$

where

$$\zeta_n = \frac{\zeta}{\Omega_n^2} \tag{9.213}$$

The Laplace transform of Eq. (9.212) is given by Eqs. (4.4) and (4.5) if we identify $\Psi_n(\tau)$ with $x(t)$, ζ with ζ_n, ω_n with Ω_n^2, and we set $F(s)/m = \hat{G}_n(s)$. Thus, if the Laplace transform of $\Psi_n(\tau)$ is $\hat{\Psi}_n(s)$, then in the Laplace domain we have

$$\hat{\Psi}_n(s) = \frac{\hat{G}_n(s) + (s + 2\zeta_n\Omega_n^2)\Psi_n(0) + \dot{\Psi}_n(0)}{s^2 + 2\zeta_n\Omega_n^2 s + \Omega_n^4} \tag{9.214}$$

where the overdot indicates the derivative with respect to τ and the modal initial conditions are given by $\Psi_n(0)$ and $\dot{\Psi}_n(0)$. Making use of Eq. (9.205), these initial conditions are determined from

$$w(\eta,0) = \sum_{n=1}^{\infty} \Psi_n(0)W_n(\eta)$$

$$\dot{w}(\eta,0) = \sum_{n=1}^{\infty} \dot{\Psi}_n(0)W_n(\eta) \tag{9.215}$$

If we multiply each of Eqs. (9.215) by $W_m(\eta)$, integrate over the interval $0 \leq \eta \leq 1$, and make use of Eq. (9.77) with $M_o = J_o = 0$, we obtain the modal initial conditions

$$\Psi_n(0) = \frac{1}{N_n} \int_0^1 W_n(\eta)w(\eta,0)d\eta \quad n = 1,2,\ldots$$

$$\dot{\Psi}_n(0) = \frac{1}{N_n} \int_0^1 W_n(\eta)\dot{w}(\eta,0)d\eta \quad n = 1,2,\ldots \tag{9.216}$$

Assuming that $0 < \zeta_n < 1$, the inverse transform of Eq. (9.214) is obtained from Eq. (4.8a) as

$$\Psi_n(\tau) = \Psi(0)e^{-\zeta\tau}\cos(\Omega_{nd}^2\tau) + \frac{\dot{\Psi}(0) + \zeta\Psi(0)}{\Omega_{nd}^2}e^{-\zeta\tau}\sin(\Omega_{nd}^2\tau)$$

$$+ \frac{1}{\Omega_{nd}^2}\int_0^\tau e^{-\zeta\tau'}\sin(\Omega_{nd}^2\tau')G_n(\tau - \tau')d\tau' \tag{9.217}$$

where $\Psi_n(\tau)$ is the nth modal amplitude and

$$\Omega_{nd}^2 = \Omega_n^2\sqrt{1 - \zeta_n^2} \tag{9.218}$$

is the damped natural frequency coefficient associated with the nth mode.

Forced Response in the Damped Case

Upon using Eqs. (9.205), (9.209), and (9.217), the displacement of the beam is

$$w(\eta,\tau) = \sum_{n=1}^\infty \frac{W_n(\eta)}{N_n}\left[\frac{1}{\Omega_{nd}^2}\int_0^\tau\left\{\int_0^1 e^{-\zeta\tau'}\sin(\Omega_{nd}^2\tau')g(\eta,\tau - \tau')W_n(\eta)d\eta\right\}d\tau'\right.$$

$$+ e^{-\zeta\tau}\cos(\Omega_{nd}^2\tau)\int_0^1 W_n(\eta)w(\eta,0)d\eta$$

$$\left.+ \frac{e^{-\zeta\tau}\sin(\Omega_{nd}^2\tau)}{\Omega_{nd}^2}\int_0^1 W_n(\eta)\{\dot{w}(\eta,0) + \zeta w(\eta,0)\}d\eta\right] \tag{9.219a}$$

The first term inside the brackets is the forced response in the nth mode and the second and third terms are the responses of the nth mode to the initial conditions. Making use of Eqs. (9.67) and (9.199), we can rewrite Eq. (9.219a) in terms of the dimensional quantities as

$$w(x,t) = \sum_n \frac{W_n(x/L)}{LN_n}\left[\frac{L^4}{\omega_{dn}t_o^2EI}\int_0^t\left\{\int_0^L e^{-\zeta't'}\sin(\omega_{dn}t')f_d(x,t - t')W_n(x/L)dx\right\}dt'\right.$$

$$+ e^{-\zeta't}\cos(\omega_{dn}t)\int_0^L W_n(x/L)w(x,0)dx$$

$$\left.+ \frac{e^{-\zeta't}\sin(\omega_{dn}t)}{\omega_{dn}}\int_0^L W_n(x/L)\left\{\frac{\partial w(x,0)}{\partial t} + \zeta'w(x,0)\right\}dx\right] \tag{9.219b}$$

where

$$\omega_{dn} = \omega_n\sqrt{1 - \zeta_n^2} \quad \text{and} \quad \zeta' = \zeta/t_o$$

Forced Response in the Undamped Case

When the damping is absent, Eq. (9.219a) simplifies to

$$w(\eta,\tau) = \sum_{n=1}^{\infty} \frac{W_n(\eta)}{N_n} \left[\frac{1}{\Omega_n^2} \int_0^\tau \left[\int_0^1 g(\eta,\tau - \tau')W_n(\eta)\sin(\Omega_n^2\tau')d\eta \right] d\tau' \right.$$

$$\left. + \cos(\Omega_n^2\tau) \int_0^1 W_n(\eta)w(\eta,0)d\eta + \frac{\sin(\Omega_n^2\tau)}{\Omega_n^2} \int_0^1 W_n(\eta)\dot{w}(\eta,0)d\eta \right] \quad (9.220a)$$

In terms of the dimensional quantities, we rewrite Eq. (9.219b) as

$$w(x,t) = \sum_{n=1}^{\infty} \frac{W_n(x/L)}{LN_n} \left[\frac{L^4}{\omega_n t_o^2 EI} \int_0^t \left\{ \int_0^L \sin(\omega_n t')f_d(x,t - t')W_n(x/L)dx \right\} dt' \right.$$

$$\left. + \cos(\omega_n t) \int_0^L W_n(x/L)w(x,0)dx + \frac{\sin(\omega_n t)}{\omega_n} \int_0^L W_n(x/L)\frac{\partial w(x,0)}{\partial t} dx \right] \quad (9.220b)$$

Equation (9.219a) describes the forced response of the system governed by Eqs. (9.198) to (9.201) and the specified initial conditions, and they have been expressed as a sum of the responses of the individual modes. In practice, the infinite sum shown in Eq. (9.219a) is replaced by a finite sum, depending on the frequency content of the excitation.

EXAMPLE 9.5 Impulse response of a cantilever beam

We shall determine the displacement response of an undamped cantilever beam that is initially at rest and subjected to an impulse force at $\eta = \xi = x_1/L$.

In this case, $c = 0$ in Eq. (9.197) and $\zeta = 0$ in Eq. (9.198). Since the beam is initially at rest, the initial conditions are

$$w(\eta,0) = \dot{w}(\eta,0) = 0 \qquad (a)$$

and the forcing is expressed as

$$g(\eta,\tau) = g_o\delta(\eta - \xi)\delta(\tau) \qquad (b)$$

Upon substituting Eqs. (a) and (b) into Eq. (9.220a), and carrying out the integration, we find that

$$w(\eta,\tau) = g_o \sum_{n=1}^{\infty} \frac{W_n(\eta)W_n(\xi)}{N_n\Omega_n^2} \sin(\Omega_n^2\tau) \qquad (c)$$

Upon using Eqs. (9.220b), we rewrite Eq. (c) in terms of the dimensional quantities as

$$w(x,t) = \frac{f_o L^3}{t_o^2 EI} \sum_{n=1}^{\infty} \frac{W_n(x/L)W_n(x_1/L)}{N_n\omega_n} \sin(\omega_n t) \qquad (d)$$

where f_o is the magnitude of the impulse at $x = x_1$ in N·s.

The values of Ω_n are determined by Eq. (9.108) and they are also given in Table 9.2. The modes $W_n(\eta)$ are given by Eq. (9.109). The beam displacement

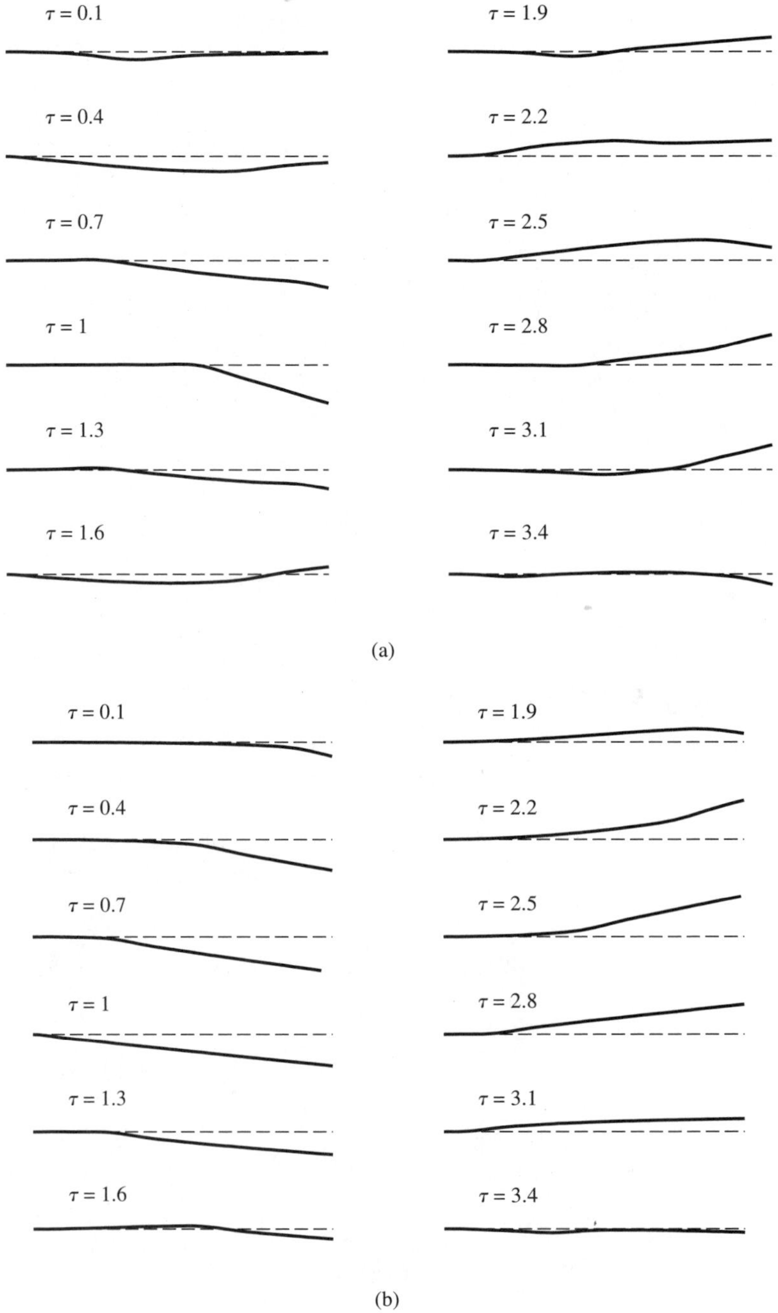

FIGURE 9.28
Normalized response of a
cantilever beam subjected to
an impulse force: (a) $\eta_1 =$
0.4 and (b) $\eta_1 = 1$.

$\tau = 0.1$

$\tau = 0.4$

$\tau = 0.7$

$\tau = 1$

$\tau = 1.3$

$\tau = 1.6$

$\tau = 1.9$

$\tau = 2.2$

$\tau = 2.5$

$\tau = 2.8$

$\tau = 3.1$

$\tau = 3.4$

(a)

$\tau = 0.1$

$\tau = 0.4$

$\tau = 0.7$

$\tau = 1$

$\tau = 1.3$

$\tau = 1.6$

$\tau = 1.9$

$\tau = 2.2$

$\tau = 2.5$

$\tau = 2.8$

$\tau = 3.1$

$\tau = 3.4$

(b)

given by Eq. (c) is plotted in Figure 9.28 for impulses applied at $\eta_1 = 0.4$ and $\eta_1 = 1$. In each case, the displacement pattern is shown at different instances of the nondimensional time τ. Each beam displacement was obtained after truncating the summation to 11 modes in Eq. (c).

EXAMPLE 9.6 Air-damped micro-cantilever beam

Consider a silicon cantilever beam that is undergoing forced harmonic oscillations at a frequency ω. In addition, it is assumed that the length L of the beam can vary between $10~\mu\text{m} < L < 1$ mm and that the width $b = L/10$ and the height $h = L/100$. Under certain assumptions, it has been shown[37] that, for these ranges of L, b, and h, the air damping is the dominant form of damping and it is reasonable to assume that the damping force is proportional to the transverse velocity of the beam; that is, from Eq. (9.41c)

$$f_{nc} = -c\,\frac{\partial w}{\partial t} \tag{a}$$

where

$$c = 1.5\mu + 0.375\pi b\sqrt{2\rho_a\mu\omega} \tag{b}$$

and $w = w(x,t)$ is the displacement of the beam, ρ_a is the density of air (1.3 kg/m^3), and μ is the viscosity of air (1.8 $\times$ 10^{-5} Pa·s). For $|w| < L/1000$ and for $\omega < 100{,}000$ rad/s, the Reynolds number for the airflow at the free end of the beam is less than 1 and the given damping model is applicable. We shall determine the form of the forced response in this case when a harmonic forcing is applied at $\eta = \xi$. We assume that the initial conditions are zero; that is,

$$w(\eta,0) = 0 \quad \text{and} \quad \dot{w}(\eta,0) = 0 \tag{c}$$

To determine the form of the response, we make use of Eq. (9.219). For the given damping coefficient c, the quantity ζ is evaluated from Eq. (b) and Eqs. (9.199) to be

$$\zeta = \frac{(1.5\mu + 0.375\pi b\sqrt{2\rho_a\mu\omega})L^3}{2c_b r m_o} \tag{d}$$

For harmonic forcing at $\eta = \xi$, the forcing function $g(\eta,\tau)$ is given by

$$g(\eta,\tau) = g_o\delta(\eta - \xi)\sin(\Omega^2\tau) \tag{e}$$

---

[37]H. Hosaka and K. Itao, "Theoretical and Experimental Study on Airflow Damping of Vibrating Microcantilevers," *J. Vibration Acoustics,* Vol. 121, pp. 64–69 (1999).

Substituting Eq. (e) into Eq. (9.219a), using the modes given by Eq. (9.109), and making use of Eq. (c) we obtain

$w(\eta,\tau)$

$$= g_o \sum_{n=1}^{\infty} \frac{W_n(\eta)}{N_n \Omega_{nd}^2} \int_0^\tau e^{-\zeta\tau'} \sin(\Omega_{nd}^2 \tau') \sin[\Omega^2(\tau - \tau')] \left\{ \int_0^1 \delta(\eta - \xi) W_n(\eta) d\eta \right\} d\tau'$$

$$= g_o \sum_{n=1}^{\infty} \frac{W_n(\eta) W_n(\xi)}{N_n \Omega_{nd}^2} \int_0^\tau e^{-\zeta\tau'} \sin(\Omega_{nd}^2 \tau') \sin[\Omega^2(\tau - \tau')] d\tau' \qquad (f)$$

Making use of Eqs. (5.6) to (5.8) and recognizing that the sinusoidal excitation is present for all time—that is, we can ignore the transient portion of the solution in Eq. (5.7)—we find that

$$w(\eta,\tau) = g_o \sum_{n=1}^{\infty} \frac{W_n(\eta) W_n(\xi)}{N_n \Omega_{nd}^2} H_n(\Omega) \sin(\Omega^2 \tau - \theta_n(\Omega)) \qquad (g)$$

where we have used Eq. (9.218) and

$$H_n(\Omega) = \left[\left(1 - \frac{\Omega^4}{\Omega_n^4} \right)^2 + \left(2\zeta_n \frac{\Omega^2}{\Omega_n^2} \right)^2 \right]^{-1/2}$$

$$\theta_n(\Omega) = \tan^{-1} \frac{2\zeta_n \Omega^2 / \Omega_n^2}{1 - \Omega^4 / \Omega_n^4}$$

EXAMPLE 9.7 Frequency-response functions of a beam

Consider the arrangement shown in Figure 9.29 in which a cantilever beam is excited by using a shaker at location $x = x_1$ on a beam. We assume that the beam is initially at rest, that the force imparted to the beam is measured by using a force transducer, and that the beam response is measured at location $x = x_2$ by using an accelerometer. We shall construct a frequency-response function based on the accelerometer measurement and the input force measurement.

The measured force applied at $x = x_1$ is represented by

$$f_d(x,t) = f(t)\delta(x - x_1) \qquad (a)$$

and the acceleration measured at $x = x_2$ is represented by

$$a(t) = \frac{\partial^2 w(x_2,t)}{\partial t^2} \qquad (b)$$

In terms of the nondimensional variables $\tau = t/t_o$ and $\eta = x/L$, Eqs. (a) and (b) are written as, respectively,

$$f_d(\eta,\tau) = f(\tau)\delta(\eta - \eta_1)$$

$$a(\tau) = \frac{1}{t_o^2} \frac{\partial^2 w(\eta_2,\tau)}{\partial \tau^2} \qquad (c)$$

FIGURE 9.29
Cantilever beam excited by a
shaker at $x = x_1$ and the
resulting acceleration
measured at $x = x_2$.

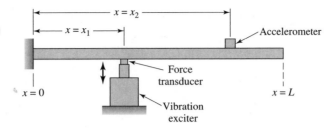
Accelerometer

Force
transducer

$x = 0$

$x = L$

Vibration
exciter

Let $A(s)$ and $F(s)$ represent the Laplace transforms of $a(\tau)$ and $f(\tau)$, respectively. Then, the transfer function $G_{21}(s)$ is constructed as

$$G_{21}(s) = \frac{A(s)}{F(s)} \tag{d}$$

The required frequency-response function is obtained from Eq. (d) by setting $s = j\Omega^2$; that is,

$$G_{21}(j\Omega^2) = \frac{A(j\Omega^2)}{F(j\Omega^2)} \tag{e}$$

To use the material in Section 9.4 to construct $G_{21}(j\Omega^2)$, we start by taking the Laplace transform of the second of Eqs. (c) to obtain

$$A(s) = \frac{s^2}{t_o^2}\,\hat{W}(\eta_2,s) \tag{f}$$

We now take the Laplace transform of Eq. (9.205) and obtain

$$\hat{W}(\eta,s) = \sum_{n=1}^{\infty} \hat{\Psi}_n(s)W_n(\eta) \tag{g}$$

Upon substituting Eq. (g) into Eq. (f), we arrive at

$$A(s) = \frac{s^2}{t_o^2}\sum_{n=1}^{\infty} \hat{\Psi}_n(s)W_n(\eta_2) \tag{h}$$

To obtain an expression for $\hat{\Psi}(s)$ we use Eq. (9.199) to obtain

$$g(\eta,\tau) = \frac{L^4}{EI}f_d(\eta,\tau) = \frac{L^4}{EI}f(\tau)\delta(\eta - \eta_1) \tag{i}$$

Upon substituting Eq. (i) into Eq. (9.209), we find that

$$\begin{aligned}
G_n(\tau) &= \frac{1}{N_n}\int_0^1 W_n(\eta)g(\eta,\tau)d\eta \\
&= \frac{L^4 f(\tau)}{N_n EI}\int_0^1 W_n(\eta)\delta(\eta - \eta_1)d\eta \\
&= \frac{L^4}{N_n EI}f(\tau)W_n(\eta_1)
\end{aligned} \tag{j}$$

Upon taking the Laplace transform of Eq. (j) we obtain

$$\hat{G}_n(s) = \frac{L^4}{N_n EI} F(s) W_n(\eta_1) \tag{k}$$

We now substitute Eq. (k) into Eq. (9.214) with the initial conditions set to zero, and arrive at

$$\hat{\Psi}_n(s) = \frac{L^4}{N_n EI} \frac{F(s) W_n(\eta_1)}{s^2 + 2\zeta_n \Omega_n^2 s + \Omega_n^4} \tag{l}$$

Upon substituting Eq. (l) into Eq. (h) and using Eq. (d), we obtain

$$G_{21}(s) = \frac{A(s)}{F(s)} = \frac{L^4}{t_o^2 EI} \sum_{n=1}^{\infty} \left[\frac{W_n(\eta_1) W_n(\eta_2)}{N_n} \right] \left[\frac{s^2}{s^2 + 2\zeta_n \Omega_n^2 s + \Omega_n^4} \right] \tag{m}$$

Then, the frequency-response function given by Eq. (e) is

$$G_{21}(j\Omega^2) = \frac{A(j\Omega^2)}{F(j\Omega^2)}$$

$$= \frac{c_b^2 r^2}{EI} \sum_{n=1}^{\infty} \left[\frac{W_n(\eta_1) W_n(\eta_2)}{N_n} \right] \left[\frac{-\Omega^4}{\Omega_n^4 - \Omega^4 + 2j\zeta_n \Omega_n^2 \Omega^2} \right] \tag{n}$$

We see from Eq. (m) that if we were to reverse the drive location, which was at $\eta = \eta_1$, with the measurement location, which was at $\eta = \eta_2$, the frequency response function would still be the same. This is an important property of linear systems and is referred to as the *reciprocity property*.

9.5   SUMMARY

In this chapter, free and forced oscillations of slender elastic beams modeled by linear systems were considered. It was illustrated how the governing equations of motion are obtained from the extended Hamilton's principle. This approach can also be used for determining the governing equations for other continua such as strings and bars. Free oscillations of beams were treated at length for different boundary conditions and varying beam properties. The effect of axial force and elastic foundations were also taken into account. The forced response of the beam was determined by using the orthogonality properties of the modes.

APPENDIX A

Laplace Transform Pairs

DEFINITION OF LAPLACE TRANSFORM

The Laplace transform is defined as

$$L[g(t)] = G(s) = \int_0^\infty e^{-st} g(t)dt \tag{A.1}$$

where the variable s is a complex variable represented as $s = \sigma + j\omega$, where $j = \sqrt{-1}$. In writing this integral transform definition, it is assumed that the time-dependent function $g(t)$ is defined for all values of time $t > 0$ and that this function is such that this integral exists; that is,

$$\int_0^\infty |g(t)| \, e^{-at} \, dt < \infty \tag{A.2}$$

where a is a positive real number. This restriction means that a function $g(t)$ that satisfies Eq. (A.2) does not increase with time more rapidly than the exponential function e^{-at}. In addition, the function $g(t)$ is required to be piecewise continuous. For the functions $g(t)$ considered in this book, these conditions are satisfied.

If the time-dependent function is a displacement response $x(t)$, which has units of a meter (m), then the corresponding Laplace transform $X(s)$ has units of meter-seconds (m·s) when t has the units of time. In this case, the variable s has units of 1/time so that the product st is nondimensional.

EVALUATION OF LAPLACE TRANSFORMS

Next, the evaluation of the Laplace transform of the function

$$g(t) = \cos(\Omega t) \tag{A.3}$$

considered in Section 4.1 is revisited. By using the definition (A.1), the Laplace transform of Eq. (A.3) is

$$L[g(t)] = G(s) = \int_0^\infty \cos(\Omega t)e^{-st}dt$$

$$= \frac{e^{-st}}{s^2 + \Omega^2}[-s\cos(\Omega t) + \Omega\sin(\Omega t)]\Big|_0^\infty$$

$$= \frac{s}{s^2 + \Omega^2} \tag{A.4}$$

which is tabulated as entry 18 of Table A. This integration can also be carried out symbolically by using MATLAB, MAPLE, or Mathematica.

Consider the Laplace transform of the first and second time derivatives of the function $h(t)$, respectively; that is,

$$\frac{dh}{dt} \quad \text{and} \quad \frac{d^2h}{dt^2}$$

Then, from Eqs. (A.1),

$$L\left[\frac{dh}{dt}\right] = \int_0^\infty \frac{dh}{dt}e^{-st}dt$$

$$= e^{-st}h(t)\Big|_0^\infty + s\int_0^\infty h(t)e^{-st}$$

$$= e^{-st}h(t)\Big|_0^\infty + sH(s)$$

$$= -h(0) + sH(s) \tag{A.5}$$

where integration by parts was used. For the second derivative of $h(t)$, we have

$$L\left[\frac{d^2h}{dt^2}\right] = \int_0^\infty \frac{d^2h}{dt^2}e^{-st}dt$$

$$= e^{-st}\frac{dh}{dt}\Big|_0^\infty + s\int_0^\infty \frac{dh}{dt}e^{-st}dt$$

$$= -\frac{dh}{dt}\Big|_{t=0} + s[sH(s) - h(0)]$$

$$= s^2H(s) - sh(0) - \dot{h}(0) \tag{A.6}$$

where integration by parts and Eq. (A.5) were used. The results given by Eqs. (A.5) and (A.6) are summarized along with higher-order derivatives in entry 2 of Table A. The evaluation of the inverse Laplace theorem is not straightforward, and for details of this, the reader is referred to Widder.[1]

---

[1]David Widder, *The Laplace Transform*, Princeton University Press, Princeton, NJ (1941).

	$G(s)$	$g(t)$	Description
TABLE A Laplace Transform Pairs			
1	$G(s/a)$	$ag(at)$	Scaling of variable
2	$s^n G(s) - \sum_{k=1}^{n} s^{n-k} g^{k-1}(0)$	$g^n(t) = \dfrac{d^n g}{dt^n}$	nth-order derivative, $n = 1, 2, \ldots$
3	$e^{-t_o s} G(s)$	$g(t - t_o)\, u(t - t_o)$	Time shifting
4	$G(s)H(s)$	$\displaystyle\int_0^t g(\eta) h(t-\eta)\,d\eta = \int_0^t g(t-\eta) h(\eta)\,d\eta$	Convolution
5	e^{-st_o}	$\delta(t - t_o)$	Delta function
6	$\dfrac{e^{-st_o}}{s}$	$u(t - t_o)$	Unit step function
7	$\dfrac{1}{s - a}$	e^{at}	Exponential
8	$\dfrac{1 - e^{-st_o}}{s}$	$u(t) - u(t - t_o)$	Rectangular pulse of duration t_o

9	$\dfrac{(1 - e^{-t_o s})^2}{s}$	$u(t) - 2u(t - t_o) + u(t - 2t_o)$	

10	$\dfrac{\omega_o (1 + e^{-\pi s/\omega_o})}{s^2 + \omega_o^2}$	$\sin(\omega_o t)[u(t) - u(t - \pi/\omega_o)]$	Half sine wave of frequency ω_o

11	$G(s) = \dfrac{1}{t_o s^2}(1 - e^{-st_o})$ $\times (1 - e^{-st_1})$	$g(t) = (t/t_o)[u(t) - u(t - t_o)] +$ $[u(t - t_o) - u(t - t_1)] +$ $(-t/t_o + 1 + t_1/t_o)$ $[u(t - t_1) - u(t - t_1 - t_o)]$	

12	$G(s) = \dfrac{1}{t_o s^2}(1 - e^{-st_o})^2$	$g(t) = (t/t_o)[u(t) - u(t - t_o)] +$ $(-t/t_o + 2)[u(t - t_o)] -$ $u(t - 2t_o)]$	

13	$\dfrac{e^{-t_o s} + t_o s - 1}{t_o s^2}$	$(1 - t/t_o)[u(t) - u(t - t_o)]$	

(continued)

	$G(s)$	$g(t)$	Description
14	$\dfrac{1}{s^2 + 2\zeta\omega_n s + \omega_n^2}$	$\dfrac{1}{\omega_d} e^{-\zeta\omega_n t} \sin(\omega_d t)$	Impulse response of single degree-of-freedom system $\omega_d = \omega_n \sqrt{1 - \zeta^2}$
15	$\dfrac{\omega_n^2}{s(s^2 + 2\zeta\omega_n s + \omega_n^2)}$	$1 - \dfrac{\omega_n}{\omega_d} e^{-\zeta\omega_n t} \sin(\omega_d t + \varphi)$	$g(t)$ is step response of single degree-of-freedom system $\varphi = \cos^{-1}\zeta \quad \zeta < 1$
16	$\dfrac{s}{s^2 + 2\zeta\omega_n s + \omega_n^2}$	$-\dfrac{\omega_n}{\omega_d} e^{-\zeta\omega_n t} \sin(\omega_d t - \varphi)$	
17	$\dfrac{s + 2\zeta\omega_n}{s^2 + 2\zeta\omega_n s + \omega_n^2}$	$\dfrac{\omega_n}{\omega_d} e^{-\zeta\omega_n t} \sin(\omega_d t + \varphi)$	Response of single degree-of-freedom system to initial displacement
18	$\dfrac{s}{s^2 + \omega^2}$	$\cos(\omega t)$	Special case of 16
19	$\dfrac{\omega}{s^2 + \omega^2}$	$\sin(\omega t)$	Special case of 14
20	$\dfrac{s}{s^2 - \omega^2}$	$\cosh(\omega t)$	
21	$\dfrac{\omega}{s^2 - \omega^2}$	$\sinh(\omega t)$	
22	$\dfrac{1}{s^n}$	$\dfrac{t^{n-1}}{(n-1)!}$	$n = 1, 2, \ldots$
23	$\dfrac{2a^3}{s^4 - a^4}$	$\sinh(at) - \sin(at)$	Term in solution to Euler beam equation
24	$\dfrac{2a^2 s}{s^4 - a^4}$	$\cosh(at) - \cos(at)$	Term in solution to Euler beam equation
25	$\dfrac{2as^2}{s^4 - a^4}$	$\sinh(at) + \sin(at)$	Term in solution to Euler beam equation
26	$\dfrac{2s^3}{s^4 - a^4}$	$\cosh(at) + \cos(at)$	Term in solution to Euler beam equation
27	$\dfrac{1}{(s + a)(s + b)}$	$\dfrac{1}{(b - a)}[e^{-at} - e^{-bt}]$	Generalization of 14
28	$\dfrac{s}{(s + a)(s + b)}$	$\dfrac{1}{(b - a)}[be^{-bt} - ae^{-at}]$	Generalization of 16
29	$a_1 G_1(s) + a_2 G_2(s)$	$a_1 g_1(t) + a_2 g_2(t)$	Linearity theorem
30	$\lim_{s \to \infty} [sG(s)]$	$\lim_{t \to 0^+} [g(t)]$	Initial value theorem assuming that $\lim_{s \to \infty} [sG(s)]$ exists
31	$\lim_{s \to 0} [sG(s)]$	$\lim_{t \to \infty} [g(t)]$	Final value theorem assuming that $\lim_{t \to \infty} [g(t)]$ exists
32	$G(s - a)$	$g(t)e^{at}$	s-domain shifting

USE OF PARTIAL FRACTIONS

The method of partial fractions is frequently used to reduce higher order polynomials to lower order polynomials whose form corresponds to one of those appearing in Table A. To illustrate this method, consider the following equation that describes the motion of a thin elastic beam on an elastic foundation that is subjected to an axial load and a harmonically excited external force:[2]

$$\frac{d^4y}{dx^4} + 2\beta \frac{d^2y}{dx^2} + (k - \Omega^4)y = f(x) \tag{A.7}$$

Using Eq. (A.6) and entry 2 of Table A, the Laplace transform of Eq. (A.7) is

$$Y(s) = \frac{1}{D(s)} [(s^3 + 2\beta s)y(0) + (s^2 + 2\beta)y'(0) + sy''(0) + y'''(0) + F(s)] \tag{A.8}$$

where the prime denotes the derivative with respect to x,

$$D(s) = s^4 + 2\beta s^2 + k - \Omega^4 = (s^2 - \delta^2)(s^2 + \epsilon^2)$$
$$\delta^2 = -\beta + \sqrt{\beta^2 + \Omega^4 - k}$$
$$\epsilon^2 = \beta + \sqrt{\beta^2 + \Omega^4 - k} \tag{A.9}$$

and we assume that $\beta^2 + \Omega^4 - k > 0$. Note that $2\beta = \epsilon^2 - \delta^2$.

Consider a ratio of two polynomials $N(s)/D(s,n)$, where n is the order of the polynomial. The first step in the method of partial fractions is to factor D into the product of lower order polynomials $D(s,n_1)D(s,n_2)...D(s,n_m)$ and then find an equivalent sum of polynomial ratios, such that

$$\frac{N}{D} = \sum_{k=1}^{m} \frac{A_k s^{n_k - 1}}{D(s,n_k)} \tag{A.10}$$

where A_k is determined by equating the coefficients of the like powers of s on both sides of Eq. (A.10). We illustrate this procedure for the first term on the right-hand side of Eq. (A.8), where we see that $m = 2$ and $n_1 = n_2 = 2$. Thus,

$$\frac{s^3 + 2\beta s}{(s^2 - \delta^2)(s^2 + \epsilon^2)} = \left[\frac{A_1 s}{(s^2 - \delta^2)} + \frac{A_2 s}{(s^2 + \epsilon^2)} \right]$$
$$= \frac{A_1 s(s^2 + \epsilon^2) + A_2 s(s^2 - \delta^2)}{(s^2 - \delta^2)(s^2 + \epsilon^2)}$$
$$= \frac{(A_1 + A_2)s^3 + s(A_1\epsilon^2 - A_2\delta^2)}{(s^2 - \delta^2)(s^2 + \epsilon^2)} \tag{A.11}$$

Upon equating the coefficients of the s^3 term and the s term in Eq. (A.11), we find that

$$(A_1 + A_2) = 1$$
$$A_1\epsilon^2 - A_2\delta^2 = 2\beta$$

[2]See Section 9.5.

Solving for A_1 and A_2, we obtain

$$A_1 = \frac{\epsilon^2}{\epsilon^2 + \delta^2}$$

$$A_2 = \frac{\delta^2}{\epsilon^2 + \delta^2} \tag{A.12}$$

On substituting Eqs. (A.12) in Eq. (A.11), we obtain

$$\frac{s^3 + 2\beta s}{(s^2 - \delta^2)(s^2 + \epsilon^2)} = \frac{1}{\epsilon^2 + \delta^2}\left[\frac{s\epsilon^2}{(s^2 - \delta^2)} + \frac{s\delta^2}{(s^2 + \epsilon^2)}\right] \tag{A.13}$$

Using entries 18 and 20 from Table A, the inverse Laplace transform of Eq. (A.13) is

$$L^{-1}\left[\frac{s^3 + 2\beta s}{(s^2 - \delta^2)(s^2 + \epsilon^2)}\right] = \frac{1}{\epsilon^2 + \delta^2} L^{-1}\left[\frac{s\epsilon^2}{(s^2 - \delta^2)} + \frac{s\delta^2}{(s^2 + \epsilon^2)}\right]$$

$$= \frac{1}{\epsilon^2 + \delta^2}[\delta^2 \cos \epsilon x + \epsilon^2 \cosh \delta x] \quad \text{(A.14)}$$

When $\beta = k = 0$, $\epsilon = \delta = \Omega$, and Eq. (A.13) simplifies to

$$\frac{s^3}{s^4 - \Omega^4} = \frac{s^3}{(s^2 - \Omega^2)(s^2 + \Omega^2)}$$

$$= \frac{1}{2}\left[\frac{s}{(s^2 - \Omega^2)} + \frac{s}{(s^2 + \Omega^2)}\right] \tag{A.15}$$

The inverse Laplace transform of Eq. (A.15) is obtained using entries 18 and 20 from Table A to arrive at

$$L^{-1}\left[\frac{s^3}{(s^2 - \Omega^2)(s^2 + \Omega^2)}\right] = \frac{1}{2} L^{-1}\left[\frac{s}{(s^2 - \Omega^2)} + \frac{s}{(s^2 + \Omega^2)}\right]$$

$$= \frac{1}{2}[\cos \Omega x + \cosh \Omega x]$$

Thus, we have verified entry 26 of Table A. This result could have also been obtained directly from Eq. (A.14).

In a similar manner, the inverse Laplace transform of the remaining terms on the right-hand side Eq. (A.8) are obtained. The result is:

$$y(x) = y(0)U_o(x) + y'(0)R_o(x)$$

$$+ y''(0)S_o(x) + y'''(0)T_o(x) + \int_0^x f(\eta)T_o(x - \mu)d\eta \tag{A.16}$$

where we have used entry 4 in Table A and

$$U_o(x) = \frac{1}{\delta^2 + \epsilon^2}[\delta^2 \cos(\epsilon x) + \epsilon^2 \cosh(\delta x)]$$

$$R_o(x) = \frac{1}{\delta^2 + \epsilon^2}\left[\frac{\delta^2}{\epsilon}\sin(\epsilon x) + \frac{\epsilon^2}{\delta}\sinh(\delta x)\right]$$

$$S_o(x) = \frac{1}{\delta^2 + \epsilon^2}[-\cos(\epsilon x) + \cosh(\delta x)]$$

$$T_o(x) = \frac{1}{\delta^2 + \epsilon^2}\left[-\frac{1}{\epsilon}\sin(\epsilon x) + \frac{1}{\delta}\sinh(\delta x)\right] \qquad (A.17)$$

The method of partial fractions[3] is quite often needed for the design of control systems.

There is a one-to-one transformation from $g(t)$ to $G(s)$, which enables us to establish a catalog of the different functions one is likely to use in Vibrations and Control. One such catalog is Table A. A large compendium of Laplace transforms and their inverse transforms has been collected by Roberts and Kaufman.[4]

[3]See, for example, R. V. Churchill, *Operational Mathematics,* McGraw-Hill, NY (1958).

[4]G. E. Roberts and H. Kaufman, *Table of Laplace Transforms,* W. B. Saunders Co, Philadelphia (1966).

APPENDIX B

Fourier Series

TABLE B Fourier Series[1]	Fourier Series	Waveform
	a) Square wave $$f(t) = \frac{4}{\pi} \sum_{i=1,3,5,\ldots} \frac{1}{i} \sin(2i\pi t/T)$$	
	b) Sawtooth $$f(t) = \frac{1}{2} + \frac{1}{\pi} \sum_{i=1} \frac{1}{i} \sin(2i\pi t/T)$$	
	c) Sawtooth $$f(t) = \frac{1}{2} - \frac{1}{\pi} \sum_{i=1} \frac{1}{i} \sin(2i\pi t/T)$$	
	d) Triangular wave $$f(t) = \frac{1}{2} - \frac{4}{\pi^2} \sum_{i=1} \frac{1}{(2i-1)^2} \cos\left((2i-1)\pi t/T\right)$$	
	e) Rectified sine wave $$f(t) = \frac{2}{\pi} + \frac{4}{\pi} \sum_{i=1} \frac{1}{1 - 4i^2} \cos(2i\pi t/T)$$	
	f) Half sine wave $$f(t) = \frac{1}{\pi} + \frac{1}{2} \sin(\pi t/T) - \frac{2}{\pi} \sum_{i=2,4,6,\ldots} \frac{\cos(i\pi t/T)}{i^2 - 1}$$	
	g) Trapezoidal $$f(t) = \frac{1}{\alpha^2} \sum_{i=1,3,5,\ldots} \frac{\sin i\pi\alpha}{(\pi i)^2} \sin(i\pi t/T)$$	
	h) Pulse train $(\alpha = t_d/T)$ $$f(t) = \alpha\left[1 + 2\sum_{i=1} \frac{\sin(i\pi\alpha)}{(i\pi\alpha)} \cos(2i\pi t/T)\right]$$	

[1]In the notation of Section 5.9, $2i\pi t/T = i\Omega_o\tau$, $T = 2\pi/\omega_o$, $\Omega_o = \omega_o/\omega_n$, $\tau = \omega_n t$, and $\omega_n = \sqrt{k/m}$]. The amplitudes of all waveforms vary from either 0 to 1 or −1 to 1.

APPENDIX C

Decibel Scale

The decibel is a unit of measurement for vibrations (and other phenomena), and is it defined[1] as

$$L_P = 10 \log_{10} \frac{P}{P_{ref}} \text{ dB} \qquad (C.1)$$

or

$$L_A = 20 \log_{10} \frac{A}{A_{ref}} \text{ dB} \qquad (C.2)$$

where P is a power or a power-like quantity, P_{ref} is a reference quantity having the same engineering units as P, A is an amplitude-like quantity and A_{ref} is a reference quantity having the same engineering units as A. Examples of P are electrical power in watts, system energy, acoustic intensity in watts/m^2; examples of A are voltage, displacement, velocity, and acceleration. The designation dB does not pertain to a particular physical quantity and therefore, it is not a physical unit in the ordinary sense. It simply indicates the relative magnitudes of two like quantities as shown in Eqs. (C.1) and (C.2).

The main reason for the introduction and use of the decibel is to compress logarithmically very large and very small numbers into a more manageable scale and to provide a convenient manner in which to talk about them. From Eq. (b), we see that for amplitude-like quantities each factor of 10 increase with respect to the reference quantity corresponds to 20 dB, whereas a decrease by a factor of 10 corresponds to -20 dB. Thus, 60 dB means that an amplitude-like quantity is 1000 times larger than the reference quantity and -60 dB indicates that it is 1000 times smaller. Two ratios that are of special interest are $A/A_{ref} = \sqrt{2}$ and $A/A_{ref} = 1/\sqrt{2}$, which correspond to 3 dB and -3 dB, respectively. For power-like quantities, $P/P_{ref} = 2$ and $P/P_{ref} = 1/2$ also correspond to 3 dB and -3 dB, respectively.

---

[1]"Acoustics—Expressions of Physical and Subjective Magnitudes of Sound or Noise in the Air," ISO 131, International Standards Organization, Geneva, Switzerland (1979).

EXPRESSING ERRORS IN DB

Frequently, errors are expressed in dB. Consider the usual definition of the percentage error ϵ of an amplitude-like quantity A with respect to a reference quantity A_{ref}:

$$\epsilon = 100 \frac{A - A_{ref}}{A_{ref}} = 100 \left(\frac{A}{A_{ref}} - 1 \right)\% \tag{C.3}$$

or

$$\frac{A}{A_{ref}} = 1 + \frac{\epsilon}{100} \tag{C.4}$$

If the ratio A/A_{ref} is expressed as Δ dB, then from Eq. (C.2)

$$\Delta = 20 \log_{10} \frac{A}{A_{ref}} \quad \text{dB} \tag{C.5}$$

or

$$\frac{A}{A_{ref}} = 10^{\Delta/20} \tag{C.6}$$

Therefore, from Eqs. (C.3) through (C.6), we obtain,

$$\epsilon = 100(10^{\Delta/20} - 1) \tag{C.7}$$

or

$$\Delta = 20 \log_{10} \left(1 + \frac{\epsilon}{100} \right) \tag{C.8}$$

Typical equivalent values for ϵ and Δ are given in Table C. Note that $\pm\Delta$ dB does not, in general, correspond to $\pm\epsilon\%$, although for $\Delta \leq 1$ dB, they are fairly close.

TABLE C

Relationship Between an Error Expressed in dB to One Expressed as a Percentage

Δ (dB)	ϵ(%)
0.01	0.12
−0.01	−0.12
0.10	1.16
−0.10	−1.14
0.50	5.93
−0.50	−5.59
1.00	12.20
−1.00	−10.87
1.50	18.85
−1.50	−15.86
2.00	25.89
−2.00	−20.57
3.00	41.25
−3.00	−29.21

APPENDIX D

Direct Methods to Determine Response to Harmonic Excitation

We present two time-domain methods for determining the particular solution of linear, second-order ordinary differential equations with constant coefficients.

METHOD 1

The governing equation of a single degree-of-freedom system subjected to a harmonic forcing of the form

$$f(\tau) = A \cos \Omega\tau + B \sin \Omega\tau \tag{D.1}$$

is given by

$$\frac{d^2x}{d\tau^2} + 2\zeta \frac{dx}{d\tau} + x = A \cos \Omega\tau + B \sin \Omega\tau \tag{D.2}$$

The solution to Eq. (D.2) can be assumed to be of the form

$$x(\tau) = C_1 \cos \Omega\tau + C_2 \sin \Omega\tau \tag{D.3}$$

where C_i are to be determined.

Upon substituting Eq. (D.3) into Eq. (D.2) and collecting terms, we obtain

$$[C_1(1 - \Omega^2) + 2\zeta\Omega C_2]\cos \Omega\tau + [C_2(1 - \Omega^2) - 2\zeta\Omega C_1]\sin \Omega\tau = A \cos \Omega\tau + B \sin \Omega\tau \tag{D.4}$$

Since Eq. (D.4) has to be valid for all values of τ, we equate the respective coefficients of $\sin(\Omega\tau)$ and $\cos(\Omega\tau)$ and obtain the following two equations:

$$(1 - \Omega^2)C_1 + 2\zeta\Omega C_2 = A$$
$$-2\zeta\Omega C_1 + (1 - \Omega^2)C_2 = B \tag{D.5}$$

567

Upon solving for C_1 and C_2 from Eqs. (D.5), we obtain

$$C_1 = \frac{1}{D(\Omega)} \left[(1 - \Omega^2)A - 2\zeta\Omega B \right]$$

$$C_2 = \frac{1}{D(\Omega)} \left[(1 - \Omega^2)B + 2\zeta\Omega A \right] \tag{D.6}$$

where

$$D(\Omega) = (1 - \Omega^2)^2 + (2\zeta\Omega)^2 \tag{D.7}$$

We now substitute Eqs. (D.6) into Eq. (D.3) to arrive at

$$x(\tau) = \frac{A}{D(\Omega)} [(1 - \Omega^2)\cos \Omega\tau + 2\zeta\Omega \sin \Omega\tau]$$

$$+ \frac{B}{D(\Omega)} [(1 - \Omega^2)\sin \Omega\tau - 2\zeta\Omega \cos \Omega\tau] \tag{D.8}$$

Although Eq. (D.8) is the solution to Eq. (D.2), we can convert it to a convenient form by defining the angle θ as a function of Ω as follows:

$$\tan \theta(\Omega) = \frac{2\zeta\Omega}{1 - \Omega^2} \tag{D.9a}$$

and, therefore,

$$\sin \theta(\Omega) = \frac{2\zeta\Omega}{\sqrt{D(\Omega)}}$$

$$\cos \theta(\Omega) = \frac{1 - \Omega^2}{\sqrt{D(\Omega)}} \tag{D.9b}$$

Upon substituting Eqs. (D.9b) into Eq. (D.8), we obtain

$$x(\tau) = \frac{A}{\sqrt{D(\Omega)}} \left[\frac{(1 - \Omega^2)}{\sqrt{D(\Omega)}} \cos \Omega\tau + \frac{2\zeta\Omega}{\sqrt{D(\Omega)}} \sin \Omega\tau \right]$$

$$+ \frac{B}{\sqrt{D(\Omega)}} \left[\frac{(1 - \Omega^2)}{\sqrt{D(\Omega)}} \sin \Omega\tau - \frac{2\zeta\Omega}{\sqrt{D(\Omega)}} \cos \Omega\tau \right]$$

$$= \frac{A}{\sqrt{D(\Omega)}} [\cos \theta(\Omega) \cos \Omega\tau + \sin \theta(\Omega) \sin \Omega\tau]$$

$$+ \frac{B}{\sqrt{D(\Omega)}} [\cos \theta(\Omega) \sin \Omega\tau - \sin \theta(\Omega) \cos \Omega\tau]$$

$$= \frac{A}{\sqrt{D(\Omega)}} \cos(\Omega\tau - \theta(\Omega)) + \frac{B}{\sqrt{D(\Omega)}} \sin(\Omega\tau - \theta(\Omega)) \tag{D.10}$$

METHOD 2

In some cases, the harmonic forcing excitation is expressed in the complex exponential form

$$f(\tau) = F_o e^{j\Omega\tau} \tag{D.11}$$

where, $j = \sqrt{-1}$. Then, the governing equation of the single degree-of-freedom system takes the form

$$\frac{d^2x}{d\tau^2} + 2\zeta\frac{dx}{d\tau} + x = F_o e^{j\Omega\tau} \tag{D.12}$$

The solution to Eq. (D.12) can be assumed to be of the form

$$x(\tau) = C_3 e^{j\Omega\tau} \tag{D.13}$$

where, in general, C_3 is a complex quantity. Upon substituting Eq. (D.13) into Eq. (D.12), we obtain

$$-\Omega^2 C_3 e^{j\Omega\tau} + 2\zeta\Omega j C_3 e^{j\Omega\tau} + C_3 e^{j\Omega\tau} = F_o e^{j\Omega\tau} \tag{D.14}$$

which, upon solving for C_3, yields

$$C_3 = \frac{F_o}{1 - \Omega^2 + 2\zeta\Omega j} \tag{D.15}$$

Then, from Eqs. (D.13) and (D.15), the response is

$$x(\tau) = \frac{F_o e^{j\Omega\tau}}{1 - \Omega^2 + 2\zeta\Omega j} \tag{D.16}$$

Equation (D.16) can also be written as

$$x(\tau) = \frac{F_o e^{j(\Omega\tau - \theta(\Omega))}}{\sqrt{D(\Omega)}} \tag{D.17}$$

where $D(\Omega)$ is given by Eq. (D.7) and $\theta(\Omega)$ is given by Eq. (D.9).

APPENDIX E

Matrices

DEFINITIONS

A *matrix* is a rectangular array of numbers consisting of m rows and n columns. Such an array is called an $(m \times n)$ matrix, and is denoted as

$$[A] = \begin{bmatrix} a_{11} & a_{12} & \cdots & a_{1n} \\ a_{21} & a_{22} & \cdots & a_{2n} \\ \vdots & & & \\ a_{m1} & a_{m2} & \cdots & a_{mn} \end{bmatrix} \qquad (E.1)$$

where the a_{ij} are called the elements of the array. The first subscript denotes the row and the second subscript denotes the column in which the element appears. When the number of rows equals the number of columns—that is, when $m = n$—the matrix is called a *square matrix* of order n.

The transpose of a matrix is obtained by interchanging the rows and columns and is denoted with a superscript T. Thus, the transpose of the $(m \times n)$ matrix $[A]$ is

$$[A]^T = \begin{bmatrix} a_{11} & a_{12} & \cdots & a_{1n} \\ a_{21} & a_{22} & \cdots & a_{2n} \\ \vdots & & & \\ a_{m1} & a_{m2} & \cdots & a_{mn} \end{bmatrix}^T = \begin{bmatrix} a_{11} & a_{21} & \cdots & a_{m1} \\ a_{12} & a_{22} & \cdots & a_{m2} \\ \vdots & & & \\ a_{1n} & a_{2n} & \cdots & a_{mn} \end{bmatrix} \qquad (E.2)$$

where $[A]^T$ is now an $(n \times m)$ matrix. A *symmetric matrix* is a square matrix in which $a_{ij} = a_{ji}$. Thus, for a symmetric matrix

$$[A] = [A]^T \qquad (E.3a)$$

A *skew-symmetric matrix* is a square matrix in which $a_{ij} = -a_{ji}$. Thus,

$$[A] = -[A]^T \qquad (E.3b)$$

A *column matrix* is a matrix with only one column; that is, an $(m \times 1)$ matrix. We denote this matrix as

$$\{a\} = \begin{Bmatrix} a_{11} \\ a_{21} \\ \vdots \\ a_{m1} \end{Bmatrix} = \begin{Bmatrix} a_1 \\ a_2 \\ \vdots \\ a_m \end{Bmatrix} \tag{E.4}$$

A column matrix is frequently referred to as a *column vector.* A *row matrix* is a matrix with only one row; that is, a $(1 \times m)$ matrix. We denote this matrix as

$$\{b\} = \{a\}^T = \{a_1 \quad a_2 \quad \cdots \quad a_m\} \tag{E.5}$$

A diagonal matrix is a square matrix in which all the elements are zero except those on the principal diagonal; that is,

$$[A] = \begin{bmatrix} a_{11} & 0 & \cdots & 0 \\ 0 & a_{22} & \cdots & 0 \\ \vdots & & & \\ 0 & 0 & \cdots & a_{nn} \end{bmatrix} \tag{E.6}$$

A special case of a diagonal matrix is the *identity matrix,* which is a diagonal matrix in which all the elements along the principal diagonal equal 1. This matrix is denoted as $[I]$ and is given by

$$[A] = \begin{bmatrix} 1 & 0 & \cdots & 0 \\ 0 & 1 & \cdots & 0 \\ \vdots & & & \\ 0 & 0 & \cdots & 1 \end{bmatrix} \tag{E.7}$$

The *null matrix* is a matrix whose every element is zero.

EQUALITY OF MATRICES

Two matrices $[A]$ and $[B]$, having the same order, are equal if and only if $a_{ij} = b_{ij}$ for every i and j.

ADDITION AND SUBTRACTION OF MATRICES

The sum or the difference of two matrices $[A]$ and $[B]$, having the same order, is denoted by

$$[C] = [A] \pm [B] \tag{E.8}$$

and is given by the sum or difference of each corresponding element; that is, $c_{ij} = a_{ij} \pm b_{ij}$ for every i and j.

MULTIPLICATION OF MATRICES

The multiplication of two matrices $[A]$ and $[B]$ is possible only if the number of columns of $[A]$ is equal to the number of rows of $[B]$; that is, if $[A]$ is an $(m \times k)$ matrix and $[B]$ is a $(k \times n)$ matrix. Such matrices are said to be *conformable.* Then the product

$$[C] = [A][B] \tag{E.9}$$

is an $(m \times n)$ matrix, where

$$c_{ij} = \sum_{p}^{k} a_{ip} b_{pj}$$

PROPERTIES OF MATRIX PRODUCTS

It is mentioned that, in general, the product $[A][B] \neq [B][A]$. However, for conformable matrices, multiplication is associative; that is,

$$([C][A])[B] = [C]([A][B]) \tag{E.10}$$

And it is distributive; that is,

$$([A] + [B])[C] = [A][C] + [B][C] \tag{E.11}$$

From the definition of the transpose of a matrix, it follows that

$$([A][B])^T = [B]^T [A]^T \tag{E.12}$$

Also, the following product plays an important role in Chapter 7. Let $\{x\}$ be an $(n \times 1)$ column vector and $[A]$ be a square matrix of order n. Then the product

$$c = \{x\}^T [A]\{x\} \tag{E.13}$$

is a scalar, since the product $\{x\}^T[A]$ results in a $(1 \times n)$ matrix and, therefore, the product $\{x\}^T[A]\{x\}$ results in a (1×1) matrix, or a scalar.

MATRIX INVERSE

The *inverse* of a square matrix $[A]$, which is denoted as $[A]^{-1}$, is defined as that matrix for which

$$[A]^{-1}[A] = [A][A]^{-1} = [I] \tag{E.14}$$

The inverse of a matrix exists if

$$\det[A] \neq 0$$

where $\det[A]$ denotes the determinant of $[A]$.

EIGENVALUES OF A SQUARE MATRIX

The eigenvalues of a square matrix $[A]$ are given by the solutions of

$$\det[[A] - \lambda[I]] = 0 \tag{E.15}$$

A matrix of order n has n eigenvalues.

Index